Statistical Inference for Stochastic Processes

This is a volume in
PROBABILITY AND MATHEMATICAL STATISTICS

A Series of Monographs and Textbooks

A complete list of titles in this series appears at the end of this volume.

Statistical Inference for Stochastic Processes

ISHWAR V. BASAWA

Department of Mathematics
La Trobe University
Australia

and

B.L.S. PRAKASA RAO
Indian Statistical Institute
New Delhi Centre,
India

ACADEMIC PRESS London New York Toronto Sydney San Francisco 1980

A Subsidiary of Harcourt Brace Jovanovich, Publishers

ACADEMIC PRESS INC. (LONDON) LTD.
24/28 Oval Road,
London NW1 7DX

United States Edition Published by
ACADEMIC PRESS INC.
111 Fifth Avenue
New York, New York 10003

British Library Cataloguing in Publication Data

Basawa, I V
Statistical inference for stochastic processes.
(Probability and mathematical statistics).
1. Stochastic processes
I. Prakasa Rao, B. L. S. II. Series
519.2 QA274 79–50533

ISBN 0–12–080250–3

Printed in Great Britain by Whitstable Litho Ltd., Whitstable, Kent

Contents

PART I SPECIAL MODELS

Preface

The current decade has witnessed an exponential growth of literature on stochastic processes. The preoccupation with model-building has resulted in a large number of theoretical models which seem to lack immediate appeal from a statistical point of view. As a result there appears to be an ever-widening gap between theory and applications in the area of stochastic modelling. The aim of this monograph is to attempt to reduce this gap by directing the interest of future researchers to the inference aspects of stochastic processes. Our choice of topics and approaches used has been partly constrained by the current trends in the literature and partly by our own interest in the area. We have attempted to put on record the work depicting the current state of the art in the general area of inference for stochastic processes as it appears to us.

Many of the results presented here will be appearing in book form for the first time. This is a research monograph written for specialists in the common area of stochastic processes and theoretical statistics; hopefully, it will appeal to both of these broad groups. Advanced graduate and M.Sc. level students should find the book useful; a seminar-type course based on a selected list of topics from the monograph seems feasible for students working towards their Ph.D.s in the North American system.

The topics in the book have been divided into three parts. Part I discusses some standard (special) models in six chapters, in addition to the introductory Chapter 0. In this part, the main ideas and methods are emphasized, avoiding technicalities as far as possible (within the constraint of maintaining a reasonable level of accuracy). In places, heuristic derivations and sketches of proofs rather than rigorous proofs are given in order not to interrupt the flow of the main arguments. The readers of Part I are expected to be familiar with *classical stochastic processes* at the level of Cox and Miller (1965) or Karlin and Taylor (1975) and with *Mathematical Statistics* at the level of Cramer (1946). Some familiarity with weak convergence (Billingsley, 1968), and martingale limit results (summarized in Appendix 1) will be an advantage.

Part II consists of three chapters on the theory of inference for general processes in discrete and continuous time including diffusion processes. Part III surveys in three chapters recent results on Bayesian, non-parametric and sequential procedures. The treatment of the subject in Parts II and III is more rigorous than that in Part I, and the theory is emphasized here rather than the methods. Familiarity with the *theory of stochastic processes* roughly at the level of Gikhman and Skorokhod (1969) and with *probability theory* at the level of Breiman (1968) is desirable for a smooth reading of most of the topics in Parts II and III.

All the chapters (except Chapters 0 and 1) contain, at the end, "Complements" which are intended to lead the interested reader to various special problems, extensions, etc. for which space could not be found in the text. Appropriate references to most of these excursions are given where solutions could be found. Bibliographical notes at the end of each chapter provide directly related references rather than attempting to trace the origin of individual results. Special topics and results relevant to some of the material in the text are briefly discussed in four appendices at the end of the book.

Any individual chapter can be read independently of the other chapters; the notation, and contents of each chapter are essentially self-contained. Some frequently used notation and abbreviations are listed on page xiii.

It is a pleasure to extend our thanks to the Departments of Statistics at La Trobe and Sheffield Universities, Department of Operations Research at Cornell University, and the Indian Statistical Institute, New Delhi Centre, for the support received during the various stages of the preparation of the manuscript. We would like to record our appreciation of Ms Kirsty Healey of the Academic Press London for her efficient work in getting the manuscript ready for the printers.

We are greatly indebted to all the authors whose work is discussed in this monograph and we extend our special thanks to those who kindly supplied us with the preprints and reprints of their work.

February 1979.

Ishwar V. Basawa
B. L. S. Prakasa Rao

List of Notation

$L_n(\theta;\, x^n)$ or $L_{x^n}(\theta)$	likelihood function based on the sample $x^n = (x_1, x_2, \ldots, x_n)$
$\mathscr{X}$, Ω, $\mathscr{Y}$	sample space
Θ	parameter space
$\mathscr{F}$, $\mathscr{A}$, $\mathscr{B}$	σ-algebras
$\mathbb{R}^p$	p-dimensional Euclidean space
$\mathbb{R}$, $\mathbb{R}^1$	real line
E	expectation
var	variance
cov	covariance
$\boldsymbol{\alpha}$	column vector
$\boldsymbol{\alpha}'$	transpose of $\boldsymbol{\alpha}$
$\mathbf{V}_n = O_p(1)$	for every $\varepsilon > 0$, there exists $A_\varepsilon > 0$ and $N_\varepsilon > 0$ such that for $n > N_\varepsilon$, $P[\mathbf{V}_n \mathbf{V}'_n > A_\varepsilon] < \varepsilon$.
$V_n = o_p(1)$	$\mathbf{V}_n$ tends to zero in probability as $n \to \infty$
χ_A	indicator function of set A
$\xrightarrow{\mathscr{L}}$	convergence in law, distribution, weak convergence
$P(\ )$	probability of the event in parens
$\xrightarrow{p}$	convergence in probability

$\xrightarrow{\text{a.s.}}$	convergence almost surely
$p-\lim$	limit in probability
$N(\mu, \sigma^2)$	normal distribution with mean μ and variance σ^2
$N_p(\boldsymbol{\mu}, \Sigma)$	p-dimensional multivariate normal distribution with mean vector $\boldsymbol{\mu}$ and covariance matrix Σ
$f_n \cong g_n$	$\frac{f_n}{g_n} \to 1$ as $n \to \infty$
$\|\cdot\|$	ordinary Euclidean norm on $\mathbb{R}^p$
$\langle \mathbf{x}, \mathbf{y} \rangle$ or $(\mathbf{x}, \mathbf{y})$	inner product of $\mathbf{x}$ and $\mathbf{y}$ in $\mathbb{R}^p$
$\mu \ll \nu$	μ is absolutely continuous with respect to ν
$\mu \sim \nu$	μ and ν are mutually absolutely continuous
det (A)	determinant of a square matrix A
i.i.d..	independent and identically distributed
MLE*	maximum likelihood estimator
LEE*	likelihood equation estimator
AMLE	approximate maximum likelihood estimator
MMLE	modified maximum likelihood estimator
UMP	uniformly most powerful
UMPU	uniformly most powerful unbiased
DF	distribution free
S–E	spherically exchangeable
H–L	Hodges–Lehmann
UMPDF	uniformly most powerful distribution free
MPDF	most powerful distribution free
SPRT	sequential probability ratio test
GSPRT	generalized sequential probability ratio test
GSP	generalized stochastic process
LAN	local asymptotic normality
SII	stationary independent increment

(*In a majority of cases treated in the book an MLE and a consistent LEE are identical. For this reason the two terms will frequently be used synonymously in the text.)

To
Veena, Ravi and Neel (IVB)
and
Saradamba and Rama Moorthy (BPR)

Chapter 0

Introductory Examples of Stochastic Models

Stochastic models are being increasingly used in scientific research in an ever-widening spectrum of disciplines. In this introductory chapter we describe some of their applications. The examples given below are by no means exhaustive nor are they intended to be a representative selection of possible fields of application or the processes to be discussed in this monograph. Their purpose is merely to highlight the diversity of applications.

EXAMPLE 1. A RANDOM WALK MODEL FOR NEURON FIRING

The neuron fires when the membrane potential reaches a critical threshold value, say C. Excitatory and inhibitory impulses are the inputs for the neuron; these impulses arrive according to a Poisson process. Each excitatory impulse increases, and each inhibitory impulse decreases, the membrane potential by a random quantity x with a common probability density $p(x)$. After each firing the membrane potential is reset at zero and the process is repeated. A typical realization of the model is given in Fig. 1.

Let $\{Y_1, Y_2, \ldots\}$ denote the times at which the neuron fires. The process of

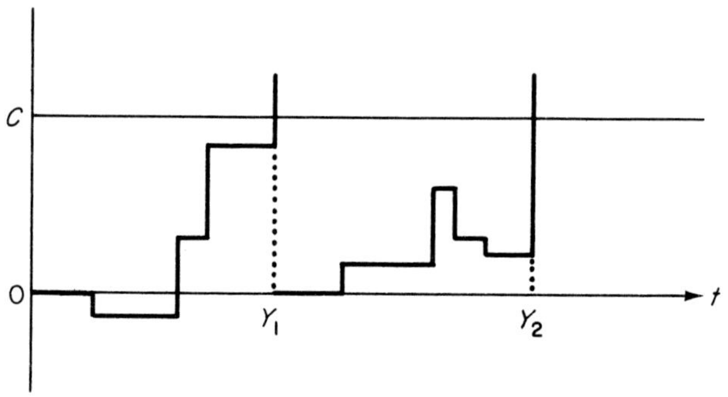

Fig. 1. Realization of a random walk model for neuron firing.

inter-spike intervals $\{Y_1, Y_2 - Y_1, Y_3 - Y_2, \ldots\}$ is of interest to the neurologist. Various questions such as the moments, correlation structure, and joint distributions of these intervals are relevant. This model is an example of a *point process*. Statistical questions related to some simple point processes are discussed in Chapter 6.

EXAMPLE 2. CHAIN BINOMIAL MODELS IN EPIDEMIOLOGY

Suppose at time $t = 0$ there are $S(0)$ susceptibles and $I(0)$ infectives. After a certain latent period of the infection, considered here as a unit of time, some of the susceptibles are infected; thus, at time $t = 1$, the initial $S(0)$ susceptibles split into two groups: those who are infected, $I(1)$ in number say, and the remaining susceptibles say, $S(1)$ in number. The process continues until

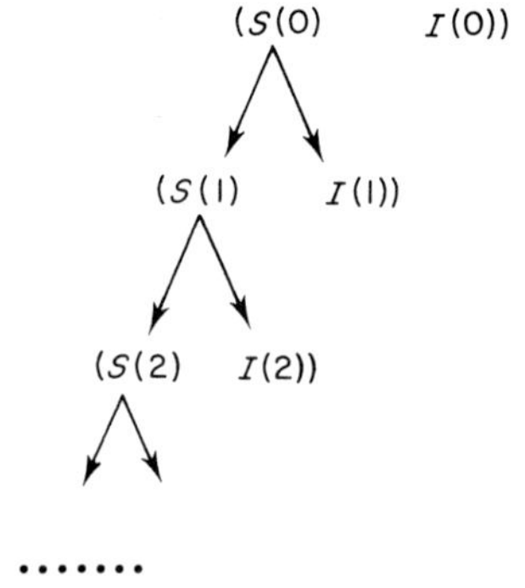

Fig. 2. Chain binomial model.

there are no more infectives in the population. The scheme is indicated in Fig. 2. Note that, in general, $S(t)=S(t+1)+I(t+1)$, $(t=0, 1, 2, \ldots)$. Assuming that the probability of a susceptible being infected is p we have, for the Greenwood Model,

$$P(S(t+1)=s(t+1)|S(t)=s(t))=\binom{s(t)}{s(t)-s(t+1)}p^{s(t)-s(t+1)}(1-p)^{s(t+1)}.$$

The process $\{S(t),\ t=0, 1, 2, \ldots\}$ is a Markov chain with the transition probabilities given above. Given $S(0)$, these transition probabilities specify the process $\{S(t)\}$ completely.

In the Greenwood Model, the number of infectives at time t is assumed not to influence the probability of infection during $(t, t+1)$. However, we can remove this restriction by viewing $\{(S(t), I(t))\}$ as a bivariate Markov chain with transition probabilities

$$P(I(t+1)=i(t+1),\ S(t+1)=s(t)-i(t+1)|I(t)=i(t),\ S(t)=s(t))$$

$$=\binom{s(t)}{s(t+1)}(1-p)^{i(t)s(t+1)}\{1-(1-p)^{i(t)}\}^{s(t)-s(t+1)}.$$

This now becomes the Reed–Frost Model. Markov chains provide a powerful tool in model building. The present example is a typical application of Markov chains in epidemiology. Problems of inference for discrete time Markov chains are discussed in Chapter 4.

EXAMPLE 3. A POPULATION GROWTH MODEL

Suppose an organism produces a random number, say Y, of offspring with $p_k=P(Y=k)$, $(k=0, 1, 2, \ldots)$, $\Sigma_0^\infty\ p_k=1$. Each offspring in turn produces organisms independently according to the same distribution $\{p_k\}$. If $Z(t)$ denotes the population size at the tth generation, $(t=0, 1, 2, \ldots)$ it is seen that $\{Z(t)\}$ is a Markov chain with transition probabilities given by

$$P(Z(t)=j|Z(t-1)=i)=P(Y_1+Y_2+\ldots+Y_i=j)$$

where $Y_1, Y_2, \ldots$ are independent identically distributed random variables with probability distribution $\{p_k\}$. The model just described is the so-called Galton–Watson branching process which is studied extensively in the literature. This model is an important example of a non-ergodic type process for which certain difficulties in extending the usual large-sample methods of inference arise. These are discussed in detail in Chapter 2.

EXAMPLE 4. A SPATIAL MODEL FOR PLANT ECOLOGY

Consider a lattice of sites (points) labelled by integer pairs (i, j) over some region. Let X_{ij} denote the random variable associated with the plant located at the site (i, j). If we are interested in a study of the pattern of infection in an array of plants we may set $X_{ij}=1$ if the plant located at (i,j) is infected, and zero otherwise. It may be reasonable to assume that the conditional distribution of X_{ij} given all other site values depends only on the four nearest neighbour sites, namely, on $X_{i-1,j}$, $X_{i+1,j}$, $X_{i,j-1}$, and $X_{i,j+1}$. This is a natural extension of the Markovian assumption to the spatial process $\{X_{ij}\}$. Spatial processes based on the "nearest-neighbour" formulation (as in this example) are of a recent origin. Work on statistical procedures relevant for such models is still at a rudimentary stage at the time of writing. Spatial models will not be discussed in this monograph.

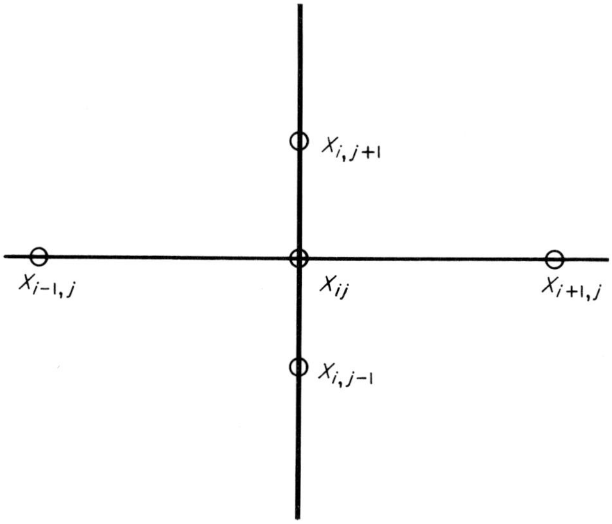

Fig. 3. Spatial ecological model.

EXAMPLE 5. A CLUSTER PROCESS FOR POPULATION SETTLEMENTS

An important area of research in geography is the study of spacings between settlements of human population and of sizes of settlements. It is frequently assumed that the settlements are located according to a spatial

Poisson process; thus one assumes that

P(one settlement in an area $\delta A) = \lambda\delta A + o(\delta A)$
P(two or more settlements in $\delta A) = o(\delta A)$,
P(no settlements in $\delta A) = 1 - \lambda\delta A + o(\delta A)$.

Then, it follows that the number of settlements in any given area A is a Poisson random variable with mean λA. Given the locations of the settlements the size of any settlement is assumed to follow a logarithmic series growth distribution. One then obtains a negative binomial distribution for the size of the population in a given area. This model is extensively used in geography, but will not be discussed any further in this monograph.

EXAMPLE 6. A MODEL IN POPULATION GENETICS

Consider a population of $2N$ genes each of which belongs to one of the two genotypes, say A and B. Let $X(t)$ denote the proportion of type A genes in the ith generation. Assuming that the total number of genes remains the same from generation to generation, and neglecting selection and mutation effects, the genes in the $(t+1)$th generation may be assumed to be a random sample of size $2N$ of genes in the tth generation. The sequence of random variables $\{X(t)\}$, $t=1, 2, \ldots$, forms a Markov chain. Conditionally on $X(t-1)=x$, $2NX(t)$ will be a binomial random variable with success probability x and index $2N$. By taking suitable limits one can approximate this Markov chain by a continuous time Markov process with a continuous state space $[0, 1]$. Such an approximation is an example of a diffusion process. Diffusion processes are discussed in detail in Chapter 9.

EXAMPLE 7. A STORAGE MODEL

Let $X(t)$ denote the annual random input during the years $(t, t+1)$, M the annual non-random release at the end of each year, and $Z(t)$ the content of the dam after the release. Then, it can be seen that

$$Z(t+1) = min\{Z(t) + X(t), K\} - min\{Z(t) + X(t), M\}$$

where K is the capacity of the dam, and $t=0, 1, 2, \ldots$. If the inputs $\{X(t)\}$ are assumed to be independent random variables, the sequence $\{Z(t)\}$ will form a Markov chain. If, however, $\{X(t)\}$ is assumed to follow a Markov chain, $\{X(t), Z(t)\}$ can be viewed as a bivariate Markov chain. Statistical questions specifically related to storage models as such will not be covered in the book; however, discrete Markov chains will be studied in Chapter 4.

EXAMPLE 8. A COMPOUND POISSON MODEL FOR INSURANCE RISK

Suppose an insurance company receives claims from its clients in accordance with a Poisson process with intensity λ. Assume that the amounts Y_k, $(k=1, 2, \ldots)$ of successive claims are independent random variables with a common distribution function $F(y)$. Then, the total amount $X(t)$ of claims arising in $(0, t)$ is given by $X(t)=Y_1+Y_2+\ldots+Y_{N(t)}$ where $N(t)$ is a Poisson random variable with mean λt. The process $\{X(t), t\geqq 0\}$, therefore, is a compound Poisson process. An interesting statistical application of compound Poisson processes is given in Chapter 6.

EXAMPLE 9. SYSTEM RELIABILITY MODELS

Let T denote the time to failure of a unit (light bulb, fuse, machine part etc.). The reliability of a unit at time t is defined as $R(t)=P(T>t)$. Consider a system consisting of several such units. A "series-system" is such that it fails if *at least one* unit fails, while a "parallel-system" can fail only when *all* the units in it fail. Statistical analysis of a simple reliability model is briefly discussed in Chapter 6.

EXAMPLE 10. A MODEL FOR CELL KINETICS

The life cycle of a cell, called the mitotic cycle, can be divided into four phases, say A_1, A_2, A_3 and A_4. The last phase A_4 is called mitosis after which the cell divides. Suppose that each cell performs a random walk $\{X(n)\}$, $(n=0, 1, 2, \ldots)$ on the integers $\{0, 1, 2, \ldots, a\}$ with the following transition probabilities:

$$P\{X(n+1)=j|X(n)=i\}=\begin{cases} p_i(n) & \text{if } j=i+1 \\ 1-p_i(n) & \text{if } j=i \\ 0 & \text{otherwise} \end{cases}$$

for $i=1, 2, \ldots, a-1$. Let $a=\Sigma_{i=1}^{4} a_i$ and assume that the event $\{X(n)\leqq\Sigma_{i=1}^{j} a_i\}$ is equivalent to the event that the cell is in one of the phases $A_1, \ldots, A_j$ at time n. A cell which reaches the last state at time n will, at its next transition remain at a with probability $1-p_a(n)$ or with probability $p_a(n)$ split into a random number of new cells which enter state 1; these new cells move through the cycle again

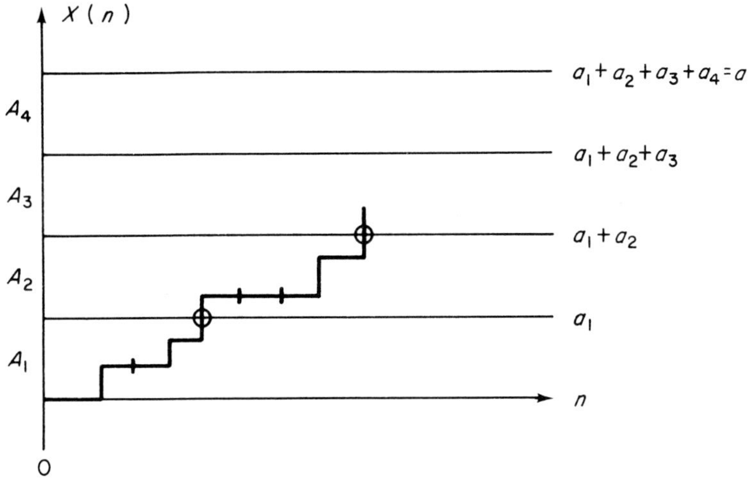

Fig. 4. A realization of the cell proliferation model.

following the random walk just described. Suppose we assume

$$p_i(n)=\begin{cases}\pi_1 & \text{for} \quad 1\leqq i\leqq a_1 & (A_1)\\ \pi_2 & \quad a_1+1\leqq i\leqq a_1+a_2 & (A_2)\\ \pi_3 & \quad a_1+a_2+1\leqq i\leqq a_1+a_2+a_3 & (A_3)\\ \pi_4 & \quad a_1+a_2+a_3+1\leqq i\leqq a_1+a_2+a_3+a_4 & (A_4)\end{cases}$$

If T_j, $(j=1,\ldots,4)$ denotes the duration of phase A_j it can be seen that (T_j-a_j) has a negative binomial distribution with mean $a_j\pi_j^{-1}(1-\pi_j)$. In this model the quantity T_j is of major interest.

While this model will not be discussed any further in this book, discrete Markov chains are studied in Chapter 4.

EXAMPLE 11. QUEUEING MODELS FOR TELEPHONE CALLS

Suppose calls arrive at a telephone exchange according to a Poisson process. Durations of successive calls may be assumed to be independent and to follow a negative exponential distribution. The capacity of the exchange may be limited to say, K calls at any given time. Various questions such as the queue-size at any particular time, expected waiting time of a call, etc. are of a practical interest. There is an immense amount of literature on queueing models and their various applications. Statistical questions related to some simple queueing models are dealt with in Chapter 5.

EXAMPLE 12. CLUSTERING–SPLITTING MODEL FOR ANIMAL BEHAVIOUR

Anthropologists have been observing how clusters of animals such as chimpanzees come together and form larger clusters, and the clusters split up into smaller groups. A simple model for this behaviour which can be applied to other phenomena of changes in the groupings is the following. Suppose we have n individuals forming $X(t)$ clusters at time t. $\{X(t), t \geqq 0\}$ is assumed to be a continuous time Markov chain on a finite state space $\{1, 2, \ldots, n\}$. Suppose the transition probabilities are as follows. In time $(t, t+\delta t)$,

$$P(X(t) \to X(t)+1) = \mu(n - X(t))\delta t + o(\delta t)$$
$$P(X(t) \to X(t)-1) = \lambda(X(t)-1)\delta t + o(\delta t),$$

and all other transitions have probability $o(\delta t)$. λ is the meeting-rate parameter, and μ the splitting rate parameter. $\{X(t)\}$ is then a logistic birth and death type process. Continuous time Markov chains are studied in Chapter 5.

EXAMPLE 13. PREDICTION OF ECONOMIC TIME SERIES

Let $X(t)$ denote the price of a certain commodity at time t, $(t=0, 1, 2, \ldots)$. Suppose we wish to fit a linear model for $X(t)$, for example

$$X(t) + \alpha_1 X(t-1) + \alpha_2 X(t-2) + \ldots + \alpha_p X(t-p)$$
$$= Z(t) + \beta_1 Z(t-1) + \beta_2 Z(t-2) + \ldots + \beta_q Z(t-q)$$

where $\alpha_j (j=1, 2, \ldots, p)$ and $\beta_k (k=1, 2, \ldots, q)$ are the real parameters to be estimated and the Z's are independent, identically distributed, unobservable random variables. This is an example of a mixed *autoregressive and moving average* model of order p and q. Given the prices at n time points, $\{X(0), X(1), \ldots, X(n-1)\}$, one may wish to predict future prices. Problems of estimation and prediction for linear models of this type will be studied in Chapter 3.

EXAMPLE 14. SIGNAL ESTIMATION

Suppose $X(t)$ is a *signal* satisfying a system equation

$$X(t+1)=aX(t)+\xi(t)$$

where a is a real parameter and $\xi(t)$ represents error. The true signal $X(t)$ is unobservable, but $y(t)$ is observed, where

$$y(t)=X(t)+Z(t),$$

$Z(t)$ being the *noise*. Having observed $\{y(0), y(1), \ldots, y(n)\}$ one may wish to estimate the true signal $X(n+1)$. This is a typical problem encountered in communication engineering and is discussed in Chapter 3.

BIBLIOGRAPHICAL NOTES

Fienberg (1974) contains a survey of models for neuron-firing; see also Kryukov (1976) for a discussion of the model in Example 1 with a time-varying threshold. Bailey (1975) discusses the epidemiological models of the type described in Example 2. Harris (1963) and Athreya and Ney (1972) may be referred for a study of branching processes. A review of spatial models is given by Besag (1974); also, see Cliff and Ord (1975). Ewens (1969), and Crow and Kimura (1970) discuss several models in genetics. A good introduction to the theory of dams is contained in the monograph by Moran (1959), and Seal (1969) discusses models of risk business. Barlow and Proschan (1965), Gnedenko *et al.* (1965), and Mann *et al.* (1974) contain models for reliability. See Jagers (1975) for cell proliferation models; the model described in Example 10 is discussed by Hopper and Brockwell (1978). Prabhu (1965) provides an introduction to queueing models. The model described in Example 12 is discussed by Morgan (1976). Box and Jenkins (1976) may be referred to for time series models.

Chapter 1

Basic Principles and Methods of Statistical Inference

1. INTRODUCTION

In this chapter, some of the basic concepts and procedures of statistical inference will be reviewed briefly. For more information, the reader should consult a book on Theoretical Statistics, for example Cox and Hinkley (1974).

2. THE LIKELIHOOD FUNCTION AND SUFFICIENT STATISTICS

Let $x^n = (x_1, x_2, \ldots, x_n)$ denote a realization of n consecutive observations from a discrete time stochastic process $\{X_k\}$, $(k = 1, 2, \ldots)$. Assume that the vector $X^n = (X_1, X_2, \ldots, X_n)$ has a probability density $p_{X^n}(x^n)$ with respect to a σ-finite positive product measure μ_n. If P_n is the basic probability measure defined on $\mathcal{B}(X^n)$, the σ-field generated by X^n, the density $p_{X^n}(x^n)$ is the Radon–Nikodym derivative of P_n with respect to μ_n at the point x^n where it is assumed that P_n is absolutely continuous with respect to μ_n. Thus,

$$p_{X^n}(x^n) = \frac{dP_n}{d\mu_n}(x^n).$$

Suppose the density p is indexed by a parameter θ, say $p_{X^n}(x^n; \theta)$ where θ takes values in a *parameter-space* Ω. Often we are required to choose a particular density from the family of possible densities $\{p_{X^n}(x^n; \theta), \theta \in \Omega\}$ such that the chosen density is the best one (in some sense) to "explain" the observed x^n. For fixed x^n, therefore, we are led to compare $p_{X^n}(x^n; \theta)$ for various values of θ. $p_{X^n}(x^n; \theta)$ viewed as a function of θ is called the likelihood function at x^n, and is denoted by $L_{x^n}(\theta)$ to emphasize that θ, and not x^n, is the variable.

The likelihood function for a realization from a continuous time stochastic process can be defined in an analogous way (see Chapter 8).

Example 1. Suppose $x^n = (x_1, x_2, \ldots, x_n)$ is a realization from a Galton–Watson branching process with $X(0) = 1$, and binomial offspring distribution with parameter p and a known index m. It is easily verified that the likelihood function of p based on x^n is given by

$$L_{x^n}(p) = C(x^n) p^{\Sigma_{ij} j n_{ij}} (1-p)^{\Sigma_{ij}(im-j) n_{ij}}$$

where $C(x^n)$ is free from p and n_{ij} denotes the number of transitions $i \to j$ in the sample x^n. The expression for the likelihood is obtained by noting that the process $\{X_k\}$ is a Markov chain on the state-space $\{0, 1, 2, \ldots\}$ with the transition probabilities

$$p_{ij} = \binom{im}{j} p^j (1-p)^{im-j}, \; j = 0, 1, \ldots, im, \; i = 0, 1, \ldots \quad .$$

Here, the measure μ_n is the counting measure.

Example 2. Consider a first-order moving average process defined by

$$X_k = Y_k - \rho Y_{k-1}, \quad (k = 1, 2, \ldots)$$

where $\{Y_k\}$ is a sequence of independent identically distributed $N(\mu, 1)$ variables; ρ is known and μ an unknown parameter. The likelihood function of μ based on the realization $x^n = (x_1, x_2, \ldots, x_n)$ is given by

$$L_{x^n}(\mu) = C \exp\{-\tfrac{1}{2} \sum_{i=1}^{n} \sum_{j=1}^{n} \sigma^{ij}(x_i - a\mu)(x_j - a\mu)\}$$

where $a = 1 - \rho$ and σ^{ij} is the (ij)th element of the inverse of the dispersion matrix of $X^n = (X_1, \ldots, X_n)$;

$$\operatorname{cov}(X_l, X_m) = \begin{cases} 1+\rho^2 & \text{for } l = m \\ -\rho & \text{for } |l-m| - 1 \\ 0 & \text{for } |l-m| > 1 \end{cases}$$

The likelihood here is obtained by noting that the vector X^n has a multinormal (multivariate normal) density with respect to the Lebesgue measure.

Example 3. Let $\{X(t),\ t \geqq 0\}$ be a Brownian motion with $X(0)=0$, $EX(t)=\mu t$, and $\operatorname{Var} X(t)=t$. Let the realization be $x^{t_0}=\{x(t),\ 0 \leqq t \leqq t_0\}$. If $P_\mu^{t_0}$ denotes the probability measure defined by $(X(t),\ 0 \leqq t \leqq t_0)$ when the drift parameter is μ, the Radon–Nikodym derivative of $P_\mu^{t_0}$ with respect to $P_0^{t_0}$ is given by

$$\frac{dP_\mu^{t_0}}{dP_0^{t_0}}(x^{t_0})=\exp\{\mu x(t_0)-\frac{1}{2}\mu^2 t_0\}$$

(see Chapter 9). The likelihood function therefore is given by the same expression considered as a function of μ. Here the dominating measure chosen is $P_0^{t_0}$.

The likelihood function forms the basis of many statistical procedures and plays a central role in the theory of inference.

Given a realization $X^n=(X_1, \ldots, X_n)$ from a stochastic process let $L_{X^n}(\theta)$ be the likelihood function where θ is a real parameter in the density of X^n. Let $T(X^n)$ be a statistic, i.e. a real valued function of X^n. T is said to be a sufficient statistic for θ if the conditional density of X^n given T is free from θ. It can be shown that this definition is equivalent to the following definition based on a factorization criterion: T is sufficient for θ if

$$L_{x^n}(\theta)=H(x^n)\cdot G(\theta, T(x^n)), \quad \text{for almost all } x^n,$$

where H depends only on x^n and G depends only on θ and T. The concept of sufficiency is useful in reducing the data X^n in a convenient form without losing relevant information about θ contained in the likelihood function.

Example 4. Refer to Example 1. It is seen that $T=(\Sigma_{ij} i n_{ij}, \Sigma_{ij} j n_{ij})$ is sufficient for p.

Example 5. For the model in Example 2 the statistic $T=\Sigma_{i=1}^n (\sigma^i X_i)$, where $\sigma^i=\Sigma_j \sigma^{ij}$, is sufficient for μ.

Example 6. An inspection of the likelihood function in Example 3 shows that $T=X(t_0)$ is sufficient for μ.

3. FREQUENCY APPROACH

The frequency approach to statistical inference can broadly be described as the one wherein the performance of any statistical method is to be evaluated on the basis of hypothetical repetitions of the experiment under the same conditions. Any criterion of optimality is thus based on the repeated sampling principle rather than on good performance for a single observed sample x^n. More precisely, one considers the sample space $\mathscr{X}^n$, the space of all possible values of x^n, and generally speaking one seeks procedures that are good on the average over $\mathscr{X}^n$.

Example 7. Consider the problem of testing the null hypothesis $H:X$ has density $p_1(x)$ against $K:X$ has density $p_2(x)$ with respect to a common dominating measure μ. The Neyman–Pearson Lemma gives the following procedure: "Reject H only when $\{p_2(x)/p_1(x)\}$ is larger than a certain predetermined number". If one follows this rule repeatedly for each $x \in \mathscr{X}$ then the Neyman–Pearson Lemma asserts that such a procedure maximizes the long-run proportion of correct rejections of H, i.e. rejection of H when in fact it is not true.

We shall be using the frequency approach in a majority of problems dealt with in this monograph.

4. THE BAYESIAN APPROACH

Suppose a random vector X^n has a density $p_{x^n}(x^n|\theta)$ for each given value of θ in Ω where Ω is a subset of the real line. In some cases it is conceivable that θ is chosen from the set Ω according to some density on Ω say $p(\theta)$, which is called a prior density. Thus the parameter, say Θ can be viewed as a random variable. After observing $X^n = x^n$, one can compute the conditional density of Θ given x^n say $p_{\Theta|X^n}(\theta|x^n)$ using Bayes' rule, as

$$P_{\Theta|X^n}(\theta|x^n) = \frac{p_{X^n|\Theta}(x^n|\theta)p_\theta(\theta)}{\int_\Omega p_{X^n|\Theta}(x^n|\theta)p_\Theta(\theta)\mathrm{d}v(\theta)}, \; \theta \in \Omega$$

where v is the dominating measure corresponding to the density $p_\Theta(\theta)$. In other words, the posterior density,

$$p_{\Theta|X^n}(\theta|x^n) \propto L_\theta(x^n) \cdot p_\Theta(\theta)$$

where $L_\theta(x^n) = p_{X^n|\Theta}(x^n|\theta)$ is the likelihood function. The posterior density then provides a basis for inference concerning θ. It should be noted that Bayesian inference is relevant to a single observed sample x^n, and the sample-space $\mathscr{X}^r$ is not involved in the formulation of any optimality criterion. A crucial requirement of this approach is, however, being able to arrive at a sensible prior density based on some facts regarding θ. As long as the prior density has some objective interpretation the Bayesian approach can be justified and is widely accepted in the literature. However, many

Bayesians prefer to view $p_\Theta(\theta)$ as a means of quantifying one's subjective beliefs about θ prior to the experiment. Since any two given statisticians may have differing beliefs about θ, it is difficult to arrive at a procedure acceptable to both – and what is more important, acceptable to others, on some sensible objective grounds.

Example 8. Consider the branching process model described in Example 1. Suppose p has prior density which is beta with parameters α and β. Then, the posterior density of p given the sample x^n is proportional to

$$\{L_{x^n}(p)\}\{p^{\alpha-1}(1-p)^{\beta-1}\}.$$

It is therefore clear that this posterior density is again beta with parameters.

$$\alpha'=\alpha+\sum_{ij} jn_{ij}=\alpha+x_1+\ldots+x_n$$

and

$$\beta'=\beta+\sum_{ij}(im-j)n_{ij}=\beta+m(x_0+x_1+\ldots+x_{n-1})-(x_1+\ldots+x_n)$$

where

$$x_0=1.$$

A prior density which leads to a posterior density of the same form (such as the one used in Example 8) is called a *conjugate prior*. Such priors are mathematically convenient and are in wide use, mainly for that reason.

5. ASYMPTOTIC INFERENCE

5.1 Maximum Likelihood Estimation

Let $X^n=(X_1, X_2, \ldots, X_n)$ denote a sequence of random variables forming a stochastic process, and possessing a density $p_n(x^n; \theta)$ with respect to a σ-finite product measure $\mu_n=\pi_1^n v_k$ where $x^n=(x_1, \ldots, x_n)$, and $\theta \in \Omega$, a subset of the real line. Suppose the conditional density of X_k given the past $(X_j, 1 \leqq j \leqq k-1)=X^{k-1}$ is given by $p_k(x^k|x^{k-1};\ \theta)=\{p_k(x^k; \theta\}\{p_{k-1}(x^{k-1}; \theta)\}^{-1}$, where

$$p_{k-1}(x^{k-1};\ \theta)=\int p_k(x^k;\ \theta)\mathrm{d}v_k(x_k).$$

The likelihood function of θ at x^n is given by

$$L_n(\theta;\ x^n)=\prod_{k=1}^{n} p_k(x^k|x^{k-1};\ \theta)$$

where $p_1(x_1|x_0;\ \theta)=p_1(x_1;\ \theta)$. In many problems the maximum likelihood estimate $\hat{\theta}_n(x^n)$ of θ, i.e. the value of θ which maximizes $L_n(\theta;\ x^n)$, is obtained as a root of the likelihood equation, $\mathrm{d}\log L_n/\mathrm{d}\theta=0$.

Let

$$S_n(\theta)=\sum_{k=1}^{n}\{S_k(\theta)-S_{k-1}(\theta)\}=\sum_{k=1}^{n}U_k(\theta)=\frac{\mathrm{d}\log L_n(\theta;\ X^n)}{\mathrm{d}\theta}$$

with $S_0(\theta)=0$. If F_k denotes the σ-field generated by $(X_1,\ \ldots,\ X_k)$ it is easy to verify under an appropriate differentiability condition that $E(S_n(\theta)|F_{n-1})=S_{n-1}(\theta)$ almost surely, and that $E(S_n(\theta))=0$. Thus, $\{(S_n(\theta),\ F_n),\ n=1, 2, \ldots\}$ is a zero-mean martingale. One can, therefore, use the Martingale Limit Theorems to establish consistency of $\hat{\theta}_n(X^n)$ and derive its asymptotic distribution.

Example 9. Consider the branching process example in Example 1. Noting that $\Sigma_{ij}jn_{ij}=\Sigma_1^n x_k$ and $\Sigma_{ij}in_{ij}=\Sigma_1^n x_{k-1}$, we see that the maximum likelihood estimate of the offspring mean, $\theta=mp$, is given by $\hat{\theta}_n(x^n)=(\Sigma_1^n x_k)\,(\Sigma_1^n x_{k-1})^{-1}$. We will show in the next chapter that $\hat{\theta}_n(X^n)$ is a strongly consistent estimator of θ and that

$$\left(\sum_1^n X_{k-1}\right)^{1/2}(\hat{\theta}_n(X^n)-\theta)\rightarrow N(0,\ mp(1-p))$$

in distribution, as $n\rightarrow\infty$, both results holding on the set of non-extinction.

Recall that an estimator $T_n(X^n)$ is said to be consistent for θ if $T_n\rightarrow\theta$ in probability and strongly consistent if this convergence is almost sure.

5.2 Asymptotic Sufficiency of the Maximum Likelihood estimator

Expand by Taylor's series $\log L_n(\theta;\ X^n)$ in θ at $\hat{\theta}_n$ to get

$$\log L_n(\theta;\ X^n)=\log L_n(\hat{\theta}_n;\ X^n)+\frac{(\hat{\theta}_n-\theta)^2}{2}\left[\frac{\mathrm{d}^2\log L_n(\theta;\ X^n)}{\mathrm{d}\theta^2}\right]_{\hat{\theta}_n}+o_p\ (1)$$

so that

$$L_n(\theta;\ X^n)=L_n(\hat{\theta}_n;\ X^n)\exp\left\{\frac{1}{2}\left[\frac{\mathrm{d}^2\log L_n(\theta;\ X^n)}{\mathrm{d}\theta^2}\right]_{\hat{\theta}_n}(\hat{\theta}_n-\theta)^2+o_p\ (1)\right\}.$$

If $I_n(\theta)=E(S_n^2(\theta))$ is the Fisher information in X^n, it often happens that

$$I_n^{-1}(\theta)\left[\frac{\mathrm{d}^2\log L_n(\theta;\ X^n)}{\mathrm{d}\theta^2}\right]_{\hat{\theta}}\rightarrow -1 \text{ in probability} \tag{1}$$

as $n \to \infty$. Such a convergence result holds, for example, when the observations are independent identically distributed, or for certain stochastic processes which can be loosely described as ergodic type. We then obtain

$$L_n(\theta;\, X^n) = L_n(\hat{\theta}_n;\, X^n)\ \exp\left\{-\frac{I_n(\theta)}{2}\,(\hat{\theta}_n-\theta)^2+o_p\,(1)\right\}.$$

This shows that the maximum likelihood estimator $\hat{\theta}_n$ satisfies the factorization criterion of sufficiency asymptotically. In this sense $\hat{\theta}_n$ is asymptotically sufficient for θ when condition (1) holds. In some cases, however, (1) does not hold, and instead we have

$$I_n^{-1}(\theta)\left[\frac{\mathrm{d}^2 \log L_n(\theta;\, X^n)}{\mathrm{d}\theta^2}\right]_{\theta} \to -W \text{ in probability} \tag{1'}$$

where $W > 0$ is a non-degenerate random variable measurable with respect to the tail σ-field of the sequence $\{X_k\}$. In such cases $\hat{\theta}_n$ is no longer asymptotically sufficient for θ.

Example 10. Refer to the branching process model discussed in Examples 1 and 9. It is shown in the next chapter that

$$-\left[\frac{\mathrm{d}^2 \log L_n(\theta,\, X^n)}{\mathrm{d}\theta^2}\right]_{\hat{\theta}_n} = \sigma^{-2}(\hat{\theta}_n)\sum_1^n X_{k-1}$$

and condition (1) is not satisfied instead (1′) is satisfied where W is given by $X_n\theta^{-n} \to W$ almost surely. Thus, the maximum likelihood estimate in this case is not asymptotically sufficient for θ. In fact, $(\hat{\theta}_n,\, \Sigma_1^n X_{k-1})$ is jointly sufficient for θ, and it is not possible to ignore $\Sigma_1^n X_{k-1}$ even asymptotically.

5.3 Asymptotic Efficiency of Estimators

Once again consider a discrete-time stochastic process $\{X_k\}$. Formulation for the continuous-time processes is analogous. We follow the notation used in the previous sections. We say that an estimator $T_n(X^n)$ of θ is asymptotically efficient if

$$\{I_n(\theta)\}^{1/2}(T_n-\theta-I_n^{-1}(\theta)S_n(\theta)) \to 0 \text{ in probability} \tag{2}$$

as $n \to \infty$. See Rao (1973) for a discussion of this criterion. Essentially the criterion states that an estimator T_n is efficient if $I_n^{1/2}(\theta)(T_n-\theta)$ and $I_n^{-1/2}(\theta)S_n(\theta)$ are asymptotically linearly related. Martingale theory is often useful in obtaining the asymptotic distribution of $I_n^{-1/2}(\theta)S_n(\theta)$.

Let $U_k(\theta)=S_k(\theta)-S_{k-1}(\theta)$ and $\xi_n(\theta)=\Sigma_1^n\ E(U_k^2(\theta)\,|\,F_{k-1})$. In the case of independence $\xi_n(\theta)=I_n(\theta)$. In general, it is seen that $E(\xi_n(\theta))=I_n(\theta)$. $\xi_n(\theta)$ can thus be viewed as a kind of conditional Fisher information in X^n. Heyde (1975) suggests replacing the Fisher information $I_n(\theta)$ in Rao's criterion (2) by its conditional analogue $\xi_n(\theta)$. Let us denote this modified criterion by (3). For many types of stochastic processes the ratio $\xi_n(\theta)/I_n(\theta)$ converges in probability to a random variable, say W. When W is degenerate the criteria (2) and (3) are asymptotically equivalent; however, when W is non-degenerate neither criterion implies the other. (Also, see Chapter 7 for further discussion).

The maximum likelihood estimator $\hat{\theta}_n$ satisfies (2) when W is degenerate (see Rao, 1973). If W is non-degenerate, $\hat{\theta}_n$ does not satisfy (2) but satisfies (3). An estimator T_n satisfying (2) is such that

(i) it is asymptotically sufficient;

(ii) considered as a test-statistic it leads to an efficient test. (see 5.4 for test-efficiency criteria).

When W is non-degenerate, an estimator satisfying (3) (in particular, the maximum likelihood estimator, $\hat{\theta}_n$) generally does not possess either of the desirable properties (i) and (ii). It therefore remains to find suitable grounds to justify (3). See Chapter 7 for one such justification.

Example 11. For the branching process discussed in the previous example with offspring distribution belonging to a power series family, we have

$$\xi_n(\theta) = \sigma^{-2}(\theta) \sum_{1}^{n} X_{k-1} \quad \text{(see next chapter).}$$

In particular, if the offspring distribution is geometric with

$$P(Z_1 = j) = \theta^{-1}(1-\theta^{-1})^{j-1}, \cdot (j = 1, 2, \ldots)$$

it turns out that $\xi_n(\theta) I_n^{-1}(\theta)$ converges in probability to W which has an exponential density with unit parameter. It will be shown in the following chapter that

$$I_n^{1/2}(\theta)(\hat{\theta}_n - \theta) \rightarrow t(2)$$

i.e. Student's t distribution with 2 degrees of freedom; $I_n^{-1/2}(\theta) S_n(\theta) \rightarrow$ double exponential random variable, in distribution, and consequently criterion (2) is not satisfied. However, it can be shown that criterion (3) is satisfied. This example will be further discussed in Chapter 2.

5.4 Efficiency Criteria for Tests

We shall introduce here only two *local* criteria. Consider the problem of testing a null hypothesis H: $\theta = \theta_0$ against an alternative K: $\theta > \theta_0$. Let

$\phi_n(X^n)$ denote any size-α test based on the sample $X^n=(X_1, \ldots, X_n)$ and $\beta_{\phi_n}(\theta)$ its power function. Thus, $\beta_{\phi_n}(\theta)=E_\theta\{\phi_n(X^n)\}$ with $\beta_{\phi_n}(\theta_0)=\alpha$ the size of the test. The usual optimality criterion is based on selecting a test ϕ_n for which $\beta_{\phi_n}(\theta)$ is large. For an asymptotic criterion we consider a sequence of tests $\{\phi_n\}$ as n becomes large. The so-called local criterion requires good performance of $\beta_{\phi_n}(\theta)$ at a sequence of parameter values θ_n converging to the null hypothesis value θ_0 as n becomes large. In this context, the following two criteria are relevant.

Pitman Power Criterion. Consider a sequence of alternatives $\theta_n=\theta_0+I_n^{-1/2}(\theta_0)\delta$, $\delta>0$. Then,

$$\gamma_p(\theta_0;\ \delta,\ \alpha)=\lim_{n\to\infty}\ \beta_{\phi_n}(\theta_n),$$

when it exists, is called the *Pitman power* corresponding to the test ϕ_n. A test that maximizes the Pitman power will be optimum according to the Pitman power criterion.

Local Power Criterion.

$$\gamma_L(\theta_0;\ \alpha)=\lim_{n\to\infty}\ I_n^{-1/2}(\theta_0)\left(\frac{\mathrm{d}\beta_{\phi_n}(\theta)}{\mathrm{d}\theta}\right)_{\theta_0},$$

when it exists, is called the *local power* corresponding to the test ϕ_n. A test which maximizes the local power is said to be optimum according to the local power criterion. It can be verified, under appropriate regularity conditions, that

$$\lim_{\delta\downarrow 0}\left\{\frac{\gamma_p(\theta_0;\ \delta,\ \alpha)-\gamma_p(\theta_0;\ 0,\ \alpha)}{\delta}\right\}=\gamma_L(\theta_0;\ \alpha).$$

6. NON-PARAMETRIC METHODS

So far we have assumed that given a sample $x^n=(x_1, \ldots, x_n)$ from a stochastic process we know its density (Radon–Nikodym derivative) say $p_n(x^n;\ \theta)$, apart from an unknown parameter θ (which may be a vector). Our object has been to estimate θ, or to test hypotheses regarding θ. As in the case of independent and identically distributed observations we may, in practice, have a less well-defined model than this. For instance, suppose x^n is a sample from a stationary Markov process with an unknown stationary density denoted as $p(x)$. The problem is then to estimate $p(x)$ or to test hypotheses regarding $p(x)$; for instance, we may be interested in a "goodness-of-fit" test of the form $p(x)=p_0(x)$, where $p_0(x)$ is a known density. Methods similar to those used in the case of independent and identically distributed observations may be applied here; however the performance of these methods needs to be explored in the presence of dependence. Density estimation and hypotheses testing for some simple types of processes will be discussed in Chapter 11.

7. SEQUENTIAL METHODS

In the fixed sample procedures one has a sample $(x_1, \ldots, x_n)$ or $\{x(t), 0<t<T\}$ with a predetermined n or T. Efficiency of estimates or tests can be increased substantially if the sample-size n (or T) is not fixed and instead the decision to terminate or continue the observation of the process at any stage is allowed to depend on the results of the observations made up to that stage. In the latter procedure, n (or T) may be viewed as a random variable. Sequential estimation and tests for both discrete and continuous time stochastic processes are discussed in Chapter 12. These results are essentially the extensions of similar results applicable for independent observations.

BIBLIOGRAPHICAL NOTES

Cox and Hinkley (1974) provide a comprehensive survey of various concepts and principles of inference, and the book contains extensive references to the literature. Specific references for Bayesian, non-parameteric, and sequential methods will be given in Chapters 10, 11, and 12 respectively.

Chapter 2

Branching Processes

1. INTRODUCTION

This chapter is concerned with inference problems for Galton–Watson process (discrete time) and Markov-branching processes (continuous time).

2. THE GALTON–WATSON PROCESS

2.1 Preliminaries

Let $\{Z_0=1, Z_1, Z_2, \ldots\}$ be a Galton–Watson process with offspring distribution $\{p_j\}$, where

$$p_j=P(Z_1=j), (j=0, 1, 2, \ldots).$$

Let $\{X_{ki}, k=0, 1, \ldots; i=0, 1, \ldots\}$ be a sequence of independent identically distributed random variables with the probability distribution $\{p_j\}$. Then, the process $\{Z_k\}$ is characterized by the following recursive relation

$$Z_{n+1}=\sum_{i=1}^{Z_n} X_{ni}. \tag{1}$$

Let $E(Z_1)=\theta$, and var $(Z_1)=\sigma^2$. It is easy to verify that the conditional mean and variance are given by

$$E(Z_{n+1}|Z_n)=Z_n\theta$$

and (2)

$$\operatorname{var}(Z_{n+1}|Z_n)=Z_n\sigma^2.$$

These considerations suggest the following autoregressive-type model for $\{Z_k\}$

$$Z_{n+1}=\theta Z_n+Z_n^{1/2}U_{n+1} \tag{3}$$

where the error random variables $\{U_k\}$ possess the following properties:

(i) $E(U_k)=0$ for all $k=0, 1, 2, \ldots$;

(ii) $\operatorname{var}(U_k)=\sigma^2$ for all k;

(iii) U_k is uncorrelated with each of $\{U_{k-1}, \ldots, U_0\}$ and is also uncorrelated with Z_{k-1}, for each k.

These properties for $\{U_k\}$ can easily be established using (1) and (2).

One problem that suggests itself is that of estimating θ and σ^2 from a realization $\{Z_0=1, Z_1, \ldots, Z_n\}$. Since the error term in (3) satisfies the classical assumptions regarding least-squares theory one may consider the "least-squares" estimator of θ which is obtained by minimizing the error sum of squares, $\Sigma_{k=0}^n U_k^2$, with respect to θ. This readily yields the estimator

$$\hat{\theta}_L=\left(\sum_1^n Z_{k-1}\right)^{-1}\left(\sum_1^n Z_k\right). \tag{4}$$

The variance estimator is obtained in the usual way as

$$\tilde{\sigma}^2=\frac{1}{n}\sum_1^n \{(Z_k-\hat{\theta}_L Z_{k-1})^2 Z_{k-1}^{-1}\}. \tag{5}$$

The estimator $\hat{\theta}_L$ was originally suggested by Harris (1948) who used a different approach in its derivation.

2.2 Maximum Likelihood Estimation of the Offspring Mean

We now assume throughout that Z_1 has a power-series distribution,

$$P(Z_1=r)=\frac{a_r\lambda^r}{A(\lambda)}, \quad (r=0, 1, 2, \ldots), \lambda>0, \tag{6}$$

where $A(\lambda)=\Sigma_{r=0}^\infty a_r\lambda^r$. The probability generating function of this distribution is given by $A(s\lambda)/A(\lambda)$ from which the mean and the variance of Z_1

can be deduced as

$$\theta = \lambda A'(\lambda)/A(\lambda), \quad \text{and} \quad \sigma^2 = \lambda \frac{\mathrm{d}\theta}{\mathrm{d}\lambda}. \tag{7}$$

Since $\{Z_0 = 1, Z_1, \ldots, Z_n\}$ is a realization from a Markov chain with the transition probabilities

$$p(Z_k|Z_{k-1}) \propto \lambda^{Z_k}\{A(\lambda)\}^{Z_{k-1}},$$

the likelihood function based on this realization is given by

$$L = \prod_{k=1}^{n} p(Z_k|Z_{k-1}) \propto \{A(\lambda)\}^{-\Sigma_1^n Z_{k-1}} \lambda^{\Sigma_1^n Z_k}.$$

From (7), we have $\mathrm{d}\theta/\mathrm{d}\lambda = \sigma^2/\lambda > 0$, and hence θ is a strictly increasing function of λ. From (8) we have

$$\frac{\mathrm{d}\log L}{\mathrm{d}\theta} = \frac{1}{\sigma^2} \sum_{k=1}^{n} (Z_k - \theta Z_{k-1}). \tag{9}$$

The maximum likelihood estimator of θ is thus

$$\hat{\theta}_n = \left(\sum_1^n Z_{k-1}\right)^{-1} \left(\sum_1^n Z_k\right) \tag{10}$$

which is the same as the least-squares estimator obtained earlier. It is easily verified that $\hat{\theta}_n$ maximizes log L. In general, however, the leat-squares estimator is not necessarily the same as the maximum likelihood estimator.

2.3 Asymptotic Properties of the Maximum Likelihood Estimator

From the previous subsection we have

$$S_n(\theta) = \frac{\mathrm{d}\log L}{\mathrm{d}\theta} = \frac{1}{\sigma^2(\theta)} \sum_{k=1}^{n} (Z_k - \theta Z_{k-1}) = \sum_{k=1}^{n} U_k(\theta),$$

say, for the power-series offspring distribution. The conditional Fisher information is given by

$$\xi_n(\theta) = \sum_{k=1}^{n} E(U_k^2(\theta)|F_{k-1})$$

$$= \sum_{k=1}^{n} \text{var}\,(U_k(\theta)|F_{k-1}) \tag{11}$$

$$= \frac{1}{\sigma^2(\theta)} \sum_{1}^{n} Z_{k-1},$$

where F_{k-1} is the σ-field generated by $(Z_{k-1}, Z_{k-2}, \ldots, Z_0)$. The Fisher information is given by

$$I_n(\theta) = E\{\xi_n(\theta)\} = \frac{1}{\sigma^2(\theta)} \left(\frac{\theta^n - 1}{\theta - 1} \right), \tag{12}$$

in which we have assumed that $\theta > 1$ (super-critical). The ratio $\xi_n(\theta)/I_n(\theta)$ plays a useful role in the asymptotic theory as already indicated in Chapter 1. It is well known (see, for instance, Athreya, and Ney, 1972) that

$$Z_n/\theta^n \rightarrow W \text{ almost surely, as } n \rightarrow \infty, \tag{13}$$

where W is a non-negative random variable. This fact can easily be deduced by noting that $\{Z_n/\theta^n\}$ is a non-negative martingale sequence and by applying the Martingale Convergence Theorem (see Feller, 1966). (13) yields, via the Toeplitz Lemma (see Loéve, 1963)

$$\zeta_n(\theta)/I_n(\theta) \rightarrow W \text{ almost surely as } n \rightarrow \infty. \tag{14}$$

It follows immediately from (14) that

$$\sum_{1}^{n} Z_k \sim W\theta^{n+1}/(\theta - 1) \text{ and } \sum_{1}^{n} Z_{k-1} \sim W\theta^n/(\theta - 1), \text{ for large } n,$$

and consequently $\hat{\theta}_n \sim \theta$ for large n on the set of non-extinction $\{W > 0\}$. Here the symbol $\sim$ means that the ratio converges to 1 almost surely. Thus, we have shown that

$$\hat{\theta}_n \rightarrow \theta \text{ almost surely as } n \rightarrow \infty \text{ on } \{W > 0\}, \tag{15}$$

i.e. the maximum likelihood estimator $\hat{\theta}_n$ is strongly consistent.

The asymptotic theory based on a single long realization breaks down on the set of extinction $\{W = 0\}$. This set consists of sub-critical (i.e. $\theta \leqq 1$) as well as some super-critical (with probability π, where π is the probability of extinction) realizations. On this set of extinction, one may consider several independent realizations observed until extinction, and study the performance of the estimator as the number of such realizations increase. Such an asymptotic theory follows essentially from the standard theory for the independent identically distributed case and hence will not be pursued here any further.

In order to derive the asymptotic distribution of $\hat{\theta}_n$ on the set $\{W>0\}$ we proceed as follows.

We first need the joint asymptotic distribution of $(\xi_n^{-1/2}(\theta)S_n(\theta), \xi_n(\theta)/I_n(\theta)$. We have

$$\xi_n^{-1/2}(\theta)S_n(\theta)=\sigma^{-1}\left(\sum_1^n Z_{k-1}\right)^{-1/2}\sum_1^n(Z_k-\theta Z_{k-1})$$
$$=\left(\sum_1^n Z_{k-1}\right)^{-1/2\Sigma_1^n}\sum_{i=1}^{Z_{k-1}}\xi_i \tag{16}$$

where ξ_i are independent identically distributed random variables with zero mean and variance one. We can therefore use the central limit theorem for random sums (Billingsley, 1968, Theorem 17.2) to conclude that

$$(\zeta_n^{-1/2}(\theta)S_n(\theta), \zeta_n(\theta)/I_n(\theta)\to(Z, W) \text{ in distribution as } n\to\infty \tag{17}$$

on the set $\{W>0\}$, where Z is a standard normal variable independent of W.

Since $\xi_n^{-1/2}(\theta)S_n(\theta)=\xi_n^{1/2}(\theta)(\hat{\theta}_n-\theta)$, we readily obtain the result

$$\xi_n^{1/2}(\theta)(\hat{\theta}_n-\theta)\to Z, \text{ in distribution, as } n\to\infty, \tag{18}$$

However, with the usual non-random norming, we get

$$I_n^{1/2}(\theta)(\hat{\theta}_n-\theta)\to ZW^{-1/2}, \text{ in distribution.} \tag{19}$$

Thus, the asymptotic distribution of the maximum likelihood estimator $\hat{\theta}_n$ is no longer normal. In particular, if W is exponential with mean unity, which occurs when the offspring distribution is geometric, i.e. when

$$P(Z_1=j)=\theta^{-1}(1-\theta^{-1})^{j-1}, \quad j=1, 2, \ldots,$$

then

$$I_n^{1/2}(\theta)(\hat{\theta}_n-\theta)\to t(2), \text{ in distribution,} \tag{20}$$

where $t(2)$ is the Student's t random variable with two degrees of freedom.

Since $S_n(\theta)=\sigma^{-2}(\theta)(\Sigma_1^n Z_{k-1})(\hat{\theta}_n-\theta)$, we have

$$\log L(\theta)=\left(\sum_1^{n-1} Z_k\right)\{\hat{\theta}_n a(\theta)-b(\theta)\}+c_n(Z_1,\ldots, Z_n). \tag{21}$$

where $a'(\theta)=\sigma^{-2}(\theta)$ and $b'(\theta)=\theta\sigma^{-2}(\theta)$, the prime denoting the derivative. It is clear from (21), via the factorization theorem, that $(\hat{\theta}_n, \Sigma_1^{n-1}Z_k)$ is jointly sufficient for θ. Expanding $\log L(\theta)$ by Taylor's series at $\hat{\theta}_n$ we have

$$\log L(\theta)=\log L(\hat{\theta}_n)+\frac{(\hat{\theta}_n-\theta)^2}{2}\left(\frac{d^2 \log L_n(\theta)}{d\theta^2}\right)_{\theta_n}+o_p(1).$$

Since $(-d^2 \log L_n(\theta)/d\theta^2)_{\hat{\theta}_n} = \sigma^{-2}(\hat{\theta}_n)\Sigma_1^n Z_{k-1} = \xi_n(\hat{\theta}_n)$, we finally obtain the asymptotic expansion of the likelihood function

$$L(\theta) = L(\hat{\theta}_n) \exp\{-\tfrac{1}{2}\zeta_n(\hat{\theta}_n)(\hat{\theta}_n - \theta)^2 + o_p(1)\}. \tag{22}$$

Since W is non-degenerate it is not possible to replace $\xi_n(\hat{\theta}_n)$ by a constant, even asymptotically. It is, therefore, clear that $\hat{\theta}_n$ is not asymptotically sufficient for θ.

We now turn to the question of asymptotic efficiency of $\hat{\theta}_n$. First, recall Rao's definition of efficiency given in the previous chapter: An estimator T_n is efficient for θ if $I_n^{1/2}(\theta)(T_n - \theta)$ and $I_n^{-1/2}(\theta)S_n(\theta)$ are asymptotically linearly related. In the present case, (17) yields the following results

$$I_n^{1/2}(\theta)(\hat{\theta}_n - \theta) \to ZW^{-1/2}, \text{ in distribution,}$$

as seen earlier, and

$$I_n^{-1/2}(\theta)S_n(\theta) \to ZW^{1/2}, \text{ in distribution.} \tag{23}$$

Consequently, the maximum likelihood estimator $\hat{\theta}_n$ is not asymptotically efficient in the usual sense. If, in particular, Z_1 has the geometric distribution as mentioned earlier, we have seen that

$$I_n^{1/2}(\theta)(\hat{\theta}_n - \theta) \to t(2), \text{ in distribution.}$$

It is easy to verify that, for this special case,

$$I_n^{-1/2}(\theta)S_n(\theta) \to V \text{ in distribution,} \tag{24}$$

where V has the double-exponential density given by

$$f_V(v) = \frac{1}{\sqrt{2}} e^{-\sqrt{2}|v|}, \quad -\infty < v < \infty.$$

We have already mentioned in Chapter 1 that $\hat{\theta}_n$ satisfies Heyde's definition of efficiency. This fact can easily be verified by noting that for the case of the power-series offspring distribution we have

$$\xi_n^{-1/2}(\theta)S_n(\theta) = \xi_n^{1/2}(\theta)(\hat{\theta}_n - \theta). \tag{25}$$

See Chapter 7 for a further discussion of the estimation-efficiency problem.

2.4 Asymptotic Tests for the Mean Offspring (Local Criterion)

Consider the problem of testing $H{:}\theta = \theta_0$ against the alternative $K{:}\theta > \theta_0$, and close to θ_0. Let ϕ_n be a size-α test of H against K based on the sample $\{Z_1, \ldots, Z_n\}$. The power function, $\beta_{\phi_n}(\theta)$, of ϕ_n is defined as

$$\beta_{\phi_n}(\theta) = E_\theta(\phi_n). \tag{26}$$

Since ϕ_n has size α, we have

$$\beta_{\phi_n}(\theta_0)=\alpha. \tag{27}$$

A first-order criterion of asymptotic efficiency is to maximize the derivative of the power function at the point θ_0, i.e. $\beta'_{\phi_n}(\theta_0)$ (See Rao, 1973). We can see, via the Neyman–Pearson Lemma, that the test

$$\phi_{0n}=\begin{cases}1 & \text{if } I_n^{-1/2}(\theta_0)S_n(\theta_0)\geqq k_n\\ 0 & \text{otherwise}\end{cases} \tag{28}$$

maximizes $\beta'_{\phi_n}(\theta_0)$. Thus

$$\beta'_{\phi_n}(\theta_0)\leqq\beta'_{\phi_{0n}}(\theta_0) \tag{29}$$

where ϕ_n is any other size-α test.

From (29) and using (17) it is not difficult to deduce the following inequality

$$\limsup_{n\to\infty}\{I_n^{-1/2}(\theta_0)\beta'_{\phi_n}(\theta_0)\}\leqq E_{\theta_0}\{W^{1/2}(2\pi)^{-1/2}\exp(-\tfrac{1}{2}a^2W^{-1})\} \tag{30}$$

where the constant a is determined such that

$$P_{\theta_0}(ZW^{1/2}\geqq a)=\alpha \tag{31}$$

where Z is a standard normal variate independent of W (see Basawa and Scott, 1976 for details).

Inequality (30) can be used as a basis for an efficiency criterion as follows.

Definition. The *local asymptotic efficiency* of any sequence $\{\phi_n\}$ of size-α tests is defined by

$$e(\{\phi_n\})=(2\pi)^{1/2}[E_{\theta_0}\{W^{1/2}\exp(-\tfrac{1}{2}a^2W^{-1})\}]^{-1} \liminf_{n\to\infty}\{I_n^{-1/2}(\theta_0)\beta'_{\phi_n}(\theta_0) \tag{32}$$

where a is given by (31).

Clearly $0\leqq e(\{\phi_n\})\leqq 1$. In particular, if $e(\{\phi_n\})=1$, $\{\phi_n\}$ is asymptotically efficient. It is obvious from the derivation of this criterion that ϕ_{0n} defined in (28) is asymptotically efficient.

As an example consider the special case with geometric offspring distribution. Suppose we wish to compare the performance of the statistics $I_n^{1/2}(\theta_0)(\hat{\theta}_n-\theta_0)$ and $\xi_n^{1/2}(\theta_0)(\hat{\theta}_n-\theta_0)$ for testing $\theta=\theta_0$ against $\theta>\theta_0$. The corresponding test functions are given by

$$\phi_{1n}=\begin{cases}1 & \text{if } I_n^{1/2}(\theta_0)(\hat{\theta}_n-\theta_0)\geqq b\\ 0 & \text{otherwise}\end{cases} \tag{33}$$

and

$$\phi_{2n}=\begin{cases}1 & \text{if } \xi_n^{1/2}(\theta_0)(\hat{\theta}_n-\theta_0)\geqq c\\ 0 & \text{otherwise}\end{cases} \tag{34}$$

where b and c are determined by

$$P(t(2)\geqq b)=\alpha,$$

where $t(2)$ is Student's t variate with two degrees of freedom, and

$$P(Z\geqq c)=\alpha,$$

Z being a standard normal variate.

The constant α in (32) is determined similarly (via (31)) using the double exponential distribution obtained in (24).

Numerical computations of $e(\{\phi_{1n}\})$ and $e(\{\phi_{2n}\})$ show that the tests ϕ_{1n} and ϕ_{2n} based on the maximum likelihood estimator $\hat{\theta}_n$ are not locally efficient ϕ_2 being more efficient than ϕ_1. The test ϕ_{0n} given in (28) with k_n given by a is locally efficient as noted earlier. In the classical situation (independent case) the test ϕ_0, ϕ_1 and ϕ_2 are asymptotically equivalent (see Rao, 1962).

2.5 The Pitman Criterion for Test-efficiency

We continue the discussion of test-efficiency for the problem of Section 2.2.4 and briefly describe here an alternative criterion of local efficiency. Consider the alternative sequence θ_n where $\theta_n=\theta_0+\delta I_n^{-1/2}(\theta_0)$, $\delta>0$, and $I_n(\cdot)$ is the Fisher information. As $n\to\infty$, it is clear that $\theta_n\to\theta_0$ The Pitman power of any size-α ϕ_n is defined to be $\lim_{n\to\infty}\beta_{\phi_n}(\theta_n)$, when the limit exists

By the Neyman–Pearson Lemma, the most powerful (MP) test of θ_0 against θ_n is given by

$$\phi_{3n}=\begin{cases}1 & \text{if } \Lambda_n\geqq\text{constant,}\\ 0 & \text{otherwise}\end{cases} \tag{35}$$

where

$$\Lambda_n=\log\left\{\frac{L(\theta_n)}{L(\theta_0)}\right\}. \tag{36}$$

Clearly,

$$\beta_{\phi_{3n}}(\theta_n)\geqq\beta_{\phi_n}(\theta_n),\quad \text{for any } n, \tag{37}$$

where ϕ_n is any other size-α test. In particular, letting $n\to\infty$, we find that ϕ_3

has the maximum Pitman power. Computations of the Pitman powers for ϕ_0, ϕ_1, and ϕ_2 introduced in Section 2.4 show, on comparison with the Pitman power for ϕ_3, that *none* of the former tests are efficient in the sense of maximizing Pitman power. See Feigin (1978) and Sweeting (1978) for details.

In the classical situation with independent and identically distributed observations it is well known that the tests ϕ_k, $(k=0, 1, 2, 3)$ are all asymptotically equivalent in the sense of having identical Pitman powers.

2.6 Further Topics

Bayesian Estimator of the Mean Offspring. One example of a Bayesian estimator appears in Section 1.5, Example 8, of Chapter 1. The following is another example. Assume that the offspring distribution is Poisson with mean λ. Let the prior density of λ be gamma with parameters p and q, i.e.

$$p(\lambda)=\frac{e^{-\lambda q}\lambda^{p-1}q^p}{\Gamma(p)}, \quad (p>0,\ q>0).$$

The likelihood function is

$$L \propto \exp\left(-\lambda\sum_1^n Z_{k-1}\right)\lambda^{\sum_1^n Z_k}$$

The posterior density of λ given $(Z_1, \ldots, Z_n)$, is therefore proportional to

$$\exp\left(-\lambda\left(q+\sum_1^n Z_{k-1}\right)\right)\lambda^{p+\sum_1^n Z_k-1}.$$

Consequently, the posterior density is again gamma with parameters $p+\Sigma_1^n Z_k$ and $q+\Sigma_1^n Z_{k-1}$. The mean of the posterior density is the Bayes estimator under quadratic loss, and is given by

$$\hat{\lambda}_B=\frac{p+\sum_1^n Z_k}{q+\sum_1^n Z_{k-1}}. \tag{38}$$

It is clear that $\hat{\lambda}_B$ is asymptotically equivalent to the maximum likelihood estimator $\hat{\lambda}=\Sigma_1^n Z_k/\Sigma_1^n Z_{k-1}$ on the set of non-extinction.

Conditional Inference. Let $(Z_0=1, Z_1, Z_2, \ldots, Z_n)$ be a realization of a branching process with offspring distribution

$$P(Z_1=j)=\theta^{-1}(1-\theta^{-1})^{j-1}.j=1, 2, \ldots, (\theta>1).$$

It is known that conditionally on $W=w$, the random variables $\{(Z_j-Z_{j-1}),$

$j=1, 2, \ldots\}$ are independent Poisson random variables with means $w(\theta^j-\theta^{j-1})$. The conditional likelihood function can therefore be written down and it is seen that the conditional likelihood is of the exponential family form, with two parameters, α and β, where $\alpha=\log\theta$, and $\beta=\log(w(\theta-1)/\theta)$. Note that we are treating w as an unknown (non-random) parameter, and after reparametrizing we get the exponential family model with the new parameters α and β.

If we conduct inference about α and β (or about the parameter α alone treating β as a nuisance parameter) with reference to the conditional likelihood we do not encounter any of the special difficulties we had in the previous sections where we used the unconditional likelihood function. Thus, in particular, we find that $\Sigma_{j=1}^n j(Z_j-Z_{j-1})$ and $\Sigma_{j=1}^n(Z_j-Z_{j-1})$ are jointly minimal sufficient for the two unknown parameters α and β. Also, for testing $\alpha=\alpha_0$ and treating β as a nuisance parameter we find that the conditional test-statistic $\{\Sigma_{j=1}^n j(Z_j-Z_{j-1})|\Sigma_{j=1}^n(Z_j-Z_{j-1})\}$ is optimal (for any n) in the sense of being uniformly most powerful unbiased (See Lehmann, 1959).

Further properties of this conditional test as $n\to\infty$ are discussed by Basawa (1980).

3. THE MARKOV BRANCHING PROCESS

3.1 Preliminaries

A Markov branching process is a continuous time branching process which can be described as follows: let $X(t)$ be the population size at time t with $X(0)=1$. Each individual has an exponentially distributed lifetime with parameter λ, and upon death is replaced by a number of offspring according to the distribution $\{p_i, i=0, 1, 2, \ldots\}$. It is assumed that individuals are mutually independent as regards death and reproduction. $\{X(t), t\geqq 0\}$ is then a continuous time Markov process with the state-space $\{0, 1, 2, \ldots\}$. The infinitesimal transition probabilities of the process are as follows

$$P(X(t+h)=j|X(t)=i)=\begin{cases} i\lambda p_{j-i+1}h+o(h), j=1-1, i+1, i+2, \ldots \\ 1-i\lambda h+o(h), j=i \\ o(h) \text{ otherwise.} \end{cases}$$

If $p_2=1$, this process reduces to a birth process; and if $p_0+p_2=1$, it reduces to the birth-and-death process with λp_0 and λp_2 as the birth and death parameters respectively. The birth-and-death process will be studied in more detail in Chapter 5.

Since the death of an individual in the process does not necessarily correspond to a jump in the $X(t)$ process we shall assume $p_1=0$ without loss of generality. With this assumption deaths correspond to the jumps in the $X(t)$ process. Suppose the process is observed continuously over a period $[0, T]$. Let $N_i(T)$ denote the number of jumps of size $i-1$ in $[0, T]$, and $N(T)=\Sigma_{i=0,2}^{\infty} N_i(T)$, the total number of jumps. Moreover, suppose $0=\tau_0<\tau_1<\tau_2, \ldots$ denote the epochs at which jumps take place. Then, the embedded realization $\{(\tau_0, X(\tau_0)), (\tau_1, X(\tau_1)), \ldots (\tau_{N(T)}, X(\tau_{N(T)})), \tau_{N(T)+1}>T\}$ is equivalent (probabilistically) to the observed sample $\{X(t), 0\leqq t\leqq T\}$. The likelihood function based on the embedded sample is seen to be

$$L_T=\lambda^{N(T)}\exp\left[-\lambda\sum_{j=1}^{N(T)} X(\tau_{j-1})(\tau_j-\tau_{j-1})\right]$$

$$\exp\left[-\lambda X(\tau_{N(T)})(T-\tau_{N(T)})\right]\prod_{i=0,2}^{\infty} p_i^{N_i(T)} \tag{39}$$

$$=\lambda^{N(T)}e^{-\lambda S(T)}\prod_{i=0,2}^{\infty} p_i^{N_i(T)}$$

where

$$S(T)=\int_0^T X(t)\mathrm{d}t=\sum_{j=1}^{N(T)} X(\tau_{j-1})(\tau_j-\tau_{j-1})+X(\tau_{N(T)})(T-\tau_{N(T)}). \tag{40}$$

The maximum likelihood estimates of λ and p_i are readily found to be

$$\hat{\lambda}=N(T)/S(T), \text{ and } \hat{p}_i=N_i(T)/N(T), \quad i=0, 2, 3, \ldots \tag{41}$$

An estimator of the offspring mean $\theta=\Sigma ip_i$ can easily be constrcuted as

$$\hat{\theta}=\Sigma i\hat{p}_i=\frac{X(T)+N(T)-1}{N(T)}. \tag{42}$$

We shall now derive the asymptotic distributions of $\hat{\lambda}$ and $\hat{\theta}$ on the set of non-extinction, $\{X(t)\to\infty\}$, assuming $\theta>1$ (super-critical case).

3.2 Consistency

Let $Y_j=X(\tau_{j-1})(\tau_j-\tau_{j-1})$, $(j=1, 2, \ldots)$. Then the $Y_1, Y_2, \ldots$ are independent exponential random variables with mean λ^{-1}. By the strong law of large numbers we have $1/n\Sigma_1^n Y_j\to\lambda^{-1}$, almost surely, as $n\to\infty$, and hence, by extending this to the random sum in the usual way, we have

$$\frac{1}{N(T)}\sum_{1}^{N(T)} Y_j\to\lambda^{-1}, \text{ almost surely,} \tag{43}$$

on the set $\{X(t)\to\infty\}$ since on this set $N(T)\to\infty$ almost surely.

Clearly, $X(\tau_{N(T)})(T-\tau_{N(T)}) \leqq Y_{N(T)}+1$ and hence

$$\frac{1}{N(T)} X(\tau_{N(T)})(T-\tau_{N(T)}) \to 0 \text{ almost surely.} \tag{44}$$

In fact, it can be shown via the Borel–Cantelli Lemma, that

$$\left(\frac{1}{N(T)}\right)^{1/2} X(\tau_{N(T)})(T-\tau_{N(T)}) \to 0, \text{ almost surely.} \tag{45}$$

By (43) and (44) via (40) we immediately deduce that

$$\hat{\lambda} \to \lambda, \text{ almost surely, as } T \to \infty \tag{46}$$

on the set $\{X(t) \to \infty\}$.

To prove consistency of $\hat{\theta}$ we proceed as follows. We have the representation

$$X(T) = \sum_{i=1}^{N(T)} \xi_i$$

where $\xi^i + 1$, (i=1, 2, ...) are independent identically distributed according to $\{p_j\}$. Thus, by applying the strong law of large numbers for the random sum as before, we have

$$\frac{X(T)}{N(T)} \to \theta - 1, \text{ almost surely,}$$

on the set $\{X(t) \to \infty\}$, and consequently,

$$\hat{\theta} \to \theta \text{ almost surely as } T \to \infty. \tag{47}$$

3.3 Asymptotic Distributions

From (40) we have

$$S(T) = \sum_{j=1}^{N(T)} Y_j + Y^*$$

where $Y_j = X(\tau_{j-1})(\tau_j - \tau_{j-1})$ are independent identically distributed exponential random variables with mean λ^{-1} and $Y^* = X(\tau_{N(T)})(T-\tau_{N(T)})$.

We have, by the central limit theorem,

$$V_n = \frac{1}{\sqrt{n}} \sum_{j=1}^{n} \left(Y_j - \frac{1}{\lambda}\right) + \frac{Y^*}{\sqrt{n}} \to N\left(0, \frac{1}{\lambda^2}\right)$$

in distribution as $n \to \infty$, since $Y^*/\sqrt{n} \to 0$ in probability (via Chebyshev's inequality). Consequently

$$\frac{\lambda S(T)-N(T)}{\lambda(N(T))^{1/2}}\to N\left(0,\ \frac{1}{\lambda^2}\right),\ \text{in distribution as } T\to\infty$$

by the central limit theorem for random sums (cf. Billingsley, 1968, Theorem 17.2). Since

$$S^{1/2}(T)(\hat{\lambda}-\lambda)=\frac{N(T)-\lambda S(T)}{\sqrt{S(T)}}$$
$$=\left\{\frac{N(T)-\lambda S(T)}{\lambda\sqrt{N(T)}}\right\}\left\{\frac{\lambda\sqrt{N(T)}}{\sqrt{S(T)}}\right\}$$

it follows that

$$S^{1/2}(T)(\hat{\lambda}-\lambda)\to N(0,\ \lambda) \text{ in distribution as } T\to\infty \tag{48}$$

where we have used the fact that $N(T)/S(T)=\hat{\lambda}\to\lambda$ in probability. The random norm $S^{1/2}(T)$ in (48) can be replaced by a non-random quantity say $C^{1/2}(T)$, but then the asymptotic distribution will be a variance mixture of normal. Thus, if $S(T)/C(T)$ converges to a random variable $\eta>0$ in probability (this result can be extablished, for instance, by taking $C(T)=E(S(T))$), the result in (48) can be written alternatively as

$$\lambda^{-1/2}C^{1/2}(T)(\hat{\lambda}-\lambda)\to Z\eta^{-1/2},\ \text{in distribution} \tag{49}$$

where Z is a standard normal variable and is independent of η.

We now turn to $\hat{\theta}$. First, recall that $X(T)=\Sigma_{i=1}^{N(T)}\xi_i$ where ξ_i are independent identically distributed random variables with mean $\theta-1$ and variance σ^2 which is assumed finite. Using the random sum central limit theorem once again we have

$$\frac{X(T)-N(T)(\theta-1)}{(N(T))^{1/2}}\to N(0,\ \sigma^2) \text{ in distribution as } T\to\infty.$$

This immediately gives the result

$$S^{1/2}(T)(\hat{\theta}-\theta)\to N(0,\ \sigma^2/\lambda). \tag{50}$$

Using the non-random norm $C(T)=E(S(T))$ the result in (50) can be written alternatively as

$$\lambda^{1/2}\sigma^{-1}C^{1/2}(T)(\hat{\theta}-\theta)\to Z\eta^{-1/2} \text{ in distribution, as } T\to\infty.$$

BIBLIOGRAPHICAL NOTES

Most of the results presented in Section 2.2 are contained in the works of Dion (1974, 1975), Heyde (1975), and Basawa and Scott (1976). For further details of results given in Section 3 see Athreya and Keiding (1977), and Feigin (1975, 1976). Heyde and Seneta (1972) (cf. Quine, 1976) give the estimators of offspring and immigration means for a subcritical Galton–Watson process with immigration. Such a process is known to possess a stationary distribution and consequently it does not present any new theoretical difficulties as far as the asymptotic theory goes.

COMPLEMENTS

1. Let $\{X_k\}$, $k=0, 1, 2, \ldots$ be a simple Branching process with $E(X_1)=\mu>1$, and var $(X_1)=\sigma^2<\infty$. $\hat{\mu}=(\Sigma_{k=1}^{n} X_k)/(\Sigma_{k=0}^{n-1} X_k)$ and $\tilde{\mu}=X_n/X_{n-1}$ are the Harris and the ratio estimates of μ from the sample $(X_0=1, X_1, \ldots, X_n)$. Compare the asymptotic efficiencies of $\hat{\mu}$ and $\tilde{\mu}$.

(Heyde, 1975)

2. Let $(X_0=1, X_1, \ldots, X_n)$ be a realization from a simple Branching process with $\mu>1$ and $\sigma^2<\infty$. Set $\hat{p}_r=(\Sigma_{m=0}^{n-1} X_{mr})/(\Sigma_{m=0}^{n-1} X_m)$ where X_{mr} denotes the number of individuals in the mth generation who give rise to r offspring. If $\{p_r\}$, $r=0, 1, 2, \ldots$, is the offspring distribution show that

$$\left(\sum_{k=0}^{n} X_k\right)^{1/2} (\hat{p}_r-p_r) \longrightarrow N(0, p_r(1-p_r)), \text{ as } n\to\infty.$$

(Dion, 1972)

3. Let $\{X_0=k, X_1, \ldots\}$ be a Galton–Watson process with initial population-size k. Suppose $\mu<1$ so that the probability of extinction is unity. Let T denote the time-to-extinction, and suppose that we observe $(X_0, X_1, \ldots, X_T)$. Assuming a power-series offspring distribution find the maximum likelihood estimate of μ and investigate its properties as $k\to\infty$.

(Becker, 1974)

4. Let $(X_1, X_2, \ldots, X_n)$ be independent identically distributed random variables with $f(s, \theta)$ as the generating function of the common distribution. The functional form of the generating function f is assumed known while θ is an unknown real parameter. Let $\hat{\theta}$ be an estimate of θ (based on $(X_1, \ldots, X_n)$) such that $n^{1/2}(\hat{\theta}-\theta)$ is asymptotically normal. Let

$$\hat{\pi}= \inf \{s{:}f(s, \hat{\theta})=s,\ s\geqq 0\}.$$

Then, show under appropriate assumptions that $n^{1/2}(\hat{\pi}-\pi)$ is asymptotically normal, where

$$\pi=\inf \{s{:}f(s, \theta)=s,\ s\geqq 0\},$$

(Stigler, 1971)

5. $\{X_0=1, X_1, \ldots, X_n\}$ is a realization from a super-critical Galton–Watson (simple Branching) process with an offspring distribution $\{p_r\}$ with generating function $f(s)$. Let $\hat{p}_r(n)=(\Sigma_{m=0}^{n-1} X_{mr}/\Sigma_{m=0}^{n-1} X_m)$, where X_{mr} is the number of individuals in the mth

generation who produce exactly r offspring. Let $\hat{f}(s, n)=\Sigma_{r=0}^{\infty}\hat{p}_r(n)s^r$, $(n=1, 2, \ldots)$ be an estimate of $f(s)$ based on $\{X_1, \ldots, X_n\}$. Define

$$\hat{\pi}(n)=\inf \{s{:}\hat{f}(s, n)=s,\ 0\leqq s<1\}.$$

$\hat{\pi}(n)$ is an estimate of the extinction probability π, Show that

$$S_n^{1/2}(\hat{\pi}(n)-\pi)\rightarrow N(0, v), \text{ in distribution,}$$

where

$$S_n=\sum_0^n X_k, \text{ and } V=\frac{f(\pi^2)-\pi^2}{(1-f'(\pi))^2}.$$

(Use the result given in Complement 2).

6. Let $\{X_k\}$ be a simple Galton–Watson process with immigration. θ and λ denote the means of the offspring distribution and the immigration distribution respectively and assume $\theta<1$ (sub-critical). It is required to estimate θ and λ from a realization $\{X_1, \ldots, X_n\}$. Define $\bar{X}=(1/n)\Sigma_1^n X_1$, and

$$R=\frac{\sum_1^n (X_i-\bar{X})(X_{i-1}-\bar{X})}{\sum_1^n (X_{i-1}-\bar{X})^2},$$

the mean and the serial correlation respectively. Show that $\bar{X}(1-R)$ and R are consistent estimates of λ and θ respectively. Also, establish the asymptotic normality of these estimates.

(Heyde and Seneta, 1972)

7. Consider a Bellman–Harris process (or age dependent branching process) $\{X_t, t\geqq 0\}$ with $\{p_i,\ i=0, 1, 2, \ldots\}$ as offspring distribution and G the life-length distribution function. The process is observed over a fixed interval $(0, T_0)$ and let $N_{T_0}(i)$ denote the number of jumps of size $(i-1)$ in $(0, T_0)$ and $N_{T_0}=\Sigma_i N_{T_0}(i)$. Assume, for simplicity, $p_1=0$. The offspring probabilities p_i can then be estimated by $\hat{p}_i=N_{T_0}(i)/N_{T_0}$. Show that conditionally on $X_t\rightarrow\infty$ as $t\rightarrow\infty$, the estimates $\hat{p}_i$ are consistent for p_i and that

$$N_{T_0}^{1/2}(\hat{p}_i-p_i)\rightarrow N(0, p_i(1-p_i)) \text{ as } T_0\rightarrow\infty.$$

in distribution with asymptotic covariances $-p_ip_j$. Construct an estimate of the mean offspring $\theta=\Sigma_i ip_i$ from those of p_i as $\hat{\theta}=\Sigma_i i\hat{p}_i$. Establish the asymptotic properties of $\hat{\theta}$ conditional on $X_t\rightarrow\infty$.

(Athreya and Keiding, 1977)

8. Show that the estimates $\hat{\lambda}$ and $\hat{\theta}$ for the Markov branching process discussed in Section 3 are asymptotically independent.

Chapter 3

Simple Linear Models

1. INTRODUCTION

In this chapter problems of prediction (i.e. estimation of future random variables), filtering (i.e. estimation of random variables in the presence of "noise" or super-imposed error), and parameter estimation for some simple discrete-time linear stochastic processes will be studied.

A discrete-time process $\{X_k\}$ is said to be a general linear process if X_k has the representation

$$X_k = \sum_{s=0}^{\infty} g_s Z_{k-s} \tag{1}$$

where $\{Z_k\}$ are independent identically distributed random variables and

$$\sum_{s=0}^{\infty} g_s^2 < \infty. \tag{2}$$

Suppose $E(Z_k)=0$ and $\mathrm{var}(Z_k)=\sigma^2$. The equality in (1) is to be interpreted in the sense of mean-square convergence. Thus, if $X_{kN} = \sum_{s=0}^{N} g_s Z_{k-s}$, we have

$$E(X_k - X_{kN})^2 \to 0, \text{ as } N \to \infty. \tag{3}$$

By Wold's Decomposition Theorem (see Doob, 1953, Chapter XII, Section 4) any regular stationary process without a singular component can be written as a linear process. Thus, the class of processes described by (1) is much larger than it might first appear. We shall, however, study only simple linear models, such as

(i) autoregressive,
(ii) moving average,
(iii) autoregressive with moving average residuals,
(iv) autoregressive with superimposed error (i.e. noise).

First, we turn to the prediction problem.

2. PREDICTION

2.1 Preliminaries: Minimum Mean-square Prediction

Consider a weakly stationary process (not necessarily linear) $\{Y_k\}$ with $E(Y_k)=0$, and cov $(Y_j, Y_k)=E(Y_jY_k)=\gamma(|j-k|)$. Suppose we wish to predict Y_{n+1} by a function of observations $\{Y_n, Y_{n-1}, \ldots, Y_0\}$ say $\hat{Y}_{n+1}$. The mean-square criterion requires choosing the function $\hat{Y}_{n+1}$ such that the mean-square error of prediction

$$E(Y_{n+1}-\hat{Y}_{n+1})^2 \tag{4}$$

is minimized. We will show that the choice

$$\hat{Y}_{n+1}=E(Y_{n+1}|Y_n, Y_{n-1}, \ldots, Y_0) \tag{5}$$

minimizes (4). Consider

$$\begin{aligned} E(Y_{n+1}-\hat{Y}_{n+1})^2 &= E[(Y_{n+1}-E(Y_{n+1}|Y_n, \ldots, Y_0))-(\hat{Y}_{n+1}-E(Y_{n+1}|Y_n, \ldots, Y_0)]^2 \\ &= E[E\{(Y_{n+1}-E(Y_{n+1}|Y_n, \ldots, Y_0)) \\ &\quad -(\hat{Y}_{n+1}-E(Y_{n+1}|Y_n, \ldots, Y_0)\}^2|Y_n, \ldots, Y_0] \\ &= E[E\{(Y_{n+1}-E(Y_{n+1}|Y_n, \ldots, Y_0))^2|Y_n, \ldots, Y_0\}] \\ &\quad + E[E\{(\hat{Y}_{n+1}-E(Y_{n+1}|Y_n, \ldots, Y_0))^2|Y_n, \ldots, Y_0\}] \end{aligned}$$

since the cross-product term vanishes.

Since the first term in the preceding expression is free from $\hat{Y}_{n+1}$ and the second term is non-negative it is clear that the mean-square error is minimized by the choice $\hat{Y}_{n+1}=E(Y_{n+1}|Y_n, \ldots, Y_0)$. From now on we shall use the notation $\hat{Y}_{n+1}$ for the optimum predictor $E(Y_{n+1}|Y_n, \ldots, Y_0)$.

If the $\{Y_n\}$ are normally distributed it is well known that (see, for instance, Rao, 1973) $E(Y_{n+1}|Y_n, \ldots, Y_0)$ is a linear function of $\{Y_n, \ldots, Y_0\}$. For a non-normal process, however, this conditional expectation is in general a non-linear function. But, for a variety of linear process (not necessarily normal) it turns out that the best predictor $\hat{Y}_{n+1}$ is linear in $\{Y_n, \ldots, Y_0\}$. This will be shown in Section 2.2.

2.2 Iterative Procedure for Prediction in Linear Processes

An *autoregressive process* $\{X_n\}$ of pth order satisfies the difference equation

$$X_n + b_1 X_{n-1} + \ldots + b_p X_{n-p} = Z_n \tag{6}$$

where the Z_n are independent identically distributed random variables with $E(Z_n)=0$. By repeated substitution for $X_{n-1}, \ldots, X_{n-p}$ in (6) it can be seen that X_n can be written as in (1).

If, on the other hand, $\{X_n\}$ satisfies

$$X_n = Z_n + C_1 Z_{n-1} + \ldots + C_q Z_{n-q} \tag{7}$$

the process is called a *moving average process* of order q.

An autoregressive process with moving average residuals is a combination of (6) and (7) where $\{X_n\}$ satisfies

$$X_n + b_1 X_{n-1} + \ldots + b_p X_{n-p} = Z_n + C_1 Z_{n-1} + \ldots + C_q Z_{n-q}. \tag{8}$$

Note that if in (8) we set $q=0$ we get (6) and $p=0$ gives (7).

We now require the best predictor, $\hat{X}_{n+1} = E(X_{n+1}|X_n, \ldots, X_0)$ of X_{n+1} from the sample $\{X_n, X_{n-1}, \ldots, X_0\}$. For the sake of illustration take $p=q=3$ in equation (8). Thus, at time $n+1$, we have

$$X_{n+1} + b_1 X_n + b_2 X_{n-1} + b_3 X_{n-2} = z_{n+1} + C_1 Z_n + C_2 Z_{n-1} + C_3 Z_{n-2}. \tag{9}$$

Taking conditional expectations given information up to time n we have

$$\hat{X}_{n+1} + b_1 X_n + b_2 X_{n-1} + b_3 X_{n-2} = C_1 Z_n + C_2 Z_{n-1} + C_3 Z_{n-1}. \tag{10}$$

Subtracting (10) from (9) gives

$$X_{n+1} - \hat{X}_{n+1} = Z_{n+1}.$$

Substituting back these values of $\{Z_n\}$ in (9) we get a recursive relation

$$\begin{aligned} &\hat{X}_{n+1} + C_1 \hat{X}_n + C_2 \hat{X}_{n-1} + C_3 \hat{X}_{n-2} \\ &= (C_1 - b_1) X_n + (C_2 - b_2) X_{n-1} + (C_3 - b_3) X_{n-2}. \end{aligned} \tag{11}$$

(11) enables us to compute $\hat{X}_{n+1}$ recursively from $\{X_n, \ldots, X_0\}$ and $\{\hat{X}_n, \hat{X}_{n-1}, \ldots, \hat{X}_0\}$. This procedure requires the initial value $\hat{X}_0$ which is usually taken as zero. It is easily verified that $\hat{X}_{n+1}$ is a linear function of $\{X_n, X_{n-1}, \ldots, X_0\}$.

2.3 Linear Prediction in a Weakly Stationary Process

Let us return to the problem of Section 2.1. We have a weakly stationary process $\{Y_n\}$ with $E(Y_k)=0$ and $E(Y_jY_k)=\gamma(|j-k|)$. Since $E(Y_{n+1}|Y_n, \ldots, Y_0)$ is not simple, in general, we wish to find a best *linear* preditor of Y_{n+1} based on $\{Y_n, \ldots, Y_0\}$. Thus, we have to determine the coefficients $(l_0, l_1, \ldots, l_n)$ such that $Y_{n+1}^L = \Sigma_{s=0}^n l_s Y_{n-s}$ is the best predictor of Y_{n+1} in the sense that

$$\phi = E(Y_{n+1} - \hat{Y}_{n+1}^L)^2$$

is minimized.

Setting the partial derivative of ϕ with respect to l_s equal to zero one gets the "normal" equations in the matrix notation:

$$\begin{bmatrix} \gamma(0) & \gamma(1) & \ldots & \gamma(\mathrm{n}) \\ \gamma(1) & \gamma(0) & \ldots & \gamma(n-1) \\ \ldots & \ldots & \ldots & \ldots \\ \gamma(n) & \gamma(n-1) & \ldots & \gamma(0) \end{bmatrix} \begin{bmatrix} l_0 \\ l_1 \\ \vdots \\ l_n \end{bmatrix} = \begin{bmatrix} \gamma(1) \\ \gamma(2) \\ \vdots \\ \gamma(n+1) \end{bmatrix}. \qquad (12)$$

Let $(\hat{l}_0, \ldots, \hat{l}_n)$ be the solution of (12). Then $\hat{Y}_{n+1}^L = \Sigma_{s=0}^n \hat{l}_s Y_{n-s}$. The best linear predictor $\hat{Y}_{n+1}^L$ can, in fact, be seen to be the projection of Y_{n+1} on the linear space spanned by $\{Y_n, \ldots, Y_0\}$.

3. FILTERING PROBLEM

3.1 Projection Theorem (Linear Filtering)

Let X_n denote the true signal at time n, η_n the noise and $Y_n = X_n + \eta_n$, the observed signal. Suppose we wish to estimate X_{n+1} as a function f of observed signals up to time n, $\{Y_n, Y_{n-1}, \ldots, Y_0\}$. It can easily be seen that the optimum function f such that $E(X_{n+1} - f)^2$ is minimized is given by $f = E(X_{n+1}|Y_n, \ldots, Y_0)$. Since this conditional expectation, in general, is difficult to find, we may consider the simpler problem, as we did in the previous section, of finding the best filter, $f_L = \Sigma_{s=0}^n l_s Y_{n-s}$, based on $(Y_n, \ldots, Y_0)$ such that $E(X_{n+1} - f_L)^2$ is minimized. We shall show that the optimum f_L is the projection of X_{n+1} on the linear space spanned by $(Y_n, Y_{n-1}, \ldots, Y_0)$. Assume $EX_n = 0$, $E\eta_n = 0$, so that $EY_n = 0$. Let $\mathscr{Y}_n$ be the linear space spanned by $(Y_n, Y_{n-1}, \ldots, Y_0)$, i.e. the space of all linear functions of $Y_n, \ldots, Y_0$. Define the inner product between any two elements u and v of $\mathscr{Y}_n$ as $E(uv)$. Consider,

$$E(X_{n+1} - f_L)^2 = E\{(X_{n+1} - \mathrm{proj}\ (X_{n+1}|\mathscr{Y}_n)) - (f_L - \mathrm{proj}\ (X_{n+1}|\mathscr{Y}_n))\}^2 \qquad (13)$$

where $\mathrm{proj}(X_{n+1}|\mathscr{Y}_n)$ denotes projection of X_{n+1} on $\mathscr{Y}_n$. By the definition of projection, we have

$$E\{(X_{n+1} - \text{proj}\ (X_{n+1}|\mathscr{Y}_n))u\} = 0$$

where u is any element of $\mathscr{Y}_n$. Since both f_L and proj $(X_{n+1}|\mathscr{Y}_n)$ are elements of $\mathscr{Y}_n$, their difference $f_L - \text{proj}\ (X_{n+1}|\mathscr{Y}_n)$ is again an element of $\mathscr{Y}_n$. Therefore

$$E\{(X_{n+1} - \text{proj}\ (X_{n+1}|\mathscr{Y}_n))(f_L - \text{proj}\ (X_{n+1}|\mathscr{Y}_n))\} = 0.$$

Using this result in (13) we have

$$E(X_{n+1} - f_L)^2 = E(X_{n+1} - \text{proj}\ (X_{n+1}|\mathscr{Y}_n))^2 + E(f_L - \text{proj}\ (X_{n+1}|\mathscr{Y}_n))^2.$$

Since the first term of the preceding expression is free from f_L and the second term is non-negative, it follows that the mean-square error is minimized for the choice

$$f_L = \text{proj}\ (X_{n+1}|\mathscr{Y}_n).$$

In the particular case (for instance, the normal process) when $E(X_{n+1}|Y_n, \ldots, Y_0)$ is a linear function it is clear that proj $(X_{n+1}|Y_n, \ldots, Y_0) = E(X_{n+1}|Y_n, \ldots, Y_0)$. In general, however, these two expressions are not the same.

3.2 Linear Filtering for a Linear Process

(Kalman-Bucy Filter)

Consider the linear (autoregressive) model

$$X_{n+1} = aX_n + \zeta_n. \tag{14}$$

Suppose that the process $\{X_n\}$ cannot be observed directly, but instead $\{Y_n\}$ can be observed where

$$Y_n = X_n + \eta_n. \tag{15}$$

The problem is that of estimating X_{n+1} by a linear function, $\hat{X}^L_{n+1}$, of $\{Y_n, Y_{n-1}, \ldots, Y_0\}$ such that the mean-square error is minimized. From the previous section we know that the optimum $\hat{X}^L_{n+1}$ is given by the projection of X_{n+1} on the linear space spanned by $\{Y_n, Y_{n-1}, \ldots, Y_0\}$. In this section we give an iterative procedure of determining $\hat{X}^L_{n+1}$. The solution is known as the Kalman–Bucy filter.

We first need the following assumptions regarding the error sequences $\{\xi_n\}$ and $\{\eta_n\}$:

(i) $E(\xi_n) = 0$, $E(\xi_n X_n) = 0$, $E(\xi_n^2) = \sigma_\xi^2$, and $E(\xi_n \xi_m) - 0$, for $n \neq m$,

(ii) $E(\eta_n) = 0$, $E(\eta_n X_n) = 0$, $E(\eta_n^2) = \sigma_\eta^2$, and $E(\eta_n \eta_m) = 0$, for $n \neq m$,

(iii) $E(\xi_n \eta_m)=0$ for $n=m$ as well as for $n \neq m$,

(iv) $E(X_0)=0$, $E(X_0^2)=\sigma_{x0}^2$, $E(X_0 \xi_n)=0$, and $E(X_0 \eta_n)=0$, for all n.

Transform the observations $\{Y_0, Y_1, \ldots, Y_n\}$ to a new set $\{Y_0^*, Y_1^*, \ldots, Y_n^*\}$ by means of an orthogonal transformation as follows:

$$Y_0^* = Y_0$$

$$Y_k^* = Y_k - \text{proj }(Y_k | \mathcal{Y}_{k-1}),\ k=1, 2, \ldots, n.$$

Then, by construction, $E(Y_r^* Y_s^*)=0$ for $r \neq s$, $E(Y_0^*)=E(Y_k^*)=0$ for each k. Also, it will be seen that

$$\text{var }(Y_k^*)=E(Y_k^{*2})=P(k)+\sigma_\eta^2 \tag{16}$$

where

$$P(k)=E(X_k - \hat{X}_k^L)^2, \tag{17}$$

is the mean-square error. We can write $\mathcal{Y}_n$ as

$$\mathcal{Y}_n = \mathcal{Y}_{n-1} + (Y_n^*) \tag{18}$$

where $\mathcal{Y}_{n-1}$ and (Y_n^*) are orthogonal (uncorrelated). Define A_{n-1} as the linear space spanned by

$$\{X_0, \xi_0, \ldots, \xi_{n-2}, \eta_0, \ldots, \eta_{n-1}\}.$$

Clearly, $\mathcal{Y}_{n-1}$ is contained in the space A_{n-1}. Since ξ_n is orthogonal to each element of A_{n-1}, it will be orthogonal to each element of $\mathcal{Y}_{n-1}$. Consequently,

$$\text{proj }(\xi_n | \mathcal{Y}_{n-1})=0. \tag{19}$$

By the same arguments we also have

$$\text{proj }(\eta_n | \mathcal{Y}_{n-1})=0. \tag{20}$$

Now, using (18) we have

$$\begin{aligned} \hat{X}_{n+1}^L &= \text{proj }(X_{n+1} | \mathcal{Y}_n) = \text{proj }(X_{n+1} | (Y_n^*)) + \text{proj }(X_{n+1} | \mathcal{Y}_{n-1}) \\ &= K_n Y_n^* + a \text{ proj }(X_n | \mathcal{Y}_{n-1}) + \text{proj }(\xi_n | \mathcal{Y}_{n-1}) \\ &= K_n Y_n^* + a\hat{X}_n^L, \end{aligned} \tag{21}$$

where K_n is determined from the fact that $(X_{n+1} - \hat{X}_{n+1}^L)$ is orthogonal to Y_n. Since

$$\begin{aligned} Y_n^* &= Y_n - \text{proj }(Y_n | \mathcal{Y}_{n-1}) \\ &= Y_n - \text{proj }(X_n | \mathcal{Y}_{n-1}) - \text{proj }(\eta_n | \mathcal{Y}_{n-1}) \\ &= Y_n - \hat{X}_n^L, \end{aligned}$$

we get after substitution in (21), the recursive relation

$$\hat{X}_{n+1}^{L}=(a-K_{n})\hat{X}_{n}^{L}+K_{n}Y_{n}. \tag{22}$$

The usual initial condition assumed is $X_0^L=E(X_0)=0$. It now remains to determine K_n. It will be seen that the condition $E\{X_{n+1}-\hat{X}_{n+1}^L)Y_n\}=0$ gives

$$K_n=\frac{aP(n)}{P(n)+\sigma_\eta^2} \tag{23}$$

where $P(n)$ is the mean-square error defined in (17). Using (22) and noting that $P(n)=E\{(X_n-\hat{X}_n^L)X_n\}$, we get after some straightforward manipulations, the recursive equation for $P(n)$ given by

$$P(n)=a(a-K_{n-1})P(n-1)+\sigma_\xi^2 \tag{24}$$

with the initial condition

$$P(0)=\sigma_{x0}^2.$$

Equations (22), (23) and (24) constitute the Kalman–Bucy solution to the filtering problem for the linear process defined by (14) and (15).

4. PARAMETER ESTIMATION

4.1 Autoregressive Process

Consider a first-order scalar autoregressive process $\{X_n\}$ defined recursively by

$$X_n=\rho X_{n-1}+Z_n, \tag{25}$$

where $\{Z_n\}$ are independent identically distributed random variables with $E(Z_n)=0$ and var $(Z_n)=\sigma^2<\infty$. Suppose $|\rho|<1$. An estimator of ρ based on the realization $(X_1, \ldots, X_n)$ is the well-known serial correlation

$$\hat{\rho}_n=\left(\sum_1^n X_{k-1}^2\right)^{-1}\left(\sum_1^n X_k X_{k-1}\right) \tag{26}$$

where the initial value X_0 is assumed to be known. It is obvious that $\hat{\rho}_n$ is the least-square estimator of ρ obtained by minimizing the error sum of squares

$$\sum_1^n (X_k-\rho X_{k-1})^2.$$

If the Z_k are normally distributed it can be verified that $\hat{\rho}_n$ is the maximum likelihood estimator of ρ. In this section we shall establish the consistency and asymptotic normality of $\hat{\rho}_n$ without making the normality assumption about $\{Z_k\}$. These results can be used to obtain confidence intervals and tests of significance for ρ.

Consistency. Note that $\hat{\rho}_n$ is the solution of the normal equation

$$S_n = \sum_1^n (X_k - \rho X_{k-1}) X_{k-1} = 0. \tag{27}$$

It can be seen that $\{S_n, n=1, 2, \ldots\}$ is a zero-mean martingale. By an application of the law of large numbers for martingales (see Feller, 1966) we find that

$$\frac{S_n}{n} \to 0, \text{ in probability, as } n \to \infty. \tag{28}$$

Also, since

$$\frac{1}{n} \sum_1^n X_{k-1}^2 \to \sigma^2 \sum_{s=0}^{\infty} \rho^{2s} = \frac{\sigma^2}{(1-\rho^2)}, \text{ in probability,} \tag{29}$$

by the law of large numbers for Markov processes (see Billingsley, 1961) the result

$$\hat{\rho}_n \to \rho, \text{ in probability,} \tag{30}$$

follows. A direct proof is available in Anderson (1971).

Asymptotic Normality. We will now show that

$$\sqrt{n}(\hat{\rho}_n - \rho) \to N(0, 1-\rho^2), \text{ in distribution.} \tag{31}$$

Note that

$$\sqrt{n}(\hat{\rho}_n - \rho) = \left(\frac{1}{\sqrt{n}} S_n\right) \left(\frac{1}{n} A_n\right)^{-1} \tag{32}$$

where

$$S_n = \sum_1^n X_k X_{k-1} - \rho \sum_1^n X_{k-1}^2 = \sum_1^n X_{k-1} Z_k \tag{33}$$

and

$$A_n = \sum_1^n X_{k-1}^2.$$

We have seen in (29) that $n^{-1}\ A_n \to \sigma^2/(1-\rho^2)$ in probability. The result in (31) will be proved if we show that

$$\frac{1}{\sqrt{n}} S_n \to N\left(0,\ \frac{\sigma^4}{1-\rho^2}\right) \tag{34}$$

Since var $(n^{-1/2} S_n) = 1/n\, \sigma^2 \Sigma_1^n E(X_{k-1}^2) \to \sigma^4 (1-\rho^2)^{-1}$ it follows on application of central limit theorem for martingales (see Billingsley, 1961) that the result in (34) holds.

For a direct proof of the asymptotic normality see Anderson (1971).

Generalization to First-order Vector Autoregressive Process. Consider the vector autoregressive process

$$\mathbf{X}_n = \mathbf{A}\mathbf{X}_{n-1} + \mathbf{Z}_n \tag{35}$$

where $\mathbf{X}$ and $\mathbf{Z}$ are p-component random vectors and $\mathbf{A}$ is a non-random $p \times p$ matrix. The condition for stationarity now becomes: the latent roots of the matrix $\mathbf{A}$ are all less than one in absolute value. The error vector $\mathbf{Z}_n$ satisfies $E(\mathbf{Z}_n) = 0$, $E(\mathbf{Z}_n \mathbf{Z}_n') = \mathbf{\Sigma}$, where the prime denotes the transpose. Also $\{\mathbf{Z}_n\}$ are assumed to be independent identically distributed random vectors.

The stationary solution of (35) is

$$\mathbf{X}_n = \sum_{s=0}^{\infty} \mathbf{A}^s \mathbf{Z}_{n-s},$$

and then, $E(\mathbf{X}_n \mathbf{X}_n') = \sum_{s=0}^{\infty} \mathbf{A}^s \mathbf{\Sigma} (\mathbf{A}')^s = \mathbf{F}$, say. (36)

The least-square normal equation for estimating $\mathbf{A}$ is

$$\left(\sum_{k=1}^{n} \mathbf{X}_{k-1} \mathbf{X}_{k-1}'\right) \mathbf{A}' = \sum_{k=1}^{n} \mathbf{X}_{k-1} \mathbf{X}_k' \tag{37}$$

which corresponds to (26). Let $\hat{\mathbf{A}}$ be the solution of (37). We then obtain the vector analogues of (30) and (31) as

$$\hat{\mathbf{A}} \to \mathbf{A} \text{ in probability,} \tag{38}$$

and,
$$\sqrt{n}(\hat{\mathbf{A}} - \mathbf{A}) \to N_{p^2}(\mathbf{0},\ \mathbf{V}), \text{ in distribution,} \tag{39}$$

where $\mathbf{V} = F^{-1} \otimes \mathbf{\Sigma}$, $\otimes$ denoting the direct product.

rth Order Autoregressive (Scalar) Process. Consider the process

$$X_n = \sum_{s=1}^{r} \beta_s X_{n-s} + Z_n. \tag{40}$$

This can be written as a first-order vector (with r components) autoregressive process given by

$$\mathbf{X}_n^* = \mathbf{B}^* \mathbf{X}_{n-1}^* + \mathbf{Z}_n^* \tag{41}$$

where

$$\mathbf{X}_n^* = \begin{bmatrix} X_n \\ X_{n-1} \\ \vdots \\ X_{n-r+1} \end{bmatrix}, \quad \mathbf{Z}_n^* = \begin{bmatrix} Z_n \\ 0 \\ \vdots \\ 0 \end{bmatrix}, \text{ and}$$

$$\mathbf{B}^* = \begin{bmatrix} \beta_1 & \beta_2 & \beta_3 & \ldots, & \beta_{r-1} & \beta_r \\ 1 & 0 & 0 & \ldots & 0 & 0 \\ 0 & 1 & 0 & \ldots & 0 & 0 \\ & \ldots & \ldots & \ldots & \ldots & \\ 0 & 0 & 0 & \ldots & 1 & 0 \end{bmatrix}$$

The stationarity condition now becomes: the roots of the polynomial equation

$$\lambda^r - \beta_1 \lambda^{r-1} - \ldots - \beta_r = 0$$

are all less than one in absolute value. If β' denotes the first row of $\mathbf{B}^*$, the least-square estimating equation for β is given by

$$\left(\sum_{k=1}^{n} \mathbf{X}_{k-1}^* \mathbf{X}_{k-1}^{*\prime} \right) \hat{\boldsymbol{\beta}} = \sum_{k=1}^{n} \mathbf{X}_{k-1}^* \mathbf{X}_k'. \tag{42}$$

From (38) and (39) we obtain the results:

$$\hat{\boldsymbol{\beta}} \to \boldsymbol{\beta} \text{ in probability} \tag{43}$$

and

$$\sqrt{n}(\hat{\boldsymbol{\beta}} - \boldsymbol{\beta}) \to N_r(\mathbf{0}, \sigma^2 \mathrm{F}^{-1}), \tag{44}$$

where F is the stationary dispersion, $E(\mathbf{X}_n^* \mathbf{X}_{n-1}^{*\prime})$. An estimator of σ^2 is

$$\hat{\sigma}^2 = 1/(n-r)\Sigma_{k=1}^{n} (X_k - \hat{\beta}_1 X_{k-1} - \hat{\beta}_2 X_{k-2} - \ldots \hat{\beta}_r X_{k-r})^2.$$

4.2 Moving Average Process

Consider the simplest case of a first order moving average process defined by

$$X_n = Z_n + \alpha Z_{n-1} \tag{45}$$

where $\{Z_n\}$ is a sequence of independent identically distributed random variables with $E(Z_n)=0$ and var $(Z_n)=1$. The problem is that of estimating α from a realization $\{X_1, X_2, \ldots, X_n\}$. A simple moment estimator can be obtained as follows:

since $E(X_k X_{k+1}) = \alpha$, we may consider

$$\hat{\alpha} = \frac{1}{(n-1)} \sum_{1}^{n-1} X_k X_{k+1} \tag{46}$$

as an estimator of α. Clearly $\hat{\alpha}$ is an unbiased estimator of α, and

$$\text{var}\ (\hat{\alpha}) \equiv \frac{1+\alpha^2\gamma+\alpha^4}{(n-1)} \tag{47}$$

where $\gamma = E(Z_n^4)$, (assumed to be finite).

Since var $(\hat{\alpha}) \to 0$ as $n \to \infty$, and $E(\hat{\alpha}) = \alpha$, it follows via Chebyshev's inequality that $\hat{\alpha}$ is a consistent estimator of α. The asymptotic distribution of $\hat{\alpha}$ can be obtained using the central limit theorem for m-dependent processes (see Anderson, 1971, Chapter 7). A sequence of random variables $\{Y_k\}$ is said to be *m-dependent* if there exists a non-negative integer m such that every subset $\{X_{t_1}, \ldots, X_{t_s}\}$ is independent of any other subset $\{X_{t_k}, X_{t_{k+1}}, \ldots\}$, where $t_1 < \ldots < t_s < t_k < t_{k+1} < \ldots$, and $(t_k - t_s) > m$. Note that $\{X_k\}$ defined by (45) is a one-dependent process. Now define $f(X_k, X_{k+1}) = X_k X_{k+1}$. Then the process $\{f(X_k, X_{k+1}), k=1, 2, \ldots\}$ is a two-dependent process. By the central limit theorem for this latter process we have

$$\frac{1}{\sqrt{n}} \sum_{1}^{n} (X_k X_{k+1} - \alpha) \to N(0, 1+\alpha^2\gamma+\alpha^4) \tag{48}$$

and consequently one obtains the results

$$(n-1)^{1/2}(\hat{\alpha}-\alpha) \to N(0, 1+\alpha^2\gamma+\alpha^4). \tag{49}$$

Similar methods can be used for higher order moving average processes. As an illustration, consider the second-order process defined by

$$X_n = Z_n + \alpha_1 Z_{n-1} + \alpha_2 Z_{n-2}. \tag{50}$$

Since $\quad E(X_k X_{k+1}) = \alpha_1 + \alpha_1\alpha_2$

and $\quad E(X_k X_{k+2}) = \alpha_2 \quad (51)$

we obtain the estimators of α_1 and α_2 after equating the sample moments with the corresponding expectations, as

$$\hat{\alpha}_1 = \left(1 + \frac{1}{n-2} \sum_1^{n-2} X_k X_{k+2}\right)^{-1} \left(\frac{1}{n-1} \sum_1^{n-1} X_k X_{k+1}\right)$$

and (52)

$$\hat{\alpha}_2 = \frac{1}{n-2} \sum_1^{n-2} X_k X_{k+2}.$$

The asymptotic distributions of $\hat{\alpha}_1$ and $\hat{\alpha}_2$ can be deduced via the central limit theorem for m-dependent processes and using Crámer's (1946) techniques for deriving the asymptotic distributions of functions of sample moments. This will not be pursued here any further, and is left as an exercise.

The maximum likelihood method can also be used for estimating the parameters of a moving average process (see Box and Jenkins, 1976, Chapter 7). This, however, does not lead to a simple solution. Even in the simplest case of the first-order normal process an iterative procedure has to be used. Two more alternative methods are due to Durbin and Walker both of which are described by Anderson (1971).

4.3 Iterative Procedure for Estimating the Parameters of an Autoregressive Process with Moving Average Residuals

Consider the autoregressive model with moving average residuals discussed in Section (2.2) with $p=1$, and $q=1$. Given a sample $\{X_0, X_1, \ldots X_n\}$, where

$$X_{k+1} - \alpha X_k = Z_{k+1} - \beta Z_k, \ (|\alpha| < 1, |\beta| < 1),$$

we are required to estimate the unknown parameters α and β. Here $\{Z_k\}$ are unobserved errors, independent and identically distributed with $E(Z_k)=0$ for each k. A natural criterion is to choose the estimates by least-squares method, i.e. minimize the error sum of squares, $\Sigma_{k=0}^{n} Z_k^2$ with respect to the parameters α and β. A computational difficulty with this approach is that (ΣZ_k^2) is a non-linear function of α and β and consequently the least-squares estimators cannot be given in a closed form. An iterative procedure can, however, be used to simplify computations. The method to be described here essentially entails approximating the error sum of squares by a linear function of α and β via Taylor's expansion around some trial value (α_0, β_0) of (α, β).

It is obvious that

$$Z_k = \sum_{r=0}^{k} (\beta^r X_{k-r}) - \alpha \sum_{r=0}^{k-1} (\beta^r X_{k-r-1})$$

after setting $X_{-1}=X_{-2}=\ldots=0$. For given $(X_0, \ldots, X_k)$, the error Z_k is thus a function (non-linear) of α and β, and let us denote this by $Z_k(\alpha, \beta)$. Let (α_0, β_0) be some initial estimate of (α, β). Expanding $Z_k(\alpha, \beta)$ around (α_0, β_0) up to the second term we get the linear approximation for $Z_k(\alpha, \beta)$ as follows:

$$Z_k(\alpha, \beta) \approx Z_k(\alpha_0, \beta_0) - \sum_{i=1}^{2} (\gamma_i - \gamma_i(o)) y_{ik}, \ (\mathrm{k}=0, 1, \ldots, n).$$

where $\gamma_1=\alpha$, $\gamma_2=\beta$, and $y_{1k}= -(\partial Z_k(\alpha, \beta)/\partial\alpha)\alpha_0$, $y_{2k} = -(\partial Z_k(\alpha, \beta)/\partial\beta)\beta_0$.

These (n+1) linear equations approximating the $(n+1)$ error terms $(Z_0, \ldots, Z_n)$ can be written compactly in matrix notation as follows:

$$\mathbf{Z}(\alpha_0, \beta_0) = \mathbf{Y}\,\boldsymbol{\gamma}^* + \mathbf{Z}(\alpha, \beta),$$

where $\mathbf{Z}'(\alpha, \beta) = (Z_0(\alpha, \beta), Z_1(\alpha, \beta), \ldots, Z_n(\alpha, \beta))$,

$$\mathbf{Y}' = \begin{pmatrix} y_{10} & y_{11} & \cdots & y_{1n} \\ y_{20} & y_{21} & \cdots & y_{2n} \end{pmatrix},$$

and $\boldsymbol{\gamma}^{*\prime} = (\alpha-\alpha_0, \beta-\beta_0)$, the prime denoting the transpose.

Treating $\mathbf{Z}(\alpha, \beta)$ as the error term and $\mathbf{Z}(\alpha_0, \beta_0)$ as the observation, we can now find an estimate of $\boldsymbol{\gamma}^*$ by minimizing $(Z(\alpha_0, \beta_0) - \mathbf{Y}\boldsymbol{\gamma}^*)'$ $(\mathbf{Z}(\alpha_0, \beta_0) - \mathbf{Y}, \boldsymbol{\gamma}^*)$. Assuming that $\mathbf{Y}'\mathbf{Y}$ has an inverse the least-square estimate of $\boldsymbol{\gamma}^*$ s seen to be

$$\hat{\boldsymbol{\gamma}}^* = (\mathbf{Y}'\mathbf{Y})^{-1}\mathbf{Y}'\mathbf{Z}(\alpha_0, \beta_0)..$$

Thus, the estimate of $\boldsymbol{\gamma}$ given by the *first* iteration is

$$\hat{\boldsymbol{\gamma}}(1) = \boldsymbol{\gamma}_0 + \boldsymbol{\gamma}^*,$$

where $\boldsymbol{\gamma}_0' = (\alpha_0, \beta_0)$. One can now use $\hat{\boldsymbol{\gamma}}(1)$ as the new initial estimate and repeat the above procedure to get an improved estimate $\hat{\boldsymbol{\gamma}}(2)$ on the second iteration. This process is continued until the estimates stabilize.

5. FURTHER TOPICS

5.1 Determination of the Order of an Autoregressive Process

For an rth order autoregressive process, $X_n = \Sigma_{s=1}^{r} \beta_s X_{n-s} + Z_n$, we have seen that the least-square estimator $\hat{\boldsymbol{\beta}}$ of the vector of coefficients $\boldsymbol{\beta}$ is given by the solution of the equations in (42). We also deduced the result

$$\sqrt{\mathrm{n}}(\hat{\boldsymbol{\beta}} - \boldsymbol{\beta}) \rightarrow \mathrm{N}_r(0, \sigma^2\mathbf{F}^{-1}).$$

Suppose we now wish to test the hypothesis that the order of the process is m ($<r$); more specifically we wish to test H:$\beta_{m+1}= \ldots =\beta_r=0$. The likelihood ratio test of this hypothesis leads to the statistic

$$T=\frac{\hat{\boldsymbol{\beta}}'_{(2)}(\mathbf{F}_{22}-\mathbf{F}_{21}\mathbf{F}_{11}^{-1}\mathbf{F}_{12})\hat{\boldsymbol{\beta}}'_{(2)}}{\hat{\sigma}^2} \tag{53}$$

where $\mathbf{F}$ is partitioned into m and r-m rows and columns as

$$\mathbf{F}=\left[\begin{array}{cc}\mathbf{F}_{11} & \mathbf{F}_{12} \\ \hline \mathbf{F}_{21} & \mathbf{F}_{22}\end{array}\right] \text{ and } \hat{\boldsymbol{\beta}} \text{ is similarly partitioned as } \left[\begin{array}{c}\hat{\boldsymbol{\beta}}_{(1)} \\ \hline \hat{\boldsymbol{\beta}}_{(2)}\end{array}\right],$$

and the prime denotes the transpose. The large values of T correspond to the critical region. Under the hypothesis H, the statistic T is asymptotically distributed as $\chi^2(r-m)$.

An alternative simpler procedure of determining (guessing rather) the order is to plot the residual sum of squares (after inserting the least-squares estimators) against the order of the process. The residual sum of squares always decreases as the order increases progressively; however, in many cases it is possible to see where the curve flattens out such that, for instance, the difference between mth and $(m+1)$th order is negligible. Also, $\hat{\sigma}^2$ can be plotted against the order. The latter criterion appears to be more sensible, since $\hat{\sigma}^2$ first decreases and then increases as the order is increased.

Also, see Complement 3 at the end of the chapter for a third method.

5.2 Explosive Autoregressive Process

For the first order autoregressive process

$$X_n=\rho X_{n-1}+Z_n$$

studied earlier we had assumed $|\rho|<1$ in order to get a limiting distribution for X_n. The case $|\rho|>1$ is referred to as explosive or unstable. It can be shown that the serial correlation $\hat{\rho}_n$ given by (25) continues to be consistent for ρ. For the asymptotic normality, however, we need to use a random norm, $(\Sigma_1^n X_{k-1}^2)^{1/2}$, instead of the usual norm $n^{1/2}$, and also assume Z_n to be normal. Thus,

$$\left(\sum_1^n X_{k-1}^2\right)^{1/2}(\hat{\rho}_n-\rho)\rightarrow N(0,\sigma^2) \tag{54}$$

provided the errors $\{Z_n\}$ are normally distributed. If Z_n are not normal, $(\Sigma_1^n X_{k-1}^2)^{1/2}(\hat{\rho}_n-\rho)$ has a non-normal limiting distribution.

Anderson (1959) provides proofs of these assertions. Feigin (1975) derives these results using martingale theory.

5.3 Spatial Autoregressive Process

Let $(X_1, \ldots, X_n)$ denote random variables corresponding to n sites (points) on a plane. Suppose that the conditional density of the X_i given all other Xs is normal with common variance σ^2 and the mean given by $\Sigma\beta_{ij}x_j$ where $\beta_{ij}=\beta_{ji}$, $\beta_{ii}=1$, and $\beta_{ij}=0$ if the sites i and j are not neighbours of each other where "neighbourhood" is defined in some suitable manner. The likelihood function corresponding to this model is given by

$$L_1 = \frac{|\mathbf{B}|^{1/2}(b^{11})^{n/2}}{(2\pi\sigma^2)^{n/2}} \exp\left\{-\frac{b^{11}}{2}\sigma^2\mathbf{X}'\mathbf{B}\mathbf{X}\right\} \tag{55}$$

where $\mathbf{X}'=(X_1, \ldots, X_n)$, $\mathbf{B}$ is the $n\times n$ matrix whose diagonal elements are unity and off-diagonal (i, j)th elements $-\beta_{ij}$, and b^{11} is the (1, 1)th element of $\mathbf{B}^{-1}$. Note that $b^{11}=b^{ii}$ for any i.

An alternative model for spatial autoregression is the following

$$X_i=\Sigma\beta_{ij}X_j+Z_i \tag{56}$$

where Z_i are independent normal variables with mean zero and variance σ^2, and β_{ij} have the same meaning as before. The likelihood function now becomes

$$L_2=\frac{|\mathbf{B}|}{(2\pi\sigma^2)^{n/2}} \exp\left\{-\frac{1}{2\sigma^2}\mathbf{X}'\mathbf{B}'\mathbf{B}\mathbf{X}\right\}. \tag{57}$$

The maximum likelihood estimator of $\mathbf{B}$ for the latter model can be obtained by minimizing

$$-2n^{-1}\log|\mathbf{B}|+\log(\mathbf{X}'\mathbf{B}'\mathbf{B}\mathbf{X}). \tag{58}$$

The real difficulty in obtaining the estimators is that of evaluating the determinant $|\mathbf{B}|$. This is discussed by Ord (1975) who also provides a computational method. See Besag (1974), and Cliff and Ord (1975) for statistical analysis of models of this kind.

BIBLIOGRAPHICAL NOTES

Box and Jenkins (1976) discuss topics related to Sections 2 and 3. Anderson (1971) discusses the estimation problems of Section 4 rather extensively. The iterative procedure of Section 4.3 is also given in Box and Jenkins (1976).

COMPLEMENTS

1. Consider the model

$$X_t = \sum_{r=1}^{p} \beta_r X_{t-r} + \sum_{r=1}^{q} \gamma_r y_{rt} + \varepsilon_t, \quad t=1, 2, \ldots, n$$

where ε_t are independent identically distributed $N(0, \sigma^2)$ variables, and y_{rt}, $r=1, \ldots$ q given constants. Suppose $X_0, X_{-1}, \ldots, X_{-(p-1)}$ are given. Find the maximum likelihood estimates of $(\beta_1, \ldots, \beta_p)$ and $(\gamma_1, \ldots, \gamma_q)$, and establish their asymptotic properties.

(Anderson, 1971)

2. Let $\hat{\beta}_p$ be the least-square estimate of β_p in a pth-order autoregressive model (with errors not necessarily normal). Show that, under the hypothesis $\beta_p=0$, $n^{1/2}\hat{\beta}_p$ is asymptotically normal with mean zero and variance 1.

3. Consider a pth-order autoregressive model

$$X_t = \sum_{r=1}^{p} \beta_r X_{t-r} + \varepsilon_t$$

where ε_t are independent identically distributed random variables with $E(\varepsilon_t)=0$ and $V(\varepsilon_t)=\sigma^2$. Let $\hat{\beta}_r(m)$ denote the least-square estimates of β_r and $S^2(m)$ a consistent estimate of σ^2 under the assumption of an mth ($m<p$) order process (i.e. setting $\beta_{m+1}=\ldots=\beta_p=0$). Assuming that the roots of the polynomial equation associated with the mth-order difference equation are less than unity, show that the statistic

$$\frac{\sum_{j=m+1}^{p} \left(\frac{1}{\sqrt{n}} \sum_{k,l,=1}^{m} \hat{\beta}_k \hat{\beta}_l a_{l,j-k} \right)}{S^4(m)}$$

is asymptotically distributed as $\chi^2(p-m)$, where $a_{r,s}$ is the (r, s)th element of $1/n \sum_{t=1}^{n} \mathbf{x}_{t-1}\mathbf{x}_{t-1}^T$, and $\mathbf{x}_{t-1}=(x_{t-1}, x_{t-2}, \ldots, x_{t-p})$.

(This statistic can be used as an alternative test for $H{:}\beta_{m+1}=\ldots=\beta_p=0$).

(Bartlett and Diananda, 1950)

4. Let $(X_1, \ldots, X_n)$ be independent identically distributed random variables with an arbitrary distribution and assume $E(X_1^2)<\infty$. The serial correlation is defined as

$$R = \frac{\sum_{i=1}^{n-1} (X_i - \bar{X})(X_{i+1} - \bar{X})}{\sum_{i=1}^{n} (X_i - \bar{X})^2},$$

where $\bar{X} = 1/n \sum_{i=1}^{n} X_i$. Show that $n^{1/2} R$ is asymptotically distributed as a standard normal variate.

(Shanbhag and Basawa, 1972)

5. Let $\{X_t\}$, $t=0, 1, 2, \ldots$, satisfy a pth order difference equation

$$X_t = \sum_{r=1}^{p} \beta_r X_{t-r} + \varepsilon_t,$$

where the errors $\{\varepsilon_t\}$ themselves satisfy a qth order difference equation

$$\varepsilon_t = \sum_{j=1}^{q} \alpha_j \varepsilon_{t-j} + \eta_t,$$

with $\{\eta_t\}$ independent identically distributed random variables satisfying $E(\eta_t)=0$ and $E(\eta_t^2)=\sigma^2$. Investigate the problem of estimating $\beta_1, \ldots, \beta_p, \alpha_1, \ldots, \alpha_q$, and σ^2. Establish the asymptotic properties of your estimates as the sample size tends to infinity.

6. For the second-order moving-average process

$$X_n = Z_n + \alpha_1 Z_{n-1} + \alpha_2 Z_{n-2}$$

the moment estimates of α_1 and α_2 based on a realization $\{X_1, \ldots, X_n\}$ are given by equation (52). Find the asymptotic joint distribution of these estimates under appropriate regularity assumptions.

7. Consider a mixed autoregressive moving average process $\{X_k\}$ defined by

$$X_n - \rho X_{n-1} = \xi_n - \eta \xi_{n-1}, \; (-1<\rho<1, \; -1<\eta<1),$$

where $\{\xi_n\}$ are independent identically distributed normal variables with zero mean and variance unity. Describe an iterative procedure for determining the maximum likelihood estimates of ρ and η from a realization $\{X_1, \ldots, X_n\}$. Find the asymptotic covariance matrix of $\sqrt{n}(\hat{\rho}, \hat{\eta})$ where $\hat{\rho}$ and $\hat{\eta}$ are the maximum likelihood estimates of ρ and η respectively.

(Box and Jenkins, 1976).

Chapter 4

Discrete Markov Chains

1. INTRODUCTION

The topic of inference for Markov chains appears to have received much attention in the sixties and the work culminated in a comprehensive book by Billingsley (1961). In recent years several publications on the topic have appeared on special aspects. In this chapter we discuss chains with finite state-space as well as those with countable state-space.

2. FINITE MARKOV CHAINS

2.1 Maximum Likelihood Estimation

Let $\{X_k\}$, $k=0, 1, 2, \ldots$ be an ergodic (positively regular) Markov chain on the state-space, $S=\{1, 2, \ldots, m\}$ with transition probabilities

$$P_{ij}=P[X_k=j \mid X_{k-1}=i],\ i, j \in S,$$

the initial probabilities

$$p_u^{(0)}=P[X_0=u],\ u \in S,$$

and the stationary (also limiting) distribution $\{\pi_j\}$, $\pi_j>0$, $\Sigma_{j=1}^m \pi_j=1$, where π_j

satisfy

$$\pi_j = \sum_{k=1}^{m} \pi_k p_{kj},$$

and

$$\lim_{n \to \infty} p_{ij}^{(n)} = \pi_j.$$

(a) Non-Parametric Case. Let $\mathbf{x}_{n+1} = (x_0, x_1, \ldots, x_n)$ be a realization of length $(n+1)$. The likelihood function based on this sample is then given by

$$\begin{aligned} L &= p_{x0}^{(0)} \prod_{k=1}^{n} p_{x_{k-1}x_k} \\ &= p_{x_0}^{(0)} \prod_{i,j,=1}^{m} p_{ij}^{n_{ij}} \end{aligned} \tag{1}$$

where n_{ij} is the frequency of the (one-step) transitions $i \to j$ in the sample $\mathbf{x}_{n+1}$. Clearly the set of m^2 *transition frequencies* $((n_{ij}))$ forms a *sufficient statistic* for the transition matrix $\mathbf{P}$.

We need to maximize

$$\log L = \log p_x^{(0)} + \sum_{i,j} n_{ij} \log p_{ij} \tag{2}$$

with respect to p_{ij}'s subject to the restriction $\Sigma_j p_{ij} = 1$. This yields the maximum likelihood estimators

$$\hat{p}_{ij} = \frac{n_{ij}}{n_i}, \tag{3}$$

where $n_i = \Sigma_j n_{ij}$. In deriving the above estimate we have assumed that the initial probability distribution $\{p_u^{(0)}\}$ is non-informative about the transition probabilities p_{ij}. In any case, for large n the effect of the first term in (2) can be ignored.

(b) Parametric Case. Suppose p_{ij} are known functions of a certain unknown parameter $\boldsymbol{\theta} = (\theta_1, \ldots, \theta_k)$, say $p_{ij}(\boldsymbol{\theta})$, taking values in R^k. The maximum likelihood estimators of the parameter $\boldsymbol{\theta}$ can be obtained by solving the system of equations

$$\frac{\partial \log L}{\partial \theta_r} = 0, \quad r = 1, 2, \ldots, k$$

i.e.

$$\sum_{i,j=1}^{m} \left[n_{ij} \frac{\partial p_{ij}(\boldsymbol{\theta})}{\partial \theta_r} \cdot \frac{1}{p_{ij}(\boldsymbol{\theta})} \right] = 0, \quad r = 1, 2, \ldots, k \tag{4}$$

simultaneously for $\theta_1, \theta_2, \ldots, \theta_k$. The existence of non-trivial sufficient statistics for $\boldsymbol{\theta}$, however, is not so straightforward. Gani (1955) gives necessary and sufficient conditions for the existence of a sufficient statistic in the case of a single parameter θ.

2.2 Asymptotic Properties of the Estimators

Since the state-space S is finite we can consider the non-parametric case as a special case of the parametric case with $m(m-1)$ parameters $(p_{i1}, p_{i2}, \ldots, p_{im-1})$, $i=1,2,\ldots,m$. Note that $p_{im}=1-\Sigma_{j=1}^{m-1} p_{ij}$. However, it is convenient to consider the non-parametric case first. Consistency and asymptotic normality of the maximum likelihood estimators can be deduced using the analogy with the multinomial distribution.

Assume for simplicity that the process has been going on for a long period so that it is stationary. Then, we can assume that $p_u^{(0)}=\pi_u$. Then $p_u^{(k)}=P(X_k=u)=\pi_u$, for any k. It should be noted that the assumption $p_u^{(0)}=\pi_u$ contains information about the transition probabilities p_{ij} since $\{\pi_u\}$ satisfy

$$\pi_u=\sum_{k=1}^{m} \pi_u p_{ku}.$$

Consequently, $\hat{p}_{ij}=n_{ij}/n_i$ which we obtained in the previous section as the maximum likelihood estimator of p_{ij} under the assumption that the initial probability distribution $p_{u.}^{(0)}$ does not give any information about p_{ij}'s will, strictly speaking, no longer be the maximum likelihood estimator of p_{ij} in the present case. However, for large n, the log-likelihood function

$$\log L=\log \pi_{x_0}+\sum_{i,j} n_{ij} \log p_{ij}$$

will be dominated by the second term, so that we can write

$$\log L\approx\sum_{i,j} n_{ij} \log p_{ij}.$$

Therefore, as far as the asymptotic theory is concerned, it is enough to consider $\hat{p}_{ij}=n_{ij}/n_i$. First, we need the following result

$$p \lim_{n\to\infty}\left(\frac{n_i}{n}\right)=\pi_i,\ i\in S. \tag{5}$$

To prove this, set

$$Z_i^{(r)}=\begin{cases}1 & \text{if } X_r=i\\ 0 & \text{if } X_r\neq i,\end{cases}$$

then

$$n_i=\sum_{r=0}^{n-1} Z_i^{(r)}$$

Now

$$E(n_i) = \sum_{r=0}^{n-1} E(Z_i^{(r)})$$

$$= n\, P[X_r = i] = n\pi_i.$$

We note that the distribution of the recurrence-time of state i has a finite mean since we are dealing with an ergodic chain. It therefore follows from the recurrent event theory (Feller, 1957) that

$$p \lim_{n\to\infty} \left(\frac{n_i}{n}\right) = E\left(\frac{n_i}{n}\right) = \pi_i,\ i \in S.$$

We can now prove the following important result regarding the asymptotic distribution of $\hat{p}_{ij}$.

Let $\phi = (p_{11}, p_{12}, \ldots, p_{1m}, \ldots, p_{m1}, p_{m2}, \ldots, p_{mm})$ be the $m^2 \times 1$ parameter-vector, and

$$\xi = (\hat{p}_{11}, \hat{p}_{12}, \ldots, \hat{p}_{1m}, \ldots, \hat{p}_{m1}, \hat{p}_{m2}, \ldots, \hat{p}_{mm})$$

be the $m^2 \times 1$ vector of estimators, where $\hat{p}_{ij} = n_{ij}/n_i$.

Then,

$$\sqrt{n}(\xi - \phi) \to N_{m^2}(0, \Sigma), \text{ in distribution,} \tag{6}$$

where the components of the matrix Σ are given by

$$\sigma_{ij,i'j'} = \frac{1}{\pi_i}\{\delta_{ii'}(\delta_{jj'}p_{ij} - p_{ij}p_{i'j'})\} \tag{7}$$

where δ_{uv} is the Kronecker delta.

We shall only give the main steps in the proof which is essentially straightforward.

For fixed i, the vector $(n_{i1}, n_{i2}, \ldots, n_{im})$ behaves like a multinomial vector with cell probabilities $(p_{i1}, p_{i2}, \ldots, p_{im})$ and (random) index n_i.

Now, for large n, n_i may be replaced by a fixed quantity $[n\pi_i]$, the integral part of $E(n_i) = n\pi_i$. The random variables

$$U_{ij} = \frac{n_{ij} - [n\pi_i]p_{ij}}{\sqrt{n}\ \pi_i}$$

are then jointly asymptotically normal (see Cramer, 1946) with zero means and

$$\operatorname{cov}(U_{ij}, U_{i'j'}) = \frac{\delta_{ii'}p_{ij}(\delta_{jj'} - p_{i'j'})}{\pi_i}. \tag{8}$$

(Notice that U_{ij} are independently distributed for different i). It can be verified that

$$\frac{n_{ij}-[n\pi_i]p_{ij}}{\sqrt{n}}-\frac{n_{ij}-n_ip_{ij}}{\sqrt{n}}\to 0, \text{ as } n\to\infty, \text{ in probability.}$$

Thus, the $U_{ij}^*=(n_{ij}-n_ip_{ij})/\sqrt{n}\ \pi_i$ have the same asymptotic joint distribution as the U_{ij}. Now let

$$V_{ij}=\sqrt{n}(\hat{p}_{ij}-p_{ij})=\frac{U_{ij}^*\ \pi_i}{(n_i/n)}.$$

Since $(n_i/n)\overset{p}{\to}\pi_i$, V_{ij} and U_{ij}^* have the same asymptotic distribution, the desired limit result in (6) follows.

The asymptotic distribution of the maximum likelihood estimator in the parametric case can be stated as follows:

Let $\hat{\theta}$ be a consistent root of the likelihood equation,

$$\sum_{i,j}\frac{n_{ij}}{p_{ij}(\theta)}\ \frac{\mathrm{d}p_{ij}(\theta)}{\mathrm{d}\theta}=0.$$

Suppose the following regularity conditions are satisfied:

(i) $p_{ij}(\theta)=p_{ij}(\eta)$ for all i, j implies that $\theta=\eta$,

(ii) $p_{ij}(\theta)$ admit continuous second derivatives with respect to θ at the true value θ_0,

(iii) at least one $(\mathrm{d}p_{ij}(\theta)/\mathrm{d}\theta)_{\theta=\theta_0}$ is non-zero.

Then

$$\sqrt{n}(\hat{\theta}-\theta_0)\to N\left(0, \frac{1}{i(\theta_0)}\right), \text{ in distribution,} \tag{9}$$

where

$$i(\theta)_0=\sum_{i,j}\frac{\pi_i(\theta_0)}{p_{ij}(\theta_0)}\left(\frac{\mathrm{d}p_{ij}(\theta)}{\mathrm{d}\theta}\right)^2_{\theta_0} \tag{10}$$

is the Fisher information. (Note that only the terms corresponding to the positive $p_{ij}(\theta)$'s enter the sum in equation (10)).

The proof again uses the analogy of the multinomial distribution. (Use Rao's results (1973) regarding the maximum likelihood estimation of multinomial distribution for the parametric case.) The result can directly be extended to the multiparameter case as follows.

Let $\boldsymbol{\theta}=(\theta_1, \ldots, \theta_k)$ take values in a k-dimensional Euclidean space. The

$(k \times k)$ Fisher information matrix $\mathbf{F} = ((\sigma_{uv}))$, where

$$\sigma_{uv} = \sum_{i,j} \left\{ \frac{\pi_i(\boldsymbol{\theta}_0)}{p_{ij}(\boldsymbol{\theta}_0)} \left(\frac{\partial p_{ij}}{\partial \theta_u} \right)_{\theta_{u0}} \left(\frac{\partial p_{ij}}{\partial \theta_v} \right)_{\theta_{v0}} \right\}, \tag{11}$$

is assumed to be non-singular.

Then,

$$\sqrt{n}(\hat{\boldsymbol{\theta}} - \boldsymbol{\theta}_0) \rightarrow N_k(0, \mathbf{F}^{-1}) \text{ in distribution.} \tag{12}$$

2.3 Asymptotic Distribution of the Serial Correlation

The serial correlation is a natural measure of dependence between successive pairs of observations. In this section we define the autocorrelation for a simple (i.e. stationary and ergodic) Markov chain on a finite state-space $S = \{E_1, E_2, \ldots, E_m\}$, say, and then suggest a natural estimator (first-order serial correlation from the sample) of this measure. The asymptotic distribution of this estimator is derived and compared with that of the maximum likelihood estimator.

Because of the stationary assumption we have

$$p_u^{(n)} = P(X_n = u) = \pi_u \text{ for all } n = 0, 1, 2, \ldots, (u \in S).$$

The autocorrelation (of first-order) associated with the simple Markov chain $\{X_k\}$ is then defined by

$$\rho = \frac{\operatorname{cov}(X_k, X_{k+1})}{V(X_k)}, \tag{13}$$

which is free from k. Suppose that the states $\{E_1, \ldots, E_m\}$ represent ordered categories. For simplicity, we associate the ranks $\{1, 2, \ldots, m\}$ with these states. Then the autocorrelation is given by

$$\rho = \frac{\sum_{i,j} ij\theta_{ij} - \left(\sum_{i,j} i\theta_{ij} \right)^2}{\sum_{i,j} i^2\theta_{ij} - \left(\sum_{i,j} i\theta_{ij} \right)^2} = g(\theta_{ij}), \text{ say,} \tag{14}$$

where $\theta_{ij} = P(X_k = i, \ X_{k+1} = j) = \pi_i p_{ij}$. Obviously, we have $\Sigma_{i,j}\theta_{ij} = 1$, $\Sigma_j \theta_{ij} = \pi_i$ and $\Sigma_i \theta_{ij} = \pi_j$.

Now, let $(x_1, x_2, \ldots, x_n)$ be a realization of length n. The (circular) serial correlation of first-order based on these observations is defined as

$$R = \frac{\sum_{r=1}^{n} (x_r - \bar{x})(x_{r+1} - \bar{x})}{\sum_{r=1}^{n} (x_r - \bar{x})^2} \tag{15}$$

where $\bar{x} = 1/n \ \Sigma_{r=1}^{n} x_r$, and x_{n+1} is replaced by x_1. Let n_{ij} denote the sample transition frequency of the system from E_i to E_j in the (modified) sample $(x_1, x_2, \ldots, x_n, x_1)$. Then, $\Sigma_{i,j} n_{ij} = n$, $\Sigma_j n_{ij} = \Sigma_k n_{ki} = n_i$, where n_i is the frequency with which state E_i occurs in the sample. It is easy to verify that

$$R = g(\hat{\theta}_{ij}), \tag{16}$$

where g is defined in the expression for ρ, and $\hat{\theta}_{ij} = n_{ij}/n$. Since $n_i/n \to \pi_i$ and $n_{ij}/n_i \to p_{ij}$ in probability, we have $n_{ij}/n \to \theta_{ij}$ in probability. By the continuity of g we can then conclude that the estimate R is a consistent estimator of ρ. For the asymptotic distribution of R we first need the following limit result.

Let $\hat{\theta}_{ij} = n_{ij}/n$, and $\mathbf{T}_n = (\hat{\theta}_{11}, \ldots, \hat{\theta}_{1m}, \ldots, \hat{\theta}_{m1}, \ldots, \hat{\theta}_{mm})$, an $m^2 \times 1$ vector. Then

$$\sqrt{n}(\mathbf{T}_n - \boldsymbol{\theta}) \to N_{m^2}(\mathbf{0}, \boldsymbol{\Sigma}) \text{ in distribution.} \tag{17}$$

where $\boldsymbol{\theta} = (\theta_{11}, \ldots, \theta_{mm})$, and the components of $\boldsymbol{\Sigma}$ are given by

$$\begin{aligned} \sigma_{ij,i'j'} &= \lim_{n\to\infty} \frac{1}{n} \operatorname{cov}(n_{ij}, n_{i'j'}) \\ &= \pi_i p_{ij}(\delta_{ii'}\, \delta_{jj'} - \pi_{i'} p_{i'j'}) \\ &\quad + p_{ij} p_{i'j'}(\pi_i c_{ji'} + \pi_{i'} c_{j'i}), \end{aligned} \tag{18}$$

where $c_{uv} = \Sigma_{k=0}^{\infty} (p_{uv}^{(k)} - \pi_v)$, and δ_{uv} is the Kronecker delta.

To prove this, define the functions,

$$Z_{ij}^{(r)} = \begin{cases} 1 & \text{if } X_r = i \text{ and } X_{r+1} = j \\ 0 & \text{otherwise,} \end{cases}$$

$(r = 1, 2, \ldots, n,\ i,j = 1, 2, \ldots, m, \text{ and } X_{n+1} \equiv X_1)$.

Applying the central limit theorem for ergodic Markov chains (Doob, 1953, p. 228 and p. 232) to the functions $Z_{ij}^{(r)}$ defined on the successive pairs of the states we readily find that the random variables

$$\frac{1}{\sqrt{n}} \sum_{r=1}^{n} (Z_{ij}^{(r)} - E(Z_{ij}^{(r)})) = \sqrt{n}(\hat{\theta}_{ij} - \theta_{ij})$$

converge in distribution to normal variates with zero means, and variances given by

$$\lim_{n\to\infty} \frac{1}{n} \operatorname{var}(n_{ij}).$$

Now, an application of the central limit theorem to every linear combination of the functions $Z_{ij}^{(r)}$ shows that any linear combination of

$\sqrt{n}(\hat{\theta}_{ij}-\theta_{ij})$ is asymptotically normal, and hence the m^2 random variables $\sqrt{n}(\hat{\theta}_{ij}-\theta_{ij})$ are asymptotically jointly normal. It now remains to show that the covariance expression is as given in (18). To this end we proceed as follows:

$$\begin{aligned}
\operatorname{cov}(n_{ij}, n_{i'j'}) &= \operatorname{cov}\left(\sum_{r=1}^{n} Z_{ij}^{(r)}, \sum_{r=1}^{n} Z_{i'j'}^{(r)}\right) \\
&= E\left\{\left(\sum_{r=1}^{n} Z_{ij}^{(r)}\right)\left(\sum_{r=1}^{n} Z_{i'j'}^{(r)}\right)\right\} - E\left(\sum_{r=1}^{n} Z_{ij}^{(r)}\right) E\left(\sum_{r=1}^{n} Z_{i'j'}^{(r)}\right) \\
&= E\left\{\sum_{r=1}^{n} Z_{ij}^{(r)} Z_{i'j'}^{(r)} + \sum_{r \neq r'} Z_{ij}^{(r)} Z_{i'j'}^{(r')}\right\} \\
&\quad - E\left(\sum_{r=1}^{n} Z_{ij}^{(r)}\right) E\left(\sum_{r=1}^{n} Z_{i'j'}^{(r)}\right).
\end{aligned}$$

Since

$$Z_{ij}^{(r)} Z_{i'j'}^{(r)} = \begin{cases} 1 & \text{if and only if } i=i' \text{ and } j=j' \\ 0 & \text{otherwise,} \end{cases}$$

we must have $E(Z_{ij}^{(r)} Z_{i'j'}^{(r)}) = \delta_{ii'}\, \delta_{jj'}\, \pi_i\, p_{ij}$. It then follows that $\operatorname{cov}(n_{ij}, n_{i'j'}) = n\delta_{ii'}\, \delta_{jj'}\, \pi_i\, p_{ij} - n^2\, \pi_i\, p_{ij}\, \pi_{i'}\, p_{i'j'} + \Sigma_{r \neq r'} E(Z_{ij}^{(r)} Z_{i'j'}^{(r')})$. It can be seen, after some algebra that the last term

$$\begin{aligned}
\sum_{r \neq r'} E(Z_{ij}^{(r)} Z_{i'j'}^{(r')}) &= \sum_{k=0}^{n-2} (n-k-1)\, \pi_i p_{ij}\, p_{ji'}^{(k)}\, p_{i'j'} \\
&\quad + \sum_{k=0}^{n-2} (n-k-1)\, \pi_{i'}\, p_{i'j'}\, p_{j'i}^{(k)}\, p_{ij}.
\end{aligned}$$

Substituting this in the expression for $\operatorname{cov}(n_{ij}, n_{i'j'})$, dividing by n, and taking limits as $n \to \infty$ we get the required result. The result in (18) is thus established.

Note that because of ergodicity, $|p_{lm}^{(k)} - \pi_m| < a\eta^k$; where $a > 0$, and $0 < \eta < 1$ are constants, (see, for instance, Kemeny and Snell, 1960, p. 71) it follows on comparing with the geometric series with common ratio η that the infinite series c_{lm} occurring in equation (18) is convergent.

The asymptotic distribution of the serial correlation R can now be shown to be normal; more specifically

$$\sqrt{n}(R-\rho) \to N(0, \sigma_R^2), \text{ as } n \to \infty, \text{ in distribution,} \tag{19}$$

where $\sigma_R^2 = \mathbf{G}^T \mathbf{\Sigma} \mathbf{G}$, $\mathbf{G}^T = (\partial g/\partial\theta_{11}, \ldots, \partial g/\partial\theta_{mm})$, a $1 \times m^2$ vector and g is the function of m^2 variables θ_{ij} as defined by the expression for ρ.

Since $R=g(\hat{\theta}_{ij})$, the stated result follows from the limit result for $\hat{\theta}_{ij}$ just established and an application of a well known convergence theorem. (cf. Rao, 1973).

If the $\{X_k\}$ are independent and identically distributed with the (unspecified) distribution $\{\pi_u, u\in S\}$, it is easily verified that

$$\sqrt{n}R\to N(0,\ 1),\ \text{in distribution.} \tag{20}$$

Parametric Case and the Maximum Likelihood Estimator. Suppose that the transition probabilities p_{ij} are functions of a single parameter say α. Let $\rho(\alpha)=g(\theta_{ij}(\alpha))=h(\alpha)$, say, be a one-to-one function of α. Then, the maximum likelihood estimator of the autocorrelation is clearly, $\hat{\rho}=h(\hat{\alpha})$, where $\hat{\alpha}$ is the maximum likelihood estimator of α obtained by solving the likelihood equation,

$$\sum_{i,j} \frac{n_{ij}}{p_{ij}(\alpha)} \frac{\mathrm{d}p_{ij}(\alpha)}{\mathrm{d}\alpha}=0.$$

Since $\sqrt{n}(\hat{\alpha}-\alpha_0)\to N(0,\ 1/i(\alpha_0))$ in distribution, where

$$i(\alpha_0)=\sum_{i,j} \frac{\pi_i(\alpha_0)}{p_{ij}(\alpha_0)} \left(\frac{\mathrm{d}p_{ij}(\alpha)}{\mathrm{d}\alpha}\right)^2_{\alpha_0}$$

we have

$$\sqrt{n}(\hat{\rho}-\rho_0)\to N(0,\ \sigma^2_{\hat{\rho}})\ \text{in distribution,} \tag{21}$$

where

$$\sigma^2_{\hat{\rho}}=\left(\frac{\mathrm{d}h(\alpha)}{\mathrm{d}\alpha}\right)^2_{\alpha_0} i(\alpha_0). \tag{22}$$

This follows by a well known convergence result (cf. Rao, 1973).

In the present parametric case (when $h(\alpha)$ is a one-to-one function of α) we have at our disposal two alternative estimators of ρ, namely R and $\hat{\rho}$. Note that although R is based on the maximum likelihood estimators $\hat{\theta}_{ij}$ of θ_{ij}, in general, it is not identical with $\hat{\rho}$. Moreover, by the maximum likelihood efficiency property we should expect

$$\sigma^2_R \geqq \sigma^2_{\hat{\rho}}. \tag{23}$$

See Basawa (1972) for further work on the efficiency problem.

The estimator R (or $\hat{\rho}$ when applicable) is useful in certain tests of hypotheses regarding the transition probabilities. This problem will be taken up in a later section (see Section 2.5).

2.4 Tests of Hypotheses (χ-squared Tests)

The results of Section (2.2) can be used to derive tests of significance for various hypotheses regarding the Markov chain under consideration.

(a) Test for a Specified Transition Matrix. Consider the hypothesis $H: p_{ij} = p_{ij}^0$, $i, j \in S$. For any fixed i, $(p_{ij}^0, j = 1, 2, \ldots, m)$ is a probability distribution. Consider the Kullback–Leibler (KL) information for discriminating between p_{ij}^0 and its maximum likelihood estimator $\hat{p}_{ij} = n_{ij}/n_i$ from a sample $\mathbf{x}_n = (x_1, x_2, \ldots, x_n)$. This is given by

$$K_n^{(i)} = \sum_{j=1}^{m} p_{ij}^0 \log \frac{p_{ij}^0}{\hat{p}_{ij}}, \quad i \in S$$

$$\approx \frac{1}{2} \sum_j \frac{(n_{ij} - n_i p_{ij}^0)^2}{n_i^2 p_{ij}^0}.$$

A weighted average of KL measures for different i is

$$\sum_{i=1}^{m} n_i K_n^{(i)} \approx \frac{1}{2} \sum_{i,j} \frac{(n_{ij} - n_i p_{ij}^0)^2}{n_i p_{ij}^0} = \frac{1}{2}\phi, \text{ say.}$$

The statistic $\Sigma_j \, (n_{ij} - n_i p_{ij}^0)^2 / n_i p_{ij}^0$ is, in fact, the χ-squared goodness-of-fit statistic for the multinomial n_{ij} (n_i fixed) with cell probabilities $n_i p_{ij}^0$ and $\Sigma_j n_{ij} = n_i$ (index). Consequently, the statistic

$$\phi = \sum_{i,j} \frac{(n_{ij} - n_i p_{ij}^0)^2}{n_i p_{ij}^0} \tag{24}$$

is a sum of m independent χ-squared statistics corresponding to the m rows of the transition matrix. Using the analogy of the multinomial distribution (as in Section 2.2) we find that ϕ is asymptotically distributed as a χ-squared random variable with degrees of freedom $m(m-1)$. Here we have assumed that all p_{ij} are positive. More generally, suppose d_i is the number of positive entries in the ith row of P, and let $D_i = \{j : p_{ij} > 0\}$. Then

$$\phi = \sum_{i \in S} \sum_{j \in D_i} \frac{(n_{ij} - n_i p_{ij}^0)^2}{n_i p_{ij}^0}$$

has a limiting χ^2 distribution with $\Sigma_{i=1}^m \, (d_i - 1) = d - m$ degrees of freedom, where d is the total number of positive entries in P.

Now, the likelihood-ratio statistic for testing $H: p_{ij} = p_{ij}^0$, $i, j \in S$ is given by

$$\lambda^* = -2 \log \lambda = -2 \log \frac{p_n(\mathbf{x}_n | p^0)}{p_n(\mathbf{x}_n | \hat{p})} = -2 \sum_{i,j} n_{ij} \log \frac{p_{ij}^0 n_i}{n_{ij}}.$$

It is seen that $\lambda^* - \phi \to 0$, in probability, under H.

The statistic ϕ is therefore a reasonable criterion for testing H. The test rejects H whenever the value of ϕ is too large (compared to the χ^2-value at a specified level of significance).

(b) Test for Independence. The hypothesis of independence of the random variables $\{X_k\}$, $k=0, 1, 2, \ldots$ within the stationary Markov dependence assumption takes the following form:

$$H{:}p_{ij} = \pi_j, \; i, j \in S,$$

where the stationary probabilities $\{\pi_j\}$ are unspecified. Under H, we have only m parameters $(\pi_1, \ldots, \pi_m)$ which are subject to a restriction, namely $\Sigma_j \pi_j = 1$. The maximum likelihood estimator of π_j (under H) is easily found to be n_j/n, where n_j=frequency of the state j in the sample $\mathbf{x}_n = (x_1, \ldots, x_n)$. The unrestricted maximum likelihood estimators of p_{ij} are, as before, n_{ij}/n_i (ignoring the initial state). The χ-squared and the likelihood-ratio criterion for testing H are respectively

$$\sum_{i,j} \frac{(n_{ij} - n_i n_j/n)^2}{n_i n_j/n} \quad \text{and} \quad 2 \sum_{i,j} n_{ij} \log \left(\frac{n_{ij}}{n_i n_j/n} \right).$$

It will be seen that the difference between these two statistics tends to zero in probability as $n \to \infty$. Again using the multinomial analogy we find that the statistic

$$\sum_{i,j} \frac{(n_{ij} - n_i n_j/n)^2}{n_i n_j/n} \tag{25}$$

has a limiting χ^2 distribution with $(m^2 - m) - (m-1) = (m-1)^2$ degrees of freedom, and the large values of the statistic correspond to rejection of H.

(c) Test for Homogeneity in Several Samples. Suppose we have r independent samples from r Markov chains with transition probabilities $p_{ij(h)}$, $h = 1, 2, \ldots, r$. Let the samples have transition frequencies $((n_{ij}(h)))$, $h = 1, \ldots, r$. Then the maximum likelihood estimator of $p_{ij(h)}$ obtained from the corresponding sample is $n_{ij}(h)/n_i(h)$. Now, under the null hypothesis $H{:}p_{ij(1)} = p_{ij(2)} = \ldots = p_{ij(r)}$, $(i, j) \in S$, i.e. the r transitions matrices are identical, it can easily be verified that the maximum likelihood estimator of the common transition probability say p_{ij} is

$$\hat{p}_{ij} = \frac{\sum_{h=1}^{r} n_{ij}(h)}{\sum_{h=1}^{r} n_i(h)}.$$

Suppose the samples are of sizes $N_1, N_2, \ldots, N_r$, where $\Sigma_{k=1}^{r} N_k = N$. Let $N \to \infty$ in such a way that $N_h/N \to \lambda_h > 0$, $(\Sigma_{h=1}^{r} \lambda_h = 1)$. Then the χ-square statistic for testing H is

$$\sum_{h=1}^{r} \sum_{i,j} \frac{(n_{ij}(h) - n_i(h)\hat{p}_{ij})^2}{n_i(h)\hat{p}_{ij}}. \tag{26}$$

This has a limiting χ^2 distribution with $rm(m-1) - m(m-1) = (r-1)m(m-1)$ degrees of freedom.

(d) Testing for the Order of Dependence. So far, we have considered first-order Markov chains. An rth-order Markov chain is defined by the following property

$$P(X_n = a_{r+1} \mid X_{n-1} = a_r, X_{n-2} = a_{r-1}, \ldots, X_{n-r} = a_1, \ldots, X_1 = x_1, X_0 = x_0)$$
$$= P(X_n = a_{r+1} \mid X_{n-1} = a_r, \ldots, X_{n-r} = a_1) = P_{a_1, \ldots, a_r; a_{r+1}}$$

Thus, the probability of the current state, given the past, depends only on the last r states. When $r = 1$ we have the usual case of first-order dependence. Any rth-order chain can, however, be transformed to an equivalent first-order Markov chain as follows.

Let $\{Y_k\}$ be the process with the state-space (r-dimensional) S^r with a typical element given by $y_n = (x_n, \ldots, x_{n+r-1})$. Suppose the transition probabilities of the X-process are all positive. Then

$$P[Y_n = (a_1, \ldots, a_r) \mid Y_{n-1} = (b_1, \ldots, b_r)]$$

$$= P[X_n = a_1, X_{n+1} = a_2, \ldots, X_{n+r-1} = a_r \mid X_{n-1} = b_1, X_n = b_2, \ldots, X_{n+r-2} = b_r]$$

$$= \begin{cases} p_{b_1, \ldots, b_r; a_r} & \text{if } a_i = b_{i+1}, i = 1, \ldots, n+r-2 \\ 0 & \text{otherwise.} \end{cases}$$

Clearly $\{Y_k\}$ now forms a first-order Markov chain with transition probabilities given by

$$p_{(a_1, \ldots, a_r), (b_1, \ldots, b_r)} = \begin{cases} p_{b_1, \ldots, b_r; a_r} & \textit{if } a_i = b_{i+1}, i = 1, \ldots, n+r-2 \\ 0 & \text{otherwise.} \end{cases}$$

Note that the number of positive entries in each row of the transition matrix $\mathbf{P}$ for $\{Y_k\}$ is m, and hence the total number of positive entries is m^{r+1} (since there are m^r rows for P).

Let $n_{a_1,\ldots,a_k}$ denote the frequency of occurrence of a run $a_1,\ldots,a_k$ in a sample of n observations. Then $n_{a_1,\ldots,a_r}$ is the frequency of the run $a_1,\ldots,a_r$, and $n_{a_1,\ldots,a_{r+1}}$ is the transition frequency of $(a_1,\ldots,a_r)$ to $(a_2,\ldots,a_{r+1})$. Thus, $n_{a_1,\ldots,a_r}$ and $n_{a_1,\ldots,a_{r+1}}$ play the same role as n_i and n_{ij} in the first-order chain. Suppose we wish to test the hypothesis H: the process is an lth order chain $(l<r)$ against the hypothesis that it is an rth order chain. Now H specifies the following:

$$p_{a_1,\ldots,a_r;a_{r+1}}=p_{a_{r-l+1},\ldots,a_r;a_{r+1}},$$

i.e. the transition probabilities are specified by m^{l+1} parameters $\{p_{a_{r-l+1},\ldots,a_r;a_{r+1}}\}$; these parameters are, however constrained by m^l row-sums being unity so that we have $m^{l+1}-m^l=m^l(m-1)$ independent parameters. Under H, the maximum likelihood estimator is given by

$$\hat{p}_{a_1,\ldots,a_r;a_{r+1}}=\frac{n_{a_{r-l+1},\ldots,a_r;a_{r+1}}}{n_{a_{r-l+1},\ldots,a_r}}.$$

The appropriate χ-squared statistic for testing H is then given by

$$\sum_{a_1,\ldots,a_{r+1}}\frac{(n_{a_1,\ldots,a_{r+1}}-n_{a_1,\ldots,a_r}\,\hat{p}_{a_1,\ldots,a_r;a_{r+1}})^2}{n_{a_1,\ldots,a_r}\,\hat{p}_{a_1,\ldots,a_r;a_{r+1}}}\tag{27}$$

and this has an asymptotic χ^2 distribution with

$m^r(m-1)-m^l(m-1)=m^l(m-1)(m^{r-l}-1)$ degrees of freedom.

(e) Testing Parametric Hypotheses. Suppose we wish to test the hypothesis H: the transition probabilities p_{ij} have specified forms, say $p_{ij}(\theta)$, where $\boldsymbol{\theta}=(\theta_1,\ldots,\theta_k)$ is an unknown parameter $\in\mathbb{R}^k$.

The appropriate statistic for this situation is

$$\sum_{i,j}\frac{(n_{ij}-n_ip_{ij}(\hat{\theta}))^2}{n_ip_{ij}(\hat{\theta})},\tag{28}$$

where $\hat{\theta}$ is the maximum likelihood estimator of θ obtained by solving the equations

$$\sum_{i,j}\frac{n_{ij}}{p_{ij}(\theta)}\frac{\partial p_{ij}}{\partial\theta_u}=0,\ u=1,\ldots,k.\tag{29}$$

Assuming that the $m^2\times k$ matrix $D=(\partial p_{ij}/\partial\theta_u)$ has rank k (so that there are no redundant parameters) and that the p_{ij} have continuous first and second partial derivatives with respect to θ_u, $u=1,\ldots,k$, it can be seen (by multinomial analogy, see Rao (1973)) that the asymptotic distribution of the statistic under consideration is χ^2 with $m(m-1)-k$ degrees of freedom. (It is assumed for simplicity that all $p_{ij}>0$. If only d elements of $\mathbf{P}$ are positive, the degrees of freedom will be $d-m-k$).

2.5 A Test of Independence Based on Serial Correlation

We shall assume that all the elements of $\mathbf{P}$ are positive and that the chain is stationary. For a wide class of transition matrices the autocorrelation can be shown to characterize statistical dependence between successive pairs of observations. Following Lehmann (1966) we have the following

Definition. A transition matrix $\mathbf{P}$ is said to be positively serially dependent if

$$\sum_{k=1}^{i}\sum_{r=1}^{j}\pi_k p_{kr} \geqq \left(\sum_{k=1}^{i}\pi_k\right)\left(\sum_{r=1}^{j}\pi_r\right), \quad \text{for all } (i,j)\in S.$$

Let F denote the family of such transition matrices. If the reverse inequality holds $\mathbf{P}$ is negatively serially dependent and we denote the corresponding family by G.

It can be shown (as in Lehmann, 1966) that $P\in F$ implies $\rho\geqq 0$ with equality holding if and only if $\{X_k\}$ are independent. Similarly, $\mathbf{P}\in G$ implies $\rho\leqq 0$ with equality holding if and only if $\{X_k\}$ are independent. Thus, whenever $\mathbf{P}\in F\cup G$, $\rho=0$ implies serial independence of $\{X_k\}$.

It is easy to verify that when there are only two states, the transition matrix $\mathbf{P}$ always belongs to $F\cup G$, so that uncorrelatedness always implies independence in this case. When the number of states is large it will be rather tedious to verify whether or not $\mathbf{P}\in F\cup G$ using the definition. Simpler sufficient conditions for positive dependence are available and they are the following:

(i) If $\mathbf{P}$ is such that $\Sigma_{r=1}^{j} p_{kr}$ is non-increasing in k for all $j\in S$, then $P\in F$.

A simpler sufficient condition for (*i*) to hold is:

(ii) If $p_{i'j}/p_{ij}$ is monotone non-decreasing in j for every $i'>i$, it follows that $\Sigma_{r=1}^{j} p_{kr}$ is non-increasing in k for all $j\in S$.

Condition (ii) is often the simplest way of determining the membership of $\mathbf{P}$ in F. Similar sufficient conditions hold for G.

Suppose that $\mathbf{P}\in F\cup G$, and we wish to test the hypothesis H:$\{X_k\}$ are independent. Testing H is then equivalent to testing $H^*:\rho=0$. A simple test statistic for testing H^* is

$$Z_n=n^{1/2}R$$

where R is the serial correlation studied in Section 1.3. Under H^*, Z_n has a limiting standard normal distribution. Large values of Z_n correspond to the rejection of H^* (and hence of H). Note that this test is considerably simpler than the χ-squared test obtained in the previous section.

2.6 Bayesian Analysis

Preliminaries. Suppose we have some prior information about the transition matrix **P** which we can represent by a prior distribution function say $H(\mathbf{P}|\psi_0)$. Let $\mathscr{H}=\{H(P|\psi);\ \psi\in\Psi\}$ where Ψ is some subset of a finite-dimensional Euclidean space. Bayesian analysis is greatly simplified if $\mathscr{H}$ is such that the posterior distribution function of **P** given the data **x**, say $H(\mathbf{P}|\psi_0, \mathbf{x})$, is again a member of the class $\mathscr{H}$ for any $\psi_0\in\Psi$. Such a family of prior distributions is said to be *closed under sampling.*

If $L(\mathbf{x}|\mathbf{P})$ is the likelihood based on the sample **x**, and $U(\mathbf{x})$ is a sufficient statistic for **P**, then we have the factorization,

$$L(\mathbf{x}|\mathbf{P})=k(U(\mathbf{x}), \mathbf{P})r(\mathbf{x}).$$

If we form a prior distribution for **P** by taking $k(U, \mathbf{P})$, (as a function of **P**) as its kernel, it can be seen that the family of such distributions is closed under sampling. This family is known as the family of natural conjugate distributions.

Assignment of Prior Distribution for P. If the parameter space for **P** consists of all $m\times m$ stochastic matrices then the class of "matrix-β" distributions appears to be sufficiently rich to express prior information about **P**. These distributions arise as special cases of the multivariate beta distribution, which is defined as a natural extension of a univariate beta distribution, by the joint density

$$f(\mathbf{p}|\mathbf{u})=\begin{cases} B(\mathbf{u})\prod_{i=1}^{m} p_i^{u_i-1}, & \mathbf{p}\in\Omega \\ 0 & \text{otherwise} \end{cases} \tag{29}$$

where $\Omega=\{\mathbf{p}=(p_1, \ldots, p_m)':\mathbf{p}\geqq 0,\ \Sigma_{i=1}^m p_i=1\}$, $\mathbf{u}=(u_1, \ldots, u_m)'>0$, and $B(\mathbf{u})=\Gamma(U)/\pi_{i=1}^m \Gamma(u_i)$, $U=\Sigma_1^m u_i$, where prime denotes the transpose.

The moments of this distribution (see Martin, 1967) are given by

$$E(p_i)=\frac{u_i}{U},\ V(p_i)=\frac{u_i(U-u_i)}{U^2(U+1)},\ \text{and}$$

$$\operatorname{cov}(p_i, p_j)=\frac{-u_iu_j}{U^2(U+1)},\quad i, j\in S.$$

Now, suppose that each row of **P** is distributed as the multivariate beta distribution, and that these are independent of each other. Then the joint density function

$$f(\mathbf{P}|\mathbf{U})=\begin{cases}\prod_{i=1}^{m}\left[B(\mathbf{u}_i)\prod_{j=1}^{m}p_{ij}^{u_{ij}-1}\right] & \text{if } \mathbf{P}\in\Omega^* \\ 0 \quad \text{otherwise}\end{cases} \tag{30}$$

where Ω^* is now the class of $m\times m$ stochastic matrices ($m\geqq 2$) and $\mathbf{U}>\mathbf{0}$ is the $m\times m$ matrix $((m_{ij}))$, $\mathbf{u}_i$ being the ith row of $\mathbf{U}$. With this prior distribution note that $p_{ij}>0$ with probability one, and hence the Markov chain is ergodic. We need to specify the matrix $\mathbf{U}$ (for instance from past data). However, it is sufficient for this purpose to have the prior knowledge regarding the means and variances given by

$$E(p_{ij})=\frac{u_{ij}}{u_{i.}},\ V(p_{ij})=\frac{u_{ij}(u_{i.}-u_{ij})}{u_{i.}^2(u_{i.}+1)}=\frac{E(p_{ij})(1-E(p_{ij}))}{u_{i.}+1},$$

where $u_{i.}=\Sigma_{j=1}^m u_{ij}$.

Now, conditional on the initial state, the likelihood function is given by

$$L(x|\mathbf{P})=\prod_{i,j}p_{ij}^{n_{ij}},$$

where n_{ij} is the transition frequency of $i\to j$ in the sample. It follows that the matrix $((n_{ij}))$ is sufficient for $\mathbf{P}$, and that the density of the natural conjugate distribution for $\mathbf{P}$ is proportional to $\Pi_{i,j=1}^m p_{ij}^{y_{ij}}$ where y_{ij} are the parameters. Thus, the matrix beta distribution is clearly the natural conjugate. It then follows that the matrix beta distribution is closed under sampling.

If $\mathbf{P}$ is assumed to have the matrix beta distribution with parameters $\mathbf{U}_0=((u_{ij}^0))>0$ as its prior, the posterior distribution is again a matrix beta distribution with parameter $\mathbf{U}^*=\mathbf{U}_0+\mathbf{N}$, where $\mathbf{N}=((n_{ij}))$.

Estimation of P. Using quadratic loss we find that the mean of the posterior distribution,

$$E(p_{ij}|\mathbf{x})=\frac{u_{ij}^0+n_{ij}}{u_{i.}^0+n_i},\ (i,j)\in S \tag{31}$$

is the Bayes estimator.

In the absence of any specified loss function, the mode of the posterior distribution is a natural choice as an estimator of **P**. Thus

$$\text{mode }(p_{ij}|\mathbf{x})=\frac{u_{ij}^0+n_{ij}-1}{u_i^0+n_i-m} \tag{32}$$

is the posterior-mode estimator of p_{ij}.

2.7 Further Topics

(a) Inference for an Absorbing Chain. Consider a Markov chain $\{X_t\}$, $t=0$, 1, 2, ... defined on the state space $S=\{0, 1, 2,\ldots, m\}$. Suppose that the state 0 is absorbing (i.e. once the state 0 is reached it is not possible to move to any other state), and the other states $\{1, 2, \ldots, m\}$ are transient (i.e. non-recurrent). Then the transition matrix $\mathbf{P}$ is of the form

$$\mathbf{P}=\begin{pmatrix} 1 & \mathbf{0}' \\ \mathbf{p} & \mathbf{Q} \end{pmatrix}, \ \mathbf{p}\neq 0 \tag{33}$$

where $\mathbf{p}'=(p_{10}, \ldots, p_{m0})$, $\mathbf{0}$ is an $m\times 1$ vector of zeros, and $\mathbf{Q}$ is an $m\times m$ matrix with elements $((p_{ij}))$, $i, j=1, 2, \ldots, m$. Note that $\Sigma_{j=1}^{m} p_{ij}=1-p_{i0}$, and hence $\mathbf{Q}$ is not a stochastic matrix. We shall assume that all the states in $\mathbf{Q}$ are irreducible and aperiodic. Let $(x_0, x_1, \ldots, x_n)$ be a realization of consecutive observations from this chain. Suppose p_{ij} depend on a vector θ of k unknown parameters $\theta_1, \ldots, \theta_k$. We now wish to estimate θ from the sample $(x_0, x_1, \ldots, x_n)$ under the condition that absorption has not taken place, i.e. $x_n\neq 0$. Now if $\mathbf{p}'^{(0)}=(p_1^{(0)}, \ldots, p_m^{(0)})$ is the initial distribution over the transient states, the probability of obtaining n observations before absorption takes place is clearly $\mathbf{p}^{(0)}\ \mathbf{Q}^n\mathbf{e}$, where $\mathbf{e}$ is an $m\times 1$ vector of 1's Therefore, the (conditional) likelihood function is given by

$$L=\frac{\prod_{i,\,j=1}^{m} p_{ij}^{n_{ij}}}{\mathbf{p}^{(0)}\mathbf{Q}^n\mathbf{e}} \tag{34}$$

where n_{ij} is the transition frequency of $i\to j$ in the sample.

Let $\rho_1, \ldots, \rho_m$ be the eigenvalues of $\mathbf{Q}$, arranged in descending order of magnitude. Let $\mathbf{w}'=(w_1, \ldots, w_m)$, and $\mathbf{v}'=(v_1, \ldots, v_m)$ be the right and left eigenvectors of $\mathbf{Q}$ respectively corresponding to the largest eigenvalue ρ_1, normalized so that $\mathbf{v}'\mathbf{e}=1$ and $\mathbf{v}'\mathbf{w}=1$. (All the components of $\mathbf{w}$ and $\mathbf{v}$ are thus positive). Darroch and Seneta (1965) have shown that

$$P(X_k=j\,|\,X_{k+n}\neq 0)\to w_j v_j \text{ as } k, n\to\infty,$$

independent of $\mathbf{p}^{(0)}$. Thus, $\pi'=(\pi_1, \ldots, \pi_m)$ where $\pi_j=w_j v_j$ is the limiting conditional distribution of X_k given that absorption does not take place for a long time afterwards.

Denote by $\mathbf{W}=((\delta_{ij}w_j))$, the diagonal matrix corresponding to $\mathbf{w}$, and let

$$\mathbf{R}=\frac{1}{\rho_1}\mathbf{W}^{-1}\mathbf{Q}\,\mathbf{W}$$

with elements given by

$$R_{ij}=\frac{1}{\rho_1}\frac{p_{ij}w_j}{w_i}, \ (i, j=1, \ldots, m).$$

It can be verified that **R** is a stochastic matrix and is irreducible and ergodic. $\boldsymbol{\pi}' = \mathbf{v}'\mathbf{W}$ is then the stationary distribution corresponding to **R**. As k and $n \to \infty$, the chain $\{X_t\}$ behaves as though $\boldsymbol{\pi}$ were its stationary distribution and **R** its transition matrix. The results obtained previously for ergodic chains can therefore be applied in the present case.

The likelihood equations are

$$\frac{\partial \log L}{\partial \theta_r} = \sum_{i,j} \frac{n_{ij}}{p_{ij}} \frac{\partial p_{ij}}{\partial \theta_r} - \frac{1}{\mathbf{p}^{(0)}\mathbf{Q}^n\mathbf{e}} \frac{\partial}{\partial \theta_r} (\mathbf{p}'^{(0)}\mathbf{Q}^n\mathbf{e}) = 0, \ r = 1, \ldots, k.$$

By differentiating the canonical form of **Q** we find that

$$\frac{\partial(\mathbf{p}'^{(0)}\mathbf{Q}^n\mathbf{e})}{\partial \theta_r} = n\rho_1^{n-1}(\mathbf{p}'^{(0)}\mathbf{W}) \frac{\partial \rho_1}{\partial \theta_r} + \rho_1^n \mathbf{p}'^{(0)} \frac{\partial \mathbf{W}}{\partial \theta_r} + o(n^{r+1}|\rho_2|^n).$$

If we neglect terms of $o(1/n)$ and $o(n^r|\rho_2|^n/\rho_1)$ in the expression for $1/n$ $(\partial \log L/\partial \theta_r)$ we obtain the modified likelihood equations

$$\frac{1}{n} \sum_{i,j} \frac{n_{ij}}{p_{ij}} \frac{\partial p_{ij}}{\partial \theta_r} - \frac{1}{\rho_1} \frac{\partial \rho_1}{\partial \theta_r} = 0, \ r = 1, \ldots, k.$$

It can be seen that if L^* is the likelihood of a sample from the ergodic chain governed by the matrix **R**, then $1/n(\partial \log L^*/\partial \theta_r) = 0$ will produce the same equations as above apart from a term of $0(1/n)$.

One therefore can use the modified likelihood equations to get the estimator θ and the large-sample properties follow as if **R** is the transition matrix (and $\boldsymbol{\pi}$ the stationary distribution). See Matthews (1970) for further details.

(b) Inverse Likelihood Estimation of States. Let $\{X_t\}$, $t = 0, 1, 2, \ldots$ be an ergodic Markov chain on a state-space $S = \{1, 2, \ldots, m\}$ and transition probabilities $((p_{ij}))$. Let $\{Y_t\}$ be a function of $\{X_t\}$ defined as follows:

$$P[Y_k = l | X_k = j] = q_{jl}, j \in S, \text{ and } l \in \bar{S} = \{c_1, \ldots, c_r\}$$

where $\bar{S}$ is a partition of S, and q_{jl} are known values, $0 \leqq q_{jl} \leqq 1$, with $\Sigma_{l=1}^r q_{jl} = 1$. Suppose the original chain $\{X_t\}$ is unobserved and instead the modified process $\{Y_t\}$ can be observed.

If we set

$$q_{jl} = \begin{cases} 1 & \text{if } j \in c_l \\ 0 & \text{otherwise,} \end{cases}$$

$\{Y_k\}$ is a grouped chain (which, in general, is not necessarily Markovian). In addition, if $\bar{S} = \{c_1, c_2\}$ with $c_1 = \{1\}$, $c_2 = \{2, \ldots, m\}$, $\{Y_k\}$ notes at each stage whether or not the state 1 of S has occurred and hence is a "*regenerative phenomenon*" (see Kingman, 1972).

It should be noted that the bivariate process $\{(X_k, Y_k), k=0, 1, 2, \ldots\}$ is Markovian with "transition probabilities"

$$P[X_k=j, Y_k=l \mid X_{k-1}=i]=p_{ij}\, q_{jl}.$$

Using a realization $\mathbf{y}=(y_0, y_1, \ldots, y_n)$ we now wish to estimate the corresponding realization $\mathbf{x}=(x_0, x_1, \ldots, x_n)$ of the X-process. We are therefore interested in estimating the posterior distribution of X given $\mathbf{y}$, $p(\mathbf{x}|\mathbf{y})$. The posterior distribution depends on the unknown transition probabilities p_{ij}. Our task is therefore to estimate p_{ij} from $\mathbf{y}$. One possibility is to estimate p_{ij} using the marginal likelihood function

$$L(\mathbf{y})=\sum_{\mathbf{x}} L(\mathbf{x}, \mathbf{y}),$$

where

$$L(\mathbf{x}, \mathbf{y})=p_{x_0} q_{x_0 y_0} \prod_{k=1}^{n} (p_{x_{k-1} x_k} q_{x_k y_k}),$$

and then substitute these estimates in the expression for $p(\mathbf{x}|\mathbf{y})$. Since the marginal Y-process generally loses Markovity the expression $L(\mathbf{y})$ is far from simple. Moreover, not all the p_{ij}'s , in general, are identified in $L(\mathbf{y})$. We suggest an alternative method which estimates $p(\mathbf{x}|\mathbf{y})$ directly.

Let $\mathcal{X}_{\mathbf{y}}$ be the set of possible realizations $\mathbf{x}$ corresponding to the given $\mathbf{y}$. For any known $\mathbf{x}$ we can estimate p_{ij} by the maximum likelihood method getting $\hat{p}_{ij}(\mathbf{x})=n_{ij}(\mathbf{x})/n_i(\mathbf{x})$, where $n_{ij}(\mathbf{x})$ is the transition frequency of $i \to j$ in $\mathbf{x}$ and $n_i(\mathbf{x})=\Sigma_j\, \mathrm{n}_{ij}(\mathbf{x})$.

Now, $p(\mathbf{x}|\mathbf{y}) \infty p(\mathbf{x}, \mathbf{y})$, the joint distribution $\propto L(\mathbf{x}, \mathbf{y})$.

Let $\hat{L}(\mathbf{x}, \mathbf{y})$ denote $L(\mathbf{x}, \mathbf{y})$ with p_{ij} replaced by $\hat{p}_{ij}(\mathbf{x})$. Then, a consistent estimate of the posterior distribution $p(\mathbf{x}|\mathbf{y})$ is

$$\hat{p}(\mathbf{x}|\mathbf{y})=\frac{\hat{L}(\mathbf{x}, \mathbf{y})}{\sum_{\mathbf{x} \in \mathcal{X}_{\mathbf{y}}} \hat{L}(\mathbf{x}, \mathbf{y})}, \quad x \in \mathcal{X}_{\mathbf{y}}, \tag{35}$$

which, in a sense, is the normalized (and estimated) likelihood.

Refer to Preston (1971) and Devore (1973) for the discussion of a related problem.

(c) Informatrics. Let $\{X_k\}$ be a stationary sequence of random variables. The entropy of the stationary distribution is

$$H_1=-\sum_{i=1}^{m} (\pi_i \log \pi_i),$$

and the entropy corresponding to the pairs of observations is

$$H_2 = -\sum_{i,j} \pi_i p_{ij} \log (\pi_i\, p_{ij}).$$

Then

$$H_2 - H_1 = -\sum_{i,j} (\pi_i\, p_{ij} \log p_{ij}) = H_{21}, \text{ say}$$

measures the average conditional uncertainty of an observation given the preceding observation. Substituting the maximum likelihood estimators $\hat{\pi}_i = n_i/n$, $\hat{p}_{ij} = n_{ij}/n_i$, we have

$$\hat{H}_{21} = \frac{1}{n}\left(\sum_{i=1}^{m} n_i \log n_i - \sum_{i,j} n_{ij} \log n_{ij} \right).$$

Note that if $\{X_k\}$ are independent, i.e. if $p_{ij} = \pi_j$, then $H_2 = 2H_1$; when $\{X_k\}$ are dependent, we have $H_1 < H_2 < 2H_1$.

Consider the quantity $2H_1 - H_2 = H_1 - H_{21}$. This clearly measures the gain in information by using dependence on the previous observation rather than not using it. Substituting the maximum likelihood estimators, we have

$$T_1 = 2n(\hat{H}_1 - \hat{H}_{21}) = 2 \sum_{i,j} n_{ij} \log \frac{n_{ij}}{n_i n_j / n} \tag{36}$$

which is the same as the likelihood ratio criterion, and hence has a limiting χ^2 distribution with $(m-1)^2$ degrees of freedom. The test-statistic T_1 is appropriate for testing independence against Markov dependence.

The procedure can be extended to test a hypothesis of $(r-1)$-dependence against r-dependence. The appropriate test-statistic is

$$T_r = 2n(\hat{H}_{r,r-1} - \hat{H}_{r+1,r}), \ (H_{10} = H_1) \tag{37}$$

where

$$H_{l,l-1} = -\sum_{a_1,\ldots,a_l} P(X_1 = a_1, \ldots, X_l = a_l) \log P(X_l = a_l | X_1 a_1, \ldots, X_{l-1} = a_{l-1})$$

and

$$\hat{H}_{l,l-1} = -\sum_{a_1,\ldots,a_l} \left(\frac{n_{a_1,\ldots,a_l}}{n} \log \left\{ \frac{n_{a_1,\ldots,a_l}}{n_{a_1,\ldots,a_{l-1}}} \right\} \right) \quad n_{a_1,\ldots,a_l}$$

denoting the frequency of the *l-tuple* $(a_1, \ldots, a_l)$ in the sample. We have used circular counting of the transitions which does not affect the asymptotic results. T_r has limiting χ^2 distribution with $m^{r-1}(m-1)^2$ degrees of freedom.

Note that this is the same test as the corresponding likelihood ratio test and is asymptotically equivalent to the chi-squared test given in Section 2.4.

Although the informatrics approach yields the same test statistics as the likelihood ratio criterion, the former provides further insight by means of the conditional entropies $H_{r,r-1}$. Plotting of $\hat{H}_{r,r-1}$ against r often gives a useful visual guide regarding the order of dependence. (See Chatfield, 1973.) Theil (1972) contains a discussion of several descriptive measures based on the entropy concept.

3. A MACRO MODEL (FINITE STATE SPACE)

The detailed information on the transitions between every pair of states (i.e. n_{ij}) may not be available in some situations. Instead, only the aggregate data such as the total number of individuals in each state at time $k=0, 1, 2, \ldots$ may be available. These aggregates are over several independent Markov chains evolving simultaneously. If the detailed transitions are available, methods similar to those in section 2.1 can be used to estimate the transition probabilities of the macro model as described in Anderson and Goodman (1957). However, if the sample consists of only the number of individuals, or the proportion of individuals in each state at varying times these methods cannot be used any longer. Instead, least-squares or minimum-χ-squared methods can be used.

Let N_k denote the total number of individuals at time k ($k=0, 1, 2, \ldots, T$), and $(y_1(k), \ldots, y_m(k))$, the proportion of individuals in the states $(1, \ldots, m)$ respectively at time k. Suppose $q_j(k)$, $(j=1, 2, \ldots, m)$ denotes the probability of an individual being in state j at time k. If p_{ij} are the transition probabilities we have

$$q_j(k)=\sum_{i=1}^{m} q_i(k-1)p_{ij}, \quad (k=1, \ldots, T), \quad (j=1, \ldots, m). \tag{38}$$

Replacing $q_j(k)$ by the estimators $y_j(k)$ in equation (38) we can consider the linear equations

$$y_j(k)=\sum_{i=1}^{m} y_i(k-1)p_{ij}+u_j(k), \quad (k=1, \ldots, T), \quad (j=1, 2, \ldots, m) \tag{39}$$

where $u_j(k)$ are error variables.

Let

$$\mathbf{y}_j' = (y_j(1), \ldots, y_j(T)),\ \mathbf{p}_j' = (p_{1j}, \ldots, p_{mj}),\ \mathbf{u}_j' = (u_j(1), \ldots, u_j(T))$$

$$\mathbf{y}' = (y_1', y_2', \ldots, y_m'),\ \mathbf{p}' = (p_1', p_2', \ldots, p_m'), \mathbf{u}' = (u_1', \ldots, u_m').$$

Also, let $\mathbf{X}_j = \begin{bmatrix} y_1(0) & y_2(0) & \ldots, & y_m(0) \\ y_1(1) & y_2(1) & \ldots, & y_m(1) \\ \cdots & \cdots & \cdots & \cdots \\ y_1(T-1) & y_2(T-1) & \ldots, & y_m(T-1) \end{bmatrix}$

and

$$\mathbf{X} = \begin{bmatrix} \mathbf{X}_1 & & 0 \\ & \ddots & \\ 0 & & \mathbf{X}_m \end{bmatrix}.$$

Then, (39) can be written as

$$\mathbf{Y} = \mathbf{X}\mathbf{p} + \mathbf{u}. \tag{40}$$

For any fixed k, $\{y_j(k), j=1, 2, \ldots, m\}$ are the proportions generated by a multinomial with means $q_j(k)$, variances $q_j(k)(1-q_j(k))/N_k$, and covariances $-q_i(k)q_j(k)/N_k$. Consider the matrix defined by

$$\mathbf{W} = \begin{bmatrix} \mathbf{W}_{11} & \mathbf{W}_{12} & \cdots & \mathbf{W}_{1m} \\ \mathbf{W}_{21} & \mathbf{W}_{22} & \cdots & \mathbf{W}_{2m} \\ \cdots & \cdots & \cdots & \cdots \\ \mathbf{W}_{m1} & \mathbf{W}_{m2} & \cdots & \mathbf{W}_{mm} \end{bmatrix} \tag{41}$$

where $$\mathbf{W}_{ii} = \text{diag}\left(\frac{q_i(1)(1-q_i(1))}{N_1}, \ldots, \frac{q_i(T)(1-q_i(T))}{N_T}\right)$$

and $$\mathbf{W}_{ij} = \text{diag}\left(-\frac{q_i(1)q_j(1)}{N_1}, \ldots, -\frac{q_i(T)q_j(T)}{N_T}\right).$$

The weighted least-square estimator of $\mathbf{p}$ using W^-, the generalized inverse of W, as the weight matrix is obtained by minimizing

$$(\mathbf{y} - \mathbf{X}\mathbf{p})'\mathbf{W}^-(\mathbf{y} - \mathbf{X}\mathbf{p}) \tag{42}$$

with respect to $\mathbf{p}$ subject to the restriction $0 \leqq p_{ij} \leqq 1$ and $\Sigma_j p_{ij} = 1$. Note that $\mathbf{W}$ involves the unknown quantities q_j which may be replaced by their estimators $y_j(k)$.

An alternative method of estimating p_{ij} is by minimizing the χ-squared type expression

$$\sum_{k=1}^{T} \sum_{j=1}^{m} \left[\frac{\{N_k y_j(k) - N_k \sum_i y_i(k-1) p_{ij}\}^2}{N_k \sum_i y_i(k-1) p_{ij}} \right] \tag{43}$$

with respect to p_{ij} subject to the restriction $0 \leqq p_{ij} \leqq 1$ and $\Sigma_j p_{ij} = 1$. It is interesting to note that this procedure leads to the same expression (42) for minimization.

Likelihood Function. The likelihood function for the model can be constructed as follows. The multinomials, $\{N_k y_j(k), j \in S\}$, $k = 1, \ldots, T$ are seen to be Markov dependent. Conditional on $\{y_j(k-1), j \in S\}$, $\{N_k y_j(k), j \in S\}$ form a multinomial distribution with probabilities $\{\Sigma_i y_i(k-1) p_{ij}, j \in S\}$ Consequently, the (approximate) likelihood function is given by

$$L \propto \prod_{k=1}^{T} \prod_{j=1}^{m} \left(\sum_i y_i(k-1) p_{ij} \right)^{N_k y_j(k)}. \tag{44}$$

The maximum likelihood estimators of p_{ij} are given by maximizing (44) subject to the usual restrictions on p_{ij}. It is seen that this approach again leads essentially to the same solution as the two approaches mentioned earlier.

Consistency and asymptotic normality (as $T \to \infty$) of the maximum likelihood estimators can be established assuming the underlying Markov chain to be positively regular.

4. GROUPED MARKOV CHAINS (FINITE STATE SPACE)

Consider a positively regular finite Markov chain $\{X_k\}$ with state-space $S = \{1, 2, \ldots, m\}$. Suppose that the chain $\{X_k\}$ is not observable directly, but a related chain $\{Y_k\}$ can be observed. The chain $\{Y_k\}$ is related to $\{X_k\}$ as follows:

$$P(Y_k = u | X_k = i, X_{k-1}, \ldots, X_0)$$

$$= P(Y_k = u | X_k = i) = q_{iu}, \qquad i \in S, u \in R \tag{45}$$

where $R = \{C_1, \ldots, C_r\}$ is a partition of the state-space S. Thus, $C_1, \ldots, C_r$ are the groups of the states in the original state-space. It is well known that the sequence $\{Y_k\}$ is no longer Markovian in general. As a special case of

(45) consider

$$q_{iu}=\begin{cases}1 & \text{if } i\in C_u\\ 0 & \text{otherwise.}\end{cases} \tag{46}$$

We then find that $\{Y_k\}$ is obtained from $\{X_k\}$ by simply grouping the states. Let p_{ij} (the transition probabilities of the original Markov chain $\{X_k\}$) depend on an unknown real parameter θ. We now consider the problem of estimating θ from a realization $\{y_1, y_2, \ldots, y_n\}$ of the grouped chain, using (46) as our grouping mechanism.

If the realization $\{x_1, \ldots, x_n\}$ was available we know that the likelihood function for this sample is

$$L_1(\theta)=\prod_{k=1}^{n} p_\theta(x_k|x_{k-1}).$$

The likelihood function based on $(y_1, \ldots, y_n)$ is given by

$$L_2(\theta)=\sum_{\mathbf{x}_n\in\mathscr{X}_{\mathbf{y}_n}} L_1(\theta) \tag{47}$$

where $\mathscr{X}_{\mathbf{y}_n}$ is the set of all realizations $\mathbf{x}_n=(x_1, \ldots, x_n)$ which correspond to the observed realization $\mathbf{y}_n=(y_1, \ldots, y_n)$.

$L_2(\theta)$, in general, does not have a simple form. However, we can write $L_2(\theta)$ as

$$\begin{aligned}L_2(\theta)&=P(Y_1=y_1, \ldots, Y_n=y_n)\\ &=P(Y_1=y_1)Pr(Y_2=y_2|Y_1=y_1)Pr(Y_3=y_3|Y_2=y_2, Y_1=y_1)\ldots\\ &\qquad P(Y_n=y_n|Y_{n-1}=y_{n-1}, \ldots, Y_1=y_1).\end{aligned} \tag{48}$$

It can then be seen that $\mathrm{d}\log L_2(\theta)/\mathrm{d}\theta$ is a zero-mean martingale. Although $\{Y_k\}$ is not a Markov chain in general, we are still able to apply martingale limit results, as given in Billingsley (1961), in the present case to deduce the asymptotic normality and consistency of the maximum likelihood estimator $\hat{\theta}$ obtained as a root of $(\mathrm{d}\log L_2(\theta))/\mathrm{d}\theta=0$. (Also, refer to Basawa *et al.* (1976) for a discussion of asymptotic properties of the maximum likelihood estimator for processes which are not necessarily Markovian). The problem of maximum likelihood estimation, therefore, is solved in principle. However, L_2 will generally have a complicated form; it is therefore worthwhile looking at a simpler, albeit less efficient, method. One such method is that of minimum χ-squared.

Let m_u, $(u=1, \ldots, r)$ denote the frequency of observations in state u in the sample $\{Y_1, \ldots, Y_n\}$. Assume, for simplicity, that $\{X_k\}$ is a stationary Markov chain. It is easily verified that $\{Y_k\}$ is also a stationary process (non-

Markovian, in general) with absolute probabilities

$$P(Y_k = u) = \pi_u^* = \sum_{j \in C_u} \pi_j \tag{49}$$

where $\{\pi_j, j \in S\}$ is the stationary distribution corresponding to $\{X_k\}$. Let us suppose that π_j and hence π_u^* are functions of an unknown parameter θ. We may now estimate θ by minimizing the χ-squared expression

$$\psi = \sum_{u=1}^{r} \frac{(m_u - n\pi_u^*(\theta))^2}{n\pi_u^*(\theta)}. \tag{50}$$

It is important to note that the asymptotic distribution of ψ is no longer χ-squared; however, it can be written as a linear combination of independent χ-squared variables. It is straightforward to establish the asymptotic normality of

$$\xi_u = \frac{m_u - n\pi_u^*(\theta)}{\sqrt{(n\pi_u^*(\theta))}} \tag{51}$$

using Doob's (1953) central limit theorem for functions defined on a finite Markov chain (also cf. Cox and Miller, 1965). Furthermore, $\xi' = (\xi_1, \ldots, \xi_r)$ has a limiting r-dimensional normal distribution, the dispersion matrix of which is *not* diagonal. By an orthogonal transformation of ξ it is possible to write ψ as a linear combination of variables which are asymptotically independent each having a χ-squared distribution with one degree of freedom.

The minimum χ-squared estimator $\hat{\theta}_n$ of θ is obtained as a root of $d\psi/d\theta = 0$, i.e.

$$\sum_{u=1}^{r} \left(\frac{m_u}{\pi_u^*(\theta)}\right)^2 \frac{d\pi_u^*(\theta)}{d\theta} = 0 \tag{52}$$

Asymptotic Properties of the Minimum χ-squared Estimator. We shall give an outline of the derivation. Let $T_u(n) = m_u/n$, $(u = 1, \ldots, r)$. The estimating equation (52) becomes

$$\phi(\mathbf{T}(n), \theta) = \sum_{u=1}^{r} \left(\frac{nT_u(n)}{\pi_u^*(\theta)}\right)^2 \frac{d\pi_u^*(\theta)}{d\theta} = 0.$$

Since $\Sigma_{u=1}^{r}\ \pi_u^*(\theta) = 1$, $\Sigma_{u=1}^{r}\ d\pi_u^*(\theta)/d\theta = 0$, and consequently we have

$$\phi(\boldsymbol{\pi}^*(\theta), \theta) = 0. \tag{53}$$

Suppose $\pi_u^*(\theta) > 0$ for all u, and that $\pi_u^*(\theta)$ are twice differentiable with respect to θ. It can then be verified that ϕ has continuous first derivatives with respect to T_u $(u = 1, \ldots, r)$, and also with respect to θ. Therefore, the estimating equation $\phi=0$ satisfies the conditions of the Implicit Function Theorem at the point $(\boldsymbol{\pi}^*(\theta), \theta)$ (see Rudin, 1964, page 196). Consequently, there exists an essentially unique root of the equation $\phi(\mathbf{T}(n), \theta)=0$, say $\hat{\theta}_n=g(T(n))$, which as $n\to\infty$ is consistent for θ; in fact, $g(\boldsymbol{\pi}^*(\theta))=\theta$, so that $\hat{\theta}_n$ is a "Fisher-consistent" estimator of θ (see Rao, 1973). It also follows by the Implicit Function Theorem that g is continuously differentiable with respect to $\mathbf{T}(n)$ in the neighbourhood of $\boldsymbol{\pi}^*(\theta)$.

To prove asymptotic normality of $\hat{\theta}_n$ we make use of the fact that

$$\sqrt{n}(\mathbf{T}(n)-\boldsymbol{\pi}^*(\theta))\to N_r(\mathbf{0}, \boldsymbol{\Sigma}) \text{ in distribution.} \tag{54}$$

where the (u, v)th element of $\boldsymbol{\Sigma}$ will be given by

$$\sigma_{uv}=\pi_v^*(\delta_{uv}-\pi_u^*)+\sum_{s=1}^{\infty}\left\{\sum_{i\in C_u}\sum_{j\in C_v}\pi_i(p_{ij}^{(s)}-\pi_j)+\sum_{j\in C_v}\sum_{i\in C_u}\pi_j(p_{ji}^{(s)}-\pi_i)\right\}. \tag{55}$$

Since $\hat{\theta}_n=g(\mathbf{T}(n))$ is continuously differentiable with respect to $\mathbf{T}(n)$, it follows by a well known convergence theorem (cf. Rao, 1973) that

$$\sqrt{n}(\hat{\theta}_n-\theta)\to N(0, \mathbf{G}'\boldsymbol{\Sigma}\mathbf{G}), \text{ in distribution,} \tag{56}$$

where $\mathbf{G}'=(\partial g/\partial\pi_1^*, \partial g/\partial\pi_2^*, \ldots, \partial g/\partial\pi_r^*)$. It should be noted that $\partial g/\partial\pi_u^*$, $(u=1, \ldots, r)$ can be computed by using the formula

$$\frac{\partial g}{\partial\pi_u^*}=\left(\frac{\partial\phi}{\partial\pi_u^*}\Big/\frac{\partial\phi}{\partial\theta}\right). \tag{57}$$

5. COUNTABLE STATE SPACE

Let $\{X_k\}$ be a Markov chain on a countable state space $S=\{1, 2, \ldots\}$ with transition probabilities

$$p_{ij}=P(X_{k+1}=j|X_k=i), \qquad i, j\in S.$$

Suppose that the chain is irreducible, aperiodic, with positively recurrent states (see Karlin and Taylor, 1975 for definitions). We then have

$$\lim_{n\to\infty} p_{ij}^{(n)}=\frac{1}{\mu_j}=\pi_j \text{ (say)}$$

where

$$\pi_j>0, \sum_{j=1}^{\infty}\pi_j=1, \quad \text{and} \quad \pi_j=\sum_{k=1}^{\infty}\pi_k p_{kj}, \tag{58}$$

μ_j being the mean-recurrence time of state j.

Suppose p_{ij} are known functions of an unknown real parameter θ. The problem is that of estimating θ from a realization $(X_1, \ldots, X_n)$ and also obtaining tests of hypotheses regarding θ. The likelihood function is given by

$$L_n(\theta) = \prod_{i,j} p_{ij}^{n_{ij}}(\theta) \tag{59}$$

where n_{ij} is the frequency of transitions from i to j in the sample $(X_1, \ldots, X_n)$. We have assumed that the initial state X_0 is fixed. The maximum likelihood estimator $\hat{\theta}_n$ of θ is obtained as a root of the equation

$$\sum_{i,j} \left(\frac{n_{ij}}{p_{ij}(\theta)}\right)\left(\frac{\mathrm{d}p_{ij}}{\mathrm{d}\theta}\right) = 0. \tag{60}$$

The product in (59) and the sum in (60) range over all the states (i, j) which occur in the sample $(X_1, \ldots, X_n)$.

The consistency and asymptotic normality of $\hat{\theta}_n$ can be established under the classical regularity conditions (see Rao, 1973) imposed on the transition probabilities $p_{ij}(\theta)$. The proofs follow the classical method for the independent case as given in Rao (1973) except that a law of large numbers and a central limit theorem for Markov chains replace the same limit results corresponding to independent observations. Billingsley (1961, Theorem 1.1) and Chung (1967, Theorems 16.1 and 16.3) provide the required law of large numbers and central limit theorem respectively. We finally find that

$$\hat{\theta}_n \to \theta \quad \text{in probability,} \tag{61}$$

and

$$\sqrt{n}(\hat{\theta}_n - \theta) \to N(0, \sigma^2(\theta)) \tag{62}$$

where

$$\sigma^2(\theta) = \left\{E\left(-\frac{\mathrm{d}^2 \log p_\theta(X_2|X_1)}{\mathrm{d}\theta^2}\right)\right\}^{-1}$$

$$= \left\{\sum_{i,j=1}^{\infty} \left(\frac{\pi_i(\theta)}{p_{ij}(\theta)}\right)\left(\frac{\mathrm{d}p_{ij}(\theta)}{\mathrm{d}\theta}\right)^2\right\}^{-1}. \tag{63}$$

Note that only the terms corresponding to the positive $p_{ij}(\theta)$ enter the sum in equation (63). These results can be extended to the multiparameter case as for the finite state-space.

Results (61) and (62) can be used in the usual manner to deduce the asymptotic distribution of the likelihood ratio test of the hypothesis, say $H: \theta \in \omega$, $\omega \subset \Omega$, Ω being the parameter space. See Billingsley (1961) for a comprehensive treatment of the asymptotic theory of inference for Markov chains with general state-space.

Note that the minimum χ-squared method of estimation for grouped chains as described in Section 4 can be extended to the countable state-space when it is partitioned into a finite number of groups. The details are left to the reader as an exercise.

BIBLIOGRAPHICAL NOTES

Most of the topics covered in this chapter are discussed by Billingsley (1961a, b). Section 2.3 is based on Basawa (1972). See Martin (1967) for further details on Section (2.6). Sections 3 and 4 are respectively based on Lee *et al.* (1970) and Toe (1976).

COMPLEMENTS

1. Let $(x_0, x_1, \ldots, x_n)$ be a realization from an ergodic finite Markov chain with unknown transition probabilities. Derive a statistic for testing the hypothesis that the Markov chain has a specified stationary distribution $\{\pi_1^0, \ldots, \pi_m^o\}$. Hence obtain a test for testing the hypothesis that the transition matrix P is doubly stochastic. (P is said to be doubly statistic if the row sums as well as the column sums equal unity).

2. Consider a Markov chain with transition probabilities

$$p_{ij} = \begin{cases} \theta_i & \text{for } j=i+1 \\ 1-\theta_i & \text{for } j=i-1, \quad i, j \in S=\{1, 2, \ldots, m\} \\ 0 & \text{otherwise.} \end{cases}$$

with $p_{10} \equiv p_{1m}$ and $p_{m,m+1} \equiv p_{m1}$. Derive a test for H: $\theta_1 = \theta_2 = \ldots = \theta_m$.

3. Consider a Markov chain with transition probabilities

$$p_{ij} = \begin{cases} \theta & \text{for } i=j \\ \eta & \text{for } i \neq j \end{cases}, \quad i, j = 1, 2, \ldots, m.$$

Derive a test of $H{:}\theta = 1/m$. (Note that this is a test of independence).

4. Let $\{X_k\}$ be a stationary Markov chain with transition matrix

$$P = \begin{pmatrix} 1-\alpha & \alpha \\ \beta & 1-\beta \end{pmatrix}.$$

If $\rho^{(r)} = \text{cov}(X_k, X_{k+r})/V(X_k)$ denotes the rth order autocorrelation, show that $\rho^{(r)} = \rho^r$, where $\rho = \rho^{(1)}$.

5. Let $\{X_k\}$ be a stationary Markov chain with state space $S=\{1, 2, 3\}$, and unknown transition probabilities $p_{ij}>0$, $i, j\in S$. Suppose while observing X_k it is not possible to distinguish between the states 1 and 2, i.e. we only observe $\{Y_k\}$ with grouped states $\{1, 2\}$ and 3. Given the five observations from $\{Y_k\}$,

$$\{1 \text{ or } 2\}, \quad 3, 3, \quad \{1 \text{ or } 2\}, \quad 3$$

estimate the posterior distribution of $\mathbf{x}=(x_1, \ldots, x_5)$.

6. Let $\{X_k\}$ be a stationary m-state Markov chain. Suppose we take r independent samples $x^{(l)}=(x_0^{(l)}, \ldots, x_n^{(l)}$, $l=1, 2, \ldots, r$ each of some (fixed) length. Let $n_{ij}(t)$ denote the number of realizations for which $x_{t-1}=i$ and $x_t=j$; $n_i(t-1)=\Sigma_{j=1}^m n_{ij}(t)$, the number of sequences for which $x_{t-1}=i$; $n_{ij}=\Sigma_{t=1}^n n_{ij}(t)$, the frequency of the transition $i\to j$ in the entire sample, and $n_i=\Sigma_{j=1}^m n_{ij}$. Find the maximum likelihood estimate $\hat{p}_{ij}$ of p_{ij} from the entire sample and obtain the asymptotic distribution of $\sqrt{r}(\hat{p}_{ij}-p_{ij})$ as the number of realizations $r\to\infty$ (n fixed). Also, find the asymptotic distribution of $\sqrt{n}(\hat{p}_{ij}-p_{ij})$ as $n\to\infty$ for r fixed.

(Anderson and Goodman, 1957)

7. Referring to Complement 6, now suppose the transition probabilities may depend on time, thus $P[X_t=j|X_{t-1}=i]=p_{ij}(t)$. Using r realizations each of same fixed length find the maximum likelihood estimates of $p_{ij}(t)$, $i, j=1, 2, \ldots, m$; $t=1, 2, \ldots, n$. Derive a test statistic for testing the hypothesis that the transition probabilities are homogeneous, i.e.

$$H: p_{ij}(1)=p_{ij}(2)=\ldots=p_{ij}(n).$$

(Anderson and Goodman, 1957)

8. Let $\{X_k\}$ be a negative binomial Markov chain with state space $S=\{0, 1, 2, \ldots\}$ and transition probabilities

$$p_{ij}(\theta)=\binom{c+i+j-1}{j}\theta^j(1-\theta)(c+i+j)\ i, j\in S.$$

where $c>0$ is known and θ an unknown parameter $(0<\theta<1)$. The stationary distribution of the chain $\{\pi_j\}$ is given by

$$\pi_j(\theta)=\binom{c+j-1}{j}\theta^j(1-\theta)^c,\ j\in S.$$

Consider a grouped chain $\{Y_k\}$ with two groups defined as $C_1=\{0\}$, $C_2=\{1, 2, \ldots\}$. Find the minimum χ-squared estimate of θ from a realization $\{Y_1, \ldots, Y_n\}$ and derive the asymptotic properties of the estimate.

(Toe, 1976)

9. Let $\{X_k\}$ be a positively regular Markov chain with a countable state space. Suppose the transition probabilities are functions of an unknown parameter θ. Consider a grouped chain $\{Y_k\}$ obtained from $\{X_k\}$, the grouped chain having a finite state space $\{C_1, \ldots, C_r\}$. It is necessary to estimate θ from a realization $\{Y_1, \ldots, Y_n\}$. Discuss the asymptotic properties of the maximum likelihood estimate and of the minimum χ-squared estimate (i.e. extend the results of Section 3 to the case when the X-chain is countable).

Chapter 5
Markov Chains in Continuous Time

1. INTRODUCTION

In this chapter problems of inference for continuous time Markov chains will be discussed. We assume the state-space to be discrete. The first two sections deal with ergodic type chains with finite and countable state-spaces respectively. We shall then discuss some recent results on birth and death processes which are of a non-ergodic type. See Chapter 2 (Section 2) for another example of a non-ergodic Markov process. Our emphasis in this chapter is on estimation rather than on testing.

2. FINITE MARKOV CHAINS

Let $\{X_t, t \geqq 0\}$ be a continuous-time Markov chain on a finite state-space $S = \{1, 2, \ldots, s\}$. We then have s types of events or units. Here X_t denotes the state of the process (i.e. the type of the unit under operation) at time t. The process $\{X_t\}$ will be completely determined by the initial probabilities $p_i(0) = P[X_0 = i]$, $i \in S$ and the transition probability functions $p_{ij}(t) = P[X_{t+u} = j | X = i]$, $i, j \in S$, which are assumed to be free from u. We further assume that

$$\lim_{t\to 0+} p_{ij}(t) = \begin{cases} 1 & \text{if } i=j \\ 0 & \text{if } i\neq j. \end{cases}$$

If we assume that all the states inter-communicate, the chain will be ergodic. Then, $p_{ij}(t)>0$ for all $i, j\in S$ and all $t>0$ and the limits $\lim_{t\to\infty} p_{ij}(t)=\pi_j$ exist. $\{\pi_j\}$ is the stationary distribution of the process with $\pi_j>0$, $\Sigma_{j=1}^{s}\pi_j=1$ and $\Sigma_{i=1}^{s}\pi_i p_{ij}(t)=\pi_j$, for all t.

For this model the transition intensities

$$\begin{aligned} q_i &= -p'_{ij}(0), \\ q_{ij} &= p'_{ij}(0),\ i\neq j, \end{aligned} \tag{1}$$

the prime denoting the derivative, exist and are such that $q_i>0$ for all i, $0\leqq q_{ij}<\infty$, and $\Sigma_{j\neq i}\, q_{ij}=q_i$. Let $\mathbf{Q}=((q_{ij}))$, with $q_{ii}=-q_i$. One then has the following constructive development of the chain in terms of the transition intensities. The system starts off at time 0 with a i-type unit with probability $p_i(0)$; the life-time of this first unit say T_1 has a density $f_{T_1}(t)=q_i e^{-q_i t}$. At time $t_1=T_1$ (when the first unit fails) a j-type unit $(i\neq j)$ is set into operation with probability q_{ij}/q_i; this second unit has a life-time T_2 distributed as $f_{T_2}(t)=q_j e^{-q_j t}$. At the epoch $t_2=T_1+T_2$ a k-type unit $(j\neq k)$ is set into operation with probability q_{jk}/q_j; this unit has a life-time T_3 distributed as $f_{T_3}(t)=q_k e^{-q_k t}$ and so on.

Let $\{X_{(1)}, X_{(2)}, \ldots,\}$ be the successive states the system passes through. Note that $X_{(k)}\neq X_{(k+1)}$, $(k=1, 2, \ldots)$. It will be seen that the bivariate discrete-time process $\{(X_{(k)}, T_k), k=1, 2, \ldots\}$ is a Markov process with the state-space $S\times[0, \infty)$, and transition probabilities given by

$$P[X_{(k+1)}=j,\ T_{k+1}>t \mid X_{(k)}=i,\ T_k=u]=\frac{q_{ij}}{q_i}e^{-q_j t},\ i\neq j. \tag{2}$$

Let $p_{ij}(t)$ be functions of an unknown parameter θ, so that q_i and q_{ij} too are functions of θ, say $q_i(\theta)$ and $q_{ij}(\theta)$. We now seek a maximum likelihood estimator of θ based on a continuous observation of $\{X_t\}$ until n_0 events have occurred. Thus, our sample is $\{(X_{(k)}, T_k), k=1, 2, \ldots, n_0-1$, and $X_{(n_0)}\}$. The likelihood function is

$$\begin{aligned} L &= p_{x_1}(\theta)(q_{x_1}(\theta)e^{-q_{x_1}(\theta)T_1})\frac{q_{x_1x_2}(\theta)}{q_{x_1}(\theta)}(q_{x_2}(\theta)e^{-q_{x_2}(\theta)T_2})\ldots\frac{q_{x_{n-1}x_n}(\theta)}{q_{x_{n-1}}(\theta)} \\ &= p_{x_1}(\theta)\left\{\prod_{\substack{i,j\\ i\neq j}} q_{ij}^{n_{ij}}(\theta)\right\}\left\{\prod_{i} e^{-q^i(\theta)\beta^i}\right\} \end{aligned} \tag{3}$$

where n_{ij} is the transition frequency of $i \to j$ in the sample $x_{(1)}, \ldots, x_{(n_0)}$, and β_i is the total observed time for which the system occupied state i. See Chapter 8 for a rigorous derivation of (3). The maximum likelihood equation for estimating θ (ignoring the effect of the initial state) is

$$\frac{\mathrm{d}\log L}{\mathrm{d}\theta} = \sum_{\substack{i,j \\ i \neq j}} \left[\frac{n_{ij} - \beta_i q_{ij}(\theta)}{q_{ij}(\theta)}\right] \frac{\mathrm{d}q_{ij}(\theta)}{\mathrm{d}\theta} = 0. \tag{4}$$

Using Billingsley's (1961) results we see that there exists a consistent root $\hat{\theta}$ of this equation such that, in distribution,

$$\sqrt{n_0}(\hat{\theta} - \theta) \to N\left(0, \frac{1}{i(\theta)}\right) \text{as } n_0 \to \infty \tag{5}$$

where

$$\begin{aligned} i(\theta) &= -\frac{1}{n_0} E\left(\frac{\mathrm{d}^2 \log L}{\mathrm{d}\theta^2}\right) \\ &= \sum_{\substack{i,j \\ i \neq j}} \frac{u_i(\theta)}{q_i(\theta) q_{ij}(\theta)} \left(\frac{\mathrm{d}q_{ij}(\theta)}{\mathrm{d}\theta}\right)^2, \end{aligned} \tag{6}$$

(See complement 1) where $\{u_i\}$ is the stationary distribution corresponding to the imbedded Markov chain $\{X_{(k)}, k = 1, 2, \ldots\}$. Note that, in general $\{u_i\}$ and $\{\pi_i\}$ are different. Extension to multiparameter case is straightforward and is left as an exercise. Likelihood-ratio tests for θ are discussed in Billingsley (1961). See Chapter 8, Section 6, for non-parametric estimation of the intensities.

3. QUEUEING MODELS

The method of the previous section can be readily extended to an infinite state-space. We shall illustrate this by considering ergodic queueing models with Poisson arrivals and exponential service-times.

When the system is in state j let λ_j and μ_j denote the arrival rate and the service rate. Set $\mu_0 = 0$ and let $\mu_j > 0$, $(j = 1, 2, \ldots)$; $\lambda_j > 0$, $(j = 0, 1, 2, \ldots)$; and assume that the mean recurrence time for each state is finite (i.e. the process is ergodic). If the queue has a limited capacity say M, then we may let $\lambda_j > 0$, $(j = 0, 1, 2, \ldots, M - 1)$, and $\lambda_j = 0$ for $j \geqq M$. The evolution of the queueing process (which is a birth and death process) can be described as follows. The system stays in state i for a time which is exponential with parameter $(\lambda_i + \mu_i)$; then a transition $i \to i + 1$ occurs (a birth occurs, or a new customer arrives) with probability $\lambda_i/(\lambda_i+\mu_i)$. If the system is in state 0 at any time, only an upward transition $0 \to 1$ can occur. We then have a Markov chain with transition intensities $q_{i,i+1}=\lambda_i$, $q_{i,i-1}=\mu_i$, and $-q_{ii}=q_i=\lambda_i+\mu_i$; all other q_{ij} are zero. Suppose that q_{ij} (i.e. λ_i and μ_i) depends on a vector of parameters θ. We seek a maximum likelihood estimator of θ based on observing the process continuously until n_0 transitions (either births or deaths) have occurred. Arguing as in the previous section, the likelihood function is found to be (ignoring the initial state)

$$L=\left(\prod_{i=0}^{\infty} \lambda_i^{u_i}\right)\left(\sum_{i=1}^{\infty} \mu_i^{d_i}\right)\left(\prod_{i=0}^{\infty} \exp\{-(\lambda_i+\mu_i)\beta_i\}\right) \tag{7}$$

where u_i is the frequency of transitions $i \to i+1$ d_i, the frequency of tran– sitions $i \to i-1$ and β_i, the observed total time spent in state i. The likelihood equations $\partial \log L/\partial\theta_r=0$, $(r=1, 2, \ldots, k)$ then yield the maximum likelihood estimators of $\theta_1, \ldots, \theta_k$.

We now consider two specific models.

3.1 Model 1: $M/M/k$ Queue

In the model with k servers, Poisson arrivals and exponential service, we have

$$\lambda_i = \lambda, \quad i=0, 1, 2, \ldots;$$
$$\mu_i = \begin{cases} i\mu, & i=0, 1, \ldots, k-1 \\ k\mu & \text{for } i \geqq k, \end{cases}$$

so that λ_i and μ_i depend on only two parameters (λ, μ). The likelihood function based on a continuous observation until n_0 transitions occur is given by

$$L=\left(\prod_{i=0}^{\infty} \lambda^{u_i}\right)\left(\prod_{i=1}^{k-1} (i\mu)^{d_i} \prod_{i=k}^{\infty} (k\mu)^{d_i}\right)\left(\prod_{i=0}^{k-1} \exp(-(\lambda+i\mu)\beta_i) \prod_{i=k}^{\infty} \exp\,(-(\lambda+k\mu)\beta_i)\right). \tag{8}$$

The maximum likelihood estimators $\hat{\lambda}$ and $\hat{\mu}$ of λ and μ respectively are

$$\hat{\lambda}=\frac{a}{T} \text{ and } \hat{\mu}=\frac{d}{\sum_{i=1}^{k-1} i\beta_i + k\sum_{i=k}^{\infty} \beta_i} \tag{9}$$

where a is the total number of arrivals, d, the total number of departures and T the total observation time. It will be seen that (assuming $\rho=\lambda/\mu<k$

for ergodicity)

$$\sqrt{n_0}\left[\begin{pmatrix}\hat{\lambda}\\ \hat{\mu}\end{pmatrix}-\begin{pmatrix}\lambda\\ \mu\end{pmatrix}\right]\to N_2\left[\begin{pmatrix}0\\ 0\end{pmatrix},\begin{pmatrix}2\lambda^2 & 0\\ 0 & 2\mu^2\end{pmatrix}\right], \text{ in distribution,} \tag{10}$$

$N_2[\cdot,\cdot]$ denotes the bivariate normal distribution.

Setting $k=1$ and $k=\infty$ one can get the appropriate results for the $M/M/1$ queue and the $M/M/\infty$ queue respectively.

3.2 Model 2: $E_k/M/1$ Queue

Consider a single server queue with the inter-arrival time distributed as gamma, with index k and parameter λ, and the service time distributed as exponential, rate μ. We can view each arrival as consisting of k stages (0, 1, 2, ..., $k-1$) in each of which the customer waits for a random time which is exponential with parameter λ. The actual arrival of the customer in the system corresponds to his reaching the kth stage. A new customer enters stage 0 just as his predecessor leaves stage $k-1$. Let Z_t denote the sum of the stages of all the customers standing in the queue at time t (including the one being served, and the one whose arrival is forthcoming). Thus, for instance, $Z_t=i=rk+l$, $(l<k)$, means that at time t there are r customers in the system and another customer is in his lth stage of arrival. When a customer departs the state of the system (corresponding to Z_t) is reduced by k (instead of 1). It is then seen that $\{Z_t,\ t\geqq 0\}$ is a Markov chain on the state-space $S=\{0, 1, 2 \ldots\}$, with the transition intensities

$$-q_{ii}=q_i=\begin{cases}\lambda & \text{for } i=0, 1, \ldots, k-1\\ \lambda+\mu & \text{for } i\geqq k\end{cases}$$

$$q_{i,i+1}=\lambda,\ i=0, 1, 2, \ldots,$$

and

$$q_{i,i-k}=\mu,\ i=k, k+1, \ldots;$$

all other q_{ij} are zero.

Suppose we observe the process $\{Z_t\}$ continuously until n_0 transitions occur. The likelihood function is

$$L=\left(\prod_{i=0}^{\infty}\lambda^{n_{i,i+1}}\right)\left(\prod_{i=k}^{\infty}\mu^{n_{i,i-k}}\right)\left(\prod_{i=k}^{\infty}\exp\left(-(\lambda+\mu)\beta_i\right)\right)\left(\prod_{i=0}^{k-1}\exp\left(-\lambda\beta_i\right)\right) \tag{11}$$

where $n_{i,i+1}$ is the transition frequency of $i\to i+1$,
$n_{i,i-k}$ is the transition frequency of $i\to i-k$,

and β_i the total observed time spent in the state i.

The maximum likelihood estimators are

$$\hat{\lambda}=\frac{\sum_{i=0}^{\infty} n_{i,i+1}}{\sum_{i=0}^{\infty} \beta_i}=\frac{a+l}{T};\ \hat{\mu}=\frac{\sum_{i=k}^{\infty} n_{i,i-k}}{\sum_{i=k}^{\infty} \beta_i}=\frac{d}{T-w}, \tag{12}$$

where a is the total number of arrivals, d the total departures, T the total observation time, w the waiting time for the arrival of the first customer, and l is the arrival stage of the forthcoming ($(a+1)$th) customer when the observation was terminated. (These quantities refer to the original process rather than to Z_t).

We see that

$$\sqrt{n_0}\left[\begin{pmatrix}\hat{\lambda}\\ \hat{\mu}\end{pmatrix}-\begin{pmatrix}\lambda\\ \mu\end{pmatrix}\right]\to N_2\left[\begin{pmatrix}0\\0\end{pmatrix},\begin{pmatrix}v_\lambda & 0\\ 0 & v_\mu\end{pmatrix}\right], \text{ in distribution,} \tag{13}$$

where v_λ and v_μ can be computed in the usual way.

4. PURE BIRTH PROCESS

Let $\{X_t,\ t\geqq 0\}$ be a pure birth process with birth-rate θ. Thus

$$P(X_{t+\delta t}=n+1|X_t=n)=n\theta\delta t+o(\delta t),$$

and

$$P(X_{t+\delta t}=n|X_t=n)=1-n\theta\delta t+o(\delta t).$$

Suppose the births occur at times $t_1<t_2<\ldots$, and let $X_0=1$ with probability 1. The intervals between births $T_k=t_k-t_{k-1}(k=1, 2, \ldots;\ t_0=0)$ are independently distributed as negative exponentials with parameters θk. The process $\{X_t\}$ is clearly a Markov process with the transition function given by

$$p_{ij}(t)=P(X_{t+u}=j|X_u=i)$$
$$=\begin{cases}\dbinom{j-1}{j-i}e^{-i\theta t}(1-e^{-\theta t})^{j-i}, & \text{for } j\geqq i\\ 0, & \text{otherwise,}\end{cases}$$

where $(i, j)\in S=\{1, 2, \ldots\}$.

We now wish to estimate θ from a sample from $\{X_t\}$. Several sampling plans are possible; we shall discuss three such plans.

4.1 Observation at Fixed Equidistant Points (Kendall, 1949)

Let T_0 be a fixed (total) time of observation. Divide T_0 into n_0 equal parts, and let $h = T_0/n_0$. Suppose we observe $\{X_t\}$ (with $X_0 = 1$) at the points 0, h, $2h, \ldots, n_0 h$ $(= T_0)$, and let $(x_{kh}, k = 0, 1, \ldots, n_0)$ be the sample. The log-likelihood function is (ignoring terms free from θ) given by,

$$\log L = -\sum_{k=1}^{n_0} x_{(k-1)h}\,\theta h + \sum_{k=1}^{n_0} (x_{kh} - x_{(k-1)h}) \log (1 - e^{-h\theta}).$$

The maximum likelihood estimator of θ is readily found to be

$$\hat{\theta} = \frac{1}{h} \log \frac{\sum_{k=1}^{n_0} x_{kh}}{\sum_{k=0}^{n_0-1} x_{kh}}, \tag{14}$$

or

$$e^{\hat{\theta} h} = \frac{\sum_{k=1}^{n_0} x_{kh}}{\sum_{k=0}^{n_0-1} x_{kh}}. \tag{15}$$

If we now let $X_0 = m$, the process $\{X_t\}$ can be viewed as a sum of m independent birth processes each with $X_0 = 1$. Denote $x_{kh}^{(r)}$ as an observation from the rth process $(r = 1, 2, \ldots, m)$ each with $X_0 = 1$. The maximum likelihood estimator $\hat{\theta}_m$ is given by

$$e^{\hat{\theta}_m h} = \frac{\sum_{r=1}^{m} \sum_{k=1}^{n_0} x_{kh}^{(r)}}{\sum_{r=1}^{m} \sum_{k=0}^{n_0-1} x_{kh}^{(r)}}. \tag{16}$$

For fixed n_0 and large m the classical asymptotic theory of maximum likelihood estimation applies and we have

$$\sqrt{m}(\hat{\theta}_m - \theta) \to N\left(0, \frac{1}{i(\theta)}\right), \text{ as } m \to \infty, \text{ in distribution} \tag{17}$$

where $i(\theta)=E(-\mathrm{d}^2 \log L/\mathrm{d}\theta^2)$. The asymptotic variance is given by

$$v(\theta)=\frac{1}{i(\theta)}=\left\{\frac{\sinh\left(\dfrac{\theta h}{2}\right)}{\dfrac{\theta h}{2}}\right\}^2\left\{\frac{\theta^2}{\exp(n_0 h\theta)-1}\right\}, \tag{18}$$

where $\sinh(u)=\dfrac{e^u-e^{-u}}{2}$.

4.2 Continuous Observation over a Fixed Period

Suppose we observe $\{X_t\}$ (with $X_0=1$) continuously over $(0, T_0)$. Recall that the intervals between births, $T_k=t_k-t_{k-1}$, $(k=1, 2, \ldots)$ are independent exponential variates with parameters $k\theta$. Suppose $N-1$ births occurred during $(0, T_0)$ at points $t_1<t_2<\ldots<t_{N-1}\leqq T_0$. Here, N is a random variable. Thus $X_{T_0}=N$, so that the epoch $t_N>T_0$. The likelihood of this sample is

$$L=\prod_{k=1}^{N-1}(k\theta\exp(-k\theta T_k))\exp(-\theta N(T_0-t_{N-1})). \tag{19}$$

The log-likelihood ignoring terms free from θ is

$$\log L=(N-1)\log\theta-\theta\sum_{k=1}^{N-1}kT_k-\theta NT_0+\theta Nt_{N-1}.$$

The maximum likelihood estimator of θ is seen to be

$$\hat{\theta}=\frac{N-1}{\sum_{k=1}^{N-1}kT_k+N(T_0-t_{N-1})}.$$

Note that the quantity $\Sigma_{k=1}^{N-1}kT_k+N(T_0-t_{N-1})$ is equal to the area under the curve in the graph of X_t against t, between 0 and T_0. We can thus write

$$\hat{\theta}=\frac{N-1}{\int_0^{T_0}X_u\mathrm{d}u}. \tag{20}$$

If we had a birth process with $X_0=m$, the maximum likelihood estimator of θ is

$$\hat{\theta}_m=\frac{N-m}{\int_0^{T_0}X_u\mathrm{d}u}. \tag{21}$$

The classical asymptotic theory for $\hat{\theta}_m$ is valid as $m \to \infty$ and T_0 fixed. We have

$$\sqrt{m}(\hat{\theta}_m - \theta) \to N(0, v^*(\theta)), \text{ in distribution,} \tag{22}$$

where the asymptotic variance $v^*(\theta)$ can be obtained from the corresponding expression $v(\theta)$ in Section 4.1 by letting $n_0 \to \infty$, $h \to 0$, and $n_0 h \to T_0$ fixed, and is given by

$$v^*(\theta) = \frac{\theta^2}{(\exp(T_0\theta) - 1)}. \tag{23}$$

The asymptotic relative efficiency of the method in 4.1 as against the method in 4.2 is given by

$$\frac{v^*(\theta)}{v(\theta)} = \left\{ \frac{\dfrac{\theta h}{2}}{\sinh \dfrac{\theta h}{2}} \right\}^2. \tag{24}$$

Asymptotic Theory for Fixed m **and** $T_0 \to \infty$. If we fix m and consider the single realization as $T_0 \to \infty$, the classical theory can no longer be applied since the observations now are dependent. The results of Billingsley (1961) too do not apply since the process is not ergodic. Keiding (1974) has shown that, for $m = 1$, and as $T_0 \to \infty$

(i) $\hat{\theta} \to \theta$, almost surely

(ii) $S_{T_0}^{1/2}(\hat{\theta} - \theta) \to N(0, \theta)$, in distribution, where $S_{T_0} = \int_0^{T_0} X_u du$, and

(iii) $(ES_{T_0})^{1/2}(\hat{\theta} - \theta) \to t(2)$ in distribution where $t(2)$ has a Student's t distribution with two degrees of freedom.

We do not provide proofs of these results here since they are special cases of the corresponding results for the birth and death process discussed in Section 5. Also, see Chapter 2 (Section 2).

4.3 Continuous Observation over a Random Period (Moran, 1951)

Suppose we observe $\{X_t, t \geqq 0\}$, $(X_0 = 1)$ continuously until N_0 births (after the first birth at time 0) have been recorded. Let these births occur at times $t_1 < t_2 < \ldots < t_{N_0}$. The likelihood function is

$$L = \prod_{k=1}^{N_0} k\theta \exp(-k\theta T_k), \tag{25}$$

where $T_k = t_k - t_{k-1}$. The maximum likelihood estimator is

$$\hat{\theta} = \frac{N_0}{\sum_{k=1}^{N_0} kT_k}.$$

Let T (random variable) denote the total time needed to collect N_0 births (after the first). Then $\Sigma_{k=1}^{N_0}(kT_k)$ is the area under the curve in the graph of X_t against t, between 0 and T, and hence $\Sigma_{k=1}^{N_0} kT_k$ can be represented by $\int_0^T X_u du$. Also, note that $N_0 = X_T - X_0 = X_T - 1$. The ML estimate $\hat{\theta}$ is thus

$$\hat{\theta} = \frac{X_T - 1}{\int_0^T X_u du} \tag{26}$$

which is the same as the estimate obtained in (4.2) except that here T is a random variable and N_0 fixed.

Now $\Sigma_{k=1}^{N_0}(kT_k)$ is distributed as a gamma variate with the characteristic function $\phi(t) = (1 - it/\theta)^{-N_0}$. Therefore $E(\hat{\theta}) = (N_0/N_0 - 1)\theta$, $(N_0 > 1)$. An unbiased estimate of θ is clearly $\tilde{\theta} = (N_0 - 1/N_0)\hat{\theta}$. The exact variance formula for $\tilde{\theta}$ is

$$V(\tilde{\theta}) = \frac{\theta^2}{(N_0 - 2)}, \quad (N_0 > 2). \tag{27}$$

Since $\Sigma_{k=1}^{N_0}(kT_k)$ is a complete sufficient statistic for θ, $\tilde{\theta}$ is the unique minimum variance unbiased estimate of θ.

Since $2\theta\Sigma_{k=1}^{N_0}(kT_k)$ is distributed as $\chi^2(2N_0)$, the $(1-\alpha)100\%$ confidence interval for θ is given by

$$\left(\frac{a}{2\sum_{k=1}^{N_0} kT_k}, \frac{b}{2\sum_{k=1}^{N_0} kT_k} \right)$$

where a and b are defined by

$$P(\chi^2(2N_0) < a) = P(\chi^2(2N_0) > b) = \frac{\alpha}{2}.$$

5. THE BIRTH AND DEATH PROCESS

Let $\{X_t, t \geqq 0\}$ be a birth and death process with λ and μ as the birth and death rates. $\{X_t\}$ is therefore a Markov chain on the state-space $S = \{0, 1, 2,$

...} with the transition functions determined by

$$P(X_{t+h}=j|X_t=i)=\begin{cases} i\lambda h+o(h), & \text{for } j=i+1 \\ 1-i(\lambda+\mu)h+o(h), & j=i \\ i\mu h+o(h) & j=i-1 \\ o(h) & \text{otherwise.} \end{cases} \tag{28}$$

Assume throughout the section that $X_0=x_0$ is fixed. The matrix of transition intensities for this Markov chain is given by

$$\mathbf{Q}=\begin{bmatrix} 0 & 0 & 0 & 0 & \cdots \\ \mu & -(\lambda+\mu) & \lambda & 0 & \cdots \\ 0 & 2\mu & -2(\lambda+\mu) & 2\lambda & \cdots \\ \cdots & \cdots & \cdots & \cdots & \cdots \\ \cdots & \cdots & \cdots & \cdots & \cdots \end{bmatrix}. \tag{29}$$

We shall first discuss an exact maximum likelihood theory using inverse sampling due to Moran (1951, 1953).

5.1 Inverse Sampling

The evolution of the process can be described as follows. The system starts off at time 0 with x_0 individuals. A transition $x_0 \to x_0+1$ (a birth) occurs with probability $p=\lambda/(\lambda+\mu)$ and a transition $x_0 \to x_0-1$ (a death) occurs with probability $1-p=\mu/(\lambda+\mu)$. The system occupies the same state x_0(i.e. no change occurs) for a random time which has an exponential distribution with parameter $x_0(\lambda+\mu)$. We wish to estimate λ and μ from a continuous observation of the process until n_0 events (births or deaths) have occurred, *conditionally on non-absorption.*

It is convenient to estimate $p=\lambda/(\lambda+\mu)$, and $\theta=(\lambda+\mu)$, rather than estimating λ and μ directly. Note that $\lambda=p\theta$ and $\mu=(1-p)\theta$, and hence the estimates of λ and μ can readily be obtained from those of θ and p.

First consider the problem of estimating θ. Let the successive transition occur at the epochs $0<t_1<t_2<\ldots<t_{n_0}$. $T_k=t_k-t_{k-1}$, $(k=1, 2, \ldots, n_0$, with $t_0=0)$ are independently distributed as exponential random variables with parameters $\theta x_{t_{k-1}}$ when $X_{t_{k-1}}=x_{t_{k-1}}$. The likelihood function (conditional on non-absorption up to t_{n_0}) is

$$L=\left(\theta^{n_0}\prod_{k=1}^{n_0} x_{t_{k-1}}\right)\left(\exp\left(-\sum_{1}^{n_0} T_k x_{t_{k-1}}\right)\right). \tag{30}$$

The maximum likelihood estimate of θ is

$$\hat{\theta}=\frac{n_0}{\sum_{k=1}^{n_0} T_k x_{t_{k-1}}}=\frac{n_0}{\int_0^{t_{n0}} X_u du}. \tag{31}$$

Since $\Sigma_{k=1}^{n_0} T_k X_{t_{k-1}}$ is distributed as a gamma variate with parameter θ and index n_0, it follows that $\tilde{\theta}=(n_0-1/n_0)\hat{\theta}$ is the unique minimum variance unbiased estimate of θ with $V(\tilde{\theta})=\theta^2/(n_0-2)$, $(n_0>2)$.

For estimating p we first represent the process on a rectangular coordinate system with epochs $(t_1, t_2, \ldots)$ on the x-axis (abscissa) and the population-sizes $(X_{t_1}, X_{t_2}, \ldots)$ on the y-axis (ordinate). The initial point of the process is $(0, x_0)$. The line at $y=0$ is an absorbing barrier. Assuming we observe until n_0 events, conditionally on non-absorption, the possible end-points are $(x=n_0, y=1)$, $(x=n_0, y=2)$, $\ldots$, $(x=n_0, y=n_0+x_0)$. Note that $Y=X_{t_{n0}}$ is a sufficient statistic for p.

The probability of the path of the random walk ending at (n_0, y) is

$$P\left(X_{t_{n0}}=y \,\middle|\, \begin{array}{l}\text{non-}\\ \text{absorption}\end{array}\right)=\left\{\binom{n_0}{\frac{1}{2}(n_0-x_0+y)}-\binom{n_0}{\frac{1}{2}(n_0+x_0+y)}\right\}p^{1/2(n_0-x_0+y)}(1-p)^{1/2(n_0+x_0-y)} \tag{32}$$

where $y=1, 2, \ldots, n_0+x_0.$ (See Feller, 1957)

The likelihood function (conditional on non-absorption) is therefore

$$L=p^{1/2(n_0-x_0+y)}(1-p)^{1/2(n_0+x_0-y)}.$$

The maximum likelihood estimate of p is $\hat{p}=(n_0-x_0+y)/2n_0$. Since the exact distribution of Y is known the exact distribution of $\hat{p}$ can be determined.

5.2 Continuous Observation Over a Fixed Period

We now derive estimates of λ and μ from a continuous realization of $\{X_t, t\geqq 0\}$ over $[0, T_0]$, with $X_0=x_0$. Let $t_1<t_2<\ldots<t_n$ be the epochs at which transitions (either births or deaths) occur. The likelihood function is

$$L=(\exp[-x_0(\lambda+\mu)t_1]x_0\xi_0)(\exp[-x_1(\lambda+\mu)t_2]x_1\xi_1)\ldots$$

$$(\exp[-x_{n-1}(\lambda+\mu)t_n]x_{n-1}\xi_{n-1})\exp\left[-x_n(\lambda+\mu)\left(T_0-\sum_1^n t_i\right)\right]$$

$$=\exp-((\lambda+\mu)S_{T_0})\prod_{i=0}^{n-1}x_i\prod_{r=0}^{n-1}\xi_r \tag{33}$$

where ξ_r is λ if the $(r+1)$th transition is a birth and μ if it is a death, and $S_{T_0}=(\Sigma_{r=1}^{n} x_{r-1}t_r + x_n(T_0-\Sigma_1^n t_r))=\int_0^{T_0} X_u du$. Ignoring the terms free from the parameters the log-likelihood is thus

$$\log L = -(\lambda+\mu)S_{T_0} + B_{T_0}\log\lambda + D_{T_0}\log\mu,$$

where B_{T_0} and D_{T_0} denote the number of births and deaths respectively in $[0, T_0]$. The maximum likelihood estimates of λ and μ are

$$\hat{\lambda}=\frac{B_{T_0}}{S_{T_0}},\ \hat{\mu}=\frac{D_{T_0}}{S_{T_0}}. \tag{34}$$

From the likelihood function it is clear that $(B_{T_0}, D_{T_0}, S_{T_0})$ is a sufficient statistic for (λ, μ). Note that the estimators of λ and μ obtained from those of θ and p in (5.1) are formally the same as the present estimators $\hat{\lambda}$ and $\hat{\mu}$.

The birth and death process with $X_0=x_0$ can be viewed as a sum of x_0 independent birth and death processes each with $X_0=1$. Consequently the asymptotic properties of $\hat{\lambda}$ and $\hat{\mu}$ follow from the classical theory as $x_0\to\infty$. Thus

$$\sqrt{x_0}\begin{bmatrix}\hat{\lambda}-\lambda\\ \hat{\mu}-\mu\end{bmatrix}\to N_2\left(\begin{bmatrix}0\\0\end{bmatrix},\begin{bmatrix}\sigma_\lambda^2 & 0\\ 0 & \sigma_\mu^2\end{bmatrix}\right) \tag{35}$$

in distribution as $x_0\to\infty$, where

$$\sigma_\lambda^2=\frac{\lambda(\lambda-\mu)}{\exp\,[(\lambda-\mu)T_0]-1}\quad\text{and}\quad\sigma_\mu^2=\frac{\mu(\lambda-\mu)}{\exp\,[(\lambda-\mu)T_0]-1}.$$

Asymptotic Theory as $T_0\to\infty$ and x_0 Fixed. We shall first show that $\hat{\lambda}$ and $\hat{\mu}$ given by (34) are strongly consistent as $T_0\to\infty$ (x_0 fixed). Consider

$$U_t(\lambda)=\frac{S_t}{\lambda}(\hat{\lambda}-\lambda)$$

$$=\frac{1}{\lambda}(B_t-\lambda S_t).$$

Thus,

$$dU_t(\lambda)=\frac{1}{\lambda}(dB_t-\lambda X_t dt).$$

Also, $E(dB_t|\mathscr{F}_t)=\lambda X_t dt$, almost surely, where $\mathscr{F}_t$ is the σ-field generated by $\{X_u, u\leqq t\}$. Consequently, $E(dU_t(\lambda)|\mathscr{F}_t)=0$, almost surely. Now, for $s<t$, we have

$$E(U_t(\lambda)-U_s(\lambda)|\mathscr{F}_s)=E\left(\int_s^t dU_k(\lambda)|\mathscr{F}_s\right)$$

$$= E\left\{\int_s^t E(\mathrm{d}U_u(\lambda)|\mathscr{F}_u)|\mathscr{F}_s\right\}$$

$$= 0, \text{ almost surely.}$$

Therefore, $\{U_t(\lambda), \mathscr{F}_t, t \geqq 0\}$ is a zero-mean martingale in continuous time.

The conditional Fisher information can be obtained as follows:

$$E\{(\mathrm{d}U_t(\lambda))^2|\mathscr{F}_t\} = \frac{1}{\lambda^2} E\{(\mathrm{d}B_t)^2|\mathscr{F}_t\}$$

$$= \frac{1}{\lambda^2} X_t \mathrm{d}t,$$

and hence $\xi_t(\lambda) = \int_0^t (X_u/\lambda)\mathrm{d}u = (S_t/\lambda)$. Thus, conditional on $S_{T_0} \to \infty$ (i.e. conditional on non-extinction) we have, by the strong law of large numbers for martingales,

$$\hat{\lambda} \to \lambda \text{ almost surely as } T_0 \to \infty.$$

The strong consistency of $\hat{\mu}$ is similarly established by considering the martingale

$$U_t(\mu) = \frac{1}{\mu}(D_t - \lambda S_t).$$

The asymptotic distributions of $\hat{\lambda}$ and $\hat{\mu}$ can be deduced as follows:

When $\lambda > \mu$ (super-critical case) it is well known (cf. Harris, 1963) that $(X_t/EX_t) \to W$ almost surely, where $P(W=0) = P(X_t \to 0) = (\mu/\lambda)^{x_0}$. Assume for simplicity $x_0 = 1$; then conditional on $W > 0$, the density of W is negative exponential given by

$$f_W(w) = \frac{\lambda - \mu}{\lambda} \exp\left[-\left(\frac{\lambda - \mu}{\lambda}\right) W\right], \; W > 0. \tag{36}$$

We shall derive the asymptotic distributions of $\hat{\lambda}$ and $\hat{\mu}$ on the set of non-extinction $\{W > 0\}$, on which $S_t \to \infty$ almost surely, as $t \to \infty$.

The essence of the derivation is to express the birth and death process as a random time transformation of an additive process and use a continuous time analogue of the random sum central limit theorem. Consider a Poisson process $\{Y_t\}$ with intensity $(\lambda + \mu)$. Corresponding to the ith jump of this process define a bivariate random variable (M_i, N_i), $(i = 1, 2, \ldots)$ taking values (1, 0) or (0, 1) with probabilities $\lambda/(\lambda + \mu)$ and $\mu/(\lambda + \mu)$ respectively. Now, define the compound Poisson process (Q_t, R_t) where $Q_t = \Sigma_1^{Y_t} M_i$ and $R_t = \Sigma_1^{Y_t} N_i$. It is then seen that (Q_t, R_t) is a process with independent increments, and consequently by the central limit theorem, we obtain, as $t \to \infty$.

$$\begin{bmatrix} (\lambda t)^{-1/2} & (Q_t - \lambda t) \\ (\mu t)^{-1/2} & (R_t - \mu t) \end{bmatrix} \to N_2\left[\begin{pmatrix} 0 \\ 0 \end{pmatrix}, \begin{pmatrix} 1 & 0 \\ 0 & 1 \end{pmatrix}\right], \tag{37}$$

in distribution. The process $\{(Q_t, R_t)\}$ is related to the birth and death process under consideration.

Now, replacing t by S_t in (37) we finally get the desired limit result:

$$\begin{bmatrix} \lambda^{-1/2} S_{T_0}^{1/2}(\hat{\lambda} - \lambda) \\ \mu^{-1/2} S_{T_0}^{1/2}(\hat{\mu} - \mu) \end{bmatrix} \to N_2\left[\begin{pmatrix} 0 \\ 0 \end{pmatrix}, \begin{pmatrix} 1 & 0 \\ 0 & 1 \end{pmatrix}\right] \tag{38}$$

as $T_0 \to \infty$ in distribution,

on the set of non-extinction.

The random norm S_{T_0} in (38) can be replaced by a constant norm $E(S_{T_0}) = (\exp(\lambda - \mu)T_0 - 1)/(\lambda - \mu)$, (assuming $x_0 = 1$), where $S_{T_0}/ES_{T_0} \to W$, almost surely, in which case the limiting distribution is no longer normal. Instead, we have as $T_0 \to \infty$, on the set of non-extinction.

$$(ES_{T_0})^{1/2} \begin{bmatrix} \lambda^{-1/2}(\hat{\lambda} - \lambda) \\ \mu^{-1/2}(\hat{\mu} - \mu) \end{bmatrix} \to \begin{bmatrix} Z_1 W^{-1/2} \\ Z_2 W^{-1/2} \end{bmatrix} \text{ in distribution,} \tag{39}$$

where Z_1 and Z_2 are independent $N(0, 1)$ variates and W has a negative exponential density given by (36). These limit results are analogous to the corresponding results for the super-critical branching process discussed in Chapter 2.

Setting $\mu = 0$ in (38) and (39) we get the corresponding limit results for the pure birth process discussed in the previous section.

BIBLIOGRAPHICAL NOTES

Sections 2 and 3 deal with ergodic Markov chains which are special cases of processes treated in Billingsley (1961) under a more general setting. See Keiding (1975) for further details regarding estimation problems for birth and death processes. It is possible to extend the methods used in this chapter to semi-Markov processes (see for instance, Moore and Pyke (1968)) by viewing the latter as two-dimensional Markov processes.

COMPLEMENTS

1. Consider an ergodic Markov chain $\{X_t, t \geqq 0\}$ on a finite state-space with the transition intensities $Q = ((q_{ij}(\theta))$. Let $\{u_i(\theta), i = 1, 2, \ldots, s\}$ be the stationary distribution

for the embedded chain $\{X_{(k)}, k=1, 2, \ldots\}$. Show that the Fisher information per observation (jump) corresponding to the estimation of θ is given by

$$i(\theta)=\sum_{\substack{i,j \\ i\neq j}} \frac{u_i(\theta)}{q_i(\theta)q_{ij}(\theta)}\left(\frac{\mathrm{d}q_{ij}(\theta)}{\mathrm{d}\theta}\right)^2$$

where $q_i(\theta)=-q_{ii}(\theta)$. Also, show that $u_i(\theta)=\pi_i(\theta)q_i(\theta)/\Sigma_i\pi_i(\theta)q_i(\theta)$ where $\{\pi_i(\theta)\}$ is the stationary distribution corresponding to $\{X_t; t\geqq 0\}$.

2. Consider a single-server queue with Poisson arrivals at rate λ and service-time distributed as gamma with index k (positive integer) and parameter μ. Let X_t denote the number of customers in the system at time t. Let $t_1 < t_2 < \ldots$ be the successive epochs of completion of service (i.e. departure epochs of customers), and denote $Z_k = X_{b_k}$. $\{Z_k\}$ is then a Markov chain. Find the maximum likelihood estimate of the traffic intensity, $\rho = k\lambda/\mu$, based on a sample of n successive observations from $\{Z_k\}$. Determine the mean and the variance of the estimate.

3. Let $\rho=\lambda/\mu$ be the traffic intensity of a single-server queue with Poisson arrivals (rate λ) and exponential service (rate μ). $\{T_k\}$, and $\{Y_k\}$ are the successive inter-arrival times and service times respectively. Find a sequential estimate $\hat{\rho}$ of ρ based on $\{T_k\}$ and $\{Y_k\}$ such that $P(|\hat{\rho}/\rho-1|\leqq\delta)\geqq(1-\alpha)$, for specified δ and α.

4. Consider a single-server queue with Poisson arrivals and exponential service times. The arrival rate is λ_i and service rate μ_i when the system is in state i, $(i=0, 1, 2, \ldots)$. Suppose $\lambda_i>0$, $i=0, \ldots, M-1$, $\lambda_j=0$ for $j\geqq M$: $\mu_i>0$, $i=1, 2, \ldots, M$ with $\mu_0=0$, where M is a given positive integer. The process is observed over a fixed period $(0, T_0)$. Find the maximum likelihood estimates of the $2M$ parameters $(\lambda_0, \ldots, \lambda_{M-1}, \mu_1, \ldots, \mu_M)$, and establish their symptotic properties.

Deduce the likelihood ratio test for testing the hypothesis $H: \lambda_0=\lambda_1=\ldots=\lambda_{M-1}$.

(Wolff, 1965)

5. Consider a semi-Markov process with finite state-space, $S=\{1, 2, \ldots, s\}$. Let $\{X_k\}$, and $\{T_k\}$, $(k=1, 2, \ldots)$ denote the successive states the system passes through, and the between-transition intervals respectively. Denote $P(X_k=j,\ T_k\leqq t|X_{k-1}=i)=Q_{ij}(t)=p_{ij}H_i(t)$, where p_{ij} is the transition probability for the embedded Markov chain $\{X_k\}$, and $H_i(t)$ is the distribution function of T_k conditional on $X_{k-1}=i$. Observe the process continuously over $(0, T_0)$ and consider the estimator

$$\hat{Q}_{ij}(t)=\hat{p}_{ij}\hat{H}_i(t),$$

where $\hat{p}_{ij}=n_{ij}/n_i$, and $H_i(t)=1/n_i\ \Sigma_{k=1}^{n_i}\ \varepsilon(t-T_{ik})$, T_{ik} being the kth interval of type i, and $\varepsilon(u)$ equals one if $u\geqq 0$ and zero otherwise; n_{ij} and n_i are the frequencies of the transition $i\to j$ and of state i in the sample $(X_1, \ldots, X_n)$, n being the total number of transitions that have occurred in $(0, T_0)$. Assuming that the embedded Markov chain is ergodic, establish the consistency and asymptotic normality of the estimates $\hat{Q}_{ij}(t)$.

(Moore and Pyke, 1968)

6. For a single-server queue with Poisson arrivals (rate λ), suppose the service times are exponential but with rates (dependent on the queue size) $i\mu$ when the queue size is i. The process is observed at successive departure epochs (until N_0 departures),

and (i) the queue size at the departure epochs, (ii) the service-times, (iii) idle periods (if any) are recorded. Find the maximum likelihood estimates of λ and μ and study their properties.

(Goyal and Harris, 1972)

7. Let $\{X_t, t \geqq 0\}$ be a semi-Markov process with a finite number of states, say m. Let $Q_{ij}(t)$ denote the transition distribution function as defined in Complement 5, i.e. $Q_{ij}(t) = p_{ij} H_i(t)$. Let

$$H_i(t) = \int_0^t \frac{e^{-\lambda_i u}(\lambda_i u)^{\alpha_i - 1}\lambda_i}{\Gamma(\alpha_i)}\, du,$$

i.e. the time-interval corresponding to the transition i to j is assumed to be gamma (λ_i, α_i). The process is observed over a fixed interval $(0, T_0)$. Find the maximum likelihood estimate of the $(m^2 + 2m - 2)$-dimensional parameter

$$\theta = (p_{11}, p_{12}, \ldots, p_{1m-1}, \ldots, p_{m,m-1}, \lambda_1, \ldots, \lambda_m, \alpha_1, \ldots, \alpha_m)$$

and derive its asymptotic properties under appropriate assumptions on p_{ij}.

Chapter 6

Simple Point Processes

1. INTRODUCTION

In this chapter problems of inference for Poisson processes, and certain Lévy processes via their imbedded compound Poisson processes, will be presented. Also, see Chapter 8, Section 4.3, for tests regarding point processes.

2. HOMOGENEOUS POISSON PROCESS

Let $\{X_t, t \geqq 0\}$ (where X_t is the number of events by time t) be a Poisson process (homogeneous) with parameter θ. We wish to estimate θ using a suitable realization from the process.

2.1 Continuous Observation over a Fixed Period

Suppose that the process is observed over a fixed interval $(0, T_0)$. Let N denote the total number of events occurring during this interval. If t_1, $t_2, \ldots, t_N$ denotes the epochs at which the events occurred, the inter-occurrence times $T_k = t_k - t_{k-1}$, $k = 1, 2, \ldots, N$ (with $t_0 = 0$) are independent identically distributed as negative exponential with the parameter θ. Assuming that $X_0=0$, the likelihood function of the realization is given by

$$L=\theta^N \exp\left(-\theta \sum_{k=1}^{N} T_k\right) \exp(-\theta(T_0-t_N))$$

$$=\theta^N \exp(-\theta T_0). \tag{1}$$

The maximum likelihood estimate of θ is given by

$$\hat{\theta}=\frac{N}{T_0}.$$

Since N has a Poisson distribution with mean θT_0 we have $E(\hat{\theta})=\theta$ and $V(\hat{\theta})=\theta/T_0$. Since N is sufficient and complete for θ it follows by the Rao-Blackwell Theorem (cf. Rao, 1973) that $\hat{\theta}$ is the minimum variance unbiased estimate.

2.2 Continuous Observation over a Random Period

Suppose we begin observation at $t=0$, and continue until N_0 events have occurred. Let these events occur at the epochs $0<t_1<t_2 \dots <t_{N_0}$. The likelihood function is then

$$L=\theta^{N_0} e^{-\theta T}, \tag{2}$$

where $T=\Sigma_{k=1}^{N_0} T_k$. The maximum likelihood estimate is $\hat{\theta}=N_0/T$.

Since T is the sum of N_0 independent exponential variates it is distributed as a gamma random variable with density

$$f_T(t)=\frac{\theta(\theta t)^{N_0-1} e^{-\theta t}}{(N_0-1)!}, \quad t>0.$$

Since $E(1/T)=\theta/(N_0-1)$, ($N_0>1$ assumed), it follows that $E(\hat{\theta})=(N_0/(N_0-1))\theta$. The bias in the ML estimate can be removed by taking $\tilde{\theta}=((N_0-1)/N_0)\hat{\theta}$ as the required estimate. Since T is a complete sufficient statistic for θ it follows that the modified maximum likelihood estimate $\tilde{\theta}$ is minimum variance unbiased, and it is seen that $V(\tilde{\theta})=\theta^2/(N_0-2)$, ($N_0>2$, assumed).

The probability density of T can also be used to construct confidence intervals for θ. Note that $2\theta T$ is distributed as $\chi^2(2N_0)$. Now, let a and b be points such that

$$P(\chi^2(2N_0)<a)=\frac{\alpha}{2}, \text{ and } P(\chi^2(2N_0)>b)=\frac{\alpha}{2},$$

where α is the given confidence coefficient. Then

$$P\left(\frac{a}{2T}<\theta<\frac{b}{2T}\right)=(1-\alpha),$$

so that, $(a/2T, b/2T)$ is a $(1-\alpha)100\%$ confidence interval for θ.

2.3 Continuous Inspection with a Given Sampling Plan

Suppose we observe the process $\{X_t\}$ at n_0 selected time points, $v_1 < v_2 < \ldots < v_{n_0}$. Let the $\{v_k\}$ be such that $\tau_k = v_k - v_{k-1}$ $(k=1, 2, \ldots, n_0, v_0=0)$ are independent and identically distributed random variables with density $f_\tau(t|\lambda)$. Let $x_{v_1}, x_{v_2}, \ldots, x_{v_{n_0}}$ be the states occupied at these instants. Note that the sampling is carried out in two stages; first n_0 observations $(\tau_1, \ldots, \tau_{n_0})$ are drawn at random from the density $f_\tau(t|\lambda)$. Then, at the points $v_1 = \tau_1$, $v_2 = \tau_1 + \tau_2, \ldots$, $v_{n_0} = \tau_1 + \ldots + \tau_{n_0}$, the process X_t is observed. We further assume that the selection distribution $f_\tau(t|\lambda)$ is independent of the Poisson process $\{X_t\}$.

Let $z_k = x_{v_k} - x_{v_{k-1}}$, $(k=1, 2, \ldots, n_0)$ denote the number of events occurring during the interval τ_k. Then, conditional on $(v_1, v_2, \ldots, v_{n_0})$ the likelihood function is given by

$$L = \prod_{k=1}^{N} \left\{ \frac{e^{-\theta\tau_k}(\theta\tau_k)^{z_k}}{z_k!} \right\}$$

$$= \theta^N e^{-\theta T} \prod_{k=1}^{N} \frac{\tau_k^{z_k}}{z_k!} \tag{3}$$

where $N = \Sigma_{k=1}^{n_0} z_k$ is the total number of events occurring by time v_{n_0}, i.e. $x_{v_{n_0}}$, and $T = \Sigma_{k=1}^{n_0} z_k = v_{n_0}$ is the total time of inspection.

The conditional (given T) maximum likelihood estimate of θ is $\hat{\theta} = N/T$, the same as the estimate obtained in (2.1). Conditional on T, the properties of $\hat{\theta}$ are the same as those in (2.1). The unconditional distribution of $\hat{\theta}$, however, depends on the distribution of N as well as the distribution of T, and the performance of $\hat{\theta}$ as an estimate will therefore depend on the choice of the selector-distribution $f_\tau(t|\lambda)$. We shall illustrate this by using two particular types of selection of observation points.

Case 1. Let $f_\tau(t|\lambda) = \lambda e^{-\lambda t}$, $t>0$. Then, T has the gamma density

$$f_T(t|\lambda) = \frac{\lambda(\lambda t)^{n_0-1} e^{-\lambda t}}{(n_0-1)!}, \quad t>0.$$

observe that

$$E(\hat{\theta}) = E(E(\hat{\theta}|T)) = \theta,$$

and

$$V(\hat{\theta}) = E(V(\hat{\theta}|T)) + V(E(\hat{\theta}|T)) = E\left(\frac{\theta}{T}\right) = \left(\frac{\lambda}{n_0 - 1}\right)\theta,$$

which shows that an estimate $\hat{\theta}$ with a desired precision can be found by suitably choosing λ. For large n_0, $V(\hat{\theta}) \cong \theta/E(T)$.

Case 2. Let $f_\tau(t|h) = \delta_{\text{th}}$, $t > 0$, $h > 0$, where δ_{th} is the Kronecker delta. This means that we observe the process at equal intervals at $v_1 = h$, $v_2 = 2h$, $\ldots$, $v_n = n_0 h$. For any given h, $T = n_0 h$ is fixed and this reduces to the method given in (2.1). Thus, the maximum likelihood estimate is $\hat{\theta} = N/n_0 h$ with $E(\hat{\theta}) = \theta$, and $V(\hat{\theta}) = \theta/n_0 h$.

A situation when the present two-stage sampling method can be used in preference to methods (2.1) and (2.2) is when the availability of observations is restricted by a counter which records the observations only at certain instants and remains locked during the rest of the time. It may then not be possible to use methods (2.1) or (2.2). For instance, if we fix N, say at N_0, the random variable T cannot be observed if the N_0th event occurs during the dead period of the counter. Similarly, if T is fixed at T_0, it would be necessary to see that T_0 coincides with an instant at which the counter is active – but this will not be easy if we have a randomly operating counter.

2.4 Discretized Realization

The realization in (2.3) was obtained by observing the process at times $v_1 < v_2 < \ldots$ such that the intervals between successive observations, $\tau_k = v_k - v_{k-1}$ formed a renewal process. It was assumed that the intervals τ_k were *observable* random variables. If these cannot be observed the two-stage procedure in (2.3) can no longer be applied.

Let $x_{v_1}, x_{v_2}, \ldots$ be the observations selected by a counter at the random points $v_1 < v_2 < \ldots$, and suppose that $\tau_k = v_k - v_{k-1}$ are independent and exponentially distributed with parameter λ. Let $z_k = x_{v_k} - x_{v_{k-1}}$ denote the number of events occurring during time τ_k. We now seek an estimate of θ based only on the observations $(z_1, z_2, \ldots, z_n)$. Since $\{X_t\}$ is a Poisson process it can be verified that the Z_k's are independent identically distributed random variables. The probability function of Z_k can be obtained as follows:

$$\begin{aligned} P(Z_k = z) &= \int_0^\infty \frac{e^{-t\theta}(t\theta)^z}{z!} \lambda e^{-\lambda t} \, dt \\ &= (1-p)p^z \end{aligned}$$

where $p=\theta/(\lambda+\theta)$. The likelihood function is

$$L=\left(\frac{\lambda}{\lambda+\theta}\right)^{n_0}\left(\frac{\theta}{\lambda+\theta}\right)^{N}. \tag{4}$$

where $N=\Sigma_{k=1}^{n_0} z_k$, the total number of events observed. This gives the maximum likelihood estimate $\hat{\theta}=\lambda N/n_0$. Since N is distributed as a negative binomial, we have

$$E(N)=\frac{n_0\theta}{\lambda}, \text{ and } V(N)=\frac{n_0\theta(\lambda+\theta)}{\lambda^2},$$

and hence

$$E(\hat{\theta})=\theta, \text{ and } V(\hat{\theta})=\frac{\theta(\lambda+\theta)}{n_0}.$$

3. NON-HOMOGENEOUS POISSON PROCESS

A non-homogeneous or time-dependent Poisson process $\{X_t, t\geqq 0\}$ (where X_t denotes the number of events by time t) with $X_0=0$ is defined by the following conditions:

(i) For any $0<u_1<u_2<u_3<\ldots$, X_{u_1}, $X_{u_2}-X_{u_1}$, $X_{u_3}-X_{u_2}$, ... are independent, i.e. $\{X_t\}$ is a process with independent increments,

(ii) $P(X_{t+\delta t}-X_t=1)=\lambda(t)\delta t+o(\delta t)$, and

(iii) $P(X_{t+\delta t}-X_t\geqq 2)=o(\delta t)$

where $o(\delta t)$ denotes any quantity which after division by δt tends to zero as $\delta t\to 0$.

Notice that if $\lambda(t)=\lambda$ this reduces to the usual definition of the homogeneous Poisson process. $\lambda(t)$ is called the *intensity function* or rate of occurrence of events.

Following are the main properties of a non-homogeneous Poisson process:

(i) Let N_1, N_2, ... denote the number of events occurring in disjoint intervals, say $[0, u_1)$, $[u_1, u_2)$, ... Then N_1, N_2, ... are independent Poisson random variables and

$$Pr(N_r=n)=\frac{\left\{\exp\left(-\int_{u_{r-1}}^{u_r}\lambda(s)\mathrm{d}s\right)\right\}\left\{\int_{u_{r-1}}^{u_r}\lambda(s)\mathrm{d}s\right\}^n}{n!} \quad n=0, 1, 2, \ldots \tag{5}$$

(ii) Let $0<t_1<t_2<\ldots$ be the epochs at which the events occur. Then the between-occurrence intervals $T_k = t_k - t_{k-1}$ $(k=1, 2, \ldots)$ are independent random variables with densities.

$$f_{T_k}(t_k) = \lambda(t_k)\exp\left(-\int_{t_{k-1}}^{t_k} \lambda(s)\mathrm{d}s\right). \tag{6}$$

Also, $\{t_1, t_2, \ldots\}$ is a Markov sequence with transition densities $p(t_k|t_{k-1}) = \lambda(t_k)\exp\left(-\int_{t_{k-1}}^{t_k} \lambda(s)\,\mathrm{d}s\right)$.

(iii) Conditional on $X_{T_0} = n$, the epochs at which events occur, i.e. $0 < t_1 < t_2 < \ldots < t_n$ $(\leqq T_0)$ are distributed as n order statistics corresponding to a random sample of n observations from the density

$$f(t) = \lambda(t) \bigg/ \int_0^{T_0} \lambda(s)\mathrm{d}s, \; 0 \leqq t \leqq T_0. \tag{7}$$

This reduces to the uniform distribution over $[0, T_0]$ when $\lambda(t) = \lambda$.

3.1 Parameter Estimation for Exponential Intensity

Let us assume that $\lambda(t)$ is of an exponential type,

$$\lambda(t) = \exp\,(h(\theta)\,k(t))\,c(\eta) \tag{8}$$

where θ and η are parameters.

Suppose we observe $\{X_t\}$ continuously over a fixed interval $(0, \tau_0)$. The likelihood function is

$$L = \left(\prod_{r=1}^{n} \lambda(t_r)\right)\exp\left(-\int_0^{\tau_0} \lambda(u)\mathrm{d}u\right)$$

$$= (c(\eta))^n \exp\left\{h(\theta)\sum_{r=1}^{n} k(t_r) - \int_0^{\tau_0} \lambda(u)\mathrm{d}u\right\} \tag{9}$$

where n is the number of events occurring by time τ_0, and $t_1, t_2, \ldots, t_n$ are the epochs at which these events occur. Clearly, $(n, \Sigma_{r=1}^{n} k(t_r))$ are sufficient statistics for (θ, η). Now, for given value of θ it is seen that n is sufficient for η; moreover for given η, the statistic $\Sigma_{r=1}^{n} k(t_r)$ is sufficient for θ. Suppose we are interested in inference regarding θ so that η is a nuisance parameter. We can therefore consider the conditional likelihood function given $n = n_0$ say. Now, conditionally on $n = n_0$, the joint density of $(0 < t_1 < t_2 < \ldots < t_{n_0} \leqq \tau_0)$ is given by

$$\frac{n_0! \prod_{r-1}^{n_0} \lambda(t_r)}{\left(\int_0^{\tau_0} \lambda(u)\mathrm{d}u\right)^{n_0}},$$

and hence the conditional likelihood is

$$L^* = \frac{\prod_{r=1}^{n_0} \lambda(t_r)}{\left(\int_0^{\tau_0} \lambda(u)\mathrm{d}u\right)^{n_0}}$$

$$= \frac{\exp\left(h(\theta)\Sigma k(t_r)\right)}{\left(\int_0^{\tau_0} \exp\left(h(\theta)k(u)\right)\mathrm{d}u\right)^{n_0}} \tag{10}$$

which is free from the nuisance parameter η.

As an example, let $\lambda(t)=\eta\mathrm{e}^{-\theta t}$, $\eta>0$, $\theta \geqq 0$. Then

$$L^* = \begin{cases} \dfrac{\theta^{n_0} \exp\left(-\theta \sum_{r=1}^{n_0} t_r\right)}{(1-\exp(-\theta\tau_0))^{n_0}}, & \text{for } \theta>0 \\ \dfrac{1}{\tau_0^{n_0}}, & \text{for } \theta=0. \end{cases} \tag{11}$$

Thus, for $\theta>0$, L^* is the likelihood function of a random sample from a truncated (at τ_0) negative exponential distribution. The likelihood equation for estimating θ is given by

$$\frac{1}{\theta} - \frac{\tau_0 \mathrm{e}^{-\theta\tau_0}}{1-\mathrm{e}^{-\theta\tau_0}} = \frac{1}{n_0} \sum_{r=1}^{n_0} t_r.$$

3.2 Test for a Trend

In the example above with $\lambda(t)=\eta\mathrm{e}^{-\theta t}$, suppose we wish to test the hypothesis H:$\theta=0$. This is equivalent to testing that the Poisson process is homogeneous, i.e. time-independent. The Neyman–Pearson Lemma (see Lehmann, 1959) for testing H:$\theta=0$ against K:$\theta>0$ gives the (best) critical region $\Sigma_{i=1}^{n_0} t_i \geqq$ constant. Under H, $\Sigma_{i=1}^{n_0} t_i$ is the sum of independent uniform $[0, \tau_0]$ random variables, and hence the statistic

$$Z = \frac{\sum_{i=1}^{n_0} t_i - \frac{1}{2} n_0 \tau_0}{\tau_0 \sqrt{\left(\frac{n_0}{12}\right)}} \tag{12}$$

is approximately standard normal for large n_0.

3.3 General Case

In general, suppose the intensity is some specified function of a vector $\boldsymbol{\theta}$ of unknown parameters, say $\lambda(t; \boldsymbol{\theta})$. The full likelihood function based on the realization $\{X_t, 0 \leqq t \leqq \tau_0\}$ is given by

$$L = \left(\prod_{k=1}^{X_{\tau 0}} \lambda(t_k)\right) \exp\left(-\int_0^{\tau_0} \lambda(s)\mathrm{d}s\right). \tag{13}$$

The maximum likelihood estimator of $\boldsymbol{\theta}$ is, therefore, obtained by maximizing.

$$\log L = -\int_0^{\tau_0} \lambda(s, \boldsymbol{\theta})\mathrm{d}s + \sum_{k=1}^{X_{\tau_0}} \log \lambda(t_k, \boldsymbol{\theta})$$

with respect to θ.

4. COMPOUND POISSON PROCESS

A compound Poisson process can be described as follows:

(i) Jumps occur according to a Poisson process with intensity, say $\lambda(t)$.

(ii) Denoting Y_i for the magnitude of the ith jump, we assume that $\{Y_i\}$ are independent, identically distributed random variables which are also independent of the Poisson process $\{X_t, t \geqq 0\}$ which produces these jumps.

The likelihood function based on $\{X_{\tau_0}; t_1, \ldots, t_{X_{\tau_0}}; Y_1, Y_2, \ldots, Y_{X_{\tau_0}}\}$ is given by

$$L = \left\{\prod_{k=1}^{X_{\tau 0}} \lambda(t_k)\right\}\left\{\exp\left(-\int_0^{\tau_0} \lambda(s)\right)\mathrm{d}s\right\}\left\{\prod_{k=1}^{X_{\tau 0}} p(Y_k)\right\}, (X_{\tau_0} \geqq 1) \tag{14}$$

where $p(Y_k)$ is the density of Y_k, the jump-size. We shall illustrate problems of inference for compound Poisson processes via those for gamma and stable processes.

4.1 Inference for Gamma and Stable Processes

Preliminaries. A pure-jump additive process (Levy process) $\{X(t), t \geqq 0\}$ with non-decreasing sample paths is characterized by its Laplace transform

$$E[\exp\{-\lambda X(t)\}] = \exp\left\{-t\int_0^{\infty} (1 - \mathrm{e}^{-\lambda u})v(\mathrm{d}u)\right\}, (\lambda > 0) \tag{15}$$

where v is a measure concentrated on $(0, \infty)$ known as the Lévy measure.

When the integral $\int_0^\infty \nu(\mathrm{d}u) < \infty$ it is seen that $\{X(t)\}$ is a compound Poisson process. In this section, we are interested in the case $\int_0^\infty \nu(\mathrm{d}u) = \infty$. In the latter case, an infinite number of jumps can occur in a finite interval. Typical examples are (i) the gamma process with $\nu(\mathrm{d}u) = pu^{-1}\mathrm{e}^{-\theta u}\mathrm{d}u$ where p and θ are the parameters and (ii) the stable process with $\nu(\mathrm{d}u) = \{\Gamma(1-\beta)\}^{-1} c\ \beta u^{-1-\beta}\ \mathrm{d}u$ where c and β are parameters. For any fixed $\varepsilon > 0$, let $N(t, \varepsilon)$ denote the number of jumps in $(0, t)$ of magnitude $\geqq \varepsilon$. Then, it can be shown, see for instance, Feller (1966), p. 539, that $\{N(t, \varepsilon), t \geqq 0\}$ is a Poisson process with intensity

$$M(\varepsilon) = \int_{u > \varepsilon} \nu(\mathrm{d}u) < \infty.$$

In particular, for the gamma process $M(\varepsilon) = p\int_\varepsilon^\infty y^{-1}\mathrm{e}^{-y}\mathrm{d}y$, (i.e. p times the exponential integral), and for the stable process $M(\varepsilon) = \{\Gamma(1-\beta)\}^{-1}c\ \varepsilon^{-\beta}$. Let $U(\varepsilon)$ denote the magnitude of a jump given that it is $\geqq \varepsilon$. Suppose we assume that $\nu(\mathrm{d}u) = g(u, \boldsymbol{\theta})\mathrm{d}u$ where $\boldsymbol{\theta}$ is a k-dimensional parameter. Then the density of $U(\varepsilon)$ is given by $\{M_\theta(\varepsilon)\}^{-1}g(u, \theta)$. Let $U_k(\varepsilon)$ denote the magnitude of the kth jump, assuming that we are only observing jumps of size $\geqq \varepsilon$. Then $\{N(t, \varepsilon), t \geqq 0;\ U_k(\varepsilon), k = 1, 2, \ldots\}$ constitutes a compound Poisson process.

We shall discuss, in what follows, problems of inference regarding $\boldsymbol{\theta}$ based on a realization of the embedded compound Poisson process described in the previous paragraph. From now onwards we shall confine ourselves to gamma and stable processes.

Gamma Process. Let

$$\phi(v) = \int_v^\infty y^{-1}\ \mathrm{e}^{-y}\mathrm{d}y, \tag{16}$$

be the exponential integral. Then, for the gamma process with parameters p and θ we have $M_\theta(\varepsilon) = p\phi(\varepsilon\theta)$. Also, the density of the random variable $U(\varepsilon)$ is

$$f_{U(\varepsilon)}(u) = \{\phi(\varepsilon\theta)\}^{-1}u^{-1}\mathrm{e}^{-\theta u},\quad u \geqq \varepsilon. \tag{17}$$

This is truncated (at ε) gamma density with zero index. Note that this density is free from p.

Without loss of generality set $X(0) = 0$ and let τ_k denote the interval between the kth and $(k-1)$th jumps of magnitude $\geqq \varepsilon$, where ε is some fixed number > 0. Suppose that N jumps of magnitude $\geqq \varepsilon$ occur in the interval $(0, T)$. Our sample now consists of $\{\tau_k, U_k(\varepsilon)\}$ $(k = 1, 2, \ldots, N)$ where N is a

Poisson random variable with mean $p\phi(\varepsilon\theta)T$. For given θ, therefore, N is sufficient for p. This suggests treating p as a nuisance parameter and estimating θ conditional on $N=n$. The relevant part of the conditional likelihood function is

$$L(\theta)=\{\phi(\varepsilon\theta)\}^{-n}\left(\prod_{k=1}^{n} u_k^{-1}\right)\exp\left(-\theta\sum_{1}^{n} u_k\right) \tag{18}$$

since conditional on $N=n$, the epochs at which jumps (of size $\geq\varepsilon$) occur have uniform distribution on $(0, T)$ and hence do not depend on the parameters. In other words, the likelihood function $L(\theta)$ is based only on $\{u_1, \ldots, u_n\}$. The maximum likelihood estimate $\hat{\theta}_n$ of θ (conditional on $N=n$) is then obtained as the solution of the equation

$$\frac{1}{n}\sum_{1}^{n} u_k=\{\theta\phi(\varepsilon\theta\}^{-1}e^{-\varepsilon\theta}.$$

Since N is a Poisson random variable with mean $p\phi(\varepsilon\theta)T$, an estimate of p is given by

$$\hat{p}=n\{T\phi(\varepsilon\hat{\theta})\}^{-1}$$

where $\hat{\theta}$ is the conditional maximum likelihood estimate obtained earlier.

If desired, p and θ can be estimated simultaneously from the joint likelihood function given by

$$L(\theta, p)=\left\{\prod_{k=1}^{n} u_k\right\}^{-1} p^n \exp\left\{-\theta\sum_{1}^{n} u_k-pT\phi(\varepsilon\theta)\right\}. \tag{19}$$

The unconditional maximum likelihood estimates of p and θ obtained from this joint likelihood are formally the same as the conditional estimates $\hat{p}$ and $\hat{\theta}$ with n replaced by the random variable N. The asymptotic properties, as $T\to\infty$, of these joint maximum likelihood estimates can be deduced in the usual way and will not be pursued here.

Now consider the problem of testing the hypothesis $H{:}\theta\leq\theta_0$ against $K{:}\theta>\theta_0$ where θ_0 is some specified value. Note that we have an unspecified nuisance parameter p. The boundary between H and K is given by $B=\{\theta_0, 0<p<\infty\}$, so that there is only one unspecified parameter p on B. Also, on B the statistic N is complete and sufficient for p. Consequently, we can consider a conditional (on N) test. Since the conditional likelihood function has monotone likelihood ratio in $\Sigma_{k=1}^{n} u_k$ (see Lehmann (1959) for definitions etc.) we finally obtain a uniformly most powerful unbiased test which rejcts when $\Sigma_{k=1}^{n} u_k$ is small. This test can be determined by using a normal approximation for $\Sigma_1^n u_k$.

In what follows we carry out a similar analysis for a stable process for which more complete and simpler results can be obtained.

Stable Process. Let τ_k and $U_k(\varepsilon)$ be as defined in the previous section. The density of $U_{k(\varepsilon)}$ for the stable process with index β, $(0<\beta<1)$, and scale parameter $c>0$ is given by

$$f_{U_k}(\varepsilon)(u)=\beta\varepsilon^{\beta}u^{-1-\beta}, u\geqq\varepsilon \tag{20}$$

which is free from c.

Using the same arguments as for the gamma process we are led to the conditional (on $N=n$) likelihood function which is based only on $\{u_1, \ldots, u_n\}$ and given by

$$L(\beta)=\beta^n\varepsilon^{n\beta}\prod_{k=1}^{n}u_k^{-1-\beta}. \tag{21}$$

The maximum likelihood estimate (conditional) of β is found to be

$$\hat{\beta}_n-\left\{\frac{1}{n}\sum_{k=1}^{n}(\log u_k-\log\varepsilon)\right\}^{-1}. \tag{22}$$

Letting $Z_k=\log u_k-\log\varepsilon$, we note that Z_k are independent negative exponential variates with mean β^{-1} so that $(\hat{\beta}_n/n)$ is distributed as a reciprocal of a gamma (n, β) variate. Thus, the density of $\hat{\beta}_n$ is free from ε. It can be seen that

$$E(\hat{\beta}_n)=\left(\frac{n}{n-1}\right)\beta \text{ and var } (\hat{\beta}_n)=\left(\frac{n}{n-1}\right)^2\frac{1}{(n-2)}\beta^2, (n>2).$$

Since within the conditional set-up $\Sigma_1^n Z_k$ is a complete sufficient statistic for β it follows that

$$\hat{\beta}_n^*=\left(\frac{n-1}{n}\right)\hat{\beta}_n$$

is the minimum variance unbiased estimate of β.

Since $2\beta\ \Sigma_1^n Z_k$ is distributed as $\chi^2(2n)$ variate a confidence interval for β with confidence coefficient $(1-\alpha)$ within the conditional set-up is given by

$$\left\{\chi_{1-\alpha/2}^2\ (2n)\Big/\left(2\sum_1^n Z_k\right),\ \chi_{\alpha/2}^2(2n)\Big/\left(2\sum_1^n Z_k\right)\right\}.$$

Now a test for $H:\beta\leqq\beta_0$ against $K:\beta>\beta_0$ (c being the unspecified nuisance parameter) can be obtained by arguing as in the previous section. We find that the uniformly most powerful umbiased test of size α rejects H when

$$2\ \beta_0\sum_{k=1}^{n}Z_k<\chi_{1-\alpha}^2\ (2n)$$

where $\chi_{1-\alpha}^2$ $(2n)$ is found by using the χ^2 table for $2n$ degrees of freedom.

The scale parameter c can be estimated by

$$\hat{c}=\{n\,\varepsilon^{\hat{\beta}}\,\Gamma(1-\hat{\beta})\}/T.$$

If required c and β can be estimated simultaneously from the full likelihood function

$$L(c,\,\beta)=\beta^{n}\left(\sum_{1}^{n} u_k^{-1-\beta}\right)c^{n}\Gamma^{n}(1-\beta)\,\exp\{-Tc\varepsilon^{-\beta}/\Gamma(1-\beta)\}$$

and the asymptotic properties of the estimates can be deduced from standard theory; the simultaneous maximum likelihood estimates of β and c are again formally the same as the conditional estimates $\hat{\beta}$ and $\hat{c}$ with n replaced by N.

5. FURTHER TOPICS

5.1 Doubly Stochastic Poisson Process

Consider a Poisson process $\{X_t,\, t\geqq 0\}$ with intensity Λ(t), where $\{\Lambda(t),\, t\geqq 0\}$ is some non-negative stochastic process. Such a process is called doubly stochastic Poisson process. For a homogeneous Poisson process the mean intensity is non-random and does not vary with time; for a non-homogeneous process however the mean intensity is allowed to depend on time but it is still assumed non-random. The doubly stochastic Poisson process provides the next step of flexibility by letting the intensity not only depend on time but also allowing it to be a stochastic process. In recent years, this model has generated considerable interest among researchers. See Snyder (1975) and Grandell (1976) for a detailed discussion of the model. In this section, we content ourselves by dealing with some simple cases.

Conditional on the process $\{\Lambda(t),\, t\geqq 0\}$, *the doubly stochastic Poisson process* $\{X_t,\, t\geqq 0\}$ is a non-homogeneous Poisson process discussed in Section 3. This fact can be used to find the likelihood function based on the realization, $\{X_t,\, 0\leqq t\leqq \tau_0\}$.

Using (13) we have the likelihood function given by

$$L=E\left\{\left(\prod_{k=1}^{X_{\tau_0}} \Lambda(t_k)\right)\exp\left(-\int_0^{\tau_0}\Lambda(s)\mathrm{d}s\right)\right\} \tag{23}$$

where t_k is the epoch at which kth event occurs. The expectation is with respect to the intensity process $\{\Lambda(t),\, 0\leqq t\leqq \tau_0\}$. Of course, the expectation in (23) is not always easy to evaluate. In the following special case this expectation can easily be found.

Let $\Lambda(t)=Zk(t)$, where Z is an exponential random variable with mean θ, and $k(t)$ is a deterministic function of t. The likelihood function using (23) is given by

$$L = E(Z^{X_{\tau 0}} \prod_{v=1}^{X_{\tau 0}} k(t_v)) \exp\left(-Z \int_0^{\tau_0} k(s)\,ds\right)\Big\}$$

$$= \theta\Gamma(X_{\tau_0}+1)\left(\theta + \int_0^{\tau_0} k(s)ds\right)^{-(X_{\tau_0}+1)} \prod_{v=1}^{X_{\tau 0}} k(t_v)). \tag{24}$$

Clearly, X_{τ_0} is a sufficient statistic for θ. The maximum likelihood estimator of θ is found to be

$$\hat{\theta} = X_{\tau 0}^{-1} \int_0^{\tau_0} k(s)ds. \tag{25}$$

Estimation of the Intensity Process. Given the realization $\{X_t,\ 0 \leqq t \leqq \tau_0\}$, suppose we wish to estimate the intensity at time τ_0, i.e. $\Lambda(\tau_0)$ such that the mean-square error of estimation is minimized. Let f denote an arbitrary function of the sample $\{X_t,\ 0 \leqq t \leqq \tau_0\}$. We then must choose f such that

$$E(\Lambda(\tau_0) - f)^2$$

is minimized. It is easily verified that the optimum f is given by the conditional expectation

$$\hat{\Lambda}(\tau_0) = E(\Lambda(\tau_0) | X_t, 0 \leqq t \leqq \tau_0). \tag{26}$$

$\hat{\Lambda}(\tau_0)$ is, in general, difficult to evaluate. One may, therefore, consider a linear estimator of $\Lambda(\tau_0)$. Thus, let f be a linear function of the sample $\{X_t,\ 0 \leqq t \leqq \tau_0\}$ in the sense that f is of the form

$$f_L = \alpha_{\tau_0} + \int_0^{\tau_0} h(\tau_0, t)dX_t \tag{27}$$

where $h(\tau_0, u)$ is called the impulse response function. The integral occurring in (27) is a counting integral defined as

$$\int_0^{\tau_0} h(\tau_0, t)dX_t = \begin{cases} \sum_{k=1}^{X_{\tau 0}} h(\tau_0, t_k) & \text{if } X_{\tau_0} \geqq 1 \\ 0 & \text{if } X_{\tau_0} = 0, \end{cases} \tag{28}$$

where t_k are the epochs at which the events occur. The problem is that of finding α_{τ_0} and $h(\tau_0, t)$ such that

$$E(\Lambda(\tau_0) - f_L)^2$$

is minimized. The optimum solution for α_{τ_0} and $h(\tau_0, t)$ are seen to be

$$\alpha_{\tau_0} = E(\Lambda(\tau_0)), \text{ and } h(\tau_0, t) = h_0(\tau_0, t),$$

where $h_0(\tau_0, t)$ satisfies the equation

$$h_0(\tau_0, t)E(\Lambda(t)) + \int_0^{\tau_0} h_0(\tau_0, s)K(s, t)\mathrm{d}s = K(\tau_0, t) \tag{29}$$

where $K(\tau_0, t)$ is the covariance function of $\{\Lambda(t), t \geqq 0\}$, i.e.

$$K(\tau_0, t) = E\{(\Lambda(\tau_0) - E(\Lambda(\tau_0)))(\Lambda(t) - E(\Lambda(t)))\}.$$

See Snyder (1975) for further details.

5.2. Renewal Process–Reliability Applications

Point and Interval Estimation. Consider independent components (units) having identically distributed failure-times $\{T_1, T_2, \ldots\}$ with the common failure-time density $f(t|\theta)$ which depends on a vector of unknown parameters $\boldsymbol{\theta}$. The process $\{X_t, t > 0\}$, where X_t denotes the number of failures by time t, is then a renewal process. One may be interested in estimating $\boldsymbol{\theta}$ or the reliability function over a fixed period say $[0, T_0]$, $R(T_0|\boldsymbol{\theta})$ where $R(T_0|\boldsymbol{\theta}) = 1 - \int_0^{T_0} f(t|\boldsymbol{\theta})\mathrm{d}t$ is the probability of failure-free operation of a unit over the period $[0, T_0]$. The data for this purpose are usually obtained from the so-called "life-tests". These take the following form: n units are put into operation. Observations are taken continuously in time and the epochs at which failures occur are noted. Several "stopping rules" are possible. If the life-test is carried out until a specified time say T_0 (before all items have failed) one has a *Type I censoring*; if the test is terminated at the epoch of rth failure ($r \leq n$) we have a *Type II censoring*. Sometimes a mixed plan may be used in which the test is carried out up to the epoch t_r of the rth failure if $t_r \leqq T_0$, and if $t_r > T_0$ the test is terminated at time T_0. Moreover we may consider these plans with replacement (i.e. each failed unit is replaced immediately by a new unit) or without replacement. The reader is referred to Gnedenko *et al.* (1969) for a detailed discussion of the various stopping rules.

Consider, for instance, a scheme with mixed stopping rule and without replacement. The likelihood of epochs of failure (these are the ordered life-times of the units all of which start operating at time 0 say), $t_1 \leqq t_2 \leqq \ldots t_m \leqq T_0$, with $m \leqq r$, is proportional to

$$L = \left(\prod_{i-1}^{m} f(t_i|\boldsymbol{\theta})\right)(1 - F(T_0|\boldsymbol{\theta}))^{n-m}, \tag{30}$$

where $F(t|\theta) = \int_0^t f(u|\theta)\mathrm{d}u$ is the life-time distribution function.

If we have Type II censoring (and without replacement) the likelihood function is proportional to

$$L^* = \left(\prod_{i=1}^{r} f(t_i|\boldsymbol{\theta}) \right) (1 - F(t_r|\boldsymbol{\theta}))^{n-t}. \tag{31}$$

Type I censoring without replacement yields the likelihood proportional to

$$L^{**} = \left(\prod_{i=1}^{d} f(t_i|\theta) \right) (1 - F(T_0|\theta))^{n-d} \tag{32}$$

where d is the number of failures observed.

It should be noted that the observations involved in the construction of these likelihood functions are not independent. We shall illustrate the method of maximum likelihood estimation by an example.

Consider n independent units each with an exponential failure-time distribution $f(t|\theta) = 1/\theta \ e^{-t/\theta}$, $t > 0$, $(\theta > 0)$. Suppose we adopt Type II censoring and that the failed units are not replaced. The likelihood function is

$$L^* = \frac{1}{\theta^r} \exp \quad - \frac{\left\{ \sum_{i=1}^{r} t_i + (n-r)t_r \right\}}{\theta}, \quad 0 \leqq t_1 \leqq \ldots \leqq t_r.$$

The maximum likelihood estimate of θ is

$$\hat{\theta} = \frac{\sum_{i=1}^{r} t_i + (n-r)t_r}{r}$$

which is clearly a sufficient statistic for θ. Writing

$$\hat{\theta} = \frac{1}{r} \sum_{i=1}^{r} (n-i+1)(t_i - t_{i-1})$$

it is seen that $r\hat{\theta}/\theta$ is distributed as a sum of r independent unit exponential random variables (and hence has a gamma distribution). We then have

$$E(\hat{\theta}) = \theta \quad \text{and} \quad V(\hat{\theta}) = \frac{\theta^2}{r}.$$

Since the distribution of $\hat{\theta}$ is complete (see Lehmann, 1959) it follows by the Rao–Blackwell Theorem that $\hat{\theta}$ is the unique minimum variance unbiased estimate of θ.

It is easily verified that $U = 2r\hat{\theta}/\theta$ has a χ^2 distribution with $2r$ degrees of freedom. This yields a $(1-\alpha)100\%$ confidence interval for θ as

$(2r\hat{\theta}/\chi^2_{1-\alpha/2}(2r),\ 2r\hat{\theta}/\chi^2_{\alpha/2}(2r))$, where $\chi^2_{1-\alpha/2}(2r)$ and $\chi^2_{\alpha/2}(2r)$ are the left and right tail values of the $\chi^2(2r)$ distribution.

Let $R(t|\theta)=e^{-t/\theta}$ be the reliability function. This being a monotonic function of θ we can get the maximum likelihood estimate of $R(t|\theta)$ for any given t, as $\hat{R}(t|\theta)=\mathrm{e}^{-t/\hat{\theta}}$, where $\hat{\theta}$ is the maximum likelihood estimate of θ.

Prediction Interval. Suppose we observe the epochs of first $(r-1)$ failures, $t_1<t_2<\ldots<t_{r-1}$, and on the basis of this sample wish to predict t_r.

Let $S_{r-1}=\Sigma_{i=1}^{r-1}\ t_i+t_{r-1}$ and $V=(t_r-t_{r-1})/S_{r-1}$. It is seen that $(r-1)(n-r+1)V$ has an F distribution with $(2,\ 2r-2)$ degrees of freedom. This can therefore be used to get a prediction interval for t_r.

Sampling with Replacement. Suppose now that we have a Type II censored sample with replacement. The likelihood of $t_1<t_2<\ldots<t_r$ is proportional to

$$L_R=\left(\frac{n}{\theta}\right)^r \exp\left(\frac{-nt_r}{\theta}\right).$$

The maximum likelihood estimate of θ is $\hat{\theta}=nt_r/r$. Note that t_i-t_{i-1}, $(i=1,\ 2,\ldots,\ r)$ with $t_0=0$, are independent exponential random variables with parameter n/θ. Since $t_r=\Sigma_{i=1}^{r}\ (t_i-t_{i-1})$, $2nt_r/\theta$ has a χ^2 distribution with $2r$ degrees of freedom, and therefore confidence intervals for θ can be readily obtained.

Testing for Constancy of the Failure-rate (Test for Exponentiality). Let $R(t)=1-F(t)$ be the reliability of a unit. The failure-rate of the unit, $\lambda(t)$, is defined as $\lambda(t)=-R'(t)/R(t)$. It is easily verified that $\lambda(t)=\lambda$ (free from t) if and only if the life-time distribution corresponding to $R(t)$ is exponential with parameter λ. This characterization property can be used to derive a test of exponentiality. Suppose n units are set into operation and the observation is carried out until the epoch of the rth failure $(r\leqq n)$. Let $t_1<t_2<\ldots<t_r$ denote the instants of failure. Consider $S_i=(n-i+1)(t_i-t_{i-1})(i=1,\ 2,\ \ldots,\ r)$, $(t_0=0)$ for "without replacement" scheme and $S_i^*=n(t_1-t_{i-1})$ for the "with replacement" plan. S_i (or S_i^*) are independent exponential random variables under the null hypothesis that $\lambda(t)=\lambda$. Let r_1 and r_2 be two positive integers such that $r=r_1+r_2$. Then the statistic

$$F=\left(r_2\sum_{i=1}^{r_1} S_i\right)\left(r_1\sum_{i=r_1+1}^{r} S_i\right) \tag{33}$$

has, under the null hypothesis, an F-distribution with degrees of freedom $(2r_1,\ 2r_2)$. The right tail of this F-distribution will serve as a critical region of the test.

A number of alternative tests of exponentiality are available (cf.

Gnedenko, 1969). Also, see Mann *et al.* (1974) for further examples of testing in reliability models including Bayesian methods.

5.3 Renewal Process – Parametric Estimation from Poisson Samples

Let $\{X_t, t \geqq 0\}$ be a renewal process where X_t denotes the number of events occurring by time t. Let $t_0 = 0 < t_1 < t_2 < \ldots$ denote the epochs at which the successive events occur. The between-event intervals $T_k = t_k - t_{k-1}$, $(k=1, 2, \ldots)$ are then independent identically distributed positive random variables with a distribution function $F(t|\theta)$, say, which depends on an unknown parameter θ. If the intervals T_k are observable and if the density $f(t|\theta) = \partial F/\partial t$ is a known function, the likelihood function can easily be constructed using, for instance, a fixed number of intervals. Suppose, however, that we are faced with either or both of the following situations: (i) the between-event intervals T_k are unobservable, and (ii) only the Laplace-Stieltjes $(L-S)$ transform $F^*(s|\theta) = \int_0^\infty e^{-st}\, dF(t)$ is known and the explicit form of the density (which may or may not exist) is unknown or difficult to evaluate. The method we suggest (Basawa, 1974) in this section provides a solution to the maximum likelihood estimation problem in this situation.

A possible alternative to observing the intervals $\{T_k\}$ is to count the number of events occurring in fixed intervals, say $(0, a)$, $(a, 2a), \ldots$ $(a > 0)$. Such a sequence of counts, for any given a, does not, however, determine the original process $\{X_t\}$ completely; besides, the joint distributions of these counts of events are not simple except in the case of a Poisson process. We shall, instead, observe the process at random observation points $\{v_0 = 0 < v_1 < v_2 < \ldots\}$, and count the number of events occurring in the random observation intervals $\xi_k = v_k - v_{k-1}$, $(k=1, 2, \ldots)$. We assume that the renewal process starts off with an event at $t_0 = 0$. ξ_k are assumed to be independent identically distributed exponential variates with the parameter 1. (Any other fixed value of the parameter, say λ, will do. However, for simplicity we take $\lambda = 1$). The sequence $\{N_k, k=1, 2, \ldots\}$, where N_k is the number of events occurring during ξ_k, is then called a Poisson count process derived from $\{X_t\}$.

Beginning from any non-zero count let Y = number of counts $\{N_r\}$ needed to get the next non-zero count, and $Z = N_Y$, the number of renewals in the Yth interval. Consider the bivariate discrete random variable (Y, Z). Denote by M_k the number of observation points $\{v_r\}$ in the kth between-event interval T_k. Since $\{v_r, r=1, 2, \ldots\}$ form a Poisson process, the counts $\{M_k\}$ $k=1, 2, \ldots$ conditional on the values of $\{T_k\}$ are independent identically distributed Poisson random variables. Unconditionally therefore the M_k are independent identically distributed with probability mass function,

$$\begin{aligned}\pi_l &= P(M_k = l)\\ &= \int_0^\infty \frac{e^{-t}t^l}{l!}\,dF(t|\theta)\\ &= \frac{(-1)^l}{l!}F^{*(l)}(1|\theta), \quad (l=0, 1, 2, \ldots)\end{aligned} \tag{34}$$

where $F^{*(l)}(1|\theta)$ is the lth derivative of the L–S transform $F^*(s|\theta)$ with respect to s at $s=1$. Now the joint distribution of Y and Z is $P(Y=y, Z=z)$

$$\begin{aligned}&= P(N_1=0, N_2=0, \ldots, N_{y-1}=0, N_y=Z)\\ &= P(M_1=y, M_2=0, \ldots, M_z=0)\\ &= \pi_y\pi_0^{z-1}, \ (y, z=1, 2, \ldots),\end{aligned} \tag{35}$$

where $\pi_0 = F^{*(0)}(1|\theta) = F^*(1|\theta)$. It should now be clear that the embedded process $\{(Y_k, Z_k), k=1, 2, \ldots\}$ is a bivariate discrete renewal process. The marginal distributions of Y and Z are

$$P(Y=y) = \frac{\pi_y}{(1-\pi_0)}, \ (y=1, 2, \ldots)$$

and

$$P(Z=z) = (1-\pi_0)\pi_0^{z-1}, \ (z=1, 2, \ldots).$$

Furthermore, each of the sequences $\{Y_k\}$, $\{Z_k\}$ is a discrete renewal process. We shall now give three sampling plans and indicate the method of construction of the likelihood function in each case.

Method 1: Fixed Number of Counts. Let $\{N_1, N_2, \ldots, N_{n_0}\}$ be a sample of n_0 Poisson counts, where n_0 is fixed. The sample "$\{N_{y_1}=z_1, N_{y_1+y_2}=z_2, \ldots, N_{y_1+\ldots+y_m}=z_m$; and all other N_r, $r<\Sigma_1^m y_i$, and the subsequent $(n_0-\Sigma_1^m y_i)$ counts are zero$\}$" is equivalent to the sample

$$\left\{(y_1, z_1), \ldots, (y_m, z_m), \left(y_{m+1} > n_0 - \sum_1^m y_i\right)\right\},$$

where m denotes the number of non-zero counts in the sample. Since m is a random variable which depends on n_0 we shall write it as $m(n_0)$. Note that the information contained in the last interval, namely $(y_{m+1} > n_0 - \Sigma_1^m y_i)$ is the same as that contained in the last incomplete interval of the renewal process when we stop sampling. Assuming the effects of ignoring this term is negligible for large n, we can write the likelihood function

$$L_1 = \prod_{i=1}^{m(n_0)} (\pi_{y_i} \pi_0^{z_i - 1}). \tag{36}$$

Here $\{(y_1, z_1), \ldots, (y_m, z_m)\}$ is the effective sample.

Method 2: Random Number of Counts. Suppose we continue sampling until we get a given number say m_0 of non-zero counts in the sample. If N_1, $N_2, \ldots, N_n$ is this sample, the sample-size n is a random variable defined as the smallest number of observation-intervals needed to produce m_0 non-zero counts. This sample is exactly equivalent to the sample $\{(y_1, z_1), \ldots, (y_{m_0}, z_{m_0})\}$, and the likelihood function is

$$L_2 = \prod_{i=1}^{m_0} (\pi_{y_i} \pi_0^{z_i - 1}). \tag{27}$$

Method 3: Conditions Likelihood Function. Suppose we observe only a given number, say m_0, of non-zero counts. Our sample is then $(z_1, \ldots, z_{m_0})$. Consequently, the conditional likelihood function is

$$L_3 = (1 - \pi_0)^{m_0} \pi_0^{\sum_{i=1}^{m_0} z_i - m_0}. \tag{28}$$

The ML estimate of $\pi_0(\theta)$ is obviously

$$\hat{\pi}_0(\theta) = 1 - \frac{m_0}{N}.$$

where $N = \sum_{i=1}^{m_0} z_i =$ total number of events in the sample. If $\pi_0(\theta)$ is a monotonic function of θ, the maximum likelihood estimate of θ can easily be obtained from that of $\pi_0(\theta)$. Because of the simplicity of L_3 it may be used to find an initial estimate of θ, especially when $\pi_l(l > 1)$ is a complicated function of θ.

Properties of Maximum Likelihood Estimates. In method 2 we have a sample of m_0 independent and identically distributed random variables from the probability distribution

$$p(y, z) = \pi_y \pi_0^{z-1}.$$

Under the standard regularity conditions on $p(y, z)$, (cf. Rao, 1973) the asymptotic properties of the maximum likelihood estimate as $m_0 \to \infty$ remain valid.

Method 1, however, needs further justification. Notice that $L_1(\theta)$ is identical in form to $L_2(\theta)$ except that in the former case $m(n_0)$ is a random variable. We must now consider the asymptotic theory as $n_0 \to \infty$.

We need a further regularity condition, namely

$$\left[-\frac{\mathrm{d}}{\mathrm{d}s}F^*(s|\theta)\right]_{s=0}=E(T_k)=\mu<\infty.$$

Then, $\sqrt{m(n_0)}(\hat{\theta}-\theta_0)\to N(0, 1/i(\theta_0))$, in distribution as $n_0\to\infty$, where $\hat{\theta}$ is the maximum likelihood estimate obtained as a root of $\mathrm{d}\log L_1/d\theta=0$, and

$$i(\theta_0)=\frac{1}{(1-\pi(0|\theta_0))}\,E\left(\frac{\mathrm{d}\,\log\,\pi(x|\theta)}{\mathrm{d}\theta}\right)^2_{\theta_0}$$

is the Fisher information. This can be established as in Rao (1973) using the following wellknown results:

(i) $1/m(n_0)\,\Sigma_{i=1}^{m(n_0)}\,\phi(Y_i, Z_i)\to E[\phi(Y, Z)]$, in probability as $n_0\to\infty$, where ϕ is any real function of Y and Z with finite expected value;

(ii) $1/\sqrt{m(n_0)}\;\;\Sigma_{i=1}^{m(n_0)}\;\;\mathrm{d}\log p(Y_i, Z_i|\theta)/\mathrm{d}\theta\to N(0, i(\theta))$ in distribution as $n_0\to\infty$.

BIBLIOGRAPHICAL NOTES

Cox and Lewis (1966) contains a comprehensive treatment of early work on statistical methods for simple point processes. A recent book on the subject is Snyder (1975) which deals with results on doubly stochastic Poisson processes among others. The work of Lévy processes discussed in Section 3 is due to Basawa and Brockwell (1978). Also see Aalen (1975, 1978) for a discussion of inference problems for more general point processes.

COMPLEMENTS

1. Suppose particles are arriving according to a Poisson process with parameter θ. Each arriving particle has a probability λ of passing through a barrier. Assuming that N_0 particles have crossed this barrier, find the maximum likelihood estimate of θ.

Suppose, instead, every hth arriving event passes through the barrier successfully; assuming that N_0 particles have crossed the barrier, find the maximum likelihood estimate of θ.

Study the properties of these estimates.

2. Suppose we observe a non-homogeneous Poisson process with rate $\lambda(t)=\exp(\Sigma_{m=0}^{r}\alpha_m t^m)$, continuously over a fixed period $[0, T_0]$. Find the maximum likelihood estimates of $(\alpha_1,\ldots,\alpha_r)$. Also, obtain a goodness-of-fit test for a pth $(p<r)$ degree polynomial rate function as against the rth degree polynomial.

(MacLean, 1974)

3. Consider a non-homogeneous Poisson process $\{X_t\}$ with rate $\lambda(t)=\mathrm{e}^{\mu+\lambda t}$, where

μ and λ are unknown parameters. Using a continuous realization over $[0, T_0]$ find the maximum likelihood estimates of μ and λ. Assuming $\lambda > 0$, show that the estimates $\hat{\lambda}$ and $\hat{\mu}$ are consistent for λ and μ. Furthermore, prove that $(\int_0^{T_0} X_u du)^{1/2} (\hat{\lambda} - \lambda)$ converges asymptotically, as $T_0 \to \infty$, to a normal random variable with mean zero and variance λ.

(Keiding, 1974)

4. Consider a non-homogeneous Poisson process with non-decreasing rate function $0 \leq \lambda(t) \leq M$, where M is a given number. Observe the process until N_0 events have occurred and find the maximum likelihood estimate of $\lambda(t)$.

Suppose instead that we observe the process over a fixed period $[0, T_0]$. Conditional on n occurrences during this time, find the maximum likelihood estimate of $\lambda(t)$. Also, find a conditional likelihood-ratio test of the hypothesis $H: \lambda(t) = \lambda$.

(Boswell, 1966)

5. Consider a doubly stochastic Poisson process $\{X_t\}$ with rate $\Lambda(t)$, where

$$\Lambda(t) = \mu + \varepsilon(t).$$

Suppose μ has a prior distribution with $E(\mu) = \mu_0$, $V(\mu) = 1$, and assume that this prior distribution is uncorrelated with the stochastic process $\{\varepsilon(t)\}$. The process $\{\varepsilon(t)\}$ is such that

$$E(\varepsilon(t)) = 0, \quad \text{and} \quad E(\varepsilon(t)\varepsilon(s)) = e^{-|s-t|}.$$

Assuming, $\Lambda(t)$ to be integrable in quadratic mean, $E(\Lambda^2(t)) < \infty$, and $\int_0^{T_0} E(\Lambda^2(t)) dt < \infty$, find the best linear Bayes estimate μ^* of μ (i.e. an estimate μ^* which is linear in X_t and such that $E(\mu^* - \mu)^2$ is a minimum) from a continuous realization of X_t over $[0, T_0]$.

(Grandell, 1976)

6. For the stable process discussed in Section 3 find the Fisher information matrix for the parameters (β, c) using the full likelihood function. Compute the asymptotic variance of the unconditional maximum likelihood estimate of β.

7. For the stable process discussed in Section 3, suppose $\hat{\beta}$ and $\hat{c}$ denote the unconditional maximum likelihood estimates computed from the full likelihood based on the observation of the process over a fixed interval $(0, T)$. For fixed T, show that as $\varepsilon \to 0$, $\hat{\beta} \to \beta$ and $\hat{c} \to c$ in probability.

(Basawa and Brockwell, 1978)

Chapter 7

Large Sample Theory for Discrete Parameter Stochastic Processes

1. INTRODUCTION

Discrete parameter stochastic processes will be discussed in this chapter. Our aim is to provide a general asymptotic theory of estimation and testing for stochastic processes without the usual assumptions about stationarity or independence of observations.

2. ESTIMATION

2.1 Preliminaries and Notations

Let $\mathbf{Y}^n=(Y_1, \ldots, Y_n)$, $n \geqq 1$ be a sequence of random vectors defined on a probability space $(\Omega, \mathscr{F}, P_\theta)$ and taking values in a measure space $(\mathscr{Y}^n, \mathscr{A}^n, \mu^n)$, μ σ-finite on $(\mathscr{Y}, \mathscr{A})$. $\mathscr{Y}$ could be IR^m and $\theta \in \Theta$, an open subset of IR^p. Let $\|\cdot\|$ denote the ordinary Euclidean norm on IR^p. Suppose that $\mathbf{Y}^n$ has density $p(\mathbf{y}^n; \theta)=p(y_1, \ldots, y_n; \theta)$ with respect to the σ-finite measure μ^n. Let θ_0 be the true value of θ. Any Borel measurable function

$$\hat{\theta}_n=\hat{\theta}_n(\mathbf{Y}^n)=\hat{\theta}_n(Y_1, \ldots, Y_n) \tag{1}$$

is called a *maximum likelihood estimator* (MLE) if

$$p(\mathbf{y}^n;\ \hat{\theta}_n) \geqq p(\mathbf{y}^n;\ \theta) \tag{2}$$

for all $\theta \in \Theta$. Suppose that the true value θ_0 of θ is an interior point of the set Θ. Let $p(y^0;\ \theta) \equiv 1$ and suppose that the conditional probability density of Y_k given $\mathbf{Y}^{k-1}$ exists for every $\theta \in \Theta$. Let

$$p_k(\theta) = p(\mathbf{y}^k;\ \theta)/p(\mathbf{y}^{k-1};\ \theta) \tag{3}$$

for every $k \geqq 1$. We shall assume throughout this chapter that the functions $p_k(\theta)$ are twice differentiable with respect to θ for all θ in a neighbourhood I of θ_0 and almost all $\mathbf{y}^k$. Further we shall assume that the support of $p(\mathbf{y}^n, \theta)$ is independent of $\theta \in I$. Define

$$\phi_k(\theta) = \log\, p_k(\theta) \tag{4}$$

and let $\dot{\phi}_k(\theta)$ be the $p \times 1$ vector whose ith component is

$$\dot{\phi}_{k,i}(\theta) = \frac{\partial}{\partial \theta_i}\, \phi_k(\theta) \tag{5}$$

and $\ddot{\phi}_k(\theta)$ be the $p \times p$ matrix whose (ij)th component is

$$\ddot{\phi}_{k,ij}(\theta) = \frac{\partial^2}{\partial \theta_i \partial \theta_j}\, \phi_k(\theta). \tag{6}$$

Let E and P denote expectation and probability with respect to θ_0; convergence of any random sequence will be studied with respect to P_{θ_0}-measure. Unless otherwise stated, limits are taken as n approaches infinity. We shall write $\dot{\phi}_k$ for $\dot{\phi}_k(\theta_0)$ and $\ddot{\phi}_k$ for $\ddot{\phi}_k(\theta_0)$.

2.2 Consistency

We formulate the regularity conditions and prove the results for the case of one dimensional parameter. Extension of the results to the multiparameter case is straightforward.

Any solution of the likelihood equation $\dot{\phi}_k(\theta) = 0$ other than a constant solution is called a *likelihood equation estimator* (LEE).

In order to simplify the notation further in the one-dimensional case, we shall write $U_k(\theta)$ for $\dot{\phi}_k(\theta)$, $V_k(\theta)$ for $\ddot{\phi}_k(\theta)$, U_k for $U_k(\theta_0)$ and V_k for $V_k(\theta_0)$. Let $L_n(\theta) = \log\, p(\mathbf{y}^n;\ \theta)$. Let $\mathscr{F}_n$ be the σ-field generated by Y_j, $1 \leqq j \leqq n$ and $\mathscr{F}_0$ be the trivial σ-field.

In additional to the assumptions made above, we shall assume that the following conditions are satisfied:

(C1) $\phi_k(\theta)$ is thrice differentiable with respect to θ for all $\theta \in I$. Let $W_k(\theta) = \dddot{\phi}_k(\theta)$ be the third derivative of $\phi_k(\theta)$ with respect to θ.

(C2) Differentiation twice with respect to θ of $p(\mathbf{y}^n; \theta)$ under the integral sign is permitted for $\theta \in I$ in $\int p\,(\mathbf{y}^n; \theta) d\mu^n(\mathbf{y}^n)$.

(C3) $E|V_k| < \infty$, $E|Z_k| < \infty$ where

$$Z_k = V_k + U_k^2. \tag{7}$$

Observe that $E[U_k| \mathscr{F}_{k-1}] = 0$ a.s. and $E[Z_k| \mathscr{F}_{k-1}] = 0$ a.s. under the condition (C2). Hence, in particular, $\{S_n = \Sigma_1^n U_k, \mathscr{F}_n, n \geqq 1\}$ is a zero-mean martingale. Condition (C3) implies that $E(U_k^2) < \infty$. Define the random variables

$$i_k(\theta_0) = \text{var}\ [U_k|\mathscr{F}_{k-1}] = E[U_k^2| \mathscr{F}_{k-1}] \tag{8}$$

and

$$I_n(\theta_0) = \sum_{k=1}^{n} i_k(\theta_0).$$

Observe that $\{S_n^* = \Sigma_1^n [V_k - E(V_k| \mathscr{F}_{k-1})], \mathscr{F}_n, n \geqq 1\}$ is a martingale and in fact $S_n^* = \Sigma_1^n V_k + I_n(\theta_0)$ since $i_k(\theta_0) = -E(V_k| \mathscr{F}_{k-1})$.

In addition to (C1)–(C3), assume that the following condition holds.

(C4) There exists a sequence of constants $K(n) \to \infty$ as $n \to \infty$ such that

(i) $\{K(n)\}^{-1} S_n \xrightarrow{p} 0$,

(ii) $\{K(n)\}^{-1} S_n^* \xrightarrow{p} 0$,

(iii) there exists $a(\theta_0) > 0$ such that for every $\varepsilon > 0$ $P[\{K(n)\}^{-1} I_n(\theta_0) \geqq 2a(\theta_0)] \geqq 1 - \varepsilon$ for all $n \geqq N(\varepsilon)$, and

(iv) $\{K(n)\}^{-1} \Sigma_{k=1}^n E|W_k(\theta)| < M < \infty$ for all $\theta \in I$ and for all n.

Theorem 2.1 *Under the conditions* (C1)–(C4), *the likelihood equation has a root* $\hat{\theta}_n$ *with* P_{θ_0}*-probability approaching one which is consistent for* θ_0 *as* $n \to \infty$.

Proof. In view of assumption (C1), one can use Taylor's expansion to write

$$\frac{\partial L_n(\theta)}{\partial \theta} = \sum_{k=1}^{n} U_k + (\theta - \theta_0) \sum_{k=1}^{n} V_k + (\theta - \theta_0) \sum_{k=1}^{n} (V_k(\theta_n^*) - V_k) \tag{9}$$

where $\theta_n^* = \theta_0 + r(\theta - \theta_0)$ with $r = r(n, \theta_0)$ satisfying $|r| \leqq 1$.

We shall now prove that there exists a consistent root of the likelihood equation

$$\frac{\cdot L_n(\theta)}{\partial \theta} = 0. \tag{10}$$

It is enough to prove that, for any $\varepsilon>0$ and $0<\delta<\delta_1(\varepsilon)$, P[there exists a root of $\partial L_n(\theta)/\partial_\theta=0$ in $(\theta_0-\delta,\ \theta_0+\delta)$]$>1-\varepsilon$ for all $n>N(\varepsilon,\delta)$. Note that $\partial L_n(\theta)/\partial_\theta$ is a continuous in I.

(C4) (i) implies that

$$\frac{1}{K(n)}\sum_{k=1}^{n} U_k \xrightarrow{p} 0. \tag{11}$$

(C4) (ii) and (iii) imply that there exists a non-random function $a(\theta_0)>0$ such that for all $\varepsilon>0$,

$$P[\{K(n)\}^{-1}|\sum_{k=1}^{n} V_k| \geqq a(\theta_0)]>1-\varepsilon \tag{12}$$

for all $n>N_1=N_1(\varepsilon)$. It is easy to see that

$$\sum_{k=1}^{n}\{V_k(\theta_n^*)-V_k\}=(\theta_n^*-\theta_0)\sum_{k=1}^{n} W_k(\theta_n^{**})$$

where $\theta_n^{**}=\theta_0+s(\theta_n^*-\theta_0)$ and $s=s(\theta_0,\ \theta_n^*)$ satisfies $|s|\leqq 1$ and hence there exists $\eta>0$ such that for all $\varepsilon>0$

$$P[\{K(n)\}^{-1}|\sum_{k=1}^{n}(V_k(\theta_n^*)-V_k)|>(1+\eta)^{-1}a(\theta_0)]$$

$$\leqq P[\{K(n)\}^{-1}|\sum_{k=1}^{n} W_k(\theta_n^{**})|>(1+\eta)^{-1}a(\theta_0)|\theta_n^*-\theta_0|^{-1}]$$

$$\leqq P[\{K(n)\}^{-1}|\sum_{k=1}^{n} W_k(\theta_n^{**})|>(1+\eta)^{-1}a(\theta_0)\{\delta_1(\varepsilon)\}^{-1}]$$

whenever $|\theta_n^*-\theta_0|<\delta_1(\varepsilon)$. But the last term is bounded by

$$\frac{M}{(1+\eta)^{-1}a(\theta_0)\{\delta_1(\varepsilon)\}^{-1}}$$

by Tchebyshev's inequality and this bound can be made smaller than ε by choosing

$$\delta_1(\varepsilon)\leqq\frac{\varepsilon a(\theta_0)}{M(1+\eta)}.$$

In view of the above remarks, there must exist an $\eta>0$ such that for every $\varepsilon>0$,

$$P[\{K(n)\}^{-1}|\sum_{k=1}^{n}(V_k(\theta_n^*)-V_k)|>(1+\eta)^{-1}a(\theta_0)]<\varepsilon \tag{13}$$

for $n > N_2(\varepsilon)$. Equations (11), (12) and (13) imply that for any $\varepsilon > 0$ and any $0 < \delta < \delta_1(\varepsilon)$,

$$P\left[\{K(n)\}^{-1}\frac{\partial L_n(\theta)}{\partial\theta}\Big|_{\theta_0-\delta} < 0 < \{K(n)\}^{-1}\frac{\partial L_n(\theta)}{\partial\theta}\Big|_{\theta_0+\delta}\right]$$
$$+P\left[\{K(n)\}^{-1}\frac{\partial L_n(\theta)}{\partial\theta}\Big|_{\theta_0-\delta} > 0 > \{K(n)\}^{-1}\frac{\partial L_n(\theta)}{\partial\theta}\Big|_{\theta_0+\delta}\right] \tag{14}$$
$$> 1-\varepsilon$$

for all $n > N(\varepsilon, \delta)$ in view of (9). (14) proves that the likelihood equation has a root $\hat{\theta}_n$ with P_{θ_0}-probability greater than $1-\varepsilon$ which is consistent for θ_0 as $n \to \infty$.

2.3 Asymptotic Normality

We now consider the question of asymptotic normality of a consistent root of the likelihood equation whose existence is assured by conditions (C1)–(C4). Instead of assuming (C4), we shall assume that the following stronger condition is satisfied.

(N) There exist sequences $\{K(n)\}$ and $\{M(n)\}$ of positive constants with $K(n) \to \infty$ and $M(n) \to \infty$ as $n \to \infty$ such that

(i) $M(n)\ \{K(n)\}^{-1}\Sigma_{k=1}^n U_k \xrightarrow{\mathcal{L}} N(0,\ i(\theta_0))$ for some non-random function $i(\theta_0) > 0$,

(ii) $\{K(n)\}^{-1}|\Sigma_{k=1}^n V_k| \xrightarrow{p} b(\theta_0)$ for some non-random function $b(\theta_0) > 0$, and

(iii) for all $\varepsilon > 0$ and for all $v > 0$, there exists $\delta = \delta(\varepsilon, v)$ and $N = N(\varepsilon, v)$ such that for all $n > N$,

$$P\left[\{K(n)\}^{-1}\left|\sum_{k=1}^{n}(V_k(\theta^*)-V_k)\right| > v,\ |\theta^*-\theta_0| < \delta\right] < \varepsilon.$$

It is clear that condition (N) implies relations (11)–(13) and hence there exists a root $\hat{\theta}_n$ of the likelihood equation with P_{θ_0}-probability approaching one which is consistent for θ_0 as $n \to \infty$. Relation (9) shows that

$$M(n)(\hat{\theta}_n - \theta_0) = -M(n)\{K(n)\}^{-1}\left\{\sum_{k=1}^{n} U_k\right\}\cdot\left[\{K(n)\}^{-1}\sum_{k=1}^{n} V_k + \{K(n)\}^{-1}\sum_{k=1}^{n}(V_k(\theta_n^*)-V_k)\right]^{-1}, \tag{15}$$

with P_{θ_0}-probability approaching one as $n \to \infty$. Consequently the asymptotic distribution of $M(n)(\hat{\theta}_n - \theta_0)$ will be normal i.e.,

$$M(n)(\hat{\theta}_n - \theta_0) \xrightarrow{\mathscr{L}} N(0, i(\theta_0)\{b(\theta_0)\}^{-2}) \tag{16}$$

provided

$$\{K(n)\}^{-1} \sum_{k=1}^{n} (V_k(\theta_n^*) - V_k) \xrightarrow{p} 0 \tag{17}$$

in view of (N) (i) and $(N)(ii)$. We shall now prove that $(N)(iii)$ implies the equation (17). We have for fixed $v > 0$, $\varepsilon > 0$

$$P\left[\{K(n)\}^{-1} \left|\sum_{k=1}^{n} (V_k(\theta_n^*) - V_k)\right| > v\right]$$

$$= P\left[\{K(n)\}^{-1} \left|\sum_{k=1}^{n} (V_k(\theta_n^*) - V_k)\right| > v, |\theta_n^* - \theta_0| < \delta(\varepsilon, v)\right]$$

$$+ P\left[\{K(n)\}^{-1} \left|\sum_{k=1}^{n} (V_k(\theta_n^*) - V_k)\right| > v, |\theta_n^* - \theta_0| \geqq \delta(\varepsilon, v)\right]$$

$$< \varepsilon + P[|\theta_n^* - \theta_0| \geqq \delta(\varepsilon, v)]$$

and the last term tends to zero, because of the consistency of θ_n^* which follows from consistency of $\hat{\theta}_n$. Hence we have the following theorem.

Theorem 2.2. *Under conditions* (C1)–(C3) *and* (N) *any consistent solution of the maximum likelihood equation is asymptotically normal.*

To deal with more general situations, we shall now strengthen conditions (C1)–(C4) to provide sufficient conditions for (N). Observe that under (C1), the condition (N)(iii) can be written as: for every $\varepsilon > 0$, for every $v > 0$, there exists $\delta = \delta(\varepsilon, v)$ and $N = N(\varepsilon, v)$ such that

$$P\left[\{K(n)\}^{-1} |\theta^* - \theta_0| \left|\sum_{k=1}^{n} W_k(\theta^{**})\right| > v,\ |\theta^* - \theta_0| < \delta\right] < \varepsilon$$

for all $n > N$ where $\theta^{**} = \theta_0 + s(\theta^* - \theta_0)$ and $|s| \leqq 1$. We modify (C4)(i) and (C4)(iii) as follows:

(C4)′ (i) $\{K(n)\}^{-1} M(n) S_n \xrightarrow{\mathscr{L}} N(0, i(\theta_0))$

(C4)′ (iii) $\{K(n)\}^{-1} I_n(\theta_0) \xrightarrow{p} i(\theta_0) = \lim_{n \to \infty} E(I_n(\theta_0))\{K(n)\}^{-1}$.

Clearly (N) is satisfied with $b(\theta_0) = i(\theta_0)$.

Convenient sufficient conditions for (C4)′(i), in the presence of (C1)–(C3), can be given by using central limit results for martingales due to Brown

(1971). Let us write $s_n^2 = E(S_n^2) = E(I_n(\theta_0))$. It follows from Brown (1971) that if

$$s_n^{-2} I_n(\theta_0) = s_n^{-2} \sum_{k=1}^{n} E(U_k^2 | \mathscr{F}_{k-1}) \xrightarrow{p} 1 \tag{18}$$

$$s_n^{-2} \sum_{k=1}^{n} E\{U_k^2 I(|U_k| \geqq \varepsilon s_n)\} \to 0 \tag{19}$$

for all $\varepsilon > 0$ where $I(\cdot)$ is the indicator function, then

$$s_n^{-1} S_n \xrightarrow{\mathscr{L}} N(0, 1). \tag{20}$$

Thus, if (C4)′ (iii) holds, then $E(I_n(\theta_0)) \sim K(n) i(\theta_0)$ and equation (18) holds. Relation (20) gives

$$\{K(n)\}^{-1/2} S_n \xrightarrow{\mathscr{L}} N(0, i(\theta_0))$$

which is (C4)′ (i) for the case $K(n) = M^2(n)$.

2.4 Efficiency

Before we study efficiency of likelihood equation estimates, we shall discuss the general concepts of asymptotic efficiency in the context of stochastic processes. Let

$$S_n(\theta) = \frac{\mathrm{d}\,.\log p(\mathbf{y}^n; \theta)}{\mathrm{d}\theta} = \sum_{k=1}^{n} U_k(\theta)$$

and

$$V_k(\theta) = \frac{\mathrm{d}U_k(\theta)}{\mathrm{d}\theta}.$$

Let us first assume that $p(\mathbf{y}^n; \theta)$ is twice differentiable with respect to θ_0 under the integral sign. Let $\mathscr{F}_k$ be the σ-field generated by $(Y_1, \ldots, Y_k)$. Then $\{S_n(\theta), \mathscr{F}_n, n \geqq 1\}$ and $\{S_n^*(\theta), \mathscr{F}_n, n \geqq 1\}$ are martingales where

$$S_n^*(\theta) = \sum_{k=1}^{n} [V_k(\theta) - E_\theta(V_k(\theta) | \mathscr{F}_{k-1})].$$

Let $J_n(\theta) = E_\theta(S_n^2(\theta)) = \Sigma_{k=1}^n E_\theta(U_k^2(\theta))$, the Fisher information in $\mathbf{Y}^n = (Y_1, Y_2, \ldots, Y_n)$ and $I_n(\theta) = \Sigma_{k=1}^n E_\theta(U_k^2(\theta) | \mathscr{F}_{k-1})$, the conditional information. Denote $U_k(\theta_0)$, $V_k(\theta_0)$, $S_n(\theta_0)$, $S_n^*(\theta_0)$, $J_n(\theta_0)$ and $I_n(\theta_0)$ by U_k, V_k, S_n, S_n^*, J_n and I_n respectively. It is clear that $E(I_n) = J_n$. Suppose that

(E1) $\quad$ (a) $J_n^{-1} I_n \xrightarrow{p} \eta$, $\quad$ (b) $-J_n^{-1} \sum_{k=1}^{n} V_k \xrightarrow{p} \eta$ $\quad$ (21)

where $\eta > 0$ is possibly a nondegenerate random variable. Let T_n be an estimator of θ based on $\mathbf{Y}^n = (Y_1, \ldots, Y_n)$.

Definition 2.1 A statistic T_n is said to be *asymptotically efficient* for θ_0 if it satisfies

$$J_n^{1/2}(T_n - \theta_0 - \beta(\theta_0, \eta)J_n^{-1}S_n) \xrightarrow{p} 0 \tag{22}$$

where $\beta(\theta, \eta)$ is a random variable depending possibly on θ and η only.

We shall first prove that likelihood equation estimator of θ is asymptotically efficient, in the sense defined above, under some regularity conditions. In addition to (E1), we shall assume that the following conditions are satisfied.

(E2) $$J_n^{-1} \sum_{k=1}^{n} E\{U_k^2 I(|U_k| \geqq \varepsilon J_n^{1/2}) | \mathscr{F}_{k-1}\} \xrightarrow{p} 0$$

for all $\varepsilon > 0$ where $I(A)$ denotes the indicator function of set A;

(E3) $p(\mathbf{y}^n; \theta)$ is thrice differentiable with respect to θ;

(E4) $E(U_k^2) < \infty$ and $E(V_k) < \infty$;

(E5) $J_n^{-1} \sum_{k=1}^{n} E|W_k(\theta)| < M < \infty$.

Conditions (E3), (E4) and (E5) have been used earlier also in connection with proving the consistency and asymptotic normality of likelihood equation estimators.

Central limit theorems for martingales due to Hall (1977) or Basawa and Scott (1977) (See Appendix 1) imply that

$$J_n^{-1/2} S_n \xrightarrow{\mathscr{L}} Z^*, \tag{23}$$

and

$$I_n^{-1/2} S_n \xrightarrow{\mathscr{L}} Z \tag{24}$$

in the presence of (E1) and (E2) where Z is a standard normal random variable and $Z^* = \eta^{1/2} Z$, η as defined in (E1) and independent of Z.

It follows from $(E1)$ that

$$J_n^{-1} \sum_{k=1}^{n} V_k \xrightarrow{p} -\eta. \tag{25}$$

By Taylor's expansion,

$$\begin{aligned} \frac{\mathrm{d} \log p(\mathbf{y}^n; \theta)}{\mathrm{d}\theta} = S_n(\theta) = S_n(\theta_0) &+ (\theta - \theta_0) \sum_{k=1}^{n} V_k(\theta_0) \\ &+ (\theta - \theta_0) \sum_{k=1}^{n} [V_k(\theta_n^*) - V_k(\theta_0)] \end{aligned} \tag{26}$$

where $\theta_n^* = \theta_0 + r_n(\theta - \theta_0)$ with $|r_n| \leqq 1$. Let $\hat{\theta}_n$ be a consistent solution of the likelihood equation $S_n(\theta)=0$. Then

$$0=S_n(\hat{\theta}_n)$$
$$=S_n(\theta_0)+(\hat{\theta}_n-\theta_0)\sum_{k=1}^{n} V_k+(\hat{\theta}_n-\theta_0)\sum_{k=1}^{n}[V_k(\theta_n^*)-V_k]$$

giving

$$J_n^{1/2}(\theta_0)(\hat{\theta}_n-\theta_0)=\frac{-J_n^{-1/2}(\theta_0)S_n(\theta_0)}{J_n^{-1}(\theta_0)\sum_{k=1}^{n} V_k+J_n^{-1}(\theta_0)\sum_{k=1}^{n}[V_k(\theta_n^*)-V_k]}. \tag{27}$$

The denominator on the right-hand side of equation (27) converges in probability to $-\eta$ by ($E5$), relation (25) and the fact that $\theta_n^*-\theta_0\xrightarrow{p}0$ since $\hat{\theta}_n\xrightarrow{p}\theta_0$. Hence

$$J_n^{1/2}[\hat{\theta}_n-\theta_0-(\eta J_n)^{-1}S_n]\xrightarrow{p}0. \tag{28}$$

Since a likelihood equation estimator $\hat{\theta}_n$ satisfies the condition in (22) with $\beta(\theta_0,\eta)=\eta^{-1}$ a.s.; it follows that $\hat{\theta}_n$ is asymptotically efficient in the sense of Definiton (2.1).

See Heyde (1978) for a justification of the criterion (28). It is shown there, for instance, that

$$\lim P_{n,\theta_0}\{J_n^{1/2}|\hat{\theta}_n-\theta_0|\leqq\delta\}\geqq\lim P_{n,\theta_0}\{J_n^{1/2}|T_n-\theta_0|\leqq\delta\}$$

where T_n is any consistent estimator satisfying certain regularity conditions and $\delta>0$.

2.5 Robustness Against Dependence of MLE's

Suppose that we have a model with independent and identically distributed (i.i.d.) real-valued random variables with the discrete distribution $\{\gamma_i(\theta):i=1,2,3,\ldots\}$ where $\theta\in\mathbb{R}^p$. In this section, we compute the MLE's of θ under the i.i.d. assumption and investigate how dependence between the observations affects large sample properties of these MLE's.

Let $\{\xi_k;\ k=1,2,3,\ldots\}$ be identically distributed random variables with state space $\mathscr{X}=\{1,2,\ldots,m\}$ or $\mathscr{X}=\{1,2,\ldots\}$ and with the distribution $\{\gamma_i(\theta):i\in\mathscr{X}\}$ where $\theta\in\Theta$ open in $\mathbb{R}^p$. If there is no further information about the dependence between the observations, a model with independent observations is often used and an MLE is computed from a likelihood function derived under this assumption. We study now the properties of the MLE when the family of one-dimensional distribution is the same but when there is in fact a dependence between the observations. We assume that the true parameter point $\theta_0 \in \Theta$ and $\gamma_i(\theta_0) > 0$ for all $i \in \mathscr{X}$. Further we shall assume that if $\gamma_i(\theta_k) \to \gamma_i(\theta_0)$ for all $i \in \mathscr{X}$ as $k \to \infty$, then $\theta_k \to \theta_0$ as $k \to \infty$. This condition is an identifiability condition.

Let $x_1, \ldots, x_n$ be observations of the process $\{\xi_k;\ 1 \leqq k \leqq n\}$. Under the assumption of independent observations the log likelihood function is

$$\log L(\theta) = \sum_i N_i(n) \log \gamma_i(\theta) \tag{29}$$

where $N_i(n)$ is the number of times $x_k = i$ *for* $1 \leqq k \leqq n$. MLE is any point $\hat{\theta} \in \Theta$ at which the likelihood function attains its supremum. For any $0 < c < 1$, c fixed but arbitrary, define $\theta^*(c)$ to be *approximate MLE* if

$$L(\theta^*(c)) \geqq c \sup_\theta L(\theta).$$

One can prove the following theorem by methods analogous to those in Rao (1958) in the independent case. We omit the proof.

Theorem 2.3. *Suppose that the stationary process $\{\xi_k\}$ is ergodic and that*

$$\sum_i \gamma_i(\theta_0) \log \gamma_i(\theta_0) > -\infty.$$

Then every approximate MLE converges to θ_0 almost surely. Further, if each $\gamma_i(\theta)$ is continuous in a neighbourhood of θ_0 then for almost every sample path and all sufficiently large n, an MLE exists.

The above theorem gives sufficient conditions for the existence of an MLE and the strong consistency of an approximate MLE. We shall now prove that, under certain regularity conditions, MLE is asymptotically normal but the approximate MLE need not be asymptotically normal. However, if we let $c \to 1$ as $n \to \infty$, then we obtain asymptotic normality. We shall call these estimators *asymptotic MLE's*.

For some classes of dependent sequences, we prove the asymptotic normality of the MLE $\hat{\theta}$, and of any asymptotic MLE, $\theta^*(c_n)$ where $\hat{\theta}$ and $\theta^*(c_n)$ are derived under the assumption that the observations are i.i.d. and $c_n \to 1$ as $n \to \infty$. In order to prove asymptotic normality, we assume that the process $\{\xi_k : k = 1, 2, \ldots\}$ is not only ergodic but also strong mixing.

Definition 2.2. The stationary process $\{\xi_k : k = 1, 2, \ldots\}$ satisfies the *strong mixing* condition if

$$|P(A \cap B) - P(A)P(B)| \leqq \alpha(s) \tag{30}$$

whenever $A \in \mathscr{B}(\xi_1, \xi_2, \ldots, \xi_k)$ and $B \in \mathscr{B}(\xi_{k+s+1}, \xi_{k+s+2}, \ldots)$ for some k, where $\alpha(s) \downarrow 0$ as $s \to \infty$; here $\mathscr{B}(\ldots)$ denotes the σ-field generated by the random variables inside parentheses.

It can easily be shown that mixing implies ergodicity (cf. Ibragimov and Linnik 1971, or Prasad, 1973). We first derive the results for processes with finite state space and then make necessary changes in assumptions for extending the results to countably infinite state space. For processes with finite state space, we use the following assumption.

(A1) (i) Each $\gamma_i(\theta)$ has continuous partial derivatives of third order in a neighbourhood of θ_0.

(ii) The $p \times m$ matrix

$$J = \left(\left(\frac{\partial\, \gamma_i(\theta_0)}{\partial\, \theta_u}\right)\right)_{p \times m}. \tag{31}$$

has rank p.

Let $g(i, \theta) = \log \gamma_i(\theta)$, $i \in \mathscr{X}$, $g_u = \partial g/\partial\theta_u$, $g_{uv} = \partial^2 g/(\partial\theta_u \partial\theta_v)$, $g_{uvw} = \partial^3 g/(\partial\theta_u \partial\theta_v \partial\theta_w)$ etc. Let $\mathbf{Y}(n) = (Y_1(n), \ldots, Y_p(n))$

where

$$Y_u(n) = n^{-1/2} \sum_{k=1}^{n} g_u(\xi_k;\, \theta_0),\ 1 \leqq u \leqq p. \tag{32}$$

For any u, v, w and any neighbourhood S *of* θ_0, define $G = G_{uvw}$ as

$$G(i) = \sup_{\theta \in S} |g_{uvw}(i;\, \theta)|. \tag{33}$$

Finally, define the matrices $\Sigma \equiv ((\sigma_{uv}))$ and $C \equiv ((c_{uv}))$ by the relations

$$\sigma_{uv} = \mathrm{E}[g_u(\xi_1;\, \theta_0)\, g_v(\xi_1;\, \theta_0)],\ 1 \leqq u,\ v \leqq p \tag{34}$$

and

$$\begin{aligned} c_{uv} = \sigma_{uv} + \sum_{k=2}^{\infty} & E[g_u(\xi_1;\, \theta_0) g_v(\xi_k;\, \theta_0) \\ & + g_u(\xi_k;\, \theta_0) g_v(\xi_1;\, \theta_0)],\ 1 \leqq u,\ v \leqq p. \end{aligned} \tag{35}$$

Under assumption (A1)(ii), it follows that Σ is positive definite (cf. Billingsley, 1961 p. 24)).

Theorem 2.4. *Suppose that the stationary process* $\{\xi_k;\, k = 1, 2, \ldots\}$ *has finite state space, satisfies the strong mixing condition with* $\Sigma_s\ \alpha(s) < \infty$ *and that* *assumption* (A1) *is satisfied. Then*

(i) $$\sqrt{n}(\hat{\theta}-\theta_0)\xrightarrow{\mathscr{L}}N(0, \Sigma^{-1} C \Sigma^{-1}) \tag{36}$$

and

(ii) $$\sqrt{n}(\theta^*(c_n)-\theta_0)\xrightarrow{\mathscr{L}}N(0, \Sigma^{-1} C \Sigma^{-1}). \tag{37}$$

Before we give a proof of this theorem, we shall state two lemmas.

Lemma 2.1. *If the stationary process* $\{\xi_k\}$ *has finite state space and satisfies the strong mixing condition with* $\Sigma_s\, \alpha(s)<\infty$ *and if the derivatives* $g_u(i;\, \theta_0)$ *exist for* $1\leqq u\leqq p$, *then* $\mathbf{Y}(u)\xrightarrow{\mathscr{L}}N(0, C)$.

Lemma 2.2. *If the stationary process* $\{\xi_k\}$ *has finite state space and is ergodic and if each* $\gamma_i(\theta)$ *has continuous partial derivatives of third-order in a neighbourhood* S of θ_0, *then*

$$\lim_{n\to\infty} n^{-1} \sum_{k=1}^{n} g_u(\xi_k;\, \theta_0)=0 \text{ a.s.},$$

$$\lim_{n\to\infty} n^{-1} \sum_{k=1}^{n} g_{uv}(\xi_k;\, \theta_0)=-\sigma_{uv} \text{ a.s.},$$

and

$$\lim_{n\to\infty} n^{-1} \sum_{k=1}^{n} G(\xi_k)=K<\infty \text{ a.s.}$$

for each u, v, w.

The first lemma follows from Theorem 18.5.4 in Ibragimov and Linnik (1971) through the Cramer–Wold device and the second from the ergodic theorem for stationary processes (cf. Doob, 1953).

Proof of Theorem 2.4. The proof of Part (i) follows as in Billingsley (1961) in view of Lemmas 2.1 and 2.2. We shall now prove part (ii). Let

$$H(\theta)=\frac{1}{n}\log L_n(\theta_u).$$

We have shown earlier that $\hat{\theta}$ exists with probability tending to one. Applying Taylor's Theorem, it follows that

$$H(\theta)=H(\hat{\theta})+\frac{1}{2}\sum_u\sum_v H_{uv}(\theta')(\theta_u-\hat{\theta})(\theta_v-\hat{\theta}_v)$$

where $||\theta' - \theta|| \leqq ||\hat{\theta} - \theta||$. Since

$$H(\theta^*(c)) \geqq \frac{\log c}{n} + H(\hat{\theta}),$$

we get

$$\frac{1}{2}\sum_u \sum_v H_{u,v}(\theta')(\theta_u^*(c) - \hat{\theta}_u)(\theta_v^*(c) - \hat{\theta}_v) \geqq \frac{\log c}{n}. \tag{38}$$

Since the partial derivatives are continuous functions of θ, it follows from the relation

$$\frac{1}{n}\sum_{k=1}^{n} g_{uv}(\xi_k;\ \theta) = \sum_{i=1}^{m} \frac{N_i(n)}{n} \frac{\partial^2 \log \gamma_i(\theta)}{\partial\theta_u \partial\theta_v}$$

that the elements of the matrix $A(\theta) = ((H_{uv}(\theta))$ are continuous functions of θ. Hence the smallest eigenvalue of $A(\theta)$ is also a continuous function of θ. From this and the fact that

$$H_{uv}(\theta_0) = \frac{1}{n}\sum_{k=1}^{n} g_{uv}(\xi_k;\ \theta_0) \rightarrow -\sigma_{uv} \text{ a.s.}$$

where Σ is positive definite, it follows that there exists a constant $B < 0$ so that

$$B||\theta^*(c) - \hat{\theta}||^2 \geqq \frac{\log c}{n}$$

with probability approaching one. Hence

$$\sqrt{n}||\theta^*(c) - \hat{\theta}|| \leqq \left(\frac{\log c}{B}\right)^{1/2}$$

with probability tending to one. Since the sequence $\theta^*(c_n)$ is such that $c_n \rightarrow 1$, it follows that

$$\sqrt{n}(\theta^*(c_n) - \hat{\theta}) \rightarrow 0.$$

Therefore

$$\sqrt{n}(\theta^*(c_n) - \theta_0) \xrightarrow{\mathscr{L}} N(0, \Sigma^{-1}\, C\, \Sigma^{-1}).$$

This completes the proof.

For processes with countably infinite state space, we replace the assumption (A1) by the following assumption (A2).

(A2) (i) Each $\gamma_i(\theta)$ has continuous partial derivatives of third order in a neighbourhood S of θ_0 and for any u, v, w

$$\sum_i \sup_{\theta' \in S} \frac{\partial \gamma_i(\theta')}{\partial \theta_u} < \infty,$$

$$\sum_i \sup_{\theta' \in S} \frac{\partial^2 \gamma_i(\theta')}{\partial \theta_u \partial \theta_v} < \infty,$$

and

$$E(\sup_{\theta' \in S} |g_{uvw}(\xi_1; \theta')|) < \infty.$$

(ii) For some $\delta > 0$,

$$E|g_u(\xi_1; \theta_0)|^{2+\delta} < \infty, \ 1 \leqq u \leqq p.$$

(iii) The $p \times p$ matrix Σ is positive definite.

Theorem 2.5. *Suppose that the stationary process $\{\xi_k\}$ has countably infinite state space, satisfies the strong mixing condition with $\Sigma_s\ \alpha(s)^{\delta/(2+\delta)} < \infty$ and that assumption (A2) is satisfied. Furthermore assume that*

$$\sum_i \gamma_i(\theta_0) \log \gamma_i(\theta_0) > -\infty.$$

Then

(i)
$$\sqrt{n}(\hat{\theta} - \theta_0) \xrightarrow{\mathscr{L}} N(0, \Sigma^{-1} C \Sigma^{-1}).$$

(ii)
$$\sqrt{n}(\theta^*(c_n) - \theta_0) \xrightarrow{\mathscr{L}} N(0, \Sigma^{-1} C \Sigma^{-1}).$$

We will sketch the proof in the case of infinite state space. When the state space of the process $\{\xi_k\}$ is infinite, the derivatives $g_u(\xi_k; \theta_0)$ are not bounded. Since the process $\{\xi_k\}$ satisfies the strong mixing condition with

$$\sum_s \alpha(s)^{\delta/(2+\delta)} < \infty$$

for some $\delta > 0$ and that

$$E\{|g_u(\xi_i; \theta_0)|^{2+\delta}\} < \infty, \ 1 \leqq u \leqq p,$$

we can use the extended central limit theorem for unbounded mixing processes (cf. Theorem 18.5.3 of Ibragimov and Linnik, 1971) to prove asymptotic normality under assumption (A2). Assumption (A2) ensures differentiations under the sum as well as the invertibility of the matrix Σ. The proof is left as an exercise.

The following counter-example shows that an approximate MLE is not necessarily asymptotically normal. The example is worked out for $p=1$. Choose θ_n^* and θ_n^{**} so that

$$H(\theta_n^*)=H(\theta_n^{**})=H(\hat{\theta})+\frac{\log c}{n}$$

where $\hat{\theta}$ is the MLE. Furthermore choose $\theta_n^* \leqq \hat{\theta}$ and $\theta_n^{**} \geqq \hat{\theta}$. As in (38), we can show that

$$\frac{1}{2}|\theta_n^*-\hat{\theta}|^2 \; A(\theta_n')=\frac{\log c}{n} \tag{39}$$

where θ_n' lies between θ_n^* and $\hat{\theta}$. Let $h(\theta)=\{2 \log c/A(\theta)\}^{1/2}$. Then

$$n^{1/2}|\theta_n^*-\hat{\theta}|=h(\theta_n').$$

Since $A(\theta)$ is a continuous function of θ with probability approaching one and since θ_n^* and $\hat{\theta}$ are consistent estimators of θ_0, it follows that $h(\theta_n')\xrightarrow{p} h(\theta_0)$. Hence asymptotic normality of $\hat{\theta}$ implies that

$$\sqrt{n}(\theta_n^*-\theta_0)$$

is asymptotically normal with mean $h(\theta_0)$. If the approximate MLE is defined as

$$\theta_n^*(c)=\begin{cases}\theta_n^* & \text{if } n \text{ is odd}\\ \theta_n^{**} & \text{if } n \text{ is even}\end{cases}$$

then $\sqrt{n}(\theta_n^*(c)-\theta_0)$ will not have any limit distribution.

We shall now give an example of a stochastic process which satisfies the assumptions stated. Conditions like those in (A1) and (A2) are used also in the independent case.

Example 2.1. Suppose that we have a stationary normal process $\{\zeta_k;\, k=1, 2, 3, \ldots\}$ with marginal distribution $N(\mu, \sigma^2)$ and that the process is only partially observable. The observed process $\{\xi_k\}$ is defined by

$$\xi_k=i \text{ if } \zeta_k \in I_i$$

where $\{I_i\}$ is a partitioning into intervals of the real line. The problem is to estimate both parameters μ and σ. It is easy to see that

$$P(\xi_k=i)\equiv\pi_i(\mu, \sigma)=\phi\left(\frac{t_i-\mu}{\sigma}\right)-\phi\left(\frac{t_{i-1}-\mu}{\sigma}\right)$$

where ϕ is the standard normal distribution and t_{i-1} and t_i are end points of the interval I_i. If the real line is divided into two intervals, then identifiability condition will not be satisfied. However, if the number of intervals is more than two, then the identifiability condition will be satisfied. It can be shown that assumptions (A1) and (A2) will be satisfied in the finite and infinite case respectively.

If the normal process $\{\zeta_k\}$ has a spectral density $f(\lambda)$ which is rational in the argument $e^{i\lambda}$, then it follows from Kolmogorov and Rozanov (1960) that $\{\zeta_k\}$ satisfies the strong mixing condition and the function $\alpha(s)$ can be chosen so that it converges to zero exponentially fast. Since $\{\xi_k\}$ is a function of the process $\{\zeta_k\}$, it satisfies the strong mixing condition with the same α-function. Hence the MLE, derived from the log-likelihood function, is strongly consistent and asymptotically normal.

2.6 MLE for Some Special Types of Dependent Processes

In the previous section, we discussed robustness of maximum likelihood estimator when the process is strong mixing and when the state space is finite or countably infinite. We shall discuss here briefly some general results when the process is stationary and ϕ-mixing.

Definition 2.3. A sequence $\{X_n\}$ of random variables is said to be *ϕ-mixing* if for each $t \geqq 1$ and, for each $n \geqq 1$

$$|P(A \cap B) - P(A)P(B)| \leqq \phi(n)P(A)$$

for every $A \in \mathscr{B}(X_1, \ldots, X_t)$ and $B \in \mathscr{B}(X_{t+n}, \ldots)$ where $\phi(n) \downarrow 0$ as $n \to \infty$.

Definition 2.4. A sequence $\{X_n\}$ of random variable is said to be *m-dependent* if there exists a non-negative integer m such that every subset $\{X_{t_1}, \ldots, X_{t_s}\}$, $t_1 < t_2 < \ldots < t_s$ is stochastically independent of any other subset $\{X_{t_k}, X_{t_{k+1}}, \ldots\}$ $t_s < t_k < t_{k+1} < \ldots$ whenever $t_k - t_s > m$.

Observe that m-dependence is a special case of ϕ-mixing.

Let $\{X_n\}$ be a stationary ϕ-mixing process. Let $p_n(x_1, \ldots, x_n; \theta)$ be the n-dimensional joint density function depending on a single unknown parameter θ. We have

$$\log p_n(x_1, x_2, \ldots, x_n; \theta) = \sum_{i=1}^{n} \log p(x_i; \theta | x_{i-1}, \ldots, x_1).$$

Let $\{l_n\}$ be a sequence of non-negative integers such that $l_n \to c < \infty$ as $n \to \infty$. It is clear that $l_n \approx c$ for large n.

An estimator $\hat{\theta}_n$ of θ which maximizes

$$\sum_{i=l_n+1}^{n} \log p(x_i;\, \theta|x_{i-1}, \ldots, x_{i-l_n})$$

in a neighbourhood of the true parameter θ_0 is called a *modified maximum likelihood estimator* (MMLE) of θ.

Suppose the following regularity conditions are satisfied:

(P1) The parameter space Θ is an open interval in the real line and the true parameter point θ_0 is an interior point of Θ.

(P2) $\mathrm{d}^i \log p_n(\theta)/\mathrm{d}\theta^i$, $i=1, 2$, exist for all $\theta\in\Theta$ and are continuous in θ almost surely, where $p_k(\theta)=p(x_k;\, \theta|x_1, \ldots, x_{k-1})$.

(P3) For all $\theta\in\Theta$ and for almost all $x_1, \ldots, x_{k-1}$,

$$\text{(i)} \qquad E_\theta\left[\frac{\mathrm{d}}{\mathrm{d}\theta} \log p_k(\theta)|\mathbf{X}^{k-1}\right]=0 \text{ a.s.}$$

and

$$\text{(ii)} \quad -E_\theta\left[\mathrm{d}^2 \log p_k(\theta)\mathrm{d}\theta^2\, \mathbf{X}^{k-1}\right]= E_\theta\left[\left\{\frac{\mathrm{d}}{\mathrm{d}\theta} \log p_k(\theta)\right\}^2|\mathbf{X}^{k-1}\right] \text{ a.s.}$$

where $\mathbf{X}^{k-1}=(X_1, \ldots, X_{k-1})$.

(P4) (i) $$I(\theta)= -\lim_{n\to\infty} n^{-1} E_\theta\left[\frac{\mathrm{d}^2}{\mathrm{d}\theta^2} \log p(X_1, \ldots, X_n;\, \theta)\right]$$

$$= \lim_{n\to\infty} n^{-1} \sum_{k=1}^{n} \sigma_k^2(\theta)<\infty$$

where $\sigma_k^2(\theta)=E_\theta[\mathrm{d}/\mathrm{d}\theta \log p_k(\theta)]^2$ and $I(\theta)>0$.

(ii) For the sequence $\{l_n\}$ chosen above,

$$\lim_{n\to\infty} E_\theta\left[\frac{\mathrm{d}}{\mathrm{d}\theta} \log p(X_{l_n+1};\, \theta|\mathbf{X}^{l_n})\right]^2 = I(\theta).$$

(P5) For every $\theta\in\Theta$, there exists a neighbourhood $V(\theta)$ of θ such that for all $\theta'\in V(\theta)$

$$\left|\frac{\mathrm{d}^2}{\mathrm{d}\theta^2} \log p_n(\theta)|_{\theta=\theta'} - \frac{\mathrm{d}^2}{\mathrm{d}\theta^2} \log p_n(\theta)\right|$$
$$\leq |\theta-\theta'|G(x_1, \ldots, x_n)$$

where

$$\lim_{n\to\infty} E_\theta[G(X_1, X_2, \ldots, X_n)]^{1+\delta}<\infty$$

for some $\delta > 0$ independent of θ.

(P6) For all $\theta \in \Theta$

$$\lim_{n\to\infty} E_\theta \left| \frac{d^2}{d\theta^2} \log p_n(\theta) \right|^{1+\delta} < \infty$$

for some $\delta > 0$ independent of θ.

Observe that no assumptions on the existence of the third order derivative of $\log p_n(\theta)$ are made here. Under Assumptions (P1)–(P6), it can be shown that there exists an MMLE which is weakly consistent. In addition to (P1)–(P6), suppose the following conditions hold.

(P7) For all $\theta \in \Theta$

$$\lim_{n\to\infty} E_\theta \left| \frac{d}{d\theta} \log p_n(\theta) \right|^{2+\delta} < \infty$$

for some $0 < \delta \leq 1$, δ independent of θ and the mixing coefficient ϕ_n satisfies the condition

$$\phi_n = 0(n^{-\alpha})$$

for some $\alpha > 1 + 2/\delta$. It can be shown that the MMLE is asymptotically normal under the conditions (P1)–(P7). We refer the reader to Prasad (1973) for detailed proofs of these results.

Example 2.2. Let $\{X_n\}$ be a sequence of random variables such that $E(X_n) = \mu$, $\mathrm{var}(X_n) = 1$ and $\mathrm{cov}(X_n, X_m) = \rho^{|n-m|}$ where $|\rho| < 1$. Let the joint density of $(X_1, \ldots, X_n)$ be $p(x_1, \ldots, x_n; \mu) = 1/(2\pi)^{n/2} (\det \Gamma_n)^{-1/2} \exp\{-\frac{1}{2}(\mathbf{x}^{(n)} - \boldsymbol{\varepsilon}^{(n)\prime})\Gamma_n^{-1}(\mathbf{x}^{(n)} - \boldsymbol{\varepsilon}^{(n)})\}$ where

$$\mathbf{x}^{(n)} = \begin{pmatrix} x_1 \\ \vdots \\ x_n \end{pmatrix} \quad \text{and} \quad \boldsymbol{\varepsilon}^{(n)} = \begin{pmatrix} E(X_1) \\ \vdots \\ E(X_n) \end{pmatrix} = \begin{pmatrix} \mu \\ \vdots \\ \mu \end{pmatrix}.$$

It is easy to see that

$$\Gamma_n^{-1} = (1-\rho^2)^{-1} \begin{bmatrix} 1 & -\rho & 0 & \ldots & 0 & 0 \\ -\rho & 1+\rho^2 & -\rho & \ldots & 0 & 0 \\ 0 & -\rho & 1+\rho^2 & \ldots & 0 & 0 \\ \ldots & \ldots & \ldots & \ldots & \ldots & \ldots \\ \ldots & \ldots & \ldots & \ldots & \ldots & \ldots \\ 0 & 0 & 0 & \ldots & 1+\rho^2 & -\rho \\ 0 & 0 & 0 & \ldots & -\rho & 1 \end{bmatrix}$$

Furthermore the process $\{X_n\}$ is stationary and ϕ-mixing with $\phi_n = \rho^n$. We leave it to the reader to check that for any sequence $l_n \to c < \infty$, the corresponding MMLE $\hat{\mu}_n$ is consistent and asymptotically normal.

3. EFFICIENT TESTS

3.1 Test Efficiency (Local Criterion)

In the spirit of asymptotic efficiency discussed in the previous section, we shall now discuss efficient tests for stochastic processes in the case of a one-dimensional parameter. We use the notation of Section 2.4.

Suppose we wish to test $H:\theta=\theta_0$ against the alternative $K:\theta>\theta_0$. Let w_n be a test (critical region) based on the sample $\mathbf{Y}^n=(Y_1,\ldots,Y_n)$. The power function of w_n is

$$\gamma_{w_n}(\theta)=\int_{w_n} p(\mathbf{y}^n;\theta)\,\mathrm{d}\mu(\mathbf{y}^n) \tag{39}$$

and w_n is a size α test if $\gamma_{w_n}(\theta_0)=\alpha$. Suppose that we can expand γ_{w_n} by the Taylor series. Then we have

$$\gamma_{w_n}(\theta)=\alpha+(\theta-\theta_0)\gamma'_{w_n}(\theta_0)+\ldots$$

If we wish to maximize the power in the neighbourhood of θ_0, then we should maximize $\gamma'_{w_n}(\theta_0)$. By the Neyman-Pearson Lemma, it can be seen that the test of the form

$$w_n^0=\{\mathbf{y}^n:S_n(\theta_0)\geqq K_n\}$$

where K_n is chosen so that the test is of given size α maximizes $\gamma'_{w_n}(\theta_0)$ i.e. $\gamma'_{w_n}(\theta_0)\leqq\gamma'_{w_n^0}(\theta_0)$ among all tests of size α. Now

$$\begin{aligned}\gamma'_{w_n}(\theta_0)&\leqq\int_{[S_n(\theta_0)\geqq K_n]} p'(\mathbf{y}^n;\theta_0)\mathrm{d}\mu(\mathbf{y}^n)\\ &=\int_{[S(\theta_0)\geqq K_n]}\{J_n^{-1/2}(\theta_0)S_n(\theta_0)\}\,J_n^{1/2}(\theta_0)p(\mathbf{y}^n;\theta_0)\mathrm{d}\mu^n(\mathbf{y}^n)\\ &=\int_{[Z_n\geqq J_n^{-1/2}K_n]} Z_nJ_n^{-1/2}\,\mathrm{d}F_{Z_n}\end{aligned}$$

where $Z_n=J_n^{-1/2}S_n$, $J_n=J_n(\theta_0)$, $S_n=S_n(\theta_0)$ and F_{Z_n} is the distribution function of Z_n. Hence

$$\overline{\lim_{n\to\infty}}\,\{J_n^{-1/2}(\theta_0)\gamma'_{w_n}(\theta_0)\}\leqq\int_{[Z^*\geqq a]} Z^*\,\mathrm{d}F_{Z^*} \tag{40}$$

where Z^* is as defined in (23) and a is such that $P(Z^* \geqq a) = \alpha$. We have assumed that conditions (E1) and (E2) of Section 2.4 are satisfied. Furthermore $Z_n \xrightarrow{\mathscr{L}} Z^*$ by relation (23). Let

$$f_{Z^*}(a) = \int_{[Z^* \geqq a]} Z^* \, \mathrm{d}F_{Z^*}. \tag{41}$$

We have proved the following lemma.

Lemma 3.1. *Under the assumptions stated above*

$$\overline{\lim_{n \to \infty}} \, \{J_n^{-1/2}(\theta_0) \gamma'_{w_n}(\theta_0)\} \leqq f_{Z^*(\theta_0)}(a). \tag{42}$$

If $\eta = 1$ a.s. *in* (E1) *of Section* 2.4 *then* $Z^* = Z$.

The result obtained above can be used as a basis for a criterion of test efficiency.

Definition 3.1. The *local asymptotic efficiency* of a test w_n of size α is defined as

$$e(w_n) = [f_{Z^*(\theta_0)}(a)]^{-1} \overline{\lim_{n \to \infty}} \, [J_n^{-1/2}(\theta_0) \gamma'_{w_n}(\theta_0)]$$

where a is such that $P[Z^*(\theta_0) \geqq a] = \alpha$. A test w_n is said to be *locally asymptotically efficient* if $e(w_n) = 1$.

Also, see Sections 4 and 5, and the preliminary discussion on test efficiency in Chapter 1. A test which is efficient according to this local criterion need not have optimum asymptotic power beyond the immediate neighbourhood of θ_0. This is related to the fact that a locally most powerful test need not be uniformly most powerful. See Chapter 2 for an application.

Suppose T_n is any statistic which estimates θ and further suppose that T_n is efficient in the sense of Definition 2.1. Then the test defined by

$$[J_n^{1/2}(\theta_0) \{\beta(\theta_0, \eta)\}^{-1} (T_n - \theta_0) \geqq K_n]$$

is efficient since $J_n^{1/2}(\theta_0)\{\beta(\theta_0, \eta)\}^{-1}(T_n - \theta_0)$ and $J_n^{-1/2}(\theta_0) S_n(\theta_0)$ will have the same asymptotic distribution.

Let $\hat{\theta}_n$ be a consistent solution of the likelihood equation. Further suppose that conditions (E1)–(E5) of Section 2.4 are satisfied. We have shown that $\hat{\theta}_n$ is efficient in the sense of Definition 2.1 with $\beta(\theta_0, \eta) = \eta^{-1}$. Hence the test defined by

$$[J_n^{1/2}(\theta_0) \eta (\hat{\theta}_n - \theta_0) \geqq K_n]$$

is efficient. But, by condition (E1),

$$J_n^{-1}(\theta_0)I_n(\theta_0)\xrightarrow{p}\eta$$

and $I_n(\theta_0)J_n^{-1/2}(\theta_0)(\hat{\theta}_n-\theta_0)$ and $J_n^{-1/2}(\theta_0)S_n(\theta_0)$ will have the same asymptotic distribution by relation (28). Hence the test defined by

$$[I_n(\theta_0)J_n^{-1/2}(\theta_0)(\hat{\theta}_n-\theta_0)\geqq K_n]$$

is an efficient test. It can be shown that the test defined by

$$[I_n^{1/2}(\theta_0)(\hat{\theta}_n-\theta_0)\geqq K_n]$$

is not efficient in general. See Chapter 2 for applications.

4. LARGE SAMPLE TESTS

4.1 Introduction

For testing hypotheses about several parameters involved in discrete parameter stochastic processes, we study test statistics based on the derivatives of the log-likelihood function and derive their asymptotic distributions under the null hypothesis both in the case of simple as well as composite hypotheses. In the case of composite hypotheses, the asymptotic distributions are obtained by substituting weakly $\sqrt{n}$ consistent estimators for unspecified parameters. Asymptotic distributions of the test statistics have also been obtained under a class of local alternatives. Results here make use of the properties of likelihood equation estimators studied in the previous section.

4.2 Preliminaries and Notation

In addition to the notation introduced in Section 2.1, we shall denote a column vector by $\boldsymbol{\alpha}$ and its transpose by $\boldsymbol{\alpha}'$. One can give regularity conditions to ensure that

$$n^{1/2}(\hat{\theta}_n-\theta_0)\xrightarrow{\mathscr{L}}N(0,\Gamma^{-1}(\theta_0)) \tag{43}$$

when $\Gamma(\theta_0)$ is a positive definite matrix. We have discussed the one-dimensional case in Section 2. The regularity conditions in the multiparameter case can be formulated analogously. Suppose that conditions similar to those in Chapter 10, Section 2, hold and

$$\frac{1}{n}\sum_{k=1}^{n} E_\theta[\dot{\phi}_k(\theta)\dot{\phi}_k'(\theta)|\mathbf{Y}^{k-1}]\to\Gamma(\theta)\text{ a.s.} \tag{44}$$

uniformly over compact sets of θ. Let

$$Z_{ni}(\theta)=n^{-1/2}\frac{\partial \log p(\mathbf{y}^n;\ \theta)}{\partial\theta}$$

$$=n^{-1/2}\sum_{k=1}^{n}\dot{\phi}_{ki}(\theta) \tag{45}$$

and

$$\mathbf{Z}_n'(\theta)=(Z_{n1}(\theta),\ldots,Z_{np}(\theta)).$$

4.3 Tests for a Simple Hypothesis

Suppose we have a sequence of null hypotheses $H_n:\theta=\theta_n=\theta_0+n^{-1/2}\delta_n$ and an alternative $H_0:\theta=\theta_0\in\Theta$ where $\delta_n\to\delta$ as $n\to\infty$.

Theorem 4.1. *Under the regularity conditions stated above*

$$\mathbf{Z}_n(\theta_n)\xrightarrow{\mathscr{L}}N(-\Gamma(\theta_0)\delta,\ \Gamma(\theta_0)). \tag{46}$$

Proof. It is easy to see that

$$\mathbf{Z}_n(\theta_n)-\dot{\mathbf{Z}}_n(\theta_0)=\left\{\sum_k\ddot{\phi}_k(\theta_n^*)\right\}\cdot n^{-1/2}(\theta_n-\theta_0) \tag{47}$$

where $\theta_n^*=\theta_0+\xi(\theta_n-\theta_0)$ where $0<\xi<1$. Since $\theta_n\to\theta_0$, it follows that

$$n^{-1}\sum_{k=1}^{n}\ddot{\phi}_k(\theta_n^*)\xrightarrow{p}-\Gamma(\theta_0) \tag{48}$$

and

$$\mathbf{Z}_n(\theta_0)\xrightarrow{\mathscr{L}}N(0,\ \Gamma(\theta_0)) \tag{49}$$

But $n^{1/2}(\theta_n-\theta_0)\to\delta$. Hence (46) holds.

As a consequence of Theorem 4.1, we have the following result.

Theorem 4.2. *Let Σ be any $p\times p$ matrix such that $\Sigma\Gamma(\theta_0)\Sigma=\Sigma$ and $Q_n(\theta)=\mathbf{Z}_n'(\theta)\Sigma\mathbf{Z}_n(\theta)$. Then $Q_n(\theta_0)$ converges in law to a central χ^2 with degrees of freedom equal to rank $(\Sigma)=$ Trace $(\Sigma\Gamma(\theta_0)$ and $Q_n(\theta_n)$ converges in law to a noncentral χ^2 with degrees of freedom equal to Rank (Σ) and noncentrality parameter $\delta'\Gamma(\theta_0)\Sigma\Gamma(\theta_0)\delta$.*

For testing H_0, one can use $Q_n(\theta_0)$ as a test statistic for a suitable choice of Σ. The asymptotic power of the test of the null hypothesis H_n against the alternative H_0 can be calculated from the asymptotic distribution of $Q_n(\theta_n)$. Note that here we are keeping the alternative hypothesis fixed (H_0) and considering a sequence of null hypotheses (H_n). (See Billingsley 1961). Choice of Σ can be made as follows. Let C be any non-singular matrix such that $C'\Gamma(\theta_0)C=\Delta$ a diagonal matrix of order p with as many non-zero diagonal elements as the rank of $\Gamma(\theta_0)$. Let ∇ be the diagonal matrix obtained by replacing the non-zero entries of Δ by their reciprocals. Let $\Sigma^*=C\nabla C'$. It can be easily seen that $\Sigma^*\Gamma(\theta_0)\Sigma^*=\Sigma^*$. Σ^* is a reflexive generalized inverse of $\Gamma(\theta_0)$ with maximum rank namely, rank of $\Gamma(\theta_0)$ (see Rao and Mitra, 1972). We have the following corollary from Theorem 4.2.

Corollary 4.1. *$Q_n(\theta_n)=\mathbf{Z}_n'(\theta_n)\Sigma^*\mathbf{Z}_n(\theta_n)$ converges in Law to a non-central χ^2 with degrees of freedom equal to Rank $(\Gamma(\theta_0))$ and non-centrality parameter $\delta'\Gamma(\theta_0)\delta$.*

Corollary 4.2. *If Σ is any reflexive generalized inverse of $\Gamma(\theta_0)$, then $Q_n(\theta_n)=\mathbf{Z}_n'(\theta_n)\Sigma\mathbf{Z}_n(\theta_n)$ converges in law to a non-central χ^2 with degrees of freedom equal to Rank (Σ) and non-centrality parameter $\delta'\,\Gamma(\theta_0)\delta$.*

4.4 Tests for a Composite Hypothesis Specifying $r(<p)$ Components

Suppose the composite hypothesis is $H_0:\theta_i=\theta_i^0$, $1\leqq i\leqq r$ $(r<p)$ and $\theta_{r+1},\ldots,\theta_p$ unspecified. Let $\Gamma(\theta)$ be partitioned as

$$\begin{pmatrix}\Gamma_{11}(\theta) & \Gamma_{12}(\theta)\\ \Gamma_{21}(\theta) & \Gamma_{22}(\theta)\end{pmatrix} \tag{50}$$

where $\Gamma_{11}(\theta)$ is the submatrix of the first r rows and r columns and $\Gamma_{22}(\theta)$ is the submatrix of the last $(p-r)$ rows and $(p-r)$ columns. Let $B(\theta)$ be any matrix such that

$$\Gamma_{12}(\theta)+B(\theta)\Gamma_{22}(\theta)=0. \tag{51}$$

The general solution of equation (51) is

$$B(\theta)=-\Gamma_{12}(\theta)\,\Gamma_{22}^-(\theta)+G-G\,\Gamma_{22}(\theta)\,\Gamma_{22}^-(\theta)$$

where $\Gamma_{22}^-(\theta)$ is any generalized inverse of $\Gamma_{22}(\theta)$ and G is an arbitrary $r\times(p-r)$ matrix. Define, for $1\leqq k\leqq n$,

$$U_{ki}(\theta)=\begin{cases}\dot\phi_{ki}(\theta)+\sum\limits_{j=r+1}^{p} b_{ij}(\theta)\dot\phi_{kj}(\theta), & 1\leqq i\leqq r\\[2ex] \dot\phi_{ki}(\theta), \quad r+1\leqq i\leqq p.\end{cases} \tag{52}$$

Here b_{ij} is (ij)th component of $B(\theta)$. Let

$${}_1\mathbf{U}_k'(\theta)=(U_{k1}(\theta),\ldots,U_{kr}(\theta)), \tag{53}$$

$$W_{ni}(\theta)=n^{-1/2}\sum_{k=1}^{n} U_{ki}(\theta) \tag{54}$$

and

$$\mathbf{W}'_n(\theta)=(W_{n1}(\theta),\ldots, W_{nr}(\theta)). \tag{55}$$

Let $H_n:\theta_n=\theta_0+n^{-1/2}\delta_n$, where the first r components of $\theta_0=(\theta_1^0,\ldots,\theta_p^0)$ are specified by H_0 and $\theta_{r+1}^0,\ldots,\theta_p^0$ are arbitrary but fixed. Further suppose that $\delta_n\to\delta$. Let δ be partitioned as $\delta'=({}_1\delta';{}_2\delta')$ where ${}_1\delta$ has the first r components of δ. It is not difficult to show under suitable regularity conditions that

$$\mathbf{W}_n(\theta_n)\xrightarrow{\mathscr{L}}N(-\Sigma(\theta_0)_1\delta,\Sigma(\theta_0)) \tag{56}$$

where

$$\Sigma(\theta)=\lim_{k\to\infty}\frac{1}{n}\sum_{k=1}^{n}\Sigma_k(\theta) \qquad \text{a.s.}$$

and

$$\Sigma_k(\theta)=E_\theta[{}_1\mathbf{U}_k(\theta)\,{}_1\mathbf{U}'_k(\theta)|\mathscr{F}_{k-1}]. \tag{57}$$

As a consequence of the result, we have the following theorem.

Theorem 4.3. *Let G be any $r\times r$ matrix such that $G\Sigma(\theta_0)G=G$ and $Q_n(\theta)=\mathbf{W}'_n(\theta)G\,\mathbf{W}_n(\theta)$. Then $Q_n(\theta_n)$ converges in law to a non-central χ^2 with degrees of freedom equal to Rank (G) (= Trace $(G\Sigma(\theta)_0)$)) and non-centrality parameter ${}_1\delta'\Sigma(\theta_0)G\Sigma(\theta_0)_1\delta$.*

If ${}_1\delta=0$, the asymptotic distribution is χ^2 with degrees of freedom equal to Rank (G) in Theorem 4.3. In order to test the sequence H_n against H_0, one can take $Q_n(\theta)$ as a test statistic. But $Q_n(\theta)$ involves nuisance parameters $\theta_{r+1},\ldots,\theta_p$. These parameters can be replaced by a sequence of weakly $\sqrt{n}$ consistent estimators. Before we study the behaviour of $Q_n(\theta)$, we shall first define $\sqrt{n}$ consistent estimators.

A sequence of estimators $\{\hat\theta_{1n},\ldots,\hat\theta_{kn}\}$ of $\{\theta_1,\ldots,\theta_k\}$ based on a sample of n observations is said to be *weakly $\sqrt{n}$ consistent* if for any set of real numbers $a_1, a_2,\ldots, a_k$ such that $\Sigma_{i=1}^k a_i^2>0$,

$$\sqrt{n}\sum_{i=1}^{k} a_i(\hat\theta_{in}-\theta_i)$$

does not converge to zero in probability.

If $\sqrt{n}(\hat\theta_{in}-\theta_i)$, $1\leqq i\leqq k$ has a limiting joint non-degenerate distribution, then the corresponding estimators form a weakly $\sqrt{n}$ consistent sequence of estimators. For example, maximum likelihood estimators $\hat{\theta}_{sn}$, $r+1 \leqq s \leqq p$ are estimators of the above type. Let

$${}_1\theta = (\theta_1, \ldots, \theta_r) \text{ and } {}_2\theta = (\theta_{r+1}, \ldots, \theta_p).$$

Let $\mathbf{W}_n(\theta) = \mathbf{W}_n({}_1\theta; {}_2\theta)$ where $\mathbf{W}_n(\theta)$ is as defined in equation (55) and ${}_2\hat{\theta}_n$ be a sequence of $\sqrt{n}$ consistent estimators. The following theorem can be proved.

Theorem 4.4. *Under the regularity conditions stated earlier*

$$\mathbf{W}_n({}_1\theta_0; {}_2\theta) - \mathbf{W}_n({}_1\theta_0; {}_2\hat{\theta}_n) \xrightarrow{p} 0. \tag{58}$$

Proof. By the definition of $U_{ki}(\theta)$ from (52) it can be easily seen that

$$E_\theta[U_{ki}(\theta) | \mathbf{Y}^{k-1}] = 0, \; 1 \leqq k \leqq n, \; 1 \leqq i \leqq p.$$

Furthermore for $1 \leqq s \leqq p$, $1 \leqq i \leqq p$ and $1 \leqq k \leqq n$

$$\begin{aligned} 0 &= \frac{\partial}{\partial \theta_s} E_\theta[U_{ki}(\theta) | \mathbf{Y}^{k-1}] \\ &= \frac{\partial}{\partial \theta_s} \int U_{ki}(\theta) p_k(\theta) \mu(\mathrm{d}y) \\ &= \int \dot{U}_{kis}(\theta) p_k(\theta) \mu(\mathrm{d}y) + \int U_{ki}(\theta) \frac{\partial p_k(\theta)}{\partial \theta_s} \mu(\mathrm{d}y) \\ &= E_\theta[\dot{U}_{kis}(\theta) | \mathbf{Y}^{k-1}] + E_\theta[U_{ki}(\theta) \dot{\phi}_{ks}(\theta) | \mathbf{Y}^{k-1}]. \end{aligned} \tag{59}$$

Hence, for $1 \leqq i \leqq r$ and $1 \leqq s \leqq p$,

$$\begin{aligned} E_\theta[\dot{U}_{kis}(\theta) | \mathbf{Y}^{k-1}] &= -E_\theta\left[\left\{\dot{\phi}_{ki}(\theta) + \sum_{j=r+1}^{p} b_{ij}(\theta) \dot{\phi}_{kj}(\theta)\right\} \dot{\phi}_{ks}(\theta) | \mathbf{Y}^{k-1}\right] \\ &= -\left\{\Gamma_{kis}(\theta) + \sum_{j=r+1}^{p} b_{ij}(\theta) \Gamma_{kjs}(\theta)\right\} \\ &= \Delta_{kis}(\theta) \text{ (say)} \end{aligned} \tag{60}$$

and for $r+1 \leqq i \leqq p$ and $1 \leqq s \leqq p$,

$$\begin{aligned} E_\theta[\dot{U}_{kis}(\theta) | \mathbf{Y}^{k-1}] &= -E_\theta[\dot{\phi}_{ki}(\theta) \dot{\phi}_{ks}(\theta) | \mathbf{Y}^{k-1}] \\ &= -\Gamma_{kis}(\theta). \end{aligned} \tag{61}$$

For any fixed $1 \leqq i \leqq r$,

$$W_{ni}({}_1\theta_0;\, {}_2\hat{\theta}_n) - W_{ni}({}_1\theta_0;\, {}_2\theta) = n^{-1/2} \sum_{k=1}^{n} \{U_{ki}({}_1\theta_0;\, {}_2\hat{\theta}_n) - U_{ki}({}_1\theta_0;\, {}_2\theta)\}$$

$$= \left\{ \sum_{s=r+1}^{p} \sqrt{n}(\hat{\theta}_{sn} - \theta_s) \right\} \cdot n^{-1} \cdot \sum_{k=1}^{n} \dot{U}_{kis}({}_1\theta_0;\, {}_2\theta)$$

$$+ n^{1/2} \sum_{s=r+1}^{p} \sum_{t=r+1}^{p} n(\hat{\theta}_{sn} - \theta_s)(\hat{\theta}_{tn} - \theta_t) \cdot n^{-1} \sum_{k=1}^{n} \ddot{U}_{kist}({}_1\theta_0;\, {}_2\theta_n^*)$$

where the derivatives $\ddot{U}_{kist}$ are evaluated at $({}_1\theta_0';\, {}_2\theta_n^{*\prime})$, ${}_2\theta_n^*$ being a vector of numbers intermediate between $\hat{\theta}_{jn}$ and θ_j for $j = r+1, \ldots, p$. The first term on the right-hand side converges to zero in probability as $n \to \infty$ by the regularity conditions, and by the fact that $\|{}_2\hat{\theta}_n - {}_2\theta\| \leqq \varepsilon$ with probability approaching one as $n \to \infty$ for every $\varepsilon > 0$. The second term converges to zero in probability by the regularity conditions and the weakly $\sqrt{n}$ consistency of ${}_2\hat{\theta}_n$.

As an application of Theorem 4.4, we have the following result.

Theorem 4.5. *If $G_n = G({}_1\theta_0;\, {}_2\hat{\theta}_n)$, $n \geqq 1$ is any sequence of matrices such that $G_n \Sigma({}_1\theta_0;\, {}_2\hat{\theta}_n) G_n = G_n$ for all n and ${}_2\hat{\theta}_n$ is a sequence of weakly $\sqrt{n}$ consistent estimators of ${}_2\theta$, then*

$$Q_n({}_1\theta_0;\, {}_2\hat{\theta}_n) = \mathbf{W}_n'({}_1\theta_0;\, {}_2\hat{\theta}_n) G_n \mathbf{W}_n({}_1\theta_0;\, {}_2\hat{\theta}_n)$$

*converges in law to a central χ^2 with degrees of freedom equal to Rank of ψ p-*lim $G_n = G$ (say).

In order to test H_n, one can use $Q_n({}_1\theta_0;\, {}_2\hat{\theta}_n)$ as a test statistic. We may compute the asymptotic power of the test of H_n against H_0 from the following theorem. We omit the proof of this theorem.

Theorem 4.6. *$Q_n({}_1\theta_n;\, {}_2\hat{\theta}_n) = \mathbf{W}_n'({}_1\theta_n;\, {}_2\hat{\theta}_n)\, G_n \mathbf{W}_n({}_1\theta_n;\, {}_2\hat{\theta}_n)$ converges in law to a non-central χ^2 with degrees of freedom equal to Rank* (G) *and non-centrality parameter ${}_1\delta' \Sigma(\theta_0) G \Sigma(\theta_0) {}_1\delta$.*

One can again choose G_n to be a reflexive generalized inverse of $\Sigma({}_1\theta_0;\, {}_2\hat{\theta}_n)$. In this case the non-centrality parameter of the limiting distribution of $Q_n({}_1\theta_n;\, {}_2\hat{\theta}_n)$ will be ${}_1\delta' \Sigma(\theta_0) {}_1\delta$ and it will have Rank $(\Sigma(\theta_0))$ as the number of degrees of freedom.

We shall now illustrate the above methods by obtaining test criteria for standard first order autoregressive models.

Example 4.1. Let $\{X_n, n \geqq 1\}$ be a stochastic process defined recursively by the relation

$$X_{n+1} = \theta_1 X_n + \theta_2 + \theta_3^{1/2} Y_{n+1}, \ n \geqq 1$$

where $\{X_1, Y_2, Y_3, \ldots\}$ is a sequence of independent random variables, the Y's being standard normal random variables. Further, we shall suppose that $|\theta_1| < 1$ and $\theta_3 > 0$. This process is called the standard first-order autoregressive model. It can be shown that the unique stationary distribution of the process is normal with mean $\theta_2/1-\theta_1$ and variance $\theta_3/1-\theta_1^2$. We shall suppose that the initial distribution is the stationary distribution. This will make the process $\{X_n, n \geqq 1\}$ a stationary process. One can verify that the regularity conditions for the validity of the test criteria hold. The transition probability density is given by

$$p(x, y; \boldsymbol{\theta}) = (2\pi\theta_3)^{-1/2} \exp\{-\frac{1}{2\theta_3}(y - \theta_1 x - \theta_2)^2\}$$

and

$$\frac{\partial \log p(x, y; \theta)}{\partial \theta_1} = x(y - \theta_1 x - \theta_2)\theta_3^{-1},$$

$$\frac{\partial \log p(x, y; \theta)}{\partial \theta_2} = (y - \theta_1 x - \theta_2)\theta_3^{-1},$$

and

$$\frac{\partial \log p(x, y; \theta)}{\partial \theta_3} = \frac{1}{2}(y - \theta_1 x - \theta_2)^2 \theta_3^{-2} - (2\theta_3)^{-1}.$$

Since $(X_{i+1} - \theta_1 X_i - \theta_2)\theta_3^{-1/2} = Y_{i+1}$ is $N(0, 1)$, the conditional covariance matrix of $(\partial \log p(x, y; \theta)/\partial \theta_j, j = 1, 2, 3)$ given $X_1 = x$ is

$$\begin{bmatrix} x^2\theta_3^{-1} & x\theta_3^{-1} & 0 \\ x\theta_3^{-1} & \theta_3^{-1} & 0 \\ 0 & 0 & (2\theta_3^2)^{-1} \end{bmatrix}.$$

Thus the unconditional covariance matrix with stationary distribution as the initial distribution is

$$\Gamma(\theta) = \begin{bmatrix} \theta_3^{-1}\{\theta_3(1-\theta^2)^{-1} + \theta_2^2(1-\theta_1)^{-2}\} & \theta_2\theta_3^{-1}(1-\theta_1)^{-1} & 0 \\ \theta_2\theta_3^{-1}(1-\theta_1)^{-1} & \theta_3^{-1} & 0 \\ 0 & 0 & (2\theta_3^2)^{-1} \end{bmatrix}.$$

Here $\Gamma(\theta)$ is non-singular and hence the regular inverse is given by

$$\Gamma^{-1}(\theta)=\begin{bmatrix}(1-\theta_1^2) & -\theta_2(1+\theta_1) & 0\\ -\theta_2(1+\theta_1) & \theta_3\theta_2^2(1+\theta_1)(1-\theta_1)^{-1} & 0\\ 0 & 0 & (2\theta_3^2)\end{bmatrix}.$$

The likelihood function of the sample of size n is

$$L_n(\theta)=\prod_{j=1}^{(n-1)}(2\pi\theta_3)^{-1/2}\exp\{(2\theta_3)^{-1}(x_{j+1}-\theta_1x_j-\theta_2)^2\}$$

and hence

$$\begin{aligned}Z_{n1}(\theta)&=(n-1)^{-1/2}\theta_3^{-1}\sum_{j=1}^{(n-1)}x_j(x_{j+1}-\theta_1x_j-\theta_2)\\&=(n-1)^{-1/2}\theta_3^{-1}A \text{ (say)},\\ Z_{n2}(\theta)&=(n-1)^{-1/2}\theta_3^{-1}\sum_{j=1}^{(n-1)}(x_{j+1}-\theta_1x_j-\theta_2)\\&=(n-1)^{-1/2}\theta_3^{-1}B \text{ (say)},\end{aligned}$$

and

$$\begin{aligned}Z_{n3}(\theta)&=(n-1)^{-1/2}(2\theta_3^2)^{-1}\sum_{j=1}^{(n-1)}(x_{j+1}-\theta_1x_j-\theta_2)-(n-1)^{1/2}(2\theta_3)^{-1}\\&=(n-1)^{-1/2}\theta_3^{-1}\{C(2\theta_3)^{-1}-\tfrac{1}{2}(n-1)\} \text{ (say)}.\end{aligned}$$

Suppose one wants to test the simple hypothesis $H_0{:}\theta=\theta_0$. Consider the quadratic form

$$\begin{aligned}Q_n(\theta)&=\mathbf{Z}_n'(\theta)\Gamma^{-1}(\theta_0)\mathbf{Z}_n(\theta)\\&=\{\theta_3^2(n-1)\}^{-1}\{(1+\theta_{10})(1-\theta_{10})^{-1}[A(1-\theta_{10})-\theta_{20}B]^2\\&\quad+\theta_{30}B^2+(\theta_{30}^2/2)(C-(n-1)\theta_3)^2\theta_3^{-2}\}.\end{aligned}$$

For testing the hypothesis H_0, $Q_n(\theta_0)$ could be used as a test criterion. From the asymptotic distribution of $Q_n(\theta_n)$, asymptotic power of the test of $H_n{:}\theta=\theta_0+n^{-1/2}\delta_n$, $\delta_n\to\delta$ as $n\to\infty$ against a fixed alternative $H_0{:}\theta=\theta_0$ could be computed.

For testing composite hypothesis $H_0{:}\theta_1=\theta_{10}$, $\theta_2=\theta_{20}$, θ_3 unspecified, the matrix B as defined in (51) will be the null matrix. Thus one has

$$W_n(\theta)=\begin{pmatrix}W_{n1}(\theta)\\ W_{n2}(\theta)\end{pmatrix}=\begin{pmatrix}\theta_3^{-1}(n-1)^{-1/2}\sum_{j=1}^{(n-1)}x_j(x_{j+1}-\theta_1x_j-\theta_2)\\ \theta_3^{-1}(n-1)^{-1/2}\sum_{j=1}^{(n-1)}(x_{j+1}-\theta_1x_j-\theta_2)\end{pmatrix}.$$

Also

$$\Sigma(\theta)=\begin{bmatrix}\theta_3^{-1}\{\theta_3(1-\theta_1^2)^{-1}+\theta_2^2(1-\theta_1)^{-2}\} & \theta_2\theta_3^{-1}(1-\theta_1)^{-1}\\ \theta_3^{-1}(1-\theta_1)^{-1}2\theta_2 & \theta_3^{-1}\end{bmatrix}$$

which is non-singular. Hence the regular inverse exists in this case and one choice of G could be $\Sigma^{-1}(\theta)$. Note that

$$\Sigma^{-1}(\theta)=(1-\theta_1^2)\begin{bmatrix}1 & -\theta_2(1-\theta_1)^{-1}\\ -\theta_2(1-\theta_1)^{-1} & \theta_3(1-\theta_1^2)+\theta_2^2(1-\theta_1)^{-1}\end{bmatrix}.$$

Thus

$$Q_n(\theta)=\mathbf{W}_n'(\theta)\Sigma^{-1}(\theta)\mathbf{W}_n(\theta).$$

Under H_0, substituting any weakly $\sqrt{n}$ consistent estimate θ_{3n} for θ_3, the quadratic form can be written as

$$Q_n(\theta_{10},\theta_{20},\hat{\theta}_{3n})=[(1-\theta_{10}^2)/\hat{\theta}_{3n}(n-1)][A^2-(2\theta_{20})(1-\theta_{10})AB$$
$$+(\hat{\theta}_{3n}(1-\theta_{10}^2)^{-1}+\theta_{20}^2(1-\theta_{10})^{-2})B^2]$$

where θ_{10}, θ_{20} and $\hat{\theta}_{3n}$ are substituted in the expressions for A and B. This is asymptotically distributed as χ^2 with two degrees of freedom and could be used as a test criterion for H_0. Moreover for a sequence of H_n:$\theta=\theta_n=\theta_0+n^{-1/2}\delta_n$ where $\delta_n\to\delta$ as $n\to\infty$ and θ_{30} is some fixed value, $Q_n(\theta_{1n},\theta_{2n},\hat{\theta}_{3n})$ is asymptotically distributed under H_0 as a non-central χ^2 with two degrees of freedom and non-centrality parameter $(\delta_1,\delta_2)\Sigma(\theta_0)$ $(\delta_1,\delta_2)'$.

In this section we have discussed some large sample tests and their properties without concerning ourselves with their effficiency question. The tests discussed are simple, arise naturally and have wide applicability.

In the next section we discuss a class of optimal asymptotic tests essentially due to Neyman (1959).

5. OPTIMAL ASYMPTOTIC TESTS

5.1 Introduction

In this section we assume that the observations $Y_1, Y_2, \ldots$ are from an arbitrary stochastic process. Suppose the conditional probability density of Y_k given Y_1, $Y_2,\ldots,Y_{k-1}$ depends on the parameters ξ and $\theta=(\theta_1,\theta_2,\ldots,\theta_k)$, $k=1,2,\ldots$ We shall derive tests for composite hypothesis H:$\xi=\xi_0$ which are optimal in a sense to be defined.

In order to use the methods presented in this section, all that is required is essentially to obtain certain derivatives of probability densities involved and to evaluate certain expectations. Naturally, this ease in constructing the test and the breadth of applicability of the methods are gained at a price: the tests are asymptotic. Thus, in particular cases there is the problem of determining whether the size of the sample contemplated is large enough to ensure a reasonable approximation to the asymptotic properties of tests.

5.2 Preliminaries and Regularity Conditions

Let $\{Y_n, n \geqq 1\}$ be an arbitrary stochastic process and let $p_k(y; \xi, \theta)$ be the conditional probability density of Y_k given $\mathbf{Y}^{k-1}$ with respect to a σ-finite measure μ. Assume that this density function depends on the parameter (ξ, θ) where $\xi \in \Lambda \subset \mathbb{R}$ where Λ is a closed interval containing the origin as an interior point, $\theta = (\theta_1, \theta_2, \ldots, \theta_p) \in \Theta \subset \mathbb{R}^p$ and is jointly measurable in y and (ξ, θ). Assume further Θ is an open subset of $\mathbb{R}^p$. The hypothesis H specifies the value of ξ as ξ_0 which we assume to be zero without loss of generality.

Let $\mathscr{Y}_n$ be the sample space of values of Y^n. It is assumed that $\mathscr{Y}_n$ and μ do not depend on (ξ, θ). We shall also assume that the conditional sample space of Y_n given $\mathbf{Y}^{n-1}$ will not depend on (ξ, θ). It is further assumed that the density functions p_k satisfy some regularity conditions stated later in this section.

Let $\{w_n\}$ be a sequence of measurable subsets of $\mathscr{Y}_n$. The sequence $\{w_n\}$ will define an asymptotic test of H corresponding to the level of significance α $(0 < \alpha < 1)$ if

$$\lim_{n} P[\mathbf{Y}^n \in w_n | H] = \alpha \tag{62}$$

for all $\theta \in \Theta$.

Let $\{\xi_n\}$ be a sequence of alternatives converging to zero and Γ denote a certain class of sequences $\{\xi_n\}$. An asymptotic test sequence $\{w_n^0\}$ satisfying relation (62) is said to be *optimal* with respect to Γ if for any sequence $\{w_n\}$ satisfying relation (62),

$$\varliminf_{n \to \infty} \{P[\mathbf{Y}^n \in w_n^0 | \xi_n, \theta] - P[\mathbf{Y}^n \in w_n | \xi_n, \theta]\} \geqq 0. \tag{63}$$

Let $U_k = U_k(y_1, \ldots, y_k; \theta)$ be a measurable function of $(y_1, \ldots, y_k)$ such that its conditional expectation and variance given $\mathbf{Y}^{k-1} = \mathbf{y}^{k-1}$ are zero and unity respectively under H for all $(y_1, \ldots, y_{k-1})$ and $\theta \in \Theta$. Under some conditions on U_k, $n^{-1/2} \sum_{k=1}^{n} U_k = Z_n(\theta)$ has an asymptotic standard normal distribution under H. If $S(\alpha)$ is an arbitrary measurable set of the real line such that

$$\frac{1}{\sqrt{2\pi}}\int_{S(\alpha)} e^{-t^2/2}\,dt=\alpha \tag{64}$$

then

$$w_n=\left[n^{-1/2}\sum_{k=1}^{n} U_k\in S(\alpha)\right] \tag{65}$$

will constitute an asymptotic test of significance level α. In order to determine U_k and $S(\alpha)$ such that the test will be optimal within a certain class, it will be necessary to compute the asymptotic power of these tests. This can be done by ensuring the asymptotic normality of $Z_n(\theta)$. Further U_k might involve the nuisance parameter θ which is unknown. This brings out the practical difficulty in obtaining the test sequence (65). This difficulty can be avoided if one can substitute a reasonable estimate $\hat{\theta}_n$ for θ and still assert the asymptotic normality of Z_n under H as well as under the alternative sequence.

Our aim is to impose suitable conditions on $Z_n(\theta)$ such that (i) it is asymptotically normal both under the hypothesis H as well as under the alternatives and (ii) it will have the same limiting distribution even after substituting any locally $\sqrt{n}$ consistent estimate $\hat{\theta}_n$ and (iii) to obtain a test, optimal within a certain class of tests, for alternatives belonging to a certain class Γ.

Let the first partial derivative of U_k with respect to θ_i and the second partial derivative of U_k with respect to θ_i and θ_j exist and denote them by U_{ki} and U_{kij} respectively for $1\leqq i, j\leqq p$. Let

$$\psi_{kj}=\left[\frac{\partial \log p_k}{\partial \theta_j}\right]_{(0,\theta)} \tag{66}$$

and

$$\psi_{k\xi}=\left[\frac{\partial \log p_k}{\partial \xi}\right]_{(0,\theta)} \tag{67}$$

We shall assume that the following regularity conditions, hereafter referred to as Cramer's conditions, are satisfied.

(C1) For all (ξ, θ), $(y_1,\ldots, y_{k-1})$,

$$\int U_k p_k(\xi, \theta)\,d\mu \tag{68}$$

exists and is differentiable under the integral sign with respect to ξ and $\theta_1,\ldots, \theta_p$ at $(0, \theta)$.

(C2) For all (ξ, θ), $(y_1, \ldots, y_{k-1})$,

$$\int U_{kj}^s p_k(\xi, \theta)\, d\mu \tag{69}$$

exists and at $(0, \theta)$ it is continuous in ξ for $1 \leqq j \leqq p$ for $s = 1,2$.

(C3) For any θ_0, there exists a function $G_{\theta_0}(y) \geqq 0$ not depending on $y_1, y_2, \ldots, y_{k-1}$ and k such that for all $\xi \in \Lambda$ and for all $(y_1, y_2, \ldots, y_{k-1})$

$$\int G_{\theta_0}(y)\, p_k(y;\, \xi, \theta_0) d\mu < M < \infty \tag{70}$$

where M does not depend on $(y_1, \ldots, y_{k-1})$. Further

$$\sum_{i=1}^{\infty} \frac{\operatorname{var}[G_{\theta_0}(Y_i) | \xi, \theta_0]}{i^2} < \infty \tag{71}$$

and for all θ in some neighbourhood of θ_0 and for all k,

$$|U_{kij}| \leqq G_{\theta_0}, \quad 1 \leqq i, j \leqq p. \tag{72}$$

(C4) There exists a function $H_{\theta_0}(y) \geqq 0$ defined on the sample space of Y_k given $\mathbf{Y}^{k-1}$ such that

$$\int H_{\theta_0}(y)\, d\mu < \infty \tag{73}$$

and

$$U_k^2\, p_k(y;\, \xi, \theta_0) \leqq H_{\theta_0}(y) \tag{74}$$

for all ξ and k.

(C5) For $j = 1, 2, \ldots, p$ and for each $\theta \in \Theta$, the covariance

$$\lambda_{kj}(\theta) = \int U_k \psi_{kj}(y, \theta)\, p_k(y;\, 0, \theta)\, d\mu \tag{75}$$

of U_k and ψ_{kj} under H has partial derivatives with respect to $\theta_1, \theta_2, \ldots, \theta_p$ at least up to the second order.

(C6) The covariance matrix of $(\psi_{k\xi}, \psi_{k1}, \ldots, \psi_{kp})$ is non-singular for all k.

In addition to the above regularity conditions, we assume further that $p_k(y;\, \xi, \theta)$ is such that the functions $\psi_{k\xi}$, ψ_{kj}, $1 \leqq j \leqq p$ also satisfy Cramer's conditions similar to those satisfied by U_k's. Even though the conditions assumed above appear formidable, they are analogous to those assumed by Cramer in his work on the theory of maximum likelihood estimation for independent and identically distributed random variables.

Let $\mu_k(\xi_n)$ denote the conditional expectation of U_k given by $\mathbf{Y}^{k-1}$ when

ξ_n is the parameter. Conditions (C1) and (C4) imply that

$$E_{\xi_n}[U_k|\mathbf{Y}^{k-1}]=\mu_k(\xi_n)\xrightarrow{\text{a.s.}}\mu_k(0)=0 \tag{76}$$

and

$$E_{\xi_n}[U_k^2|\mathbf{Y}^{k-1}]\xrightarrow{\text{a.s.}}E_0[U_k^2|\mathbf{Y}^{k-1}]=1 \tag{77}$$

as $\xi_n \to 0$. The following central limit theorem can be proved using the results of Loeve (1963).

Theorem 5.1. *Let* $\xi_n \to 0$ *such that* $\mu_k(\xi_n)\sqrt{n}<K$ a.s. *for all* k *where* K *does not depend on* n. *Then, under the conditions* (C1)–(C5), *for every* $\theta \in \Theta$,

$$Z_n=n^{-1/2}\sum_{k=1}^{n}[U_k-\mu_k(\xi_n)]\xrightarrow{\mathscr{L}}N(0,1). \tag{78}$$

As a special case of this theorem, we have the following corollary.

Corollary 5.1. *If* $\xi_n \equiv 0$, *then*

$$Z_n=n^{-1/2}\sum_{k=1}^{n}U_k\xrightarrow{\mathscr{L}}N(0,1)$$

for every $\theta \in \Theta$.

Let $\mu_k'(\xi)$ denote the derivative of $\mu_k(\xi)$ with respect to ξ. In view of conditions (C1)–(C5) and since $\mu_k(\xi_n)\to 0$ a.s., it follows that

$$\mu_k(\xi_n)=(\xi_n-\xi_0)\mu_k'(\xi_0)+(\xi_n-\xi_0)o(1)$$

where $o(1)$ tends to zero almost surely uniformly in $Y_1, Y_2, \ldots, Y_{k-1}$ and k. Hence, it follows from Theorem 5.1, that

$$n^{-1/2}\sum_{k=1}^{n}U_k-n^{1/2}(\xi_n-\xi_0)\left(\frac{1}{n}\right)\sum_{k=1}^{n}\mu_k'(\xi_0)\xrightarrow{\mathscr{L}}N(0,1).$$

5.3 An Asymptotic Test of $H:\xi=0$

Let us consider the sequence $\{w_n\}$ of critical regions given by

$$w_n=[Z_n \in S(\alpha)] \tag{79}$$

to test H provided Z_n does not involve θ. If Z_n involves θ, we would like to substitute a locally $\sqrt{n}$ consistent estimate $\hat{\theta}_n$ for θ. We shall now derive the asymptotic distribution of $Z_n(\hat{\theta}_n)$. Let $\sigma_\xi^2(X)$ denote the variance of a random variable X under (ξ, θ).

Theorem 5.2. *Let $\xi_n \to 0$ such that $\sqrt{n}\xi_n = O(1)$. Further suppose that*

$$\sum_{k=1}^{\infty} \sigma_\xi^2(U_{kj})k^{-2} < \infty \tag{80}$$

for all (ξ, θ). Then

$$Z_n(\hat{\theta}_n) - Z_n(\theta) \xrightarrow{p} 0 \tag{81}$$

provided

$$\frac{1}{n}\sum_{k=1}^{n} E_0[U_k\psi_{kj}|\mathbf{Y}^{k-1}] = 0 \tag{82}$$

for $n = 1, 2, \ldots$.

Proof. Expanding $Z_n(\hat{\theta}_n)$ around a fixed point θ by using Taylor's expansion, it follows that

$$Z_n(\hat{\theta}_n) - Z_n(\theta) = \sum_{j=1}^{p} (\hat{\theta}_{jn} - \theta_j)\sqrt{n}\left(\frac{1}{n}\sum_{k=1}^{n} U_{kj}\right)$$
$$+ \frac{1}{2\sqrt{n}} \sum_{j=1}^{p} \sum_{l=1}^{p} n(\hat{\theta}_{jn} - \theta_j)(\hat{\theta}_{ln} - \theta_l)\left(\frac{1}{n}\sum_{k=1}^{n} U_{kjl}(\theta^*)\right)$$

where θ^* lies between $\hat{\theta}_n$ and θ i.e. $\|\theta^* - \theta\| \leqq \|\hat{\theta}_n - \theta\|$.

Recall that $\hat{\theta}_{jn}$ is a *locally $\sqrt{n}$ consistent* estimate of θ_j, if there exists a constant A_j such that

$$|\hat{\theta}_{jn} - \theta_j - A_j\xi|\sqrt{n}$$

is bounded in probability for all (ξ, θ) as $n \to \infty$. Since by assmption $\sqrt{n}\xi_n$ is bounded, it follows that $\sqrt{n}(\hat{\theta}_{jn} - \theta_j)$ is bounded in probability. (C3) implies that the second term on the right hand side of equation (83) tends to zero in probability.

In view of assumptions (C1) and (C2), it is easy to check that

$$E_0[U_k\psi_{kj}|\mathbf{Y}^{k-1}] = -E[U_{kj}|\mathbf{Y}^{k-1}] \text{ a.s.} \tag{84}$$

for every $(0, \theta)$. Further more (C2) and equation (80) imply that for every (ξ, θ)

$$\frac{1}{n}\sum_{k=1}^{n} U_{kj} \xrightarrow{\text{a.s.}} 0$$

and hence in probability for every $1 \leqq j \leqq p$ by results in Loéve (1963). Hence the first term on the right-hand side of (83) tends to zero by standard arguments. This completes the proof of the theorem.

In Theorem 5.2, equation (82) may be replaced by

$$E_{\xi_n, \theta}[U_k \psi_{kj} | \mathbf{Y}^{k-1}] = 0 \tag{85}$$

for all k, ξ_n and θ. Theorem 5.2 implies that $Z_n(\hat{\theta}_n)$ will have asymptotic standard normal distribution under $(0, \theta)$ for every θ provided (82) or (85) holds.

In order to satisfy (85) i.e. in order to make U_k's uncorrelated with ψ_{kj}, $1 \leqq j \leqq p$, we substract from U_k its linear regression on ψ_{kj}, $1 \leqq j \leqq p$, and obtain

$$G_k = U_k - b_1 \psi_{k1} - \ldots - b_p \psi_{kp} \tag{86}$$

where b_j is the conditional partial regression coefficient of G_k on ψ_{kj}, $1 \leqq j \leqq p$ given $\mathbf{Y}^{k-1}$. Under the assumption that ψ_{kj}, $1 \leqq j \leqq p$ and U_k satisfy Cramer's conditions, G_k does too. Let $\sigma_k^2(G_k, \theta)$ be the conditional variance of G_k under H given $\mathbf{Y}^{k-1}$. (C6) implies that $\sigma_k^2(G_k, \theta)$ is positive. Let

$$Z_n^G(\hat{\theta}_n) = n^{-1/2} \sum_{k=1}^{n} \frac{G_k(\hat{\theta}_n)}{\sigma_k(G_k, \hat{\theta}_n)}. \tag{87}$$

By the construction of the sequence G_k, it will follow that

$$Z_n^G(\hat{\theta}_n) \xrightarrow{\mathscr{L}} N(0, 1) \tag{88}$$

under H and

$$Z_n^G(\hat{\theta}_n) - \sqrt{n}\xi_n \cdot \frac{1}{n} \sum_{k=1}^{n} \mu_k'(0) \rightarrow N(0, 1) \tag{89}$$

under the alternative sequence $\{\xi_n\}$ satisfying the conditions $\sqrt{n}\xi_n = O(1)$ and $\xi_n \rightarrow 0$. Here $\mu_k'(0)$ is the conditional covariance of G_k and $\psi_{k\xi}$ under H given $\mathbf{Y}^{k-1}$.

We define a class $C_s(\alpha)$ of tests of size α as follows.

(i) We select a sequence $\{U_k\}$ of functions such that G_k defined by (86) has non-zero conditional variance $\sigma_k^2(G_k, \theta)$ and $E_0[G_k | \mathbf{Y}^{k-1}] = 0$ and satisfies the Cramer conditions.

(ii) We choose a set $S(\alpha)$ on the real line which has its indicator function continuous almost everywhere and which satisfies (64). In such an event, the sequence $\{w_n\}$ of critical regions given by

$$w_n = [n^{-1/2} \sum_{k=1}^{n} F_k(y_1, \ldots, y_k; \hat{\theta}_n) \in S(\alpha) \tag{90}$$

is an asymptotic test of size α where

$$F_k = \frac{G_k}{\sigma_k(G_k, \theta)}. \tag{91}$$

5.4 Optimal Tests of the Class $C_s(\alpha)$

We shall first derive optimal tests belonging to class $C_s(\alpha)$ when the alternative is one-sided, namely $\xi > 0$.

Let Γ_+ be the class of sequences $\{\xi_n\}$ such that $\xi_n > 0$, $\xi_n \to 0$ and $\sqrt{n}\xi_n$ remains bounded. Further assume that $p_k(y; \xi, \theta)$ is sufficiently regular to make $\psi_{k\xi}$ and ψ_{kj}, $1 \leqq j \leqq p$ satisfy Cramer conditions. Let a_{jk}^0 be the value of a_{jk}, $1 \leqq j \leqq p$ for which the conditional variance of

$$\psi_{k\xi} - \sum_{j=1}^{p} a_{jk}\psi_{kj} \tag{92}$$

given $\mathbf{Y}^{k-1}$ is minimum and $\sigma_{k0}^{*2}(\theta)$ be the minimum conditional variance. Since the covariance matrix of

$$(\psi_{k\xi}, \psi_{k1}, \ldots, \psi_{kp})$$

is assumed non-singular by assumption (C6), there does not exist a linear relationship between $\psi_{k\xi}$ and ψ_{kj}'s almost surely and hence a_{kj}^0, $1 \leqq j \leqq p$ are well defined and $\sigma_{k0}^{*2}(\theta) > 0$. Let

$$F_k^* = \left\{\psi_{k\xi} - \sum_{j=1}^{p} a_{kj}^0 \psi_{kj}\right\} \sigma_{k0}^{*-1}(\theta). \tag{93}$$

Since $\psi_{k\xi}$ and ψ_{kj}, $1 \leqq j \leqq p$ satisfy Cramer conditions, F_k^* does too. Furthermore

$$E_0[F_k^* | \mathbf{Y}^{k-1}] = 0 \text{ a.s.}, \tag{94}$$

$$E_0[F_k^{*2} | \mathbf{Y}^{k-1}] = 1 \text{ a.s.}, \tag{95}$$

and

$$E_0(F_k^* \psi_{kj} | \mathbf{Y}^{k-1}] = 0 \text{ a.s. for } 1 \leqq j \leqq p. \tag{96}$$

We have the following theorem.

Theorem 5.3. *If the estimates $\hat{\theta}_{jn}$ of θ_j are locally root n consistent, then the test of H, given by the critical region*

$$[Z_n^* = n^{-1/2} \sum_{k=1}^{n} F_k^*(y_1, \ldots, y_k; \hat{\theta}_n) > \gamma(\alpha)], \tag{97}$$

where $\gamma(\alpha)$ is defined by

$$\frac{1}{\sqrt{2\pi}}\int_{\gamma(\alpha)}^{\infty} \exp\,(-t^2/2)\mathrm{d}t=\alpha, \tag{98}$$

is an optimal test of the class $C_s(\alpha)$ with reference to the sequence of alternatives belonging to the family Γ_+.

Proof. The sequence of critical regions given by (97) belongs to the class $C_s(\alpha)$. We shall now prove that the region given by (97) is optimal in the sense of relation (63). Since for every test sequence $w_n=[Z_n\in S(\alpha)]$, we have an equivalent test sequence $w_n'=[-Z_n\in -S(\alpha)]$ the subclass of tests which belong to $C_s(\alpha)$ and for which $\lim 1/n\Sigma_{i=1}^n\mu_i'(0)\geqq 0$ is essentially complete for $C_s(\alpha)$. Hence, without loss of generality, we can assume that

$$\lim \frac{1}{n}\sum_{i=1}^{n}\mu_i'(0)\geqq 0.$$

Let $S^*(\alpha)=(\gamma(\alpha),\ \infty)$. It can be shown as in Neyman (1959) that it is always advantageous to use $S^*(\alpha)$ over any other set $S(\alpha)$. We shall now show that the test with F_k^* and $S^*(\alpha)$ is better than the test with any other F_k and $S^*(\alpha)$ in the sense of (63). Clearly,

$$\mu_k'(0)=E_0[F_k\psi_{k\xi}|\mathbf{Y}^{k-1}] \tag{99}$$

and let

$$\mu_k^*(0)=E_0[F_k^*\psi_{k\xi}|\mathbf{Y}^{k-1}]. \tag{100}$$

By Schwartz inequality for conditional expectations,

$$\begin{aligned}
\zeta_n&=\frac{1}{n}\sum_{k=1}^{n}\mu_k'(0)\\
&=\frac{1}{n}\sum_{k=1}^{n}E_0[F_k\psi_{k\xi}|\mathbf{Y}^{k-1}]\\
&=\frac{1}{n}\sum_{k=1}^{n}\sigma_{k0}^*E_0[F_kF_k^*|\mathbf{Y}^{k-1}]\\
&\leqq\frac{1}{n}\sum_{k=1}^{n}\sigma_{k0}^*E_0^{1/2}[F_k^2|\mathbf{Y}^{k-1}]\,E_0^{1/2}[F_k^{*2}|\mathbf{Y}^{k-1}]\\
&=\frac{1}{n}\sum_{k=1}^{n}\sigma_{k0}^*\ \text{(by 95)}\\
&=\frac{1}{n}\sum_{k=1}^{n}E_0[F_k^*\psi_{k\xi}|\mathbf{Y}^{k-1}]\\
&=\zeta_n^*.
\end{aligned} \tag{101}$$

Let $\eta_n = \sqrt{n}\xi_n\zeta_n$ and $\eta_n^* = \sqrt{n}\xi_n\zeta_n^*$. Let Z_n and Z_n^* be the appropriate variables corresponding to F_k and F_k^*. Relation (90) implies that

$$Z_n - \eta_n \xrightarrow{\mathscr{L}} N(0, 1) \tag{102}$$

and

$$Z_n^* - \eta_n^* \xrightarrow{\mathscr{L}} N(0, 1),$$

under the alternative sequence $\{\xi_n\}$. But $\eta_n^* \geqq \eta_n$ a.s. Hence

$$\lim_{n\to\infty} P_{\xi_n,\theta}[Z \leqq x + \eta_n] = \lim_{n\to\infty} P[Z_n^* \leqq x + \eta_n^*]$$

$$\geqq \lim_{n\to\infty} P[Z_n^* \leqq x + \eta_n]. \tag{103}$$

Hence Z_n^* is stochastically larger than Z_n asymptotically. Therefore the critical region defined by (97) gives an optimal test in the class $C_s(\alpha)$. This completes the proof of the Theorem.

Definition 5.1. An asymptotic test $\{w_n\}$ of size $\alpha(0<\alpha<1)$ is said to be *centralized* if for all alternatives $\{\xi_n\}$ tending to zero,

$$\lim_{n\to\infty} \{P[\mathbf{Y}^n \in w_n|(\xi_n, \theta)] - P[\mathbf{Y}^n \in w_n|(0, \theta)]\} \geqq 0. \tag{104}$$

Let $\{\xi_n\}$ be such that $\xi_n \to 0$ and $\sqrt{n}\xi_n$ is bounded. Let Γ be the class of all such sequences. This class corresponds to two-sided alternatives while Γ_+ corresponds to one sided alternatives. The following theorem holds.

Theorem 5.4. *If the estimates $\hat{\theta}_{jn}$, $1 \leqq j \leqq p$ of θ_j, $1 \leqq j \leqq p$ are locally root-n consistent, then the test of $H: \xi = 0$ given by the critical region*

$$|Z_n^*| > \gamma(\alpha/2) \tag{105}$$

is an optimal unbiased test of the class $C_s(\alpha)$ with reference to the alternative sequences of the class Γ.

We omit the proof as it is similar to that of Theorem 5.3.

It should be pointed out that the test described above for a one-sided hypothesis may not be consistent in general in the sense that

$$\lim_{n\to\infty} P[\text{Reject } H_0|\theta, \xi] = a \neq 1.$$

However, under additional regularity conditions, a set of necessary and sufficient conditions for these tests to be consistent can be given.

Theorem 5.5. *Let $Z_n = n^{-1/2}\Sigma_{k=1}^{n} U_k(\mathbf{Y}^k; \hat{\theta}_n)$ be a test statistic with rejection region $(\eta(\alpha), \infty)$ of size α, where U_k, $k \geqq 1$ satisfy the regularity conditions* (Cl)–(C6) *described earlier. Further, suppose that*

(i) $\sqrt{n}(\hat{\theta}_n - \theta^*) = 0_{P_{\theta,\xi}}(1)$ *for some* $\theta^*(\xi, \theta) \in \Theta$;

(ii) *for every $\hat{\theta}$ in an open neighbourhood V of θ^**

$$\frac{1}{n}\sum_{k=1}^{n} U_{kjl}(\mathbf{Y}_k; \hat{\theta}) = 0_{P_{\theta,\xi}}(1),\ 1 \leqq j,\ l \leqq p;$$

(iii) *for* $1 \leqq j \leqq p$

$$\frac{1}{n}\sum_{k=1}^{n} U_{kjl}(\mathbf{Y}_k; \theta^*) = 0_{P_{\theta,\xi}}(1);$$

(iv) *the sequence of random variables*

$$n^{1/2}\sum_{k=1}^{n} \{U_k(\mathbf{Y}_k; \theta^*) - E_{\theta,\xi}[U_k(\mathbf{Y}_k; \theta^*)]\}$$

is asymptotically normal with mean zero and finite variance when the true parameter values are θ and ξ and

(v) *finally*

$$\lim_{n\to\alpha} n^{-1}\sum_{k=1}^{n} E_{\theta,\xi}[U_k(\mathbf{Y}_k; \theta^*)]$$

exists and is equal to τ.

Then a necessary and sufficient condition for the test to be consistent at the alternative (θ, ξ) is $\tau > 0$.

Proof. Clearly

$$Z_n(\hat{\theta}_n) - Z_n(\theta^*) = \sum_{j=1}^{p} (\hat{\theta}_{jn} - \theta_j^*)\sqrt{n}\left\{\frac{1}{n}\sum_{k=1}^{n} U_{kj}(\mathbf{Y}_k; \theta^*)\right\}$$
$$+ \frac{1}{2\sqrt{n}}\sum_{j=1}^{p}\sum_{l=1}^{p} n(\hat{\theta}_{jn} - \theta_j^*)(\hat{\theta}_{ln} - \theta_l^*)\left\{\frac{1}{n}\sum_{k=1}^{n} U_{kj}(\mathbf{Y}_k; \tilde{\theta})\right\}$$

where $\tilde{\theta}_i$ lies between $\hat{\theta}_{in}$ and θ_i^*, $1 \leqq i \leqq p$. Conditions (i)–(iii) of the theorem imply that

$$Z_n(\hat{\theta}_n) - Z_n(\theta^*) - 0_{P_{\theta,\xi}}(1) \xrightarrow{p} 0$$

under $P_{\theta,\xi}$. It is sufficient to prove that for every constant γ,

$$P_{\theta,\xi}(Z_n(\theta^*) + 0_{P_{\theta,\xi}}(1) > \gamma) \to 1$$

as $n \to \infty$ if and only if $\tau > 0$. Suppose $\tau > 0$; then condition (iv) implies that $Z_n(\theta^*) - \sqrt{n}\tau$ is asymptotically normal with mean 0 and finite variance. Hence

$$P_{\theta,\xi}[\{Z_n(\theta^*) - \sqrt{n\tau}\} + \{0_{P_{\theta,\xi}}(1) + \sqrt{n}\,\tau\} > \gamma] \to 1 \text{ as } n \to \infty$$

since $\tau > 0$. It is easy to see that for every γ

$$\lim_{n\to\infty} P_{\theta,\xi}[Z_n(\theta^*) + 0_{P_{\theta,\xi}}(1) > \gamma]$$

exists and is equal to zero if $\tau < 0$ and is less than one if $\tau = 0$. This completes the proof of the theorem.

5.5 Some Remarks and Examples

Neyman (1959) defined an estimator $\hat{\theta}_{jn}$ to be a *locally* $\sqrt{n}$ *consistent* estimator of θ if

$$\sqrt{n}|\hat{\theta}_{jn} - \theta_j - A_j \xi| \tag{106}$$

is bounded in probability for all$(\xi, \dot{\theta})$, as $n \to \infty$ where A_j is a constant. The theory developed above depends on the existence of such estimates of nuisance parameter θ. It is easy to see that the theory developed here is true even if $\hat{\theta}_{jn}$ are such that A_j's are bounded randon variables in (106) instead of constants. Such estimators are called *generalized locally* $\sqrt{n}$ *consistent.*

We shall now give a method of constructing such generalized locally $\sqrt{n}$ consistent estimators. Let $\hat{\theta}_k(\mathbf{Y}^k)$ be an estimate of θ based on the conditional distribution of Y_k given $\mathbf{Y}^{k-1}$ such that

$$E(\hat{\theta}_k|\mathbf{Y}^{k-1}) = \theta + A_k(\mathbf{Y}^{k-1};\theta, \xi)\, g(\xi) \text{ a.s.} \tag{107}$$

where $A_k(\mathbf{Y}^{k-1}\colon \theta, \xi)$ is almost surely bounded in $\mathbf{Y}^{k-1}$, θ, ξ and k and $g(\xi)$ is any function vanishing at $\xi = 0$ and is such that $\sqrt{n}\, g(\xi_n)$ is bounded as $n \to \infty$. Let

$$\sigma_k^2(\xi) = E[\hat{\theta}_k - \theta - A_k(\mathbf{Y}^{k-1}, \theta, \xi) g(\xi)]^2, \tag{108}$$

$$\bar{\hat{\theta}}_n = \frac{1}{n} \sum_{k=1}^{n} \hat{\theta}_k, \tag{109}$$

and

$$\bar{\sigma}_n^2(\xi) = \frac{1}{n} \sum_{k=1}^{n} \sigma_k^2(\xi). \tag{110}$$

Since $g(0) = 0$, it is easy to see that $\hat{\theta}_k$ is an unbiased estimate of θ under H.

It follows from extended Bienayme inequality that

$$P\left[\sqrt{n}|\bar{\theta}_n-\theta-n^{-1}\sum_{k=1}^{n} A_k\, g(\xi)|>M\right]\leqq\frac{\delta_n^2(\xi)}{M^2}. \tag{111}$$

Relation (111) implies that sufficient conditions for $\bar{\theta}_n$ to be generalized locally $\sqrt{n}$ consistent estimate of θ under the alternative sequence $\{\xi_n\}$ is

(i) $$\bar{\sigma}_n^2(0)<\sigma^2<\infty \tag{112}$$

(ii) $$\bar{\sigma}_n^2(\xi) \text{ is uniformly continuous in } \xi. \tag{113}$$

The following example illustrates the theory developed in this section.

Example 5.1. Consider the following stochastic model for epidemics. Here the sequence of random variables are the times $t_0=0\leqq T_1\leqq T_2\leqq\ldots\leqq T_N$ at which the individuals of a population of size $(N+1)$ become infected. The conditional probability density function of T_k given $T_1=t_1$, $T_2=t_2,\ldots,$ $T_{k-1}=t_{k-1}$ is assumed to be

$$p(t|t_1,\ldots,t_{k-1};\lambda,\beta)$$

$$=(N-k+1)(\lambda+\beta G_k(t))\exp\left[-(N-k+1)\int_{t_{k-1}}^{t_k}(\lambda+\beta G_k(\tau))d\tau\right] \tag{114}$$

where

$$\lambda>0,\ \beta\geqq 0,\ G_k(\tau)=\sum_{i-0}^{k-1} f(t_i,\tau),$$

f being a non-negative measurable function bounded by a constant c such that $f(t,\tau)=0$ for $t>\tau$. $f(t,\tau)$ may be interpreted as the amount of pathogenic material contributed at time $\tau>t$ by an individual infected at time t.

We develop the asymptotic theory under the following assumptions

(i) N and the number of observed variables n vary in such a way that $n\leqq N$, $n\to\infty$, $n/N=O(1)$

(ii) $(N-k+1)\displaystyle\int_{t_{k-1}}^{t} G_k(\tau)d\tau$ remains bounded uniformly in k.

The second condition ensures that the interval between two consecutive infections does not become degenerate as $k\to\infty$. Hence the epidemic will not become "explosive" even if $\beta>0$.

The hypothesis to be tested is $H{:}\beta=0$, λ being unspecified. It can be seen that the theory developed is valid even if $\xi_0=0$ is an end point of Λ

provided we take for $\varphi_{k\xi}$ the one-sided derivative with respect to ξ at that point.

The range of T_k given $T_1 = t_1, \ldots, T_{k-1} = t_{k-1}$ does not depend on (λ, β), for all k. It can be checked that

$$\varphi_{k\beta}(t) = \frac{G_k(t)}{\lambda} - (N-k+1)\int_{t_{k-1}}^{t} G_k(\tau)\mathrm{d}\tau$$

and

$$\varphi_{k\lambda}(t) = \frac{1}{\lambda} - (N-k+1)(t - t_{k+1}).$$

Under Condition (ii), $\psi_{k\beta}$ and $\psi_{k\lambda}$ satisfy Cramer conditions (C1)–(C5) since f is bounded. Under H, the conditional covariance matrix of $\psi_{k\lambda}$ and $\psi_{k\beta}$ given $t_1, \ldots, t_{k-1}$ is

$$\begin{bmatrix} \dfrac{1}{\lambda^2} & \dfrac{1}{\lambda^2} E(G_k(t)|\mathbf{T}^{k-1}) \\ \dfrac{1}{\lambda^2} E(G_k(t)|\mathbf{T}^{k-1}) & \dfrac{1}{\lambda^2} E(G_k^2(t)|\mathbf{T}^{k-1}) \end{bmatrix}.$$

Hence the test criterion of the optimal test is

$$Z_n^* = n^{-1/2} \sum_{k=1}^{n} \frac{\psi_{k\beta}(t) - E(G_k(t)|\mathbf{T}^{k-1})\psi_{k\lambda}(t)}{\frac{1}{\lambda}(\mathrm{var}(G_k(t)|\mathbf{T}^{k-1}))^{1/2}}. \tag{115}$$

It can be shown that

$$\frac{1}{n}\sum_{k=1}^{n} (N-k+1)(t_k - t_{k-1}) \tag{116}$$

is generalized locally $\sqrt{n}$ consistent estimate of $1/\lambda$ by using conditions (112) and (113). Substituting the expression in (116) for $1/\lambda$ in (115), we obtain the optimal asymptotic test given in (97).

6. FURTHER TOPICS

In Section 2, we have given regularity conditions involving the existence of third-order derivative of the log-likelihood function for exhibiting a consistent and asymptotically normal likelihood equation estimator. However, one can weaken these conditions further to obtain consistent and asymptotically normal maximum likelihood estimator. We refer the reader to Prakasa Rao (1974) for further details. Recently Crowder (1976) obtained another set of sufficient conditions for consistency and asymptotic normality of MLE for dependent observations by using central limit theorems for martingales due to Brown (1971) and Scott (1973). He considered the case when some of the parameters are "transient" in the sense that the accumulated information on them from the sample does not increase indefinitely in which case interest lies in estimating the other parameters consistently. We mention that it is not enough to know that the estimator is asymptotically normal. It is very important to know the rate of convergence for practical use. Results of this type for special classes of processes like stationary Markov processes and for independent not identically distributed observations have recently been obtained in Prakasa Rao (1973, 1975).

In Section 2.5, it is possible to derive asymptotic normality of the MLE in the finite state space case assuming only the existence of continuous partial derivatives of second order instead of third order. The technique is to establish the asymptotic normality of the random vector

$$n^{-1/2}[N_1(n)-n\gamma_1(\theta_0)), \ldots, (N_m(n)-n\gamma_m(\theta_0))]$$

as in Ranneby (1975b) and then follow Rao (1973), p 359. Tests for general composite hypotheses of the form $H{:}g_i(\theta)-0$, $1 \leqq i \leqq r (r \leqq p)$ can also be obtained by methods analogous to those given in Section 4. Details may be found in Prakasa Rao (1974).

Even though we studied optimal asymptotic tests of hypotheses involving a single parameter in Section 5, the theory can be extended to the testing of multiparameter hypotheses by methods similar to those developed in Bhat and Nagnur (1965) and Buhler and Puri (1966).

BIBLIOGRAPHICAL NOTES

Statistical inference for discrete parameter stochastic processes has been investigated earlier by Wald (1948), Silvey (1961) and more recently by Bar-Shalom (1971), Bradley (1973), Bhat (1974), Prasad (1973), Weiss (1971, 1973) and Rao (1966) among others. Billingsley (1961), Mihoc and Crain (1972), Roussas (1965, 1968), Prakasa Rao (1972, 1973) and Sarma (1968, 1973) studied the case of discrete time stationary Markov processes. Prasad (1973) studied inference problems when the process is stationary m-dependent or stationary ϕ-mixing. Grandell (1976) discussed estimation problems for the special class of doubly stochastic Poisson processes. Most of the papers cited above use strong regularity conditions on the distributions involved as the method of proof for studying the asymptotic behaviour consists in expanding the derivative of log-likelihood function by Taylor's expansion. An alternative approach for studying the asymptotic behaviour of MLE is through the study of the likelihood ratio process as a process in the parameter. This approach is discussed at first by Prakasa Rao (1968) and later by Ibragimov and Khas'minskii (1972). Inagaki and Ogata (1975b, 1976) studied the MLE for Markov dependent observations through this method. Ibragimov and Khas'minskii (1975a, b) studied the behaviour of MLE and Bayes estimators for independent but not necessarily identically distributed observations. A different approach to these problems is due to Roussas (1976) and Phillippou (1977a, b). Roussas (1972) discussed the case of discrete time stationary Markov processes extensively by using contiguity methods.

Sections 2.1–2.4 contain the work of Basawa, *et al.* (1976). Results in Section 2.5 are due to Ranneby (175b). Section 2.6 contains results and examples due to Prasad (1973). Section 3.1 discusses the work of Basawa and Scott (1977). Theory of large sample tests in Section 4 is due to Prakasa Rao (1974) and it extends the work of Sarma (1968, 1973) for discrete-time stationary Markov processes and of Rao (1948) and Wald (1943) for i.i.d. observations. Example 4.1 is due to Sarma (1973). Section 5 contains the work of Bhat and Kulkarni (1972). Example 5.1 uses a model developed by Yang (1968).

The tests described in Section 5 were first studied by Neyman (1959) in the i.i.d. case. Neyman and Scott (1965) applied their results for studies on random cloud seeding experiments. Neyman's theory has been extended by Buhler and Puri (1966) and Bartoo and Puri (1967) for independent but not necessarily identically distributed observations and by Bhat and Nagnur (1965) for testing a vector parameter hypothesis with different optimality conditions. Ray (1973) gave an example of an inconsistent $C(\alpha)$ test. Foutz (1974) obtained necessary and sufficient conditions for the consistency of $C(\alpha)$ tests for vector parameter when the observations are independent but not necessarily identically distributed.

COMPLEMENTS

1. Let $\{X_n, n \geqq 0\}$ be a stochastic process defined by the one step autoregression given by

$$X_n = \theta X_{n-1} + V_n$$

where $\theta \in \Theta$, a bounded non-degenerate interval including the origin, and V_n's independent of each other with standard normal distribution, V_n being independent of X_{n-1}. Show that the likelihood equation estimator of θ is consistent.

(Rao 1966; Basawa *et al.*, 1976)

2. Let $\{X_n, n \geqq 0\}$ be a stochastic process defined by

$$X_n = \alpha\ X_{n-1} + V_n$$

where $|\alpha| < 1$ and α is known and the sequence V_n is independent with unknown mean μ and variance 1. Show that the likelihood equation estimator of μ is strongly consistent and asymptotically normal.

(Prasad, 1973)

3. Let $\{X_n, n \geqq 1\}$ be a sequence such that

(i) $E(X_n) = \mu$, n odd, or $2\mu, n$ even,
(ii) $\text{cov}(X_n, X_m) = C$ for $n \neq m$
(iii) $\text{var}(X_n) = 2C$ for all n.

Suppose that μ is unknown, C is known, and the joint density of $(X_1, \ldots, X_n)$ is $1/(2\pi)^{n/2}(\det \Gamma_n)^{-1/2} \exp\{-\frac{1}{2}(x^{(n)} - \varepsilon^{(n)})'\Gamma_n^{-1}(x^{(n)} - \varepsilon^{(n)})\}$ where Γ_n is the variance–covariance matrix of $X^{(n)}$, $x^{(n)} = (x_1, \ldots, x_n)$ and $\varepsilon^{(n)'} = (E(X_1), \ldots, E(X_n))$. Show that the maximum likelihood estimator is consistent and asymptotically normal.

(Prasad, 1973)

4. Suppose the finite-dimensional distribution of the process $\{\xi_k\}$ is defined by the multivariate exponential distribution

$$\begin{aligned} &P[\xi_1 > x_1, \xi_2 > x_2, \ldots, \xi_l > x_l] \\ &= \exp\Big\{ -\sum_i \lambda_i x_i - \sum_{i<j} \lambda_{ij} \max(x_i, x_j) \\ &\quad - \sum_{i<j<k} \lambda_{ijk} \max(x_i, x_j, x_k) \\ &\quad \vdots \\ &\quad -\lambda_{123\ldots l} \max(x_1, x_2, x_3, \ldots, x_l)\Big\} \end{aligned}$$

where $\lambda_i, \lambda_{ij}, \ldots$ etc. are such that the process is m-dependent. Show that the MLE is strongly consistent and asymptotically normal.

(Marshal and Olkin, 1967; Ranneby, 1975)

5. Let $\{\zeta_i, i \geqq 0\}$ be independent random variables with stable distribution with characteristic exponent α, $1 < \alpha \leqq 2$ and with mean zero. Let $a_i > 0$, be such that

(i) $$\sum_{i=0}^{\infty} a_i^{\alpha} < \infty$$

(ii) $$\sum_{i=0}^{\infty} i a_i^{(\varepsilon + \alpha/(1+\alpha))} < \infty \quad \text{for some } \varepsilon > 0$$

(iii) $$\sum_s \alpha(s) < \infty \quad \text{where} \quad \alpha(s) \cong \sum_{i=s+1}^{\infty} i a_i^{2/(1+\alpha)}.$$

Define

$$\xi_k = b^{-1} \sum_{i=0}^{\infty} a_i \zeta_{k-i} + \theta$$

where $b=(\Sigma_{i=0}^{\infty} a_i^{\alpha})^{1/\alpha}$. Show that the process $\{\xi_k\}$ is strong mixing with α-function defined by (iii) and that the MLE of θ is strongly consistent and asymptotically normal.

(Chanda, 1974; Ranneby, 1975)

6. Show that it is sufficient to assume existence of moments of order two of the derivatives $g_u(\xi_1;\ \theta_0)$, $1 \leqq u \leqq p$ for the validity of Theorem 2.5 if the stationary process $\{\xi_k\}$ is ϕ-mixing.

(Ibragimov and Linnik, 1971)

7. Give a detailed proof of Theorem 2.3.

8. State and prove Theorem 2.4 assuming only the existence of continuous partial derivatives of the first order.

(Ranneby, 1975; Rao, 1973)

9. A problem in astronomy is to test whether galaxies are homogeneous with respect to the frequency of occurrence of supernovae. Assume that, if galaxies are homogeneous the number X of supernovae has a Poisson distribution and if they are not, the number X_i of supernovae observed during controlled time t_i in a randomly selected ith galaxy is a mixture of Poisson variables and its distribution is given by

$$P(X_i=K_i)=\frac{(\lambda_0 t_i)^{K_i}}{K_i!}\int_a^b \exp\{-[\lambda_0 t_i e^{x\xi^{1/2}} - K_i x\xi^{1/2}]\} \mathrm{d}F(x),$$

$$K_i=0, 1, 2, \ldots; i=1, \ldots, n.$$

where

n = number of galaxies observed,
$\log \lambda_0$ = average number of frequencies of supernovae per unit time,
ξ = variance of the logarithms of frequences of supernovae over all galaxies,

with $\int_a^b x\mathrm{d}F(x)=\int_a^b x^3\mathrm{d}F(x)=0$; $\int_a^b x^2\mathrm{d}F(x)=\int_a^b \mathrm{d}F(x)=1$ $a<b$ being two arbitrary but finite constants. The t_i are such that $n^{-1}\Sigma_{i=1}^n t_i$ has a finite positive limit as $n\to\infty$ and max $t_i<c$ for some constant $c>0$. Obtain the optimal test for testing the hypothesis H_0:$\xi=0$.

(Bartoo and Puri, 1967)

10. Extend the results of Section 2 to the multiparameter case.

11. Extend standard statistical procedures like Students t-test and F-test for one-way analysis variance for stationary strong mixing processes.

(Norman, 1971)

12. Consider the standard "regression" model where the Y_t's are independent $N(\theta_0+\phi_0 x_t, 1)$ the x_i's being given.

(i) If $x_t=t^{-\alpha}$ for some $\alpha>1/2$, show that $\hat{\theta}_n$ is mean-square consistent but $\hat{\phi}_n$ is not even consistent in probability and that ϕ is transient.

(ii) If $x_t=t^{-\alpha}$ for some $0<\alpha\leqq 1/2$, show that $\hat{\phi}_n$ is mean sqaure consistent and $\hat{\theta}_n$ is consistent in probability but ϕ is not transient.

(iii) If $x_t = 1 + t^{-\alpha}$, $\alpha > 1/2$, show that both $\hat{\theta}_n$ and $\hat{\phi}_n$ are not consistent and θ and ϕ are not transient.

Show that in all these cases $(\hat{\theta}_n, \hat{\phi}_n)$ is asymptotically normal with mean (θ_0, ϕ_0) and appropriate variance covariance matrix.

(Crowder, 1976)

13. Suppose X_{ij}, $1 \leqq j \leqq n_i$ are independent $N(\theta, \phi_1^2)$, $1 \leqq i \leqq r$, the n_i's are fixed and $r \to \infty$. Show that $\hat{\theta}_n$ is consistent in probability, where $n = \Sigma_{i=1}^{n} n_i$. If X_{ij}, $1 \leqq j \leqq n_i$ are independent $N(\phi_i, \theta^2)$, $1 \leqq i \leqq r$, show that $\hat{\theta}_n$ is not consistent.

(Crowder, 1976; Neyman and Scott, 1948)

14. Let $\{X_n, n \geqq 1\}$ be an m-dependent stationary Gaussian process with mean θ. Obtain a modification of t-test for testing the mean θ. Exhibit a test for the process with the property that if the observations are independent then this test requires asymptotically mu_α^2 more observations than the ordinary t-test where α is the size of the test and u_α is the upper α-point of the standard normal distributions.

(Albers, 1976)

Chapter 8

Large Sample Theory for Continuous Parameter Stochastic Processes

1. INTRODUCTION

In the previous chapter, large sample theory for discrete parameter stochastic processes was discussed. Here we take up the study of continuous parameter processes. Two approaches have been used up to now to study asymptotic theory of inference for continuous parameter processes. The first method consists of reduction of a continuous record of a process to a countable set of observable coordinates and then using the methods developed in the previous chapter. The second approach is to study the inference aspects using martingale theory. Continuous-time Markov processes are discussed in Chapter 5 and point processes in Chapter 6. Inference problems for diffusion processes will be studied in the next chapter.

2. OBSERVABLE COORDINATES

In contrast to the discrete parameter stochastic processes, here we have the additional problems of determining the right properties of the sample paths to be observed, finding a way of representing these observations and seeing whether the probability structure can be determined from these observations. In practice, we cannot observe the function X_t for all t such that $a \leqq t \leqq b$. Such precision can never be realized. We can only observe a finite number of functionals of the process. In general, we could have at our disposal a finite number of observables of the form

$$Z_v = \int_a^b X_t \phi_v(t) \mathrm{d}t, \quad v \geqq 1$$

where the integrals are suitably defined stochastic integrals. We may choose the system $\{\phi_v(t), v=1, 2, \ldots\}$ so that it is complete in the space of sample paths of the process. We then have a complete knowledge of the system. Another possibility is to take $\{Z_v\}$ as measurements $\{X_{t_v}\}$ at the times $\{t_v\}$ which is an everywhere dense set in the index set T of the process. This would give complete information if the sample paths are known to be continuous or stepwise continuous. The $\{Z_v\}$ are called *observable coordinates* of the process. We shall see that it is important to choose these coordinates in a way that facilitates the construction of estimates, test functions, etc.

Suppose $X(t)$ is a Gaussian process which is observed in a finite time interval $T=[a, b]$. The process is continuous in the mean with mean function $m(t)$ and covariance function $r(s, t)$. It can be shown that

$$X(t) = m(t) + \sum_{v=1}^{\infty} \lambda_v^{1/2} Y_v \phi_v(t)$$

with convergence in the mean for every $t \in T$ where λ_v, ϕ_v are respectively eigenvalues and eigenfunctions of the linear homogeneous integral equation

$$\lambda \phi(t) = \int_T r(t, s) \phi(s) \mathrm{d}s, \quad t \in T$$

and Y_v, $v \geqq 1$ are independent standard normal random variables. This can be seen from Karhunen (1947). We can take as observables the quantities

$$Z_v = \int_T X(t) \phi_v(t) \mathrm{d}t = a_v + \lambda_v^{1/2} Y_v$$

where

$$a_v = \int_T m(t) \phi_v(t) \mathrm{d}t.$$

Suppose we consider the solution of the stochastic differential equation

$$\mathrm{d}X(t) = \beta X(t) \mathrm{d}t + \mathrm{d}W(t), \; 0 \leqq t \leqq 1, \; X(0) = 0.$$

(See Appendix 2 for the definition of a stochastic differential equation). It has been shown that the solution of this differential equation is a stationary Markov process with continuous sample paths almost surely. Let $\{t_n\}$ be a denumerable set of points everywhere dense in $T=[0, 1]$. Then, with probability one, the path is completely specified when $X(t_n)=Z_n$ are known for all n.

Another important class of processes which falls into this set up is the class of point processes. Since the natural sample space of these processes consists of step functions, one can use the following system of coordinates. If a sample path has the form

$$X(t)=\begin{cases} x_0 & \text{for } a\leqq t\leqq t_1 \\ x_i & \text{for } t_i<t\leqq t_{i+1}, \quad i=1, \quad 2, \ldots, n-1 \\ x_n & \text{for } t_n<t\leqq b \end{cases}$$

where $n, t_1, \ldots, t_n, x_0, \ldots, x_n$ are random variables, we use as coordinates the sequence $n; x_0, t_1, x_1, \ldots, t_n, x_n$. In order to get the coordinates in a form similar to the earlier examples, we add an infinite sequence of real numbers. We assign these a simple probability distribution, for example that they are independent of each other and of the preceeding coordinates each having a normal distribution with mean 0 and variance 1. We assume, of course, that $P(n<\infty)=1$.

3. THE GENERAL PROBLEM

Consider a family of probability measures $\{P_\alpha:\alpha\in I\}$ on the space of observables. We saw earlier that we must differentiate between the set of sample paths and the set of observables and that in general the relation between the two probability measures so obtained may not be simple in the sense that it is not always obvious which quantities should be taken as observables. Suppose that this problem has been settled.

Let us first suppose that $I=\{0, 1\}$; we have just two probability measures P_0 and P_1 and we want to use the knowledge of observable x to choose between them. Suppose there is a set $E\subset X$, the sample space, such that $P_0(E)=1$, $P_1(E)=0$. Then we use the test

$$\delta(x)=\begin{cases} 1 & \text{if } x\in E^c \\ 0 & \text{if } x\in E \end{cases}$$

and

$$P_\alpha(\delta(x)=\alpha)=1, \ \alpha=0, 1.$$

If E exists, the two measures P_0 and P_1 are *singular* with respect to each other. The opposite case, for any set F, $P_0(F)=0$ implies $P_1(F)=0$ and vice versa, we shall call the *regular* case. In other words P_0 and P_1 are equivalent to each other. By the Radon–Nikodym Theorem, there exists a non-negative measurable function $p(x)$ such that

$$P_1(E)=\int_E p(x)P_0(\mathrm{d}x)$$

for all E in the appropriate σ-algebra. Here $p(x)$ is a Radon–Nikodym derivative of P_1 with respect to P_0. In the intermediate case, there might exist a set E with $P_0(E)=0$ and $0<P_1(E)<1$. We assume that this type of case does not arise and in an important class of problems to be discussed later, this can always be ruled out.

In the standard statistical theory, the likelihood function or a likelihood ratio plays a crucial role in problems of inference. Radon–Nikodym, derivative thus plays an important rule in problems of inference for processes.

The problem now is that of actual calculation of Radon–Nikodym derivative if the measures are equivalent or proving singularity. Suppose the observable is given as a vector $x=(x_1, x_2, \ldots)$ with real components and denote by P_α^n the marginal distribution of the first "n" coordinates (in other words, restricted to the σ-algebra $\mathscr{A}_n$ of the n-dimensional cylinders). Suppose that P_0^n and P_1^n are absolutely continuous with respect to each other for each "n". If this is not the case, then P_0 and P_1 will not be absolutely continuous with respect to each other. Let $p_n(x)$ be a Radon–Nikodym derivative of P_1^n with respect to P_0^n. $p_n(x)$ is uniquely defined almost everywhere $[P_0^n]$. Hence for any $F\in\mathscr{A}_n$,

$$\int_F p_n(x)P_0(\mathrm{d}x)=P_1(F)=\int_F p_m(x)P_0(\mathrm{d}x)$$

for every n and $m\geqq n$. Hence $\{p_n(x), \mathscr{A}_n, n\geqq 1\}$ forms a martingale and since $E|p_n|=Ep_n=1$ for all "n", it follows that $p_n(x)$ converges almost surely $[P_0]$ to some functions $p(x)$. If P_1 is absolutely continuous with respect to P_0, then

$$p(x)=\frac{\mathrm{d}P_1}{\mathrm{d}P_0}(x)$$

exists, $Ep(x)=1$ and the martingale property implies $p_n\to p$ a.s. (P_0) as well as a.s. (P_1) and also in the mean of order 1. On the other hand, the following result holds: if for any $r\geqq 1$

$$E(p_n^r)\uparrow C_r \tag{1}$$

and if $C_r < \infty$ for some $r > 1$, then P_1 is absolutely continuous with respect to P_0 and $p_n \to p$ in the sense indicated above.

The method will be as follows: Calculate $p_n(x)$. If it does not converge almost surely $[P_1]$, then absolute continuity does not hold. On the other hand, if relation (1) holds for some $r > 1$ for a finite C_r, then absolute continuity holds and the Radon–Nikodym derivative is the limit of the likelihood ratios $p_n(x)$.

The above result can also be put in a different way. We always have convergence in probability $[P_0]$ of p_n. With respect to P_1, it converges in probability to p except in the singular part of the sample space X where it tends to $+\infty$ in probability $[P_1]$.

4. TESTING HYPOTHESES

We shall now study the problems of testing statistical hypotheses for continuous parameter stochastic processes. We shall show that it is possible to transfer the ideas and methods of Neyman–Pearson theory to this case.

4.1 Simple Versus Simple

Let us consider a simple hypothesis H_0 corresponding to the probability measure P_0 against a simple alternative H_1 corresponding to P_1. As in the classical theory, we want to form a critical region W of size α such that the error of the second kind $P_1(W^c)$ is as small as possible. Here W^c denotes the complement of the set W.

By the Lebesgue Decomposition Theorem and the Radon–Nikodym Theorem, there exists a set H of P_0 of measure zero and a non-negative function $f(w)$ integrable with respect to P_0 such that, for every measurable set $E \subset \Omega$ the sample space,

$$P_1(E) = \int_E f(w) P_0(dw) + P_1(E \cap H).$$

Consider the set

$$S_k = \{w : f(w) \geq k\} \cup H$$

and determine k so that $P_0(S_k) = \alpha$, the size of the test. One can deal with the case when there is no such k without much difficulty. Suppose there is such a k.

Theorem 4.1. *The test corresponding to the critical region S_k is the most powerful test of H_0 against H_1 of level α.*

Proof. Let E be another measurable set such that $P_0(E)=\alpha$. Let $A=E\cap H^c$, $B=E\cap H$ and $F=H^c\cap(w:f(w)\geqq k)$ where H^c denotes the complement of H in Ω. Note that $P_0(F)=P_0(A)=\alpha$ since $P_0(H)=0$ and $P_0(E)=P_0(S_k)=\alpha$. Furthermore

$$\begin{aligned} P_1(A-A\cap F) &\leqq kP_0[A-A\cap F]+P_1[(A-A\cap F)\cap H] \\ &=kP_0[A-A\cap F]. \end{aligned}$$

Hence

$$\begin{aligned} P_1(F) &= P_1(F-F\cap A)+P_1(F\cap A) \\ &\geqq kP_0(F)-kP_0(F\cap A)+P_1(F\cap A) \\ &= kP_0(A)-kP_0(F\cap A)+P_1(F\cap A) \\ &\geqq P_1(A-A\cap F)+P_1(F\cap A)=P_1(A) \end{aligned}$$

and

$$P_1(S_k)=P_1(F)+P_1(H)\geqq P_1(A)+P_1(B)=P_1(E).$$

Example 4.1. We shall now derive a test for the mean value function of a normal process with known covariance function $r(s, t)$. We want to test the hypothesis

$$H_0\colon E_0(X(t))=m_0(t) \text{ against the alternative } H_1\colon E_1(X(t))=m_1(t).$$

We take as the observable coordinates of the process

$$Z_\nu=\int_T X(t)\phi_\nu(t)\mathrm{d}t \tag{2}$$

where $\{\phi_\nu(t)\}$ is the orthonormal system corresponding to the integral equation of the process. We have assumed, of course, that the covariance function $r(s, t)$ is continuous on $T\times T$ and $m_i(t)$ are square integrable.

In terms of our observables $Z=(Z_1, Z_2, \ldots)$ taking values in $\mathbb{R}^\infty$, we can write the hypothesis as

$$H_0\colon Z_\nu\sim N(a_{0\nu}, \lambda_\nu), \quad \nu=1, 2, \ldots$$

and

$$H_1\colon Z_\nu\sim N(a_{1\nu}, \lambda_\nu), \quad \nu=1, 2, \ldots$$

where

$$a_{i\nu}=\int_T m_i(t)\phi_\nu(t)\mathrm{d}t$$

and Z_ν's are independent under H_0 and H_1.

If a finite number of eigenvalues are different from zero and ϕ_n is a normalized eigenfunction corresponding to $\lambda_n = 0$, we have the singular case if

$$\int_T m_1(t)\phi_n(t)\mathrm{d}t \neq \int_T m_0(t)\phi_n(t)\mathrm{d}t.$$

In this case, the set

$$E = \{\mathbf{z}:(z_1, z_2, \ldots)|z_n = \int_T m_0(t)\phi_n(t)\mathrm{d}t\}$$

has $P_0(E) = 1$ and $P_1(E) = 0$. We have thus arrived at the extreme end of the singular case and can determine the true hypothesis from the knowledge of one realization if we disregard events of probability zero.

Let us now consider the case where $\lambda_\nu \neq 0$ for all ν so that the covariance function is positive definite. Then

$$p_n(\mathbf{z}) = \frac{\prod_{\nu=1}^{n} (2\pi\lambda_\nu)^{-1/2} \exp\left\{-\frac{1}{2\lambda_\nu}(z_\nu - a_{1\nu})^2\right\}}{\prod_{\nu=1}^{n} (2\pi\lambda_\nu)^{-1/2} \exp\left\{-\frac{1}{2\lambda_\nu}(z_\nu - a_{0\nu})^2\right\}} = \exp q_n(\mathbf{z})$$

with

$$q_n(\mathbf{z}) = \sum_{\nu=1}^{n} \left\{z_\nu\left(\frac{a_{1\nu} - a_{0\nu}}{\lambda_\nu}\right) - \left(\frac{a_{1\nu}^2 - a_{0\nu}^2}{2\lambda_\nu}\right)\right\} = \sum_{\nu=1}^{n} \zeta_\nu \text{ (say).}$$

Let us first suppose that

$$K = \sum_{\nu=1}^{\infty} \frac{(a_{1\nu} - a_{0\nu})^2}{\lambda_\nu} < \infty. \tag{3}$$

Then

$$E_0(\zeta_\nu) = -\frac{1}{2}\frac{(a_{1\nu} - a_{0\nu})^2}{\lambda_\nu} \tag{4}$$

and

$$E_1(\zeta_\nu) = \frac{1}{2}\frac{(a_{1\nu} - a_{0\nu})^2}{\lambda_\nu} \tag{5}$$

while the variance is given by

$$\operatorname{var}(\zeta_\nu) = \frac{(a_{1\nu} - a_{0\nu})^2}{\lambda_\nu} \tag{6}$$

for both the hypotheses. Hence, by relation (3), $q=\Sigma_{\nu=1}^{\infty}\zeta_\nu$ converges almost surely $[P_0]$ and $[P_1]$ so that we have the regular case and the Radon–Nikodym derivative p of P_0 with respect to P_1 is obtained as the limit of p_n. To find the most powerful test of H_0 against H_1, we apply Theorem 4.1. Critical region of the most powerful test will be of the form

$$\{\mathbf{z}|p(\mathbf{z})\geqq c\}=\{\mathbf{z}|q(\mathbf{z})\geqq c_1\}. \tag{7}$$

Let

$$f_n(t)=\sum_{\nu=1}^{n}\left\{\frac{a_{1\nu}-a_{0\nu}}{\lambda_\nu}\right\}\phi_\nu(t). \tag{8}$$

Then $q_n(\mathbf{z})$ can be written in the form

$$q_n(\mathbf{z})=\int_T f_n(t)\left\{X(t)-\frac{m_0(t)+m_1(t)}{2}\right\}\mathrm{d}t. \tag{9}$$

Observe that the sequence of f_n need not converge in general and q cannot always be written in the form (9). f_n converges in mean to some limit f if and only if

$$\sum_{\nu=1}^{\infty}\left(\frac{a_{1\nu}-a_{0\nu}}{\lambda_\nu}\right)^2<\infty \tag{10}$$

and then f satisfies the integral equation

$$\int_T r(s,\,t)f(t)\mathrm{d}t=m_1(s)-m_0(s). \tag{11}$$

In other words, this extremely regular case occurs exactly when (11) has a square integrable solution f.

Suppose (3) does not hold i.e. $K=+\infty$. In this case $q_n\to-\infty$ under H_0 and to $+\infty$ under H_1 and we have the singular case.

Example 4.2. As a different example, let us consider a Poisson process on the interval $T=(0,A)$ with positive and continuous intensities. Let us test $H_0:\lambda(t)=\lambda_0(t)$ against $H_1:\lambda(t)=\lambda_1(t)$. As the observables we take the number of events f_ν in the interval $i_\nu=((\nu/n)A,\,(\nu+1/n)A)$, $\nu=0,\,1,\,\ldots,\,n-1$. We get

$$p_n=\prod_{\nu=0}^{n-1}\frac{\left(\int_{i_\nu}\lambda_1(t)\mathrm{d}t\right)^{f_\nu}\exp\left\{-\int_{i_\nu}\lambda_1(t)\mathrm{d}t\right\}}{\left(\int_{i_\nu}\lambda_0(t)\mathrm{d}t\right)^{f_\nu}\exp\left\{-\int_{i_\nu}\lambda_0(t)\mathrm{d}t\right\}}.$$

As n increases, we get convergence almost surely $[P_0]$ and $[P_1]$ to the

Radon–Nikodym derivative

$$p=\prod_{k=1}^{N}\frac{\lambda_1(t_k)}{\lambda_0(t_k)}\exp\left\{\int_0^A(\lambda_0(t)-\lambda_1(t))\mathrm{d}t\right\}$$

where N is the number of events in $(0, A)$ and $t_1, \ldots, t_N$ are the corresponding time points. The most powerful test, therefore, is

$$\prod_{k=1}^{N}\frac{\lambda_1(t_k)}{\lambda_0(t_k)}\geqq\text{constant}.$$

Example 4.3. Suppose that in the space $L_2[a, b]$ of sample paths of some real Gaussian random process $\xi(t)$, we are given two Gaussian measures μ_1 and μ_2 with mean values equal to zero and correlation functions $r_1(s, t)$ and $r_2(s, t)$ respectively. The problem consists of choosing one of the two hypotheses H_1 and H_2 asserting that the measure $\mu_i (i=1, 2)$ corresponds to the process $\xi(t)$.

Consider the integral equation

$$\lambda\int_a^b r_2(s, t)\phi(t)\mathrm{d}t=\int_a^b r_1(s, t)\phi(t)\mathrm{d}t$$

and assume that there exists a sequence of solutions $\{\phi_k(t)\}$ with corresponding sequence of eigenvalues $\{\lambda_k\}$ tending to one. If we multiply both sides of the integral equation

$$r_2(s, t)-r_1(s, t)=\int_a^b r_1(s, u)c(u, t)\mathrm{d}u \tag{12}$$

by $\phi_k(s)$ (where $c(u, t)$ is a square integrable function) and integrate, we obtain

$$\frac{1-\lambda_k}{\lambda_k}g_k(t)=\int_a^b c(u, t)g_k(u)\mathrm{d}u$$

where

$$g_k(s)=\int_a^b r_2(s, t)\phi_k(t)\mathrm{d}t$$

i.e. $g_k(t)$ is the left eigenfunction of the kernel $c(s, t)$. Note that the sequences $\{\boldsymbol{\varphi}_k(t)\}$ and $\{g_k(t)\}$ form a bi-orthogonal system i.e.

$$\int_a^b \phi_j(\mathrm{s})\mathrm{g}_k(\mathrm{s})\mathrm{d}s=\begin{cases}0 & \text{if } j\neq k\\ a_k & \text{for } j=k\end{cases}$$

for some constant a_k. Let us normalize ϕ_k so that

$$\int_a^b \phi_j(s)g_k(s)\mathrm{d}s=1 \quad \text{for } j=k.$$

It is easy to see that the function $c(s, t)$ has the expansion

$$c(s, t)=\sum_{k=1}^{\infty} \frac{1-\lambda_k}{\lambda_k} \phi_k(s)g_k(t).$$

If we take the variables

$$X_k=\int_a^b \xi(t)\phi_k(t)\mathrm{d}t$$

as observable coordinates, they form a sequence of independent normally distributed random variables with

$$E_1(X_k)=0, \ \mathrm{var}_1(X_k)=\int_a^b\int_a^b r_1(s, t)\phi_k(s)\phi_k(t)\mathrm{d}s\mathrm{d}t=\lambda_k$$

under the hypothesis H_1 and

$$E_2(X_k)=0, \ \mathrm{var}_2(X_k)=\int_a^b\int_a^b r_2(s, t)\phi_k(s)\phi_k(t)\mathrm{d}s\mathrm{d}t=1$$

under the hypothesis H_2. Consequently, the likelihood ratio is

$$L(x_1, \ldots, x_n)=(\lambda_1\lambda_2 \ldots \lambda_n)^{1/2} \exp\left\{\frac{1}{2} \sum_{k=1}^{n} \frac{1-\lambda_k}{\lambda_k} X_k^2\right\}.$$

We assume that $\Sigma_{k=1}^{\infty}|1-\lambda_k|<\infty$. This ensures the absolute convergence of the infinite product $\Pi_{k=1}^{\infty}\lambda_k$. Further suppose that $\Sigma_{k=1}^{\infty} (1-\lambda_k)^2/\lambda_k<\infty$. This condition will ensure that the Gaussian measures are equivalent (cf. Prohorov and Rozonov, 1969). It is also easy to see, by Kolmogorov's Three-series Theorem, that the above condition ensures that $\Sigma_{k=1}^{\infty} (1-\lambda_k/\lambda_k) X_k^2$ converges almost surely under μ_1 and μ_2. Consequently, the best critical region for the hypothesis H_1 against H_2 is given by the inequality

$$\sum_{k=1}^{\infty} \frac{1-\lambda_k}{\lambda_k} X_k^2 \geqq c. \tag{13}$$

Assuming that we know the solution of (12), we define the random function $\eta(t)$ as follows:

$$\eta(t)=\int_a^b c(s, t)\xi(s)\mathrm{d}s. \tag{14}$$

If a solution exists for the integral equation

$$\int_a^b r_2(s, t)\zeta(s)\mathrm{d}s = \eta(t) \tag{15}$$

then the integral $\int_a^b \xi(s)\xi(s)\mathrm{d}s$ can be shown to be equal to $\Sigma_{k=1}^{\infty} (1-\lambda_k/\lambda_k)X_k^2$. Thus the test can be written in the form

$$\int_a^b \xi(s)\zeta(s)\mathrm{d}s \geqq c. \tag{16}$$

Hence the problem of finding a test reduces to solving equation (12). It can be solved explicitly only in exceptional cases. Šagdar (1974) presents a method of solving the equation to obtain an approximate test of the hypothesis. He finds a sequence of tests with critical region $\int_a^b \zeta^{(n)}(s)\xi(s)\mathrm{d}s \geqq c$ where $\zeta^{(n)}(s)$ is the solution of the equation

$$\int_a^b r_2(s, t)\zeta^{(n)}(s)\mathrm{d}s = \eta^{(n)}(t),$$

and

$$\eta^{(n)}(t) = \int_a^b c_n(s, t)\xi(s)\mathrm{d}s,$$

$c_n(s, t)$ being a mean-square approximation to $c(s, t)$.

4.2 Simple Versus Composite

In the preceding section we saw how to construct the most powerful region for testing a simple hypothesis against another simple one. In practice, more complicated situations might arise.

Suppose that we want to test the simple hypothesis H_0 against a family H_α, $\alpha \in A$ of simple alternatives. Here α is a real parameter which may be normed so that $\alpha = 0$ corresponds to H_0. Fix α and construct the best critical region S_α for testing H_0 against H_α on the level ε. If we get the same S for all values of α under consideration, then the test corresponding to S is uniformly most powerful. As in the classical case of independent and identically distributed observations, it is very rare for a uniformly most powerful (UMP) test to exist. It is sometimes possible to find a uniformly most powerful unbiased (UMPU) test.

Let P_α be the probability measure induced by a stochastic process under H_α. We shall assume that P_α, $\alpha \in A$ are absolutely continuous with respect to each other and that $f_\alpha(w)$ is a Radon–Nikodym derivative of P_α with respect to P_0. If the derivative $\partial f_\alpha(w)/\partial\alpha$ exists almost surely and is dominated by a function $F(w)$ integrable with respect to P_0, we take as the critical region the set

$$S=\left[w: f_\alpha(w) \geqq c_0 + c_1 \frac{\partial f_\alpha}{\partial \alpha}\Big|_{\alpha=0}\right]. \tag{17}$$

Suppose the set S does not depend on α and it is possible to choose the constants c_0 and c_1 to satisfy

$$P_0(S)=\varepsilon, \text{ and } \frac{\partial P_\alpha(S)}{\partial \alpha}\Big|_{\alpha=0}=0.$$

If E is another unbiased region of the same size, then

$$\begin{aligned} P_\alpha(S-E\cap S) &= \int_{S-E\cap S} f_\alpha(w) P_0(\mathrm{d}w) \\ &\geqq c_0 P_0(S-E\cap S) + c_1 \frac{\partial P_\alpha(S-E\cap S)}{\partial \alpha}\Big|_{\alpha=0} \\ &\geqq P_\alpha(E-E\cap S) \end{aligned}$$

and hence

$$P_\alpha(S) \geqq P_\alpha(E).$$

Thus S is the uniformly most powerful unbiased region.

Example 4.4. Let us again consider the example of Gaussian processes discussed earlier (4.1). Suppose we want to test

$$H_0: E_0[X(t)]=0 \text{ against}$$
$$H_1: E_\alpha[X(t)]=m_\alpha(t).$$

Assume that

$$\sum_{\nu=1}^{\infty} \frac{a_{\nu\alpha}^2}{\lambda_\nu} < \infty \tag{18}$$

for all α under consideration. This condition is analogous to the relation (3) and implies that we have the regular case. The most powerful critical region for any fixed α has the form

$$\lim_{n\to\infty} \sum_{\nu=1}^{n} \frac{Z_\nu a_{\nu\alpha}}{\lambda_\nu} \geqq c(\alpha)$$

Where Z_ν is as defined in (2). In order to get a UMP test we must have

$$a_{\nu\alpha} = K(\alpha) a_\nu$$

where $K(\alpha)$ is of constant sign. Suppose that this is the case. Then the most powerful test will be of the form

$$\lim_{n\to\infty} \int_a^b f_n(t)X(t)\mathrm{d}t \geqq c \tag{19}$$

if $K(\alpha) > 0$ and is of the form

$$\lim_{n\to\infty} \int_a^b f_n(t)X(t)\mathrm{d}t \leqq c \tag{20}$$

if $K(\alpha) < 0$ where $f_n(t) = \Sigma_{\nu=1}^{n} a_\nu \phi_\nu(t)/\lambda_\nu$. Hence the region defined by (19) is a uniformly most powerful one-sided test with respect to alternatives such that $K(\alpha) > 0$ and the region defined by (20) is a UMP one-sided test with respect to α such that $K(\alpha) < 0$.

Suppose the alternative hypothesis is of the form

$$H_1 : E_\alpha[X(t)] = \alpha a(t)$$

where $a(t)$ is a known function. In this case

$$\frac{\partial f_\alpha}{\partial \alpha} = \exp\left\{-\frac{1}{2}\alpha^2 \sum_{\nu=1}^{\infty} \frac{a_{\nu 0}^2}{\lambda_\nu} + \alpha \sum_{\nu=1}^{\infty} \frac{Z_\nu a_{\nu 0}}{\lambda_\nu}\right\}.$$

$$\left\{-\alpha_\nu \sum_{\nu=1}^{\infty} \frac{a_{\nu 0}^2}{\lambda_\nu} + \sum_{\nu=1}^{\infty} \frac{Z_\nu a_{\nu 0}}{\lambda_\nu}\right\}. \tag{21}$$

Let $Z = \Sigma_{\nu=1}^{\infty} Z_\nu a_{\nu 0}/\lambda_\nu$. In view of condition (18), it follows that Z is a well-defined almost surely finite random variable. In fact Z is a normally distributed random variable. Furthermore there exists a constant K such that

$$\left|\frac{\partial f_\alpha}{\partial \alpha}\right| < K e^{|Z|} \left\{\sum_{\nu=1}^{\infty} \frac{a_{\nu 0}^2}{\lambda_\nu} + |Z|\right\} \tag{22}$$

wherever $|\alpha| < 1$. Since the expression on the right-hand side of inequality (22) has finite expectation, it is easy to see that assumptions made for the existence of an UMPU test are satisfied. In fact the UMPU region will be of the form

$$\left|\lim_{n\to\infty} \int_a^b f_n(t)X(t)\mathrm{d}t\right| \geqq k. \tag{23}$$

4.3 Tests for Point Processes

We shall discuss tests for some special types of point processes in this section. Let $X(t)$ be a generalized Poisson process with probability intensity

$\lambda(t)$ or $\mu(t)$ according to whether H_0 or H_1 is true. The process is observed in the time interval $[0, T]$. Using the coordinates $(n, t_1, \ldots, t_n)$, the Radon–Nikodym derivative can easily be seen to be

$$f(w)=\frac{\mu(t_1)\ \ldots\ \mu(t_n)}{\lambda(t_1)\ \ldots\ \lambda(t_n)}\exp\left\{\int_0^T[\mu(t)-\lambda(t)]\mathrm{d}t\right\} \tag{24}$$

in the case $\lambda(t)$ is different from zero almost everywhere in the set where $\mu(t)\neq 0$. We shall suppose that this is the case. The critical region for testing H_0 against H_1 will be of the form

$$S=\left[\prod_1^n \frac{\mu(t_i)}{\lambda(t_i)}\geqq k\right], \tag{25}$$

as derived in Example 4.2. If the hypotheses are of the form,

H_0: the probability intensity is $\lambda(t)$,

and

H_1: the probability intensity is $\mu\lambda(t)$
then the critical region is of the form

$$S=[\mu^n\geqq k]$$

and if we confine ourselves to alternatives $\mu>1$ $(\mu<1)$ we get the one-sided uniformly most powerful test

$$S=[n\geqq n_0](\text{or } S=[n\leqq n_0]). \tag{26}$$

If the alternative is not one-sided, then the uniformly most powerful unbiased test exists and the critical region can be shown to be of the form

$$S=[n<n_1]\cup[n>n_2]. \tag{27}$$

Since n has a discrete distribution, it may be impossible to solve $P_0(S)=\varepsilon$ exactly so that the test is of size ε for n_0 in (26) and n_1 and n_2 in (27). However, randomized tests could be used in these cases.

Suppose the hypotheses are of the form

$$H_0{:}\ \lambda(t)=\mathrm{e}^{\alpha t}$$

and

$$H_\mu{:}\ \mu(t)=\mathrm{e}^{(\alpha+\mu)t}.$$

Then the most powerful critical region S is of the form

$$\left[\exp\left(\mu\sum_1^n t_i\right)\geqq k\right]. \tag{28}$$

For the one-sided alternative $\mu > 0$, we thus get the uniformly most powerful test

$$S = \left[\sum_{1}^{n} t_1 \geqq k'\right]. \tag{29}$$

Here $\Sigma_1^n t_i$ has a continuous distribution so that we have no difficulty in determining k for a given value of ε.

(Also, see Chapter 6 Section 3).

Suppose that we observe a nonhomogeneous Poisson process P_μ on $[0, t]$, the mean measure μ of the process assumed to be either μ_1 or μ_2 where $\mu_i[0, t] < \infty$, $i = 1, 2$. Let ν be any measure which dominates both μ_1 and μ_2 with $\nu[0, t] < \infty$ and let

$$f_1 = \frac{\mathrm{d}\mu_1}{\mathrm{d}\nu}, \; f_2 = \frac{\mathrm{d}\mu_2}{\mathrm{d}\nu}.$$

Let $P_{\mu_1}^{(t)}$ $P_{\mu_2}^{(t)}$ and $P_\nu^{(t)}$ represent Poisson processes on $[0, t]$ with mean measures μ_1, μ_2 and ν respectively. It then follows that $P_{\mu i}^{(t)} \ll P_\nu^{(t)}$, $i = 1, 2$ and

$$g_i(t) = \frac{\mathrm{d}P_{\mu i}^{(t)}}{\mathrm{d}P_\nu^{(t)}} = \left\{\prod_{j=1}^{N_t} f_i(x_j)\right\} \exp\left[-\int_0^t \{f_i(x) - 1\}\nu(\mathrm{d}x)\right], \quad i = 1, 2, \tag{30}$$

where N_t is the number of events in $[0, t]$ and x_i, $i \geqq 1$, are the epochs at which events occur. This is essentially what we did to derive (24). The most powerful critical region for testing the null hypothesis $\mu = \mu_1$ against the alternative $\mu = \mu_2$ is obtained by Neyman–Pearson Lemma and it tells us to reject the null hypothesis if

$$\prod_{i=1}^{N_t} \left\{\frac{f_2(x_i)}{f_1(x_i)}\right\} \exp\left[-\int_0^t \{f_2(x) - f_1(x)\}\nu(\mathrm{d}x)\right]$$

is large, the cut-off point being determined by the desired α level of the test. This is equivalent to rejecting if

$$W_t = \sum_{i=1}^{N_t} \log \frac{f_2(x_i)}{f_1(x_i)} \tag{31}$$

is large. If P_{μ_1} and P_{μ_2} are singular with respect to each other, Brown (1972) has shown that W_t is asymptotically normal as $t \to \infty$.

Suppose that the family of mean measures $\{\mu_\theta, \theta \in I\}$ where I is a (possibly infinite) interval, is dominated by a σ-finite measure ν and that $f_\theta(x) = (\mathrm{d}\mu_\theta/\mathrm{d}\nu)(x)$ is monotone non-decreasing in θ for each x. Suppose that $\mu_\theta(0, t) < \infty$ for all θ and t. Brown (1972) has shown that the level-α Neyman–Pearson test of $\theta = \theta_1$ versus $\theta = \theta_2$ $(\theta_1 < \theta_2)$ is also a level-α test of $\theta \leqq \theta_1$ versus $\theta > \theta_2$ and has a monotone non-decreasing power function.

Thus, if $P_{\mu_{\theta_1}} \perp P_{\mu_{\theta_2}}$, then the power of a fixed level-α Neyman–Pearson test approaches one as $t \to \infty$ uniformly for $\theta \geqq \theta_2$.

5. ESTIMATION

In this section, we shall be mainly concerned with asymptotic theory of likelihood equation estimates for continuous parameter stochastic processes. The estimation problem can be considered under two different situations. The problem of parameter estimation arises either when we have observed n independent sample paths of the process or when we have observed the process over a period $[0, T]$. The asymptotic theory arises as $n \to \infty$ in the former case and as $T \to \infty$ in the latter case.

5.1 The Likelihood Function

Let $(\Omega, \mathscr{F})$ be a measurable space and $\{P_\theta : \theta \in \Theta\}$ be a class of probability measures, $\Theta \subset \mathbb{R}^k$, k-dimensional Euclidean space. We assume that P_{θ_1} is equivalent to P_{θ_2} for all θ_1, θ_2 in Θ and $(\Omega, \mathscr{F}, P_\theta)$ is complete for all $\theta \in \Theta$. The stochastic process $X = \{X_s, s \geqq 0\}$ on $(\Omega, \mathscr{F})$ evolves according to one of the probability measures P_θ, say P_{θ_0} (i.e. θ_0 is the true value of the parameter). For each $t \geqq 0$, define

$$\mathscr{F}_t = \sigma(X_s : 0 \leqq s \leqq t,\ N)$$

the σ-field generated by sets $\{w \in \Omega : X_s(w) \in B\}$, where $s \leqq t$, B is a Borel set of $\mathbb{R}$ and N is the class of P_{θ_0}-null sets of $\mathscr{F}$. We consider processes X which have right continuous sample paths almost surely so that the increasing sequence $\{\mathscr{F}_t;\ t \geqq 0\}$ is also right continuous. Let P_θ^t be the restriction of P_θ to $\mathscr{F}_t$ and define

$$L_t(\theta) = \frac{\mathrm{d}P_\theta^t}{\mathrm{d}P_{\theta_0}^t} \tag{32}$$

as a Radon–Nikodym derivative of P_θ^t with respect to $P_{\theta_0}^t$. Thus $L_t(\theta)$ is an $\mathscr{F}_t$-measurable function and can be expressed almost surely as a function of the sample path $\{X_s(w) : 0 \leqq s \leqq t\}$. Let $L(w;\ \theta)$ be a Radon–Nikodym derivative of P_θ with respect to P_{θ_0}.

In order to calculate the likelihood function for any particular model the following two schemes can be adopted.

(i) Calculate the likelihood ratio (with respect to P_{θ_0}) for a finite set of time points $0 = t_0 < t_1 < \ldots < t_n = T$ and then let the number of time points tend to infinity so that

$$\max_{1 \leq i \leq n} (t_i - t_{i-1}) \to 0 \text{ as } n \to \infty.$$

(ii) Find a countable set of independent variables (the so-called "observable coordinates" of the process of Section 2) and calculate the likelihood based on them.

Both these methods were illustrated in the previous section. The first is the better of the two as the second is likely to be mainly restricted to Gaussian processes.

5.2 Estimation when Several Sample Paths of a Process are Available

Suppose the following conditions are satisfied:

(G1) $$\frac{\partial^\nu \log L(w;\theta)}{\partial\theta^\nu}; \ \nu = 1, 2, 3 \text{ exist almost surely.}$$

(G2) For every $\theta \in \Theta$,

$$\left|\frac{\partial L}{\partial \theta}\right| < F_1(w), \left|\frac{\partial^2 L}{\partial \theta^2}\right| < F_2(w), \left|\frac{\partial^3 \log L}{\partial \theta^3}\right| < H(w) \tag{33}$$

where $E_{\theta_0} F_1 < \infty$, $E_{\theta_0} F_2 < \infty$ and $E_\theta H < K$ for all $\theta \in \Theta$.

(G3) For every $\theta \in \Theta$,

$$0 < E_\theta \left\{\frac{\partial \log L}{\partial \theta}\right\}^2 < \infty. \tag{34}$$

Here $L(w; \theta)$ denotes the Radon–Nikodym derivative of the process when θ is the true parameter with respect to a suitable dominating measure which is assumed to exist. We are considering the case when θ is one-dimensional. The general problem can be dealt with on similar lines.

Suppose we have observed N independent sample paths of the process and we want to estimate θ. Denoting the sample paths by $w_1, \ldots, w_N$, we form the simultaneous likelihood function

$$L(w_1, \ldots, w_N; \theta) = L(w_1; \theta) \ldots L(w_N; \theta) \tag{35}$$

and it can be shown that the likelihood equation

$$\frac{\partial \log L(w_1, \ldots, w_N; \theta)}{\partial \theta} = 0 \tag{36}$$

has a solution $\theta^*(w_1, \ldots, w_N)$ which is consistent, asymptotically normal and asymptotically efficient estimate of θ as $N \to \infty$ under the conditions (G1)–(G3).

This is proved by a method similar to that in Cramér (1946) for proving consistency, asymptotic normality and asymptotic efficiency of likelihood equation estimators for independent and identically distributed random variables.

Example 5.1. As an application of the above method, let us consider the stochastic process which describes the movement of pebbles at the bottom of water channel. A stone which is observed during the time interval $[0, T]$ has two possible states: either it rests or it moves. Suppose the movement takes place without loss of time. At $t=0$, we shall assume that the stone is moving and the time points of movement $t_1, \ldots, t_n$ are distributed according to a Poisson process with probability intensity $1/\theta$. In other words, θ is the mean resting time. Suppose further that the stone moves a length of x_i at time instant t_i and length x_0 at time 0. Here the x's are independent random variables taking positive values with exponential density $\xi^{-1} e^{-x/\xi}$ for $x>0$ and 0 for $x \leqq 0$. ξ is the mean length of transport. Suppose N stones are independently observed. We want to estimate θ.
It is easy to see that the joint density of $(x_0, \ldots, x_n; t_1, \ldots, t_n)$ is

$$\xi^{-(n+1)} e^{-X/\xi} \theta^{-n} e^{-T/\theta}$$

where $X = x_0 + x_1 + \ldots + x_n$. Labelling the coordinates of the different paths with the index $i = 1, 2, \ldots, N$, we get the likelihood function

$$\xi^{-\sum_{i=1}^{N}(n_i+1)} \exp\left(-\sum_{i=1}^{N} X_i/\xi\right) \theta^{-\sum_{i=1}^{N} n_i} \exp(-NT/\theta).$$

The likelihood equation estimate is easily seen to be

$$\theta^* = \frac{NT}{\sum_{i=1}^{N} n_i}. \tag{37}$$

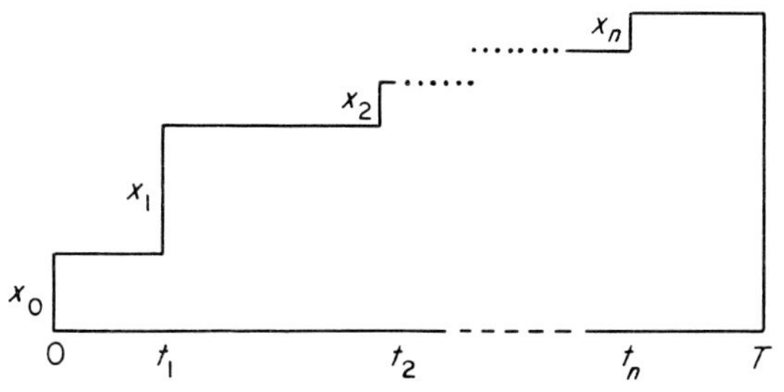

If $\Sigma_{i=1}^{N} n_i = 0$, then θ^* is infinite but when $N \to \infty$ the probability of this tends to zero. Observe that

$$N^{1/2}(\theta^* - \theta) = -\theta \frac{N^{-1/2} \sum_{i=1}^{N} (n_i - T\theta^{-1})}{N^{-1} \sum_{i=1}^{N} n_i} \tag{38}$$

where n_i has a Poisson distribution with mean $T\theta^{-1}$. Central limit theorem implies

$$N^{-1/2} \sum_{i=1}^{N} (n_i - T\theta^{-1}) \xrightarrow{\mathscr{L}} N(0, T\theta^{-1}) \tag{39}$$

and weak law of large numbers shows that

$$N^{-1} \sum_{i=1}^{N} n_i \xrightarrow{p} T\theta^{-1}. \tag{40}$$

Applying Slutsky's Theorem (cf. Cramer, 1946), relations (38)–(40) show that

$$N^{1/2}(\theta^* - \theta) \xrightarrow{\mathscr{L}} N(0, \theta^3 T^{-1}). \tag{41}$$

Since

$$E_\theta \left(\frac{\partial \log L(\theta)}{\partial \theta} \right)^2 = E_\theta \left(\frac{n}{\theta} - \frac{T}{\theta^2} \right)^2 = \frac{T}{\theta^3} \tag{42}$$

it follows that θ^* is asymptotically efficient.

Example 5.2. We shall now discuss another example which occurs in a problem with medical statistics. The problem is concerned with the investigation of the occurence of a phenomenon A in an interval which we denote by $[0, T]$ although the parameter does not denote time but location. At $t=0$, we suppose that A^* always occurs. As the sample space we take functions taking the values 0(for A^*) and 1(for A) and having the value 0 for $t=0$. As coordinates we take the numbers $n_1, n_2, t_1, t_2, \ldots, t_{n_1+n_2}$ where the ts denote the time points when the states change and n_1 denotes the number of starting points for the state A and n_2 the number of end points for A-intervals. Clearly $n_1 = n_2$ or $n_1 = n_2 + 1$ according to whether the state at $t = T$ is A^* or A. Suppose that the length of an A^* (or A) interval has the density function $\beta e^{-\beta t}$, $t > 0$(or $\alpha\, e^{-\alpha t}$, $t > 0$) with independence in the usual manner between different intervals. Let L and Λ denote the total length of A^* intervals and A intervals respectively. We leave it to the reader to check that the joint density of $(t_1, \ldots, t_{n_1+n_2})$ is given by

$$\beta^{n_1} e^{-\beta L} \alpha^{n_2}\, e^{-\alpha \Lambda}. \tag{43}$$

Repeating the experiment N times independently, we get the maximum likelihood estimates

$$\beta^* = \frac{\sum_1^N n_{1i}}{\sum_1^N L_i}; \quad \alpha^* = \frac{\sum_1^N n_{2i}}{\sum_1^N \Lambda_i}. \tag{44}$$

The quantities in the denominator are the times of risk of the event that one state changes to the other and the quantities in the numerator are the number of times this has happened.

5.3 Estimation when a Part of a Single Sample Path of a Process is Available

In the previous section, we considered the case when N independent sample paths of the process have been observed. As shown there, it is then possible to use a maximum likelihood method to obtain consistent and asymptotically efficient estimates when N tends to infinity. We shall now investigate the possibility of getting estimates by the maximum likelihood method using one sample path observed over $[0, T]$ and discuss its asymptotic properties as T approaches infinity.

We now consider a martingale (related to the likelihood function of section 5.1) with the aid of which asymptotic properties of the maximum likelihood estimate can be determined. Under the set up given in Section 5.1, let $l_t(\theta) = \log L_t(\theta)$ and assume that

$$V_t(\theta) = \frac{\partial_{l_t}(\theta)}{\partial\theta} \tag{45}$$

exists with finite expectation for all $t \geqq 0$ and $\theta \in \Theta$. Then $V_t(\theta)$ is $\mathscr{F}_t$-measurable and for $\theta \in \mathbb{R}^k$, observe that the term on the left-hand side of relation (45) is to be interpreted as a k-dimensional column vector.

In the following discussion, we shall consider only the scalar case. For any $A \in \mathscr{F}_s$, $0 \leqq s \leqq t$, we have formally,

$$\int_A V_t(\theta) \mathrm{d}P_\theta = \int_A \frac{\partial}{\partial\theta} \log L_t(\theta) \mathrm{d}P_\theta^t \tag{46}$$

$$= \int_A L_t^{-1}(\theta) \frac{\partial L_t(\theta)}{\partial\theta} \frac{\mathrm{d}P_\theta^t}{\mathrm{d}P_{\theta_0}^t} \mathrm{d}P_{\theta_0}^t \tag{47}$$

$$= \int_A \frac{\partial L_t(\theta)}{\partial\theta} \mathrm{d}P_{\theta_0}^t \tag{48}$$

$$=\frac{\partial}{\partial\theta}\int_A L_t(\theta)\mathrm{d}P^t_{\theta_0} \tag{49}$$

$$=\frac{\partial}{\partial\theta}\, P^t_\theta(A).$$

$$=\frac{\partial}{\partial\theta}\, P^s_\theta(A) \quad \text{since } A\in\mathscr{F}_s \tag{50}$$

$$=\int_A V_s(\theta)\mathrm{d}P_\theta.$$

Some of the above equalities do not hold in general. Relation (47) follows from a property of Radon–Nikodym derivative but relation (49) need not hold in general. If, for any $t\geqq 0$, relation (49) holds for all $A\in\mathscr{F}_t$, then equation (50) holds too. Assuming that equation (49) holds, the above calculations show that

$$\{V_t(\theta), \mathscr{F}_t;\, t\geqq 0\} \tag{51}$$

is a zero mean martingale with respect to P_θ for all $\theta\in\Theta$.

The likelihood equation estimate $\hat{\theta}(t)$ is defined to be the solution of the equation

$$V_t(\theta)=0. \tag{52}$$

We shall now discuss some examples of processes where the above theory can be applied usefully.

Example 5.3. (Pure Jump Process). A process is a *pure jump process* if its transitions are jumps so that between jumps it remains in a specific state. Let T_i denote the time between the $(i-1)$th and ith jump and let

$$X_j = X_{\tau_j},\ j\geqq 0 \tag{53}$$

where $\tau_j = T_1 + \ldots + T_j$, $j\geqq 1$, $\tau_0=0$. Assume that all the sample paths of $\{X_t;\, t\geqq 0\}$ are right continuous. The discrete time process $\{X_j;\, j=0, 1, \ldots\}$ is called the embedded jump process and we will consider only those processes for which the embedded jump process is a Markov chain with stationary probabilities

$$P(X_j\in B|X_{j-1}=x)=\int_B f(y|x)\mu(\mathrm{d}y) \tag{54}$$

for some measure μ. For the process to be Markov, there must exist some function λ such that

$$P(T_k\leqq t|X_{k-1}=x)=1-\exp\,(-\lambda(x)t) \tag{55}$$

and T_k, $k \geqq 1$ are independent.

For a Markov jump process, the likelihood is

$$L_t(\theta) = \prod_{j=1}^{n(t)} \{f(X_j | X_{j-1}) \lambda(X_{j-1}) \exp(-\lambda(X_{j-1}) T_j)\} \exp(-\lambda(X_n)(t - \tau_n)) \tag{56}$$

where $n = n(t)$ is the number of jumps up till time t. We assume that X_0 is non-random. Here θ can enter into f or λ or both. Equation (56) can also be written in the form

$$L_t(\theta) = \prod_{j=1}^{n(t)} \{f(X_j | X_{j-1}) \lambda(X_{j-1})\} \exp\left\{-\int_0^t \lambda(X_u) \mathrm{d}u\right\}. \tag{57}$$

If we take the birth–death process (cf. Karlin, 1966) with linear birth and death rates, then we have

$$\lambda(X_t) = (\lambda + \mu) X_t,$$

and

$$f(X_{j-1} + d | X_{j-1}) = \begin{cases} \lambda/(\lambda+\mu) & \text{if } d = 1 \\ \mu/(\lambda+\mu) & \text{if } d = -1 \\ 0 & \text{otherwise.} \end{cases} \tag{58}$$

Let

$$\theta = \begin{pmatrix} \lambda \\ \mu \end{pmatrix} \tag{59}$$

and in this case

$$L_t(\theta) = \left\{\prod_{j=1}^{n(t)} X_{j-1}\right\} \lambda^{(1/2)(n + X_n - X_0)} \mu^{(1/2)(n - X_n + X_0)} \exp\left(-(\lambda + \mu) \int_0^t X_u \, \mathrm{d}u\right)$$
$$= K \lambda^{B_t} \mu^{D_t} \exp(-(\lambda + \mu) S_t) \tag{60}$$

where K is independent of λ and μ, B_t is number of births up to time t, D_t is number of deaths up to time t and

$$S_t = \int_0^t X_u \mathrm{d}u. \tag{61}$$

Keiding (1975) has attained asymptotic properties of the maximum likelihood estimate of θ based on observation over fixed time intervals. These results are presented in Chapter 5, Section 5.

Example 5.4. (Ornstein–Uhlenbeck Process). Let us consider the stochastic differential equation (see Appendix 2).

$$dX_t = \theta X_t dt + dW_t;\ t \geqq 0 \tag{62}$$

where $\theta > 0$ and W_t is the standard Wiener process (see Chapter 9). In this case

$$\begin{aligned} V_t(\theta) &= \int_0^t X_s dX_s - \theta \int_0^t X_s^2 ds \\ &= \int_0^t X_s dW_s \text{ a.s.}[P_\theta] \end{aligned} \tag{63}$$

and

$$\begin{aligned} \hat{\theta}(t) &= \left\{\int_0^t X_s^2 ds\right\}^{-1} \left\{\int_0^t X_s dX_s\right\} \\ &= \frac{1}{2}\left\{\int_0^t X_s^2 ds\right\}^{-1} \{X_t^2 - X_0^2 - t\} \end{aligned} \tag{64}$$

on applying Itos' Lemma (cf. Mckean, 1969) to the process $U(X_t)$ where $U(x) = x^2$. Note that

$$V_t(\theta) = I_t(\theta)(\hat{\theta}(t) - \theta) \tag{65}$$

where

$$I_t(\theta) = \int_0^t X_s^2 ds.$$

To check the existence and properties of the above integrals, we follow the approach of Breiman (1968). Relation (62) shows that

$$e^{-\theta t} X_t - X_0 = \int_0^t e^{-\theta s} dW_s. \tag{66}$$

It follows, from results on stochastic integrals with respect to Wiener process, that X_t is Gaussian, $\{e^{-\theta t} X_t - X_0, \mathscr{F}_t;\ t \geqq 0\}$ is a zero mean martingale satisfying

$$E_\theta[e^{-\theta t} X_t - X_0]^2 = \int_0^t e^{-2\theta s} ds = \frac{1}{2\theta}(1 - e^{-2\theta t}) \leqq \frac{1}{2\theta} \tag{67}$$

for all $t \geqq 0$. Here $\mathscr{F}_t$ is the σ-field generated by W_s, $0 \leqq s \leqq t$. The inequality (67) ensures the existence of

$$E_\theta[I_t(\theta)] = E_\theta\left\{\int_0^t X_s^2 ds\right\}$$

so that the stochastic integral given by (63) is well-defined and it also allows us to conclude that

$$e^{-\theta t} X_t - X_0 \to Z \quad \text{a.s.}$$

by the martingale convergence theorem. Since

$$\lim_{t\to\infty} E_\theta(e^{-\theta t} X_t - X_0)^2 < \infty$$

from relation (67), Doob (1953) Chapter 4, Theorem 4.1, implies that

$$Z \sim N(0, (2\theta)^{-1}).$$

Applying the integral version of the Toeplitz Lemma to

$$e^{-2\theta t} X_t^2 \to (Z + X_0)^2 \text{ a.s.}$$

we obtain that

$$e^{-2\theta t} I_t = e^{-2\theta t} \int_0^t X_s^2 \to \frac{1}{2\theta} (Z + X_0)^2 \text{ a.s.}$$

This result will imply that

$$I_t(\theta) \to \infty \text{ a.s.} \tag{68}$$

since $P((Z + X_0)^2 = 0) = 0$. Moreover, from McKean (1969), for each θ, there exists a standard Brownian motion B such that

$$V_t(\theta) = B_{I_t}(\theta) \text{ a.s.}$$

for all $t \geqq 0$. Relations (66), and (68) show that

$$\hat{\theta}(t) - \theta_0 \to 0 \text{ a.s. } [P_{\theta_0}] \tag{69}$$

and

$$I_t^{1/2}(\theta_0)(\hat{\theta}(t) - \theta_0) \xrightarrow{\mathcal{L}} N(0, 1) \tag{70}$$

by Theorems 4.1 and 4.2 of Appendix 1 when $\theta_0 > 0$.

For the stationary case, that is $\theta_0 < 0$, Brown and Hewitt (1975) showed that

$$t^{1/2}(\hat{\theta}(t) - \theta_0) \xrightarrow{\mathcal{L}} N(0, -2\theta_0) \tag{71}$$

and

$$I_t^{1/2}(\theta_0)(\hat{\theta}(t) - \theta_0) \xrightarrow{\mathcal{L}} N(0, 1). \tag{72}$$

If $\theta_0 = 0$, it can be shown that

$$I_t^{1/2}\hat{\theta}(t) \stackrel{d}{=} \frac{1}{2}\left\{W_1^2 - 1\right\}\left\{\int_0^1 W_u^2 \, du\right\}^{-1/2} \tag{73}$$

for each t, under $[P_0]$ and

$$\hat{\theta}(t)\to 0 \text{ a.s. } [P_0]. \tag{74}$$

6. ESTIMATING THE INFINITESIMAL GENERATOR FOR A CONTINUOUS TIME FINITE STATE MARKOV PROCESS

Let $\{Z(t),\ t\geqq 0\}$ be a separable continuous time Markov process with stationary transition probabilities $P_{ij}(t)$, $i, j=1, 2, \ldots, M$. It is well known that under suitable regularity conditions, the matrix of transition probabilities $P(t)=((P_{ij}(t))$ can be expressed in the form $P(t)=\exp\{t\,Q\}$ where Q is an $M\times M$ matrix and Q is called the infinitesimal generator for the process.

In this section, we shall discuss the problem of estimating Q when it is not known. (See Chapter 5, Section 2 for the parametric version). Even though the results of this section are not in the main stream, we have decided to include them here in view of the special problems they pose in estimation. We shall only sketch the proofs.

The question of defining a maximum likelihood estimate is not a trivial one here. Customarily, we define a maximum likelihood estimate relative to a density (which depends on the parameter to be estimated) over the sample space of the experiment. In our problem, the sample space of the experiment will be a function space and one of the major problems is that of the construction of the density over the set of sample paths of the process.

6.1 Preliminaries

Let $(\Omega, \mathscr{A}, P)$ be a probability space and let $\{Z(t, w);\ t\geqq 0\}$ be a separable Markov process on that space which takes values in a finite set $\{1, 2, \ldots, M\}$. Assume that the transition probability function is stationary i.e. for any $s, t\geqq 0$,

$$P[Z(t+s)=j|Z(s)=i]=P_{ij}(t) \tag{75}$$

depends only on the time difference, t. Further assume that

$$\lim_{h\to 0}\frac{1-P_{ii}(h)}{h}=q(i) \tag{76}$$

exists for all $1\leqq i\leqq M$ and

$$\lim_{h\to 0}\frac{P_{ij}(h)}{h}=q(i, j)$$

exists for all $1 \leqq i \neq j \leqq M$. The following theorem can be proved on the basis of these assumptions. We refer the reader to Doob (1953) for a proof.

Theorem 6.1. *Let*

$$q(i, j) = \begin{cases} -q(i) = -\sum_{j \neq i} q(i, j) & \text{if } i = j \\ q(i, j) & \text{if } i \neq j \end{cases} \tag{77}$$

and $Q = (q(i, j))$. *Then*

(i) $$P(t) = \exp\{tQ\} \tag{78}$$

where $P(t)$ *is the matrix of transition probability functions and*

(ii) $$P\{Z(t) = i,\ t_0 \leqq t \leqq t_0 + \alpha | Z(t_0) = i\} = e^{-q(i)\alpha} \tag{79}$$

for all $t_0 \geqq 0$ *and* $\alpha \geqq 0$.

(iii) *If* $Z(t_0) = i$ *and* $q(i) > 0$, *then, with probability one, there exists a sample path discontinuity (of the first kind) which is a jump. For any* $0 < \alpha \leqq \infty$, *the conditional probability that the first discontinuity in* $[t_0, t_0 + \alpha)$ *is a jump to* j *given that* $Z(t_0) = i$ *and there is a discontinuity in* $(t_0, t_0 + \alpha)$ is

$$q(i, j)/q(i). \tag{80}$$

(iv) *Almost all sample paths are step functions with a finite number of jumps in any finite time interval.*

Let $\Omega' \subset \Omega$ be the subset for which sample paths of the process $\{Z(t), t \geqq 0\}$ are step functions with a finite number of jumps in any finite interval. This set has probability one in view of the previous theorem. Define

$$\tau_0(w) = 0, \tag{81}$$

$$\tau_i(w) = \begin{cases} \text{the time at which the jump occurs if } w \in \Omega' \\ +\infty \quad \text{otherwise,} \end{cases} \tag{82}$$

$$T_i(w) = \begin{cases} \tau_{i+1}(w) - \tau_i(w) \text{ if } \tau_i(w) < \infty \\ 0 \quad \text{otherwise,} \end{cases} \tag{83}$$

$$N(T, w) = \text{the largest integer } n \text{ for which } \tau_n(w) < T, \tag{84}$$

and

$$Z_i(w) = Z(\tau_i(w), w),\ i = 0, 1, 2 \ldots \tag{85}$$

It is clear that $T_i(\cdot)$ is the time spent in the ith state, $N(T, \cdot)$ is the number of jumps made by the process in $[0, T)$ and $Z_i(\cdot)$ is the state of process immediately after the ith jump. In fact, with probability one, a sample path of $\{Z(t,\cdot), 0 \leqq t \leqq T\}$ can be represented as an ordered sequence

$$((Z_0(w), T_0(w)), \ldots, (Z_{N(T,w)-1}(w), T_{N(T,w)-1}(w)), Z_{N(T,w)}(w)).$$

If

$$\{Z(t, w); 0 \leqq t \leqq T\} = ((Z_0, t_0), (Z_1, t_1), \ldots, (Z_{n-1}, t_{n-1}), Z_n),$$

then the path starts at Z_0 at time zero, remains in Z_0 for t_0 units of time, makes a jump to Z_1, remains in Z_1 for t_1 units of time, ..., jumps to Z_{n-1}, remains there for t_{n-1} units of time and then makes the final jump to Z_n and remains there at least until time T. Let

$$q'(i,j) = \begin{cases} 0 & \text{if } i=j \\ q(i,j) & \text{if } i \neq j. \end{cases} \tag{86}$$

It is easy to show that

$$P(N(T)=0, Z(0)=Z_0) = P(Z(0)=Z_0)e^{-q(Z_0)T} \tag{87}$$

and for $n>0$

$$P[N(T)=n, Z(0)=Z_0, T_0 \leqq \alpha_0, \ldots, Z(n-1)=Z_{n-1}, T_{n-1} \leqq \alpha_{n-1}; Z(n)=Z_n]$$
$$= P[Z(0)=Z_0)e^{-q(Z_n)T}.$$

$$\int_{S_n} \prod_{j=0}^{n-1} \{q'(Z_j, Z_{j+1})e^{-[q(Z_j)-q(Z_n)]t_j}dt_j\} \tag{88}$$

where

$$S_n = \left\{(t_0, t_1, \ldots, t_{n-1}): \sum_{j=0}^{n-1} t_j < T \text{ and } 0 \leqq t_j \leqq \alpha_j\right\}.$$

We shall now construct the density function. Let λ be the Lebesgue measure on $\mathbb{R}$ and C be the counting measure on $W_0 = \{1, 2, \ldots, M\}$. Let

$$W_n = \left[\prod_{j=1}^{n} (W_0 \times \mathbb{R})\right] \times W_0 \tag{89}$$

and $\sigma^{(n)}$ be the sigma-finite measure on W_n defined by

$$\sigma^{(n)} = \left[\prod_{j=1}^{n} (C \times \lambda)\right] \times C. \tag{90}$$

Let

$$W = \bigcup_{n=0}^{\infty} W_n \tag{91}$$

and for each set $E \subset W$ for which $E \cap W_n$ is $\sigma^{(n)}$-measureable define

$$\sigma^*(E) = \sum_{n=0}^{\infty} \sigma^{(n)}(E \cap W_n). \tag{92}$$

As a consequence of (iv) of Theorem 6.1, every sample path of the process $\{Z(t), 0 \leqq t \leqq T\}$ can be represented as a point in E. Note that σ^* is defined on the σ-field $\mathscr{B}^*$ which is the smallest σ-field containing all sets $E \subset W$ whose projection on $\mathbb{R}^n$ is a Borel set for each n.

Let σ be a measure on the space of all sample functions, defined for all subsets whose intersection with W is in $\mathscr{B}^*$ by the relation $\sigma(B) = \sigma^*(B \cap W)$. Note that σ is a σ-finite measure. The following theorem gives the required density function.

Theorem 6.2. *If B is a σ-measurable subset of the space of all sample functions over* $[0, T]$, *then*

$$P(B) = \int_B f_Q(v) \mathrm{d}\sigma(v) \tag{93}$$

where

$$f_Q(v) = \begin{cases} P(Z(0) = Z_0) \mathrm{e}^{q(Z_0)T} \textit{ if } v = (Z_0) \\ P(Z(0) = Z_0) \mathrm{e}^{-q(Z_n)T} \prod_{j=0}^{n-1} q'(Z_j, Z_{j+1}) \mathrm{e}^{-(q(Z_j) - q(Z_n))t_j} \\ \qquad \textit{if } v = ((Z_0, t_0), \ldots, (Z_{n-1}, t_{n-1}), Z_n) \textit{ with} \\ \qquad n > 0,\ Z_j \in W_0,\ t_j \geqq 0,\ j = 0, 1, \ldots, n-1 \textit{ and} \\ \qquad \sum_{j=0}^{n-1} t_j < T \\ 0 \qquad \textit{otherwise.} \end{cases} \tag{94}$$

This theorem follows easily from the definition of the measure σ and from relations (87) and (88).

6.2 Maximum Likelihood Estimation

Suppose k independent sample paths $v_1, \ldots, v_k$ of $\{Z(t), 0 \leqq t \leqq T\}$ are observed. The likelihood function $L_Q^{(k)}$ is given by

$$L_Q^{(k)} = \prod_{j=1}^{k} f_Q(v_j) \tag{95}$$

in view of Theorem 6.2. Let

$$N_T^{(k)}(i, j) = \text{the total number of transitions from state } i \text{ to state } j \text{ observed during } k \text{ trials,} \tag{96}$$

and

$$A_T^{(k)}(i) = \text{the total length of time that state } i \text{ is occupied during } k \text{ trials.} \tag{97}$$

Then

$$\log L_Q^{(k)} = C_k + \sum_i \sum_j N_T^{(k)}(i, j) \log q(i, j) - \sum_{i=1}^{M} A_T^{(k)}(i) q(i) \tag{98}$$

where C_k is finite almost surely and does not depend on Q. Observe that the set

$$\{N_T^{(k)}(i, j), A_T^{(k)}(i), 1 \leqq i \neq j \leqq M\} \tag{99}$$

is a sufficient statistic for Q. It is now easy to see that the maximum likelihood estimate for $q(i, j)$ is given by

$$\hat{q}_T^{(k)}(i, j) = \frac{N_T^{(k)}(i, j)}{A_T^{(k)}(i)} \tag{100}$$

if $i \neq j$ and $A_T^{(k)}(i) > 0$. If $A_T^{(k)}(i) = 0$, the maximum likelihood estimate does not exist and we adopt the convention that

$$\hat{q}_T^{(k)}(i, j) = 0 \text{ if } i \neq j \text{ and } A_T^{(k)}(i) = 0. \tag{101}$$

6.3 Large Sample Properties

We will now investigate the large sample properties of the set $\{\hat{q}_T^{(k)}(i, j), j \neq i\}$ both as $T \to \infty$ while k is fixed and as $k \to \infty$ with T fixed. The former corresponds to the case when a fixed number of sample paths can be observed over a long period of time and the latter to the case when a large number of sample paths can be observed over a fixed period of time. In the first case, we shall suppose that $k = 1$ to simplify calculations. If $k = 1$ we shall denote $N_T^{(1)}(i, j)$ by $N_T(i, j)$ and $A_T^{(1)}(i)$ by $A_T(i)$. The following theorem gives moments of $N_T^{(k)}(i, j)$ and $A_T^{(k)}(i)$ which will be useful later.

Theorem 6.3

(i) $E\{N_T(i,j)\} = q(i,j)\int_0^T P[Z(t)=i]\mathrm{d}t.$

(ii) $E\{A_T(i)\} = \int_0^T P[Z(t)=i]\mathrm{d}t.$

(iii) $$E\{N_T(i,j)N_T(r,s)\} = q(i,j)q(r,s)\int_0^T\int_0^x\{P_{si}(x-t)P[Z(t)=r] + P_{jr}(x-t)P[Z(t)=i]\}\mathrm{d}t\mathrm{d}x + \delta(i,j;r,s)q(i,j)\int_0^T P[Z(t)=i]\mathrm{d}t$$

where

$P_{ij}(t) = P[Z(t+s)=j|Z(s)=i],$

$$\delta(i,j;r,s) = \begin{cases} 1 & \textit{if } r=i \textit{ and } j=s \\ 0 & \textit{otherwise.} \end{cases}$$

(iv) $E[A_T(i)A_T(r)] = \int_0^T\int_0^x\{P_{ri}(x-t)P[Z(t)=r] + P_{ir}(x-t)P[Z(t)=i]\}\mathrm{d}t\mathrm{d}x.$

(v) $E[N_T(r,s)A_T(i)] = q(r,s)\int_0^T\int_0^x\{P_{ir}(x-t)P[Z(t)=i] + P_{si}(x-t)P[Z(t)= r]\}\mathrm{d}t\mathrm{d}x.$

(vi) $$E\{[N_T(r,s) - q(r,s)A_T(r)][N_T(i,j) - q(i,j)A_T(i)]\} = \delta(i,j;r,s)q(i,j)\int_0^T P[Z(t)=i]\mathrm{d}t.$$

Proof of this theorem is left as an exercise for the reader. We now study the easier problem of estimation when T is fixed and $k\to\infty$. Note that if $i\neq j$,

$$\hat{q}_T^{(k)}(i,j) = \begin{cases} N_T^{(k)}(i,j)/A_T^{(k)}(i) & \text{if } A_T^{(k)}(i)>0 \\ 0 & \text{if } A_T^{(k)}(i)=0. \end{cases}$$

If $P(A_T(i)=0)=i$, then $\hat{q}_T^{(k)}(i,j)=0$ with probability one for every k even if $q(i,j)>0$. In this case, the maximum likelihood estimate is a bad one. This situation occurs when we try to estimate parameters associated with transitions out of a state that can never be reached (for $P[A_T(i)=0]=1$ if and only if $P[Z(t)=i]=0$ for every t). We exclude this case from consideration. If $P[Z(t)=i]>0$, then $E(A_T(i))>0$ and hence, by Theorem 6.3,

$$\frac{E[N_T(i,j)]}{E[A_T(i)]} = q(i,j).$$

Since $N_T^{(k)}(i,j)$ and $A_T^{(k)}(i)$ are sums of independent identically distributed random variables,

$$\lim_{k\to\infty} \hat{q}_T^{(k)}(i,j) = \lim_{k\to\infty} \frac{N_T^{(k)}(i,j)/k}{A_T^{(k)}(i)/k} = q(i,j) \text{ a.s.}$$

by the strong law of large numbers. By applying a theorem of Cramer (1946) p.

254, the set of random variables

$$\{k^{1/2}(\hat{q}_T^{(k)}(i, j) - q(i, j))\}, \ i \neq j$$

has the same asymptotic distribution as the set

$$\left\{\frac{1}{E[A_T(i)]}\left(\frac{N_T^{(k)}(i, j) - q(i, j)A_T^{(k)}(i)}{k^{1/2}}\right)\right\}, \ i \neq j$$

and by the multivariate central limit theorem, the last is asymptotically normal with mean zero and covariances

$$C(i, j; k, 1) = \frac{1}{E[A_T(i)]E[A_T(k)]}E\{[N_T(i, j) - q(i, j)A_T(i)]$$

$$[N_T(k, 1) - q(k, 1)A_T(k)]\}$$

$$= \delta(i, j; k, 1)q(i, j) \Big/ \int_0^T P[Z(t) = i]\mathrm{d}t.$$

We have now proved the following theorem.

Theorem 6.4. *If there is positive probability that the ith state will be occupied at some time $t \geqq 0$, then*

$$\lim_{k \to \infty} \hat{q}_T^{(k)}(i, j) = q(i, j) \text{ a.s.}$$

and if every state has positive probability of being occupied, then the set of random variables

$$\{k^{1/2}(\hat{q}_T^{(k)}(i, j) - q(i, j))\}_{j \neq i}$$

are asymptotically normal and independent with zero mean and variance

$$q(i, j) \Big/ \int_0^T P[Z(t) = i]\mathrm{d}t.$$

In case k is held fixed ($k = 1$) and T grows large, consistency and asymptotic normality of maximum likelihood estimator holds good under some regularity conditions. The following theorem holds.

Theorem 6.5. *If $\{Z(t), t \geqq 0\}$ is positively regular, then*

$$\lim_{T \to \infty} \hat{q}_T(i, j) = q(i, j) \text{ a.s}$$

and the joint distribution of the set of random variables

$$\{T^{1/2}(\hat{q}_T(i, j) - q(i, j))\}_{\substack{i,j=1 \\ i \neq j}}^{n}$$

is asymptotically normal and independent with zero mean and variances $q(i,j)\,\rho/Q^{(i,i)}$ *where* ρ *is the product of the non-zero eigenvalues of* Q *and* $Q^{(i,i)}$ *is the* (i, i)*th cofactor of* Q.

We omit the proof of this theorem to save space; the reader is referred to Albert (1962) for a proof.

6.4 Asymptotic Efficiency

Under the conditions of Theorem 6.4,

$$k^{1/2}(\hat{q}_T^{(k)}(i, j) - q(i, j))$$

is asymptotically normally distributed with mean zero and variance

$$C_T^2 = \frac{q(i, j)}{\int_0^T P[Z(t)=i]\mathrm{d}t}.$$

It is easy to see that

$$\frac{\partial \log f_Q}{\partial q(i, j)} = \frac{1}{q(i, j)}[N_T(i, j) - q(i, j)A_T(i)]$$

and
$$\left[E\left\{\frac{\partial \log f_Q}{\partial q(i, j)}\right\}^2\right]^{-1} = q(i, j)\Big/\int_0^T P[Z(t)=i]\mathrm{d}t = C_T^2$$

so that $\hat{q}_T^{(k)}(i, j)$ is asymptotically efficient as the number k of independent sample paths observed goes to infinity.

If the observation consists of a single sample path of $Z(t)$ over the interval $(0, T)$, it is natural to define

$$\frac{1}{T}\,E\left\{\frac{\partial \log f_Q}{\partial q(i, j)}\right\}^2$$

as the amount of information per unit time and we shall call an estimate α_T^* of $q(i, j)$ asymptotically efficient as $T \to \infty$ if

$$T^{1/2}(\alpha_T^* - q(i, j))$$

is asymptotically normal with mean zero and variance

$$C^2 = \lim_{T\to\infty}\left[\frac{1}{T}\,E\left\{\frac{\partial \log f_Q}{\partial q(i, j)}\right\}^2\right]^{-1}. \tag{102}$$

It is not difficult to show that

$$\int_0^T P[Z(t)=i]\mathrm{d}t = Q^{(i,\,i)}T\rho^{-1} + o(T) \tag{103}$$

as $T \to \infty$. Since

$$\frac{1}{T}\,E\left\{\frac{\partial \log f_Q}{\partial\, q(i,j)}\right\}^2 = \int_0^T P[Z(t)=i]\mathrm{d}t/Tq(i,j), \tag{104}$$

relations (103) and (104) imply that relation (102) holds in view of the fact that

$$T^{1/2}(\hat{q}_T(i,j) - q(i,j)) \xrightarrow{\mathscr{L}} N(0,\ C^2)$$

where $C^2 = q(i,j)\rho/Q(i,i)$. Hence $\hat{q}_T(i,j)$ is asymptotically efficient as $T \to \infty$.

7. FURTHER TOPICS

Inference problems for special types of continuous parameter processes have been studied by several authors. Billingsley (1961) and Ranneby (1975b) studied maximum likelihood estimation for continuous parameter Markov chains with general state space and finite state space respectively. Keiding (1975), Athreya and Keiding (1977) and Feigin (1977) considered the case of birth and death processes as well as Markov branching processes. M. Brown (1971, 1972) has a general discussion of statistical analysis for non-homogeneous Poisson processes. Estimation problems of stationary point processes in general are discussed in Brillinger (1975).

A discussion for densities for measures generated by stochastic processes on suitable function spaces is given in Striebel (1959) and Gikhman and Skorokhod (1966).

Inference problems using contiguity methods for Markov processes, semi-Markov processes, Lévy processes of the discontinuous type, diffusion processes, and Gaussian processes with known covariance in the continuous parameter case have been discussed by Akritas (1978), Akritas and Roussas (1978), and Akritras and Johnson (1978).

BIBLIOGRAPHICAL NOTES

The basic and fundamental work in the area of statistical inference for stochastic processes was by Grenander (1951). A comprehensive survey of the results on this topic is given in Grenander (1967). General theory and examples discussed in Sections 2, 3, 4.1, 4.2, 5.1 and 5.2 are entirely due to Grenander (1951, 1967) except for Example 4.3 which is taken from Sagdar

(1974). Theory of inference for non-homogeneous Poisson processes in Section 4.3 is due to Brown (1972). Section 5.3 contains the work of Feigin (1977). Example 5.3 contains some work of Keiding (1975). Results in Section 6 are due to Albert (1962) and they are continuous time analogues of Anderson and Goodman (1957) to describe asymptotic behaviour of MLE of the transition probabilities of a Markov chain.

COMPLEMENTS

1. Let $\{X(t);\ t \geqq 0\}$ be a stationary Markov–Gaussian process with covariance function $r(s, t) = \exp\{-\beta|t-s|\}$. If the process is observed over $[0, T]$, obtain the uniformly most powerful test for testing the hypothesis $H_0: m \equiv E(X(t)) = 0$ against the one-sided alternative $H_1: m \equiv E(X(t)) > 0$.

(Grenander 1950, Rao, M.M. 1963)

2. (Continuation of Problem 1) Obtain the likelihood equation estimate of m if the process is observed for $[0, T]$. Show that it is consistent and asymptotically efficient as $T \to \infty$.

(Grenander, 1950)

3. Let $\{X_t,\ t \geqq 0\}$ be a Gaussian process with mean $E[X_t] = \mathbf{k}'\boldsymbol{\phi}(t)$ where $\mathbf{k}' = (k_1, \ldots, k_s)$, $\boldsymbol{\phi}(t) = (\phi_1(t), \ldots, \phi_s(t))'$ and $\text{cov}(X_t, X_s) = C(t, s)$ for C a non-singular continuous covariance function. Assume that ϕ_j are known functions. Obtain the likelihood equation estimate of $\mathbf{k}$ based on a single sample path of the process over $[0, T]$ and discuss its asymptotic properties as $T \to \infty$.

(Feigin, 1975)

4. Complete the discussion of the problems in Section 4.3.

(Grenander, 1950)

5. State and prove the theorem on asymptotic properties of MLE when several sample paths are available.

(Grenander, 1960)

6. Prove Theorem 6.3 giving moments of $N_T(i, j)$ and $A_T(i)$.

(Albert, 1962)

7. Let P_μ denote a non-homogeneous Poisson process on $[0, \infty)$ with mean measure μ. Show that P_{μ_1} and P_{μ_2} are singular with respect to each other i.e. $P_{\mu_1} \perp P_{\mu_2}$ if and only if at least one of the following two conditions hold:

(i) $$\int_{[|f_1/f_2 - 1| > C]} |f_1 - f_2|\,\mathrm{d}\nu = \infty \text{ for some } C > 0;$$

(ii) $$\int_{[|f_1/f_2 - 1| > C]} (f_1/f_2 - 1)^2 f_2\,\mathrm{d}\nu = \infty \text{ for all } C > 0$$

where f_i is the density of μ_i, $i=1, 2$, with respect to a measure ν dominating μ_1 and μ_2. If $P_{\mu_1} \perp P_{\mu_2}$, show that any level α Neyman–Pearson test for testing $H_0:\mu=\mu_1$ versus $H_1:\mu=\mu_2$ is consistent.

(Brown, 1972)

8. (*Continuation of* 7). If

(i) $\mu_1 \ll \mu_2$

(ii) there exists M such that $\mu_1\{x:|\log f_2(x)/f_1(x)|>M\}<\infty$,

(iii) $\int_{[|f_2/f_1-1|>C]}(f_2/f_1-1)^2 f_1 \, d\nu=\infty$ for all $C>0$ show that under P_{μ_1},

$$\frac{W_t-\varphi_{t,M}}{b_{t,M}} \xrightarrow{\mathscr{L}} N(0, 1)$$

where W_t is as defined in (31)

$$\varphi_{t,M}=\int_{S_{t,M}}\left(\log\frac{f_2}{f_1}\right)f_1 \, d\nu,$$

$$b_{t,M}^2=\int_{S_{t,M}}\left(\log\frac{f_2}{f_1}\right)^2 f_1 \, d\nu,$$

and

$$S_{t,M}=\left\{x\in[0, t]; \left|\log\frac{f_2(x)}{f_1(x)}\right|\leq M\right\}.$$

(Brown, 1972)

9. Let $P_{\mu_1}^{(t)}$ and $P_{\mu_2}^{(t)}$ be Poisson processes with $\mu_i([0, t])<\infty$, $i=1, 2$ and $f_1(x)=(d\mu_1/d\nu)(x)\leq f_2(x)=(d\mu_2/d\nu)(x)$ for all $x\in[0, t]$ where ν is a measure dominating μ_1 and μ_2. Let K be a non-negative function on $[0, t]$. Show that for all a.

$$P_{\mu_1}\left[\sum_1^{N_t} K(x_i)>a\right]\leq P_{\mu_2}\left[\sum_1^{N_t} K(x_i)>a\right],$$

where N_t is the number of events in $[0, t]$ and x_i, $i\geq 1$ are the epochs at which these events occur. Hence prove the result mentioned at the end of Section 4.

(Brown, 1972)

10. Let $\{\xi(t), t\geq 0\}$ be an irreducible separable continuous time Markov chain defined on a complete probability space $(\Omega, \mathscr{A}, P)$ and suppose that the state space is finite and the transition probability $P_{ij}(t; \theta)$ is stationary. Under suitable assumptions to be stated, show that any modified maximum likelihood estimator (MMLE) is strongly consistent. Further show that, under suitable regularity conditions, MLE exists, is unique and is a solution of the likelihood equation with probability one.

(Ranneby, 1975a)

Chapter 9
Diffusion Processes

1. INTRODUCTION

In this chapter, we shall study inference problems for an important class of processes, the class of diffusion processes which serve as probabilistic models of the physical process of diffusion.

An example of a diffusion phenomenon is the motion of small particles suspended in a homogeneous liquid which are under the influence of collisions with the molecules of the liquid in chaotic thermal motion. Under reasonable assumptions, the process $\xi(t)$ in $\mathbb{R}^3$ that describes the position of the particle at an instant t is a process with independent increments. (For definition, see Doob, 1953.) Furthermore, physical considerations imply that the process will be continuous and stationary with respect to time if the physical state of the liquid does not change with time. But every continuous homogeneous process with independent increments is a Gaussian process. At time $t=0$, suppose that $\xi(0)=0$. Further suppose that $E[\xi(t)]=t\mathbf{a}$ and $\operatorname{var}[\xi((t), \boldsymbol{\alpha})]=\boldsymbol{\alpha}' A\boldsymbol{\alpha} t$ where A is a symmetric matric of order 3×3. In the case of a homogeneous liquid, it can be seen again that $\mathbf{a}=\mathbf{0}$ and $A=cI$ where c is a nonnegative constant and I is the identity matrix of order 3×3.

A process $W(t)$ is called a *Brownian motion* or *Wiener process* if it is a stationary Gaussian process with independent increments such that $EW(t)=0$ and $\operatorname{var}[(W(t), \boldsymbol{\alpha})\}=c\boldsymbol{\alpha}'\boldsymbol{\alpha} t$ for some $c\geqq 0$.

Using a similar construction in the case of a non-homogeneous medium, we can arrive at the concept of a general diffusion process. We shall clarify the basic concepts of a diffusion process by giving an example of a one-dimensional process.

Let X_t denote the coordinate of a sufficiently small particle suspended in a liquid at an instant t. Suppose the velocity of the motion of the liquid at the point x and the instant t is equal to $a(t, x)$. Further assume that the fluctuational component of the displacement is a random variable whose distribution depends on the position x of the particle, the instant t at which the displacement is observed, and the quantity Δt which is the length of the interval of time during which its displacement is observed. We assume that the average value of this displacement is equal to zero independently of t, X_t and Δt. Thus the displacement of the particle can be written approximately in the form

$$X_{t+\Delta t} - X_t = a(t, X_t)\Delta t + \xi_{t, X_t, \Delta t}. \tag{1}$$

Here $E\xi_{t, X_t, \Delta t} = 0$. If $a(t, x) = 0$ and the distribution of $\xi_{t, X_t, \Delta t}$ is independent of x and t as in the case of Brownian motion, then $E\,\xi^2_{t, X_t, \Delta t} = l\Delta t$. Since the properties of the medium are naturally assumed to change only slightly for small changes in t and x, the process is homogeneous. Therefore, we may assume that

$$\xi_{t, X_t, \Delta t} = \sigma(t, X_t)\xi_{t, \Delta_t}$$

where $\sigma(t, x)$ characterizes the properties of the medium at the point x at the instant t and $\xi_{t, \Delta t}$ is the value of the increment that is obtained in the homogeneous case under the condition that $\sigma(t, x) = 1$. Thus $\xi_{t, \Delta t}$ must be distributed like the increment of a process of Brownian motion: $W(t+\Delta t) - W(t)$. Consequently, we can write the approximate formula

$$\begin{aligned} X_{t+\Delta t} - X_t &\sim a(t, X_t)\Delta t \\ &+ \sigma(t, X_t)[W(t+\Delta t) - W(t)]. \end{aligned} \tag{2}$$

If we replace the increments by differentials, we obtain the differential equation

$$\mathrm{d}X_t = a(t, X_t)\mathrm{d}t + \sigma(t, X_t)\mathrm{d}W(t) \tag{3}$$

which can be taken as the starting point for defining diffusion process.

We note that we cannot as yet give a precise meaning to (3) since $W(t)$ does not have a derivative in the usual sense. We can give a precise meaning to (3) by introducing the concepts of stochastic integrals and stochastic differentials. Precise definitions and the theory of these concepts are outside the scope of this book. However, we have given some of the relevant theorems and proofs in Appendix 2.

2. DIFFUSION PROCESSES

Diffusion processes are special cases of Markov processes with continuous sample functions. Depending on the classification of methods, there are two basically different approaches to the class of diffusion processes. One can either define them in terms of the conditions on the transition probabilities or one can study the state X_t itself and its variation with respect to time as was done in the previous section. It can be shown that these two approaches lead essentially to the same class of processes (cf. Arnold, 1974, p. 156). We shall first formally define a Markov process and then a diffusion process. The reader is referred to Doob (1953), Gikhman and Skorokhod (1969), Dynkin (1965) etc. for a deeper study of Markov processes.

Definition 2.1. A stochastic process $\{X_t, t_0 \leqq t \leqq T\}$ defined on a probability space $(\Omega, \mathscr{F}, P)$ with state space $\mathbb{R}^d$, is called a *Markov process* if the following Markov property is satisfied: For $n \geqq 1$, $t_0 \leqq t_1 \leqq \ldots < t_n \leqq T$ and $B \in \mathscr{B}^d$,

$$P(X_t \in B | X_{t_1}, \ldots, X_{t_n}) = P(X_t \in B | X_{t_n}) \tag{4}$$

where $\mathscr{B}^d$ is the σ-algebra of Borel sets of $\mathbb{R}^d$.

Loosely speaking, Markov property says that if the state of a system at a particular time s is known, additional information regarding the behaviour of the system at times $t < s$ (the past) has no effect on our knowledge of the probable development of the system at $t > s$ (in the future). It can be shown that for a Markov process, the past and future are statistically independent when the present is known.

Let $\{X_t, t_0 \leq t \leq T\}$ be a Markov process. There exists a conditional distribution $Q(X_s, B) = P(s, X_s, t, B)$ corresponding to the conditional probability $P(X_t \in B | X_s)$. The function $P(s, x, t, B)$ is called a *transition probability* of the Markov process X_t. The function $P(s, x, t, \cdot)$ is a probability measure on $\mathscr{B}^d$ for fixed $s \leqq t$ and $x \in \mathbb{R}^d$ and $P(s, \cdot; t, B)$ is $\mathscr{B}^d$-measurable for fixed $s \leqq t$ and $B \in \mathscr{B}^d$. Furthermore, for $t_0 \leqq s \leqq u \leqq t \leqq T$, $B \in \mathscr{B}^d$ and for almost all $x \in \mathbb{R}^d$, the function $P(s, x, t, B)$ satisfies the Chapman–Kolmogorov Equation, namely

$$P(s, x; t, B) = \int_{\mathbb{R}^d} P(u, y; t, B) P(s, x; u, dy). \tag{5}$$

A Markov process $\{X_t, t_0 \leqq t \leqq T\}$ is said to be *homogeneous* (with respect to time) if its transition probability $P(s, x; t, B)$ is stationary, that is, if the condition

$$P(s+u, x; t+u, B) = P(s, x; t, B)$$

for $t_0 \leqq s \leqq t \leqq T$ and $t_0 \leqq s+u \leqq t+u \leqq T$ holds.

It is easy to see that the Wiener process described in the introduction is a homogeneous Markov process with stationary transition probability.

Definition 2.2. A Markov process $\{X_t, t_0 \leqq t \leqq T\}$ with values in $\mathbb{R}^d$ and almost certainly continuous sample functions is called a multidimensional *diffusion process* if its transition probability $P(s, x; t, B)$ satisfies the following three conditions for every $s \in [t_0, T]$, $x \in \mathbb{R}^d$ and $\varepsilon > 0$:

(i) $$\lim_{t \downarrow s} \int_{\|y-x\| > \varepsilon} P(s, x; t, \mathrm{d}y) = 0; \tag{6}$$

(ii) there exists an $\mathbb{R}^d$-valued function $a(s, x)$ such that

$$\lim_{t \downarrow s} (t-s)^{-1} \int_{\|y-x\| \leqq \varepsilon} (y-x)\, P(s, x; t, \mathrm{d}y) = a(s, x); \tag{7}$$

(iii) there exists a $d \times d$ matrix-valued function $B(s, x)$ such that

$$\lim_{t \downarrow s} (t-s)^{-1} \int_{\|y-x\| \leqq \varepsilon} (y-x)(y-x)' P(s, x; t, \mathrm{d}y) = B(s, x). \tag{8}$$

The functions a and B are called the *coefficients* of the diffusion process. a is called the *drift vector* and B is called the *diffusion matrix*. $B(s, x)$ is symmetric and non-negative definite.

We remark that truncated moments are used in (i) and (ii) of Definition 2.2 since the regular moments need not necessarily exist. However, if

$$\lim_{t \downarrow s} \frac{1}{t-s} \int_{\mathbb{R}^d} \| y-x \|^{2+\delta} P(s, x, t, \mathrm{d}y) = 0 \tag{9}$$

for some $\delta > 0$, then condition (i) is automatically satisfied by Tchebyshev's inequality and we can choose $\mathbb{R}^d$ as the region of integration in conditions (ii) and (iii). Condition (i) means that large changes in X_t over a short period of time are improbable i.e.

$$P(\|X_t - X_s\| > \varepsilon | X_s = x) = o(t-s).$$

Let us suppose that the truncated moments in (ii) and (iii) are replaced with the usual ones. Then, it is easy to see that

$$E(X_t - X_s | X_s = x) = a(s, x)(t-s) + o(t-s) \tag{10}$$

and

$$E\{(X_t - X_s)(X_t - X_s)' | X_s = x\} = B(s, x)(t-s) + o(t-s). \tag{11}$$

Therefore, given that $X_s = x$,

$$\operatorname{cov}(X_t - X_s) = B(s, x)(t-s) + o(t-s) \tag{12}$$

where $\text{cov}(X_t - X_s)$ is the covariance matrix of $X_t - X_s$ with respect to the probability $P(s, x; t, \cdot)$. $a(s, x)$ is the mean velocity vector of the random motion described by X_t under the condition $X_s = x$, and $B(s, x)$ is the measure of the local magnitude of the fluctuation of $X_t - X_s$ about the mean value.

The decisive property of diffusion processes is that their transition probability $P(s, x; t, B)$ is, under certain regulating assumptions, uniquely determined by the drift vector and diffusion matrix. This is surprising since a and B are obtained only from the first two moments of $P(s, x; t, B)$ which do not in general define a distribution.

Uniform motion with velocity v is a diffusion process with $a \equiv v$ and $B \equiv 0$ and the Wiener process W_t is a diffusion process with drift vector $a \equiv 0$ and diffusion matrix $B \equiv I$.

The next section deals with the study of the absolute continuity of measures corresponding to processes of diffusion type relative to a Wiener measure. This is essential for the study of inference problems for diffusion processes.

3. ABSOLUTE CONTINUITY OF MEASURES FOR DIFFUSION PROCESSES

Let $(\Omega, \mathscr{F}, P)$ be a probability space and $\{\mathscr{F}_t, 0 \leqq t \leqq 1\}$ be a non-decreasing family of σ-algebras contained in $\mathscr{F}$. Let $W = (W_t, \mathscr{F}_t, P)$ be a standard Wiener process i.e., a continuous square integrable martingale with $W_0 \equiv 0$, $E[(W_t - W_s)^2 | \mathscr{F}_s] = (t-s) a \cdot s[P]$, $t \geqq s$. For a discussion about martingales, see Appendix 1. We note that the definition of the Wiener process given earlier coincides with this definition if $\mathscr{F}_t$ is taken to be the σ-algebra generated by the random variables W_s, $s \leqq t$.

Let $(C, \mathscr{B})$ denote the measurable space of continuous functions $x = (x_t, 0 \leqq t \leqq 1)$ on $[0, 1]$, $x_0 = 0$ with σ-algebra $\mathscr{B} = \sigma\{x : x_s, s \leqq 1]$. Also let $\mathscr{B}_t = \sigma\{x : x_s, s \leqq t\}$.

Further, let $\alpha_t(x)$ be jointly measurable with respect to the product σ-algebra $\tau \times \mathscr{B}$ (τ is the σ-algebra of Borel sets on $[0, 1]$ not depending on the "future" i.e. $\mathscr{B}_t$-measurable for each $0 \leqq t \leqq 1$).

Suppose $\xi = (\xi_t, \mathscr{F}_t)$ is a continuous random process defined on $(\Omega, \mathscr{F}, P)$. This implies in particular that ξ_t is $\mathscr{F}_t$-measurable and the sample functions of the process ξ belong to C with probability 1. The measure μ_ξ defined on the σ-algebra $\mathscr{B}$ of Borel sets of C by

$$\mu_\xi(B) = P\{w : \xi_t(w) \in B\}$$

is called the measure induced by the process ξ on C or in short measure of ξ.

Definition 3.1. A continuous random process $\xi=(\xi_t, \mathscr{F}_t)$ defined on $(\Omega, \mathscr{F}, P)$ is called a *process of diffusion type* if there exists a $\tau.\times \mathscr{B}$ measurable function $\alpha_t(x)$ such that

$$P\left\{\int_0^1 |\alpha_t(\xi)|\mathrm{d}t<\infty\right\}=1 \tag{13}$$

and for each $0\leqq t\leqq 1$

$$\xi_t=\int_0^t \alpha_s(\xi)\mathrm{d}s+W_t \text{ a.s.}[P]. \tag{14}$$

If we say that "ξ is a process admitting the stochastic differential"

$$\mathrm{d}\xi_t=\alpha_t(\xi)\mathrm{d}t+\mathrm{d}W_t,\ \xi_0=0 \tag{15}$$

we shall mean that ξ is a process of diffusion type. For a discussion on stochastic integrals and stochastic differential equations, see Appendix 2.

Let μ_ξ and μ_W denote the measures on the space $(C, \mathscr{B})$ corresponding to the process ξ and the Wiener process W respectively. These are also called the probability distributions of the random processes ξ and W on the functional space $(C, \mathscr{B})$.

The most important problem is to find necessary and sufficient conditions for the mutual absolute continuity or equivalence $(\mu_\xi\sim\mu_W)$ of the measures μ_ξ and μ_W and find their Radon–Nikodym derivatives when the equivalence obtains. We shall state the main results as the proofs are beyond the scope of this book.

Theorem 3.1. *If ξ is a process of diffusion type with stochastic differential* (15) *then,*

$$P\left\{\int_0^1 \alpha_t^2(\xi)\mathrm{d}t<\infty\right\}=1$$

and (16)

$$P\left\{\int_0^1 \alpha_t^2(W)\mathrm{d}t<\infty\right\}=1$$

if and only if $\mu_\xi\sim\mu_W$. In such an event the densities (the Radon–Nikodym derivatives) $(\mathrm{d}\mu_\xi/(\mathrm{d}\mu_W)(x)$ and $(\mathrm{d}\mu_W/\mathrm{d}\mu_\xi)(x)$ are given by

$$\frac{\mathrm{d}\mu_\xi}{\mathrm{d}\mu_W}(W)=\exp\left\{\int_0^1 \alpha_t(W)\mathrm{d}W_t-\frac{1}{2}\int_0^1 \alpha_t^2(W)\mathrm{d}t\right\} \text{ a.s.}[P] \tag{17}$$

and

$$\frac{d\mu_W}{d\mu_\xi}(\xi)=\exp\left\{-\int_0^1 \alpha_t(\xi)d\xi_t+\frac{1}{2}\int_0^1 \alpha_t^2(\xi)dt\right\} \text{ a.s.}[P]. \tag{18}$$

By $d\mu_\xi/d\mu_W(W)$, we denote the random variable $(d\mu_\xi/d\mu_W)(W(w))$ derived from $(d\mu_\xi/d\mu_W)(x)$ by substituting for the function x the sample path $W(w)$. $(d\mu_W/d\mu_\xi)(\xi(w))$ is similarly defined.

Theorem 3.2. *If ξ is a process of diffusion type with stochastic differential* (15), *then*

$$P\left\{\int_0^1 \alpha_t^2(\xi)dt<\infty\right\}=1 \text{ if and only if } \mu_\xi \ll \mu_W. \tag{19}$$

Here

$$\frac{d\mu_\xi}{d\mu_W}(\xi)=\exp\left\{\int_0^1 \alpha_t(\xi)d\xi_t-\frac{1}{2}\int_0^1 \alpha_t^2(\xi)dt\right\} \text{ a.s.}[P]. \tag{20}$$

It is important to note that every random process ξ for which $\mu_\xi \sim \mu_W$ is a process of diffusion type up to equivalence with stochastic differential

$$d\xi_t=\alpha_t(\xi)dt+dW_t$$

where

$$P\left\{\int_0^1 \alpha_t^2(W)dt<\infty\right\}=P\left\{\int_0^1 \alpha_t^2(\xi)dt<\infty\right\}=1.$$

An exact formulation of this result and its proof is given in Lipcer and Širjaev (1972); also see Lipcer and Širjaev (1977) for a complete discussion of the results.

The proofs of these results rely essentially on the following theorem due to Girsanov (1960).

Theorem 3.3. *Let $W=(W_t, \mathscr{F}_t, P)$ be a standard Wiener process defined on $(\Omega, \mathscr{F}, P)$ and let the process $\beta=(\beta_t, \mathscr{F}_t)$ be such that*

$$P\left\{\int_0^1 \beta_t^2(w)dt<\infty\right\}=1. \tag{21}$$

Let

$$\phi(w)=\exp\left\{\int_0^1 \beta_t(w)\mathrm{d}W_t-\frac{1}{2}\int_0^1 \beta_t^2(w)\mathrm{d}t\right\}. \tag{22}$$

If $E\phi(w)=1$, *then the randon process* $\xi=(\xi_t, \mathscr{F}_t, \tilde{P})$ *where*

$$\xi_t=-\int_0^1 \beta_s(w)\mathrm{d}s+W_t,\ 0\leqq t\leqq 1 \tag{23}$$

and

$$\tilde{P}(\mathrm{d}w)=\phi(w)P(\mathrm{d}w) \tag{24}$$

is a Wiener process (relative to $(\mathscr{F}_t, \tilde{P})$*).*

We shall now discuss briefly some results for multidimensional diffusion processes. Suppose that $\xi(t)$ is a multidimensional diffusion process with drift vector $\mathbf{a}(s, x)$ and diffusion matrix $B(s, x)$. Under some regularity conditions $\mathbf{a}(s, x)$ and $B(s, x)$ completely determine the transition probability $P(s, x; t, A)$ of the process $\xi(t)$. Assume that relation (9) holds. Then relations (10) and (11) hold by our earlier remarks. We shall first define the integral

$$\int_0^t (\mathbf{f}(s, \xi(s)), \mathrm{d}\xi(s)) \tag{25}$$

where $\mathbf{f}(s, x)$ is a continuous vector function with values in $\mathbb{R}^d$. This can be defined as the limit of the sums

$$\sum_{k=0}^{n-1} (\mathbf{f}(s_k, \xi(s_k)), [\xi(s_{k+1})-\xi(s_k)])$$

where $0=s_0<s_1<\ldots<s_n=t$. It can be seen that

$$\int_0^t (\mathbf{f}(s, \xi(s)), \mathrm{d}\xi(s))$$
$$=\int_0^t (\mathbf{f}(s, \xi(s)), \mathbf{a}(s, \xi(s)))\mathrm{d}s+\int_0^t (\mathbf{f}(s, \xi(s)), \mathrm{d}\zeta(s)) \tag{26}$$

where

$$\zeta(s)=\xi(s)-\int_0^t a(s, \xi(s))\mathrm{d}s.$$

Since $\zeta(s)$ is a martingale, one can define

$$\int_0^t (\mathbf{f}(s, \xi(s)), \mathrm{d}\zeta(s))$$

as a stochastic integral. Furthermore

$$E\left\{\int_0^t (\mathbf{f}(s, \xi(s)), d\zeta(s))\right\}=0, \tag{27}$$

and

$$E\left\{\int_0^t (\mathbf{f}(s, \xi(s)), d\zeta(s))\right\}^2$$

$$=\int_0^t E\{(B(s, \xi(s))\mathbf{f}(s, \xi(s)), \mathbf{f}(s, \xi(s))\}ds. \tag{28}$$

Theorem 3.4. *Let* $\mathbf{C}(s,x)$ *be a vector function continuous in the pair* (s,x) *and*

$$d(s,x)=(\mathbf{a}(s,x), \mathbf{C}(s,x))+\tfrac{1}{2}(B(s,x)\mathbf{C}(s,x), \mathbf{C}(s,x)). \tag{29}$$

Then

$$\exp\left\{\int_0^T (\mathbf{C}(s, \xi(s)), d\xi(s))-\int_0^T d(s, \xi(s))ds\right\} \tag{30}$$

is the density of a measure corresponding to a Markov process $\xi_1(t)$ *on* $[0, T]$ *relative to the measure corresponding to the process* $\xi(t)$ *on* $[0, T]$. *Furthermore the transition probability of this Markov process is given by*

$$P^*(t, x; s, A)$$

$$=E(\chi_A(\xi_1(s)) \exp\left\{\int_t^s (\mathbf{C}(u, \xi(u)), d\xi(u)) \right. \tag{31}$$

$$\left. -\int_t^s d(u, \xi(u))du\right\} \Bigg| \xi_1(t)=x).$$

(Here χ_A denotes the indicator function of set A)

Theorem 3.5. *If the functions* $\mathbf{C}(u, x)$ *and* $d(u, x)$ *given in Theorem* 3.4 *are continuous in* (u, x) *and* $\mathbf{a}(t, x)$, $B(t, x)$ *and*

$$\mathbf{a}^*(t, x)=\mathbf{a}(t, x)+B(t, x)\mathbf{C}(t, x)$$

satisfy Lipschitz conditions, then a process with transition probability $P^*(t, x; s, A)$ *defined by relation* (31) *is a diffusion process with drift vector* $\mathbf{a}^*(t, x)$ *and diffusion matrix* $B(t,x)$.

Theorem 3.6 *Suppose that the diffusion coefficients* $\mathbf{a}_1(t, x)$, $B_1(t, x)$, $\mathbf{a}_2(t, x)$ *and* $B_2(t, x)$ *of two diffusion process* $\xi_1(t)$ *and* $\xi_2(t)$ *on* $[0, T]$ *satisfy*

Lipschitz conditions in addition to conditions stated above. If μ_i is the measure corresponding to $\xi_i(t)$, then for the absolute continuity of μ_2 relative to μ_1, it is sufficient that $B_1(t, x) = B_2(t, x)$ and that there exists a vector function $\mathbf{C}(t, x)$ for which

$$\mathbf{a}_2(t, x) - \mathbf{a}_1(t, x) = B_1(t, x)\mathbf{C}(t, x)$$

and the distribution of $\xi_2(0)$ is absolutely continuous relative to the distribution of $\xi_1(0)$ (we denote the corresponding density by $\rho_0(x)$). Under these conditions $\mathrm{d}\mu_2/\mathrm{d}\mu_1$ is determined by

$$\frac{\mathrm{d}\mu_2}{\mathrm{d}\mu_1}(\xi_1(\cdot)) = \rho_0(\xi_1(0))\eta$$

where η is given by (30) *with $\mathbf{a}(t, x) = \mathbf{a}_1(t, x) - \mathbf{a}_2(t, x)$, $B(t, x) = B_1(t, x)$ and $\mathbf{C}(t, x)$ as defined above.*

Proofs of Theorems 3.4, 3.5 and 3.6 and further discussion on these topics are given in Gikhmon and Skorokhod (1966). The one-dimensional version of this theorem is given in Skorokhod (1965).

4. PARAMETER ESTIMATION IN A LINEAR STOCHASTIC DIFFERENTIAL EQUATION

This section is concerned with the problem of estimation for the drift parameter θ in a vector linear stochastic differential equation with constant coefficients of the form

$$\mathrm{d}\mathbf{X}_t = \theta\mathbf{X}_t\mathrm{d}t + G\mathrm{d}\mathbf{W}_t, \ t \geqq 0; \ \mathbf{X}_0 \equiv 0$$

where $\{\mathbf{W}_t, \ t \geqq 0\}$ is a standard Wiener process and G is a non-singular matrix. See Example 5.4.

4.1 Continuous Sampling

We first discuss likelihood equation estimation of the parameter based on observing $\mathbf{X}_t$ continuously throughout $0 \leqq t \leqq T$.

Let $(\Omega, \mathscr{A})$ be a measurable space and $\{\mathbf{X}_t, \ t \geqq 0\}$ be a family of random vectors defined on $(\Omega, \mathscr{A})$ with values in $\mathbb{R}^n$. Let $\mathscr{A}_t$ be the σ-algebra generated by $\{\mathbf{X}_s, \ 0 \leqq s \leqq t\}$. Let Θ be a subset of the space of square matrices of order $n \times n$ and ζ be a subset of the space $\mathscr{I}_n$ of non-singular square matrices of order $n \times n$. We assume that there exists a matrix $\theta_0 \in \Theta$, a matrix $G_0 \in \zeta$ and a probability measure P_{θ_0, G_0} on $(\Omega, \mathscr{A})$ such that the process $\{\mathbf{X}_t, t \geqq 0\}$ is a solution of the stochastic differential equation

$$d\mathbf{X}_t = \theta_0 \mathbf{X}_t d_t + G_0 d\mathbf{W}_t, \ t \geqq 0; \ \mathbf{X}_0 = 0 \tag{32}$$

with respect to P_{θ_0, G_0}.

Let $C(T, \mathbb{R}^n)$ be the space of continuous functions from $[0, T]$ to $\mathbb{R}^n$, $\mathscr{F}_T$ be the σ-algebra generated by the coordinates for $0 \leqq t \leqq T$. Let $\mu_{\theta,G}^T$ be the measure induced by the solution of the stochastic differential equation

$$d\mathbf{Z}_t = \theta \mathbf{Z}_t dt + G d\mathbf{W}_t, \ t \geqq 0; \ \mathbf{Z}_0 \equiv 0 \tag{33}$$

on $\mathscr{F}_T$ by the mapping $\pi_t(\mathbf{X}_t(w), t \geqq 0) = \mathbf{X}_t(w), 0 \leqq t \leqq T, w \in \Omega$. Let μ_*^T be the measure induced by Wiener Process on $\mathscr{F}_T$.

We first prove that it can be assumed that $GG' = E_n$ where E_n is the identity matrix of order n. Even though this might appear to be a severe restriction, it is in fact a consequence of the assumption that $\mathbf{X}_t$ can be observed continuously throughout the time interval $0 \leqq t \leqq T$.

Lemma 4.1. *For every $\theta \in \Theta$ and $G \in \zeta$, the measure $\mu_{\theta,G}^T$ is absolutely continuous with respect to the measure $\mu_{0,G}^T$ with Radon–Nikodym derivative*

$$\frac{d\mu_{\theta,G}^T}{d\mu_{0,G}^{*T}} = \exp\left\{\int_0^T \langle \theta\pi_t, (GG')^{-1} d\pi_t \rangle - \tfrac{1}{2}\int_0^T \langle \theta\pi_t, (GG')^{-1}\theta\pi_t \rangle dt\right\} \tag{34}$$

where the stochastic integral

$$\int_0^T \langle \theta\pi_t, (GG')^{-1} d\pi_t \rangle$$

is defined with respect to $\mu_{0,G}^T$.

Proof. Consider the process $\{\mathbf{Z}_t^1, 0 \leqq \mathbf{t} \leqq T\}$ defined by the stochastic differential equation

$$d\mathbf{Z}_t^1 = G d\mathbf{W}_t, \ 0 \leqq t \leqq T; \ \mathbf{Z}_0^1 \equiv 0 \tag{35}$$

with respect to μ_*^T. Let $\tilde{\mu}^T$ be the measure absolutely continuous with respect to μ_*^T on $\mathscr{F}_T$ defined by the Radon–Nikodym derivative

$$\frac{d\tilde{\mu}^T}{d\mu_*^T} = \exp\left\{\int_0^T \langle G^{-1}\theta\mathbf{Z}_t^1, d\mathbf{W}_t \rangle - \tfrac{1}{2}\int_0^T \langle G^{-1}\theta\mathbf{Z}_t^1, -G^{-1}\theta\mathbf{Z}_t^1 \rangle dt\right\}. \tag{36}$$

Theorem 3.3 asserts that the process

$$\hat{\mathbf{W}}_t = \mathbf{W}_t - \int_0^t G^{-1}\theta \mathbf{Z}_s^1 \mathrm{d}s$$

is a Wiener process with respect to $\tilde{\mu}^T$ and

$$\mathrm{d}\mathbf{Z}_t^1 = \theta \mathbf{Z}_t^1 \mathrm{d}t + G\mathrm{d}\hat{\mathbf{W}}_t;\ 0 \leqq t \leqq T;\ \mathbf{Z}_0^1 \equiv 0$$

with respect to $\tilde{\mu}^T$. Hence the measure induced by the process $\{\mathbf{Z}_t^1, 0 \leq t \leqq T\}$ on $\mathscr{F}_T$ with respect to $\tilde{\mu}^T$ is absolutely continuous with respect to the measure induced by the same process on $\mathscr{F}_T$ with respect to μ_*^T and their Radon–Nikodym derivative is

$$\exp\left\{\int_0^T \langle G^{-1}\theta\pi_t, G^{-1}\theta \mathrm{d}\pi_t\rangle - \tfrac{1}{2}\int_0^T \langle G^{-1}\theta\pi_t, G^{-1}\theta\pi_t\rangle \mathrm{d}t\right\}.$$

Since the solution of the stochastic differential equation

$$\mathrm{d}Z_t = \theta Z_t \mathrm{d}t + G\mathrm{d}W_t$$

is unique (in distribution) (cf. Gikhman and Skorokhod, 1970, Theorem 1, Chapter 2), it follows that the processes $\{\mathbf{Z}_t, 0 \leqq t \leqq T]$ and $\{\mathbf{Z}_t^1, 0 \leqq t \leqq T\}$ are the same as far as induced measures are concerned.

Lemma 4.2. *For every $(\theta, G) \in \Theta \times \zeta$ and for $0 \leqq t \leqq T$,*

$$\hat{\sigma}_t(X) \equiv \lim_{N\to\infty} \frac{1}{t} \sum_{i=1}^{2^N} (\mathbf{X}_{it2^{-N}} - \mathbf{X}_{(i-1)t2^{-N}})(\mathbf{X}_{it2^{-N}} - \mathbf{X}_{(i-1)t2^{-N}})'$$

$= GG'$*–almost surely.*

Proof. Observe that

$$G^{-1}\hat{\sigma}_t(X)G'^{-1} = \lim_{N\to\infty} \frac{1}{t} \sum_{i=t}^{2^N} (G^{-1}\pi_{it2^{-N}} - G^{-1}\pi_{(i-1)t2^{-N}})$$
$$(G^{-1}\pi_{it2^{-N}} - G^{-1}\pi_{(i-1)t2^{-N}})'$$

where $(G^{-1}\pi_t, 0 \leqq t \leqq T)$ is a standard Wiener process with respect to the measure $\mu_{0,G}^T$. But for a Wiener process $\{\mathbf{W}_t, 0 \leqq t \leqq T]$, it is known (cf. Doob, (1953 p. 395) that

$$\lim_{N\to\infty} \frac{1}{t} \sum_{i=1}^{2^N} \mathbf{W}_{it2^{-N}} - \mathbf{W}_{(i-1)t2^{-N}})(\mathbf{W}_{it2^{-N}} - \mathbf{W}_{(i-1)t2^{-N}})' = E_n \text{ a.s.}$$

Hence

$$G^{-1}\hat{\sigma}_t G'^{-1} = E_n \text{ a.s. } \mu^T_{0,G}$$

which implies that $\hat{\sigma}_t = GG'$ a.s. $\mu^T_{0,G}$. Therefore, by Lemma 4.1, $\hat{\sigma}_t = GG'$ a.s. $\mu^T_{\theta,G}$ for all $(\theta, G) \in \Theta \times \zeta$.

Lemma 4.2 implies that the matrix GG' can be computed with probability one with the help of a single trajectory of the process on any finite time interval.

In view of the above observation, there is no loss of generality in assuming that GG' is completely known whenever continuous sampling is involved. We can now study the problem of estimation of θ.

Observe that, the family of measure $\{\mu^T_{\theta,G}, \theta \in \Theta\}$ on $(C(T; \mathbb{R}^n), \mathscr{F}_T)$ is dominated by the measure $\mu^T_{0,G}$ with

$$\frac{d\mu^T_{\theta,G}}{d\mu^T_{0,G}} = \exp\left\{\int_0^T \langle \theta\pi_t, (GG')^{-1} d\pi_t\rangle - \tfrac{1}{2}\int_0^T \langle \theta\pi_t, (GG')^{-1}\theta\pi_t\rangle dt\right\} \tag{37}$$

where the measure $\mu^T_{0,G}$ is completely known since GG' is known. The Radon–Nikodym derivative in relation (37) can also be written in the form

$$\frac{d\mu^T_{\theta,G}}{d\mu^T_{0,G}} = \exp\left\{Tr\left[(GG')^{-1}\theta\int_0^T \pi_t d\pi'_t\right] - \tfrac{1}{2} Tr\left[\theta'(GG')^{-1}\theta\int_0^T \pi_t\pi'_t dt\right]\right\} \tag{38}$$

where Tr means Tracc. Since the structure of the Radon–Nikodym derivative is of exponential type, it follows that

$$\left\{\int_0^T d\pi_t \cdot \pi'_t, \int_0^T \pi_t \cdot \pi'_t dt\right\}$$

is a sufficient statistic for the family $\{\mu^T_{\theta,G}, \theta \in \Theta\}$ where the stochastic integral is defined with respect to $\mu^T_{0,G}$. If $n=1$, then

$$\left(\pi^2_T - T, \int_0^T \pi^2_s ds\right)$$

is the sufficient statistic.

The log-likelihood function is

$$L_G(\theta, \cdot) = \int_0^T \langle \theta\pi_t, (GG')^{-1} d\pi_t\rangle - \tfrac{1}{2}\int_0^T \langle \theta\pi_t, (GG')^{-1}\theta\pi_t\rangle dt$$

where the stochastic integral is defined with respect to $\mu_{0,G}^T$.

The system of likelihood equations can be obtained by computing

$$dL_G(\theta, \cdot) = Tr\left\{d\theta\left[\int_0^T \pi_t \cdot d\pi_t'\right](GG')^{-1} - d\theta\left[\int_0^T \pi_t \cdot \pi_t' dt\right]\theta'(GG')^{-1}\right\}$$

and the system can be written in the form

$$\int_0^T d\pi_t \cdot \pi_t' - \theta \int_0^T \pi_t \cdot \pi_t' dt = 0. \tag{39}$$

Any solution of the system of likelihood equations (39) is a *likelihood equation estimator* of the parameter.

We can now state the following theorem.

Theorem 4.1. *The likelihood equation estimator of θ is*

$$\hat{\theta}_G(T) = \left\{\int_0^T d\pi_t \cdot \pi_t'\right\}\left\{\int_0^T \pi_t \cdot \pi_t' dt\right\}^{-1} \tag{40}$$

where the stochastic integral $\int_0^T d\pi_t \cdot \pi_t'$ is defined with respect to $\mu_{0,G}^T$.

Proof. It is sufficient to verify that the matrix

$$\int_0^T \pi_t \cdot \pi_t' dt$$

is non-singular a.s. $\mu_{0,G}^T$. We shall treat the case when $GG' = E_n$. The general case can be treated in a similar manner. Observe that in this case $\mu_{0,G}^T \equiv \mu_*^T$. Define

$$A_0 = \left\{x \in C(T;\ \mathbb{R}^n)\text{: There exists } \mathbf{h} \in \mathbb{R}^n,\ \mathbf{h} \neq 0 \text{ such that } \left\langle\left[\int_0^T x_s \cdot x_s' ds\right]\mathbf{h},\ \mathbf{h}\right\rangle = 0\right\}$$

$$= \{x \in C(T;\ \mathbb{R}^n)\text{: There exists } \mathbf{h} \in \mathbb{R}^n,\ \mathbf{h} \neq 0 \text{ such that for all } s \in [0,\ T],\ \langle x_s,\ h\rangle = 0.$$

Let s_i, $1 \leq i \leq n+1$ be an increasing sequence of points in $[0, T]$, $s_i \neq s_j$ for $i \neq j$. It is clear that

$$A_0 \subset A_0^* = \{x \in C(T;\ \mathbb{R}^n)\text{: determinant of } [x_{s_2} - x_{s_1}, \ldots, x_{s_{n+1}} - x_{s_n}] = 0\}.$$

It is sufficient to prove that $\mu_*^T(A_0^*)=0$. But

$$\mu_*^T(A_0^*)=P_\tau(\Delta)$$

where P_τ is the joint distribution of the random vector

$$\tau=(\pi_{s_2}-\pi_{s_1}, \ldots, \pi_{s_{n+1}}-\pi_{s_n})$$

and

$$\Delta=\{(\mathbf{V}_1, \ldots, \mathbf{V}_n)\in(\mathbb{R}^n)^n\text{: determinant of }[\mathbf{V}_1, \ldots, \mathbf{V}_n]=0\}.$$

Since P_τ is a Gaussian non-singular measure, $P_\tau(\Delta)=0$ if Δ has Lebesgue measure zero. Define

$$\phi_i(\mathbf{w}_1, \ldots, \mathbf{w}_{n-1}, \lambda_1, \ldots, \lambda_{n-1})=(\mathbf{v}_1, \ldots, \mathbf{v}_n)$$

where

$$\mathbf{w}_j\in\mathbb{R}^n, \begin{pmatrix}\lambda_1\\ \vdots\\ \lambda_{n-1}\end{pmatrix}\in\mathbb{R}^{n-1}, 1\leqq j\leqq n-1 \text{ and } \mathbf{v}_i=\sum_{j=1}^{n-1}\lambda_j\mathbf{w}_j, \mathbf{v}_k=\mathbf{w}_k$$

for $k<i$ and $\mathbf{v}_k=\mathbf{w}_{k-1}$ for $k>i$. Since $\mathbf{v}_1, \ldots, \mathbf{v}_n$ are not linearly independent in $\mathbb{R}^n$, it follows that det $[\mathbf{v}_1, \ldots, \mathbf{v}_n]=0$. Hence the range of ϕ_i is contained in Δ for $1\leqq i\leqq n$ i.e. $\phi_i((\mathbb{R}^n)^{n-1}\times\mathbb{R}^{n-1})\subset\Delta$. Conversly suppose $(\mathbf{v}_1, \ldots, \mathbf{v}_n)\in\Delta$. There then exists an $i_0\in\{1, 2, \ldots, n\}$ and coefficients $\lambda_i\in\{1, \ldots, i_0-1, i_0+1, \ldots, n\}$ such that

$$\mathbf{v}_{i_0}=\sum_{i\neq i_0}\lambda_i\mathbf{v}_i.$$

Let $\mathbf{w}_r=\mathbf{v}_r$, $\mu_r=\lambda_r$ for $r<i_0$ and $\mathbf{w}_r=\mathbf{v}_{r+1}$, $\mu_r=\lambda_{r+1}$ for $r\geqq i_0$. Then

$$\phi_{i_0}(\mathbf{w}_1, \ldots, \mathbf{w}_{n-1}, \mu_1, \ldots, \mu_{n-1})=(\mathbf{v}_1, \ldots, \mathbf{v}_n).$$

Hence

$$\Delta=\bigcup_{i=1}^{n}\phi_i((\mathbb{R}^n)^{n-1}\times\mathbb{R}^{n-1}).$$

The mappings ϕ_i are differentiable and hence, by Sard's Theorem (cf. Dieudonné, 1972, p. 167), Δ has Lebesgue measure zero.

It can be shown that $\hat{\theta}_G(T)$ maximizes $L_G(\theta, \cdot)$ a.s. $\mu_{0,G}^T$ over $\theta\in\mathcal{M}_n$ the space of symmetric matrices of order $n\times n$. Furthermore, if $n=1$, then

$$\hat{\theta}_G(T)(X)=\frac{X_T^2 \quad G^2T}{2\int_0^T X_s^2 ds}.$$

We shall now study the asymptotic behaviour of the estimate $\hat{\theta}_G(T)$. It can be proved that if the true value θ_0 of θ is a real stability matrix (i.e. the eigenvalues of θ_0 have negative real parts) then $\hat{\theta}_G(T)$ is consistent and asymptotically normally distributed.

Theorem 4.2. *If Θ is the subset of stability matrices in $\mathcal{M}_n$, then*

(i) $\hat{\theta}_G(T) \xrightarrow{p} \theta$ *as* $T \to \infty$, (41)

(ii) $T^{1/2}(\hat{\theta}_G(T) - \theta) \to K^\theta$ *as* $T \xrightarrow{\mathcal{L}} \infty$ (42)

where K^θ is a Gaussian matrix with mean zero and covariance defined by

$$E(K^\theta_{ij} K^\theta_{kl}) = (GG')_{ik}(Q_\theta^{-1})_{jl} \tag{43}$$

where Q_θ is a positive definite matrix and is a solution of $\theta Q + Q\theta' = -GG'$. (44)

Note. If θ is a real stability matrix, then Lyapunov's Theorem (cf. Lee and Markus 1967, p. 198) asserts that equation (44) has a solution.

Since we will discuss asymptotic behaviour of estimates for a more general class in the next section, we omit the proof of this theorem. Proof of this theorem can also be found in Le Breton (1974). Most of the above results can be extended for vector equations of type

$$\mathrm{d}\mathbf{X}_t = \theta A(t, \mathbf{X}) + G\,\mathrm{d}\mathbf{W}_t \tag{45}$$

where A is a known function. See Example 4.1.

4.2 Discrete Sampling

In practice, even if continuous sampling is available, in order to compute $\hat{\theta}_G(T)$, the stochastic integrals

$$\int_0^T \mathrm{d}\pi_t \cdot \pi_t' \quad \text{and} \quad \int_0^T \pi_t \cdot \pi_t' \mathrm{d}t$$

have to be approximated by appropriate finite sums of the form

$$\sum_{i=1}^N (\pi_{t_i} - \pi_{t_{i-1}}) \cdot \pi'_{t_{i-1}} \quad \text{and} \quad \sum_{i=1}^N (t_i - t_{i-1}) \pi_{t_{i-1}} \cdot \pi'_{t_{i-1}}$$

where $\{t_0, t_1, \ldots, t_N\}$ defines a partition of $[0, T]$. Then a natural question is whether the resulting estimate

$$\hat{\theta}_{N,T} = \left[\sum_{i=1}^N (\pi_{t_i} - \pi_{t_{i-1}})\pi'_{t_{i-1}}\right] \cdot \left[\sum_{i=1}^N (t_i - t_{i-1}) \pi_{t_{i-1}} \cdot \pi'_{t_{i-1}}\right]^{-1} \tag{46}$$

is the best one that could be obtained in view of the discrete sampling at t_0, $t_1, \ldots, t_N$. We are then led to the study the problem of estimation of the drift parameter in the stochastic differential equation

$$\mathrm{d}\mathbf{X}_t = \theta \mathbf{X}_t \mathrm{d}t + \mathrm{d}\mathbf{W}_t;\ t \geqq 0;\ X_0 = 0$$

based on discrete observations at $t_0, t_1, \ldots, t_N$. Le Breton (1975) showed that if $\delta_N = \max_{1 \leqq i \leqq N} |t_i - t_{i-1}|$ tends to zero and $\tilde{\theta}_{N,T}$ denotes the likelihood equation estimate obtained by discrete sampling, then

$$\text{(i) } \hat{\theta}_{N,T} \xrightarrow{p} \hat{\theta}_T \quad \text{as } N \to \infty,$$

$$\text{(ii) } \tilde{\theta}_{N,T} \xrightarrow{p} \hat{\theta}_T \quad \text{as } N \to \infty, \tag{47}$$

(iii) $\delta_N^{-1/2}(\hat{\theta}_{N,T} - \hat{\theta}_T)$ and $\delta_N^{-1/2}(\tilde{\theta}_{N,T} - \hat{\theta}_T)$ are bounded in probability.

Here $\hat{\theta}_T \equiv \hat{\theta}_G(T)$. This shows that if one used the approximate likelihood equation estimator $\hat{\theta}_{N,T}$ instead of $\hat{\theta}_T$ based on continuous sampling, then the error is of order $\delta_N^{1/2}$ in probability. Hence, if the order of the variance (or asymptotic variance) of $\hat{\theta}_T$ is known, then one can get information for approximate procedure by comparing the order of the variance with δ_N. On the other hand, if discrete sampling is only available, then $\hat{\theta}_{N,T}$ is almost the "best estimate" close to the order $\delta_N^{1/2}$ and it is computable.

Example 4.1. Let us consider the diffusion process with stochastic differential

$$\mathrm{d}\xi_t = \lambda A(t, \xi)\mathrm{d}t + \mathrm{d}W_t,\ t \geqq 0,\ \xi_0 = 0, \tag{48}$$

where the unknown parameter is λ, $-\infty < \lambda < \infty$ and $A(t, \xi)$ is measurable with respect to $\sigma\{\xi(s), 0 \leqq s \leqq t\}$ for every $t \geqq 0$. Further let us assume that for every continuous function $x = (x_t)$, $t \geqq 0$, $x_0 = 0$ there exists $\varepsilon = \varepsilon(x) > 0$ such that

$$\int_0^{\varepsilon(x)} A^2(t, x)\mathrm{d}t < \infty$$

and for every λ and $t \geqq 0$,

$$P_\lambda\left\{\int_0^t A^2(s, \xi)\mathrm{d}s < \infty\right\} = P_0\left\{\int_0^t A^2(s, \xi)\mathrm{d}s < \infty\right\} = 1$$

where P_λ denotes the probability under the measure generated by the process ξ_t given λ. It follows from Theorem 3.2 that the Wiener measure μ_W is equivalent to μ_ξ^λ of the process ξ for every $-\infty < \lambda < \infty$ and

$$\frac{\mathrm{d}\mu_\xi^\lambda}{\mathrm{d}\mu_W}(t, \xi) = \exp\left\{\lambda \int_0^t A(s, \xi)\mathrm{d}\xi_s - \frac{\lambda^2}{2}\int_0^t A^2(s, \xi)\mathrm{d}s\right\}. \tag{49}$$

Hence the likelihood equation estimate of λ given the observation $\{\xi_s, 0 \leqq s \leqq T\}$ is

$$\lambda_T(\xi) = \frac{\int_0^T A(s, \xi) d\xi_s}{\int_0^T A^2(s, \xi) ds}. \tag{50}$$

Let E_λ denote the expectation under the measure μ_ξ^λ. It can be shown, under some natural conditions, that

$$E_\lambda[\lambda_T(\xi)] = \lambda + \frac{\partial}{\partial\lambda} E_\lambda\left[\int_0^T A^2(s, \xi) ds\right]^{-1} \tag{51}$$

(for example, the condition $0 < c \leqq |A(t, \xi)| \leqq C < \infty$ is sufficient for (51) to hold). As a consequence of this formula, it follows that the likelihood equation estimate is biased and the bias, in general, is

$$\frac{\partial}{\partial\lambda} E_\lambda\left[\int_0^T A^2(s, \xi) ds\right]^{-1}. \tag{52}$$

5. ASYMPTOTIC LIKELIHOOD THEORY FOR MULTIDIMENSIONAL DIFFUSION PROCESSES

Suppose we have the stochastic differential equation

$$d\xi(t) = \mathbf{a}(t, \xi(t); \boldsymbol{\alpha}) dt + A^{1/2}(t, \xi(t)) d\mathbf{W}(t), \ \xi(0) = \xi_0 \tag{53}$$

where $\mathbf{a}(t, \mathbf{x}; \boldsymbol{\alpha})$ and $A(t, \mathbf{x})$ are not random, $\mathbf{a}(t, \mathbf{x}; \boldsymbol{\alpha})$ is a known vector-valued function except for the parameter $\boldsymbol{\alpha}$ with values in $\mathbb{R}^m$, $t \in (0, T]$, $\mathbf{x} \in \mathbb{R}^m$, $\boldsymbol{\alpha}$ is a parameter with values in $\mathbb{R}^m$, $A(t, \mathbf{x})$ is a known diffusion matrix acting linearly from $\mathbb{R}^m$ into $\mathbb{R}^m$ and $\mathbf{W}(t)$ is an m-dimensional Wiener process with independent components. A solution of (53) under sufficiently general conditions will be a diffusion process with phase space $\mathbb{R}^m$.

5.1. Estimation

Let us consider the stochastic differential equation (53). We shall assume that the initial value ξ_0 is a random vector independent of $\mathbf{W}(t)$. Let $\mathcal{F}_t$ denote the smallest σ-algebra generated by ξ_0 and $W_k(s)$, $1 \leqq k \leqq m$, $s \in [0, T]$. Suppose that $A(t, \mathbf{x})$ and $\mathbf{a}(t, \mathbf{x}; \boldsymbol{\alpha})$ are such that there exists a unique solution of (53) on $[0, T]$. This solution $\xi(s)$ will be $\mathscr{F}_s$ measurable for each $s \in [0, T]$. Moreover, we shall assume that the parameter $\boldsymbol{\alpha}$ has values in some admissible set $K \subset \mathbb{R}^n$ containing the true value of $\boldsymbol{\alpha}$ as an interior point. The measure corresponding to the solution of (53) on $[0, T]$ for fixed initial value ξ_0 will be denoted by $\mu_{\boldsymbol{\alpha}}$ and we shall assume that for any $\boldsymbol{\alpha}'$, $\boldsymbol{\alpha}'' \in K \subset \mathbb{R}^n$ the measures $\mu_{\boldsymbol{\alpha}'}$ and $\mu_{\boldsymbol{\alpha}''}$ are equivalent. Let μ_0 denote the dominating measure correspond to some $\boldsymbol{\alpha}_0 \in K$. For each $\boldsymbol{\alpha} \in K$, the measures $\mu_{\boldsymbol{\alpha}}$ and μ_0 are equivalent and their Radon–Nikodym derivative can be obtained by Theorem 3.6. It is given by the relation

$$\frac{\mathrm{d}\mu_\alpha}{\mathrm{d}\mu_0}(\xi(\cdot)) = \exp\left\{\int_0^T \mathbf{C}(s, \xi(s))\mathrm{d}\xi(s) - \int_0^T d(s, \xi(s))\mathrm{d}s\right\}$$

where $\mathbf{C}(t, \mathbf{x})$ is the vector-valued function defined by the relation

$$\mathbf{a}(t, \mathbf{x}; \boldsymbol{\alpha}) - \mathbf{a}(t, \mathbf{x}; \boldsymbol{\alpha}_0) = A(t, \mathbf{x})\mathbf{C}(t, \mathbf{x}) \tag{54}$$

and

$$d(s, \mathbf{x}) = (\mathbf{a}(s, \mathbf{x}; \boldsymbol{\alpha}_0),\ \mathbf{C}(s, \mathbf{x})) + \tfrac{1}{2}\,(A(s, \mathbf{x})\mathbf{C}(s, \mathbf{x}, \mathbf{C}(s, \mathbf{x})). \tag{55}$$

Let us consider the case when the diffusion matrix $A(t, \mathbf{x})$ is nondegenerate for all t and $\mathbf{x}$ and the drift coefficient depends linearly on the parameter

$$\mathbf{a}(t, \mathbf{x}; \boldsymbol{\alpha}) = \mathbf{a}(t, \mathbf{x}) + \sum_{k=1}^{n} \alpha_k \boldsymbol{\varphi}_k(t, \mathbf{x}) \tag{56}$$

where $\boldsymbol{\varphi}_k(t, x)$, $1 \leq k \leq n$ are given functions with values in $\mathbb{R}^m$ and are linearly independent on $[0, T] \times \mathbb{R}^m$.

In this case, we have

$$\begin{aligned}\frac{\mathrm{d}\mu_\alpha}{\mathrm{d}\mu_0}(\xi(\cdot) = \exp\Bigg\{&\sum_{k=1}^{n} (\alpha_k - \alpha_k^0)\int_0^T (A^{-1}(t, \xi(t))\boldsymbol{\varphi}_k(t, \xi(t)),\ \mathrm{d}\xi(t))\\ &- \sum_{k=1}^{n} (\alpha_k - \alpha_k^0)\int_0^T [(A^{-1}(t, \xi(t))\boldsymbol{\varphi}_k(t, \xi(t)), \mathbf{a}(t, \xi(t))\\ &+ \sum_{j=1}^{n} \alpha_j^0 \boldsymbol{\varphi}_j(t, \xi(t)))]\mathrm{d}t - \tfrac{1}{2}\sum_{k=1}^{n}\sum_{j=1}^{n} (\alpha_k - \alpha_k^0)(\alpha_j - \alpha_j^0).\\ &\int_0^T (A^{-1}(t, \xi(t))\boldsymbol{\varphi}_k(t, \xi(t)),\ \boldsymbol{\varphi}_j(t, \xi(t)))\mathrm{d}t\Bigg\}.\end{aligned} \tag{57}$$

Let us write

$$L = \log\frac{\mathrm{d}\mu_\alpha}{\mathrm{d}\mu_0}(\xi(\cdot)).$$

Then

$$\frac{\partial L}{\partial \alpha_k}=\int_0^T (A^{-1}(t, \xi(t))\boldsymbol{\varphi}_k(t, \xi(t)), \mathrm{d}\xi(t)-\mathbf{a}(t, \xi(t))\mathrm{d}t) \\ -\sum_{j=1}^n \alpha_j \int_0^T (A^{-1}(t, \xi(t))\boldsymbol{\varphi}_k(t, \xi(t)), \boldsymbol{\varphi}_j(t, \xi(t)))\mathrm{d}t \tag{58}$$

for $1 \leqq k \leqq n$ and

$$\frac{\partial^2 L}{\partial \alpha_k \partial \alpha_j}=-\int_0^T (A^{-1}(t, \xi(t))\boldsymbol{\varphi}_k(t, \xi(t)), \boldsymbol{\varphi}_j(t, \xi(t)))\mathrm{d}t$$

for $1 \leqq k \leqq n$, $1 \leqq j \leqq n$. Let

$$\frac{\partial L}{\partial \alpha}=\left\{\frac{\partial L}{\partial \alpha_1}, \ldots, \frac{\partial L}{\partial \alpha_n}\right\}, \tag{59}$$

$$\boldsymbol{\eta}_T=\left\{\frac{1}{T}\int_0^T (A^{-1}(t, \xi(t))\boldsymbol{\varphi}_k(t, \xi(t)), \mathrm{d}\xi(t)-\mathbf{a}(t, \xi(t))\mathrm{d}t),\ 1 \leqq k \leqq n\right\} \tag{60}$$

and

$$B_T=\left(\left(\frac{1}{T}\int_0^T (A^{-1}(t, \xi(t))\boldsymbol{\varphi}_k(t, \xi(t)), \boldsymbol{\varphi}_j(t, \xi(t)))\mathrm{d}t\right)\right)_{n\times n}. \tag{61}$$

Let $I_T=E_{\boldsymbol{\alpha}}[B_T]$. The existence of these integrals and expectations is ensured by the mutual absolute continuity of $\mu_{\boldsymbol{\alpha}}$ and μ_0. Here $E_{\boldsymbol{\alpha}}$ denotes the expectation under the condition that the measure $\mu_{\boldsymbol{\alpha}}$ corresponds to the process $\xi(t)$. The likelihood equation can be written in the form

$$\boldsymbol{\eta}_T-B_T\boldsymbol{\alpha}=0 \tag{62}$$

and it is linear. Suppose that $\mu_{\boldsymbol{\alpha}}$ and μ_0 are mutually absolute continuous for all $T>0$. We have the following theorem on the asymptotic behaviour of solutions of likelihood equations.

Theorem 5.1. *Suppose that the following condition is satisfied in addition to all the assumptions stated above.*

$$\frac{1}{T}\int_0^T (A^{-1}(t, \xi(t))\boldsymbol{\varphi}_k(t, \xi(t)), \boldsymbol{\varphi}_j(t, \xi(t)))\mathrm{d}t \xrightarrow{p} b_{kj} \tag{63}$$

as $T\to\infty$ where $|b_{kj}|<\infty$ and $((b_{kj}))=J$ is a non-singular matrix. Then, for any $\varepsilon>0$ there is a $T_\varepsilon>0$ such that for $T>T_\varepsilon$ with probability not less than $1-\varepsilon$, the likelihood equation (62) has a unique solution $\hat{\boldsymbol{\alpha}}_T$ which is a consistent estimate of the vector $\boldsymbol{\alpha}$; the distribution of $T^{1/2}(\hat{\boldsymbol{\alpha}}_T-\boldsymbol{\alpha})$ is asymptotically normal with mean zero and covariance matrix J^{-1}.

Note that b_{kj} and J depend on the parameter $\boldsymbol{\alpha}$. For simplicity of notation we have not indicated the dependence. We shall first state a lemma which will be used in the proof of Theorem 5.1. Proof of this lemma is left as an exercise for the reader.

Lemma 5.1. *Suppose the following conditions hold as $t \to t_0$:*

(i) $D_t \xrightarrow{p} D$ *where D is a non-random non-singular matrix,*

(ii) $P(|\zeta_t^{(i)}| > M) \to 0$ *as $M \to \infty$ uniformly in t in the domain of variation of t for $1 \leqq i \leqq n$;*

(iii) $\zeta_t \xrightarrow{\mathscr{L}} \zeta$ *where $\xi_t = (\zeta_t^{(1)}, \ldots, \zeta_t^{(n)})$.*

Then

$$D_t \zeta_t \xrightarrow{\mathscr{L}} D\zeta.$$

Proof of Theorem 5.1. Let $P_{\boldsymbol{\alpha}}(\cdot)$ indicate that the probability is computed when $\boldsymbol{\alpha}$ is the true parameter. Since

$$B_T \xrightarrow{p} ((b_{kj})) \equiv J \text{ as } T \to \infty$$

and J is non-singular, given any $\varepsilon > 0$, there exists $T_{1\varepsilon} > 0$ such that

$$P_{\boldsymbol{\alpha}}(B_T \text{ is non-singular}) > 1 - \varepsilon/3$$

for $T > T_{1\varepsilon}$. Hence, with probability $1 - \varepsilon$, the likelihood equation

$$\boldsymbol{\eta}_T - B_T \boldsymbol{\alpha} = 0 \tag{64}$$

has a unique solution for $T > T_{1\varepsilon}$ and it is given by

$$\hat{\boldsymbol{\alpha}}_T = B_T^{-1} \boldsymbol{\eta}_T.$$

We shall now prove that $\hat{\boldsymbol{\alpha}}_T \xrightarrow{p} \boldsymbol{\alpha}$ as $T \to \infty$. In other words $\hat{\boldsymbol{\alpha}}_T$ is a consistent estimate of $\boldsymbol{\alpha}$. Clearly

$$\begin{aligned} &P_{\boldsymbol{\alpha}}(\|\hat{\boldsymbol{\alpha}}_T - \boldsymbol{\alpha}\| > \varepsilon) \\ &\quad = P_{\boldsymbol{\alpha}}(\|B_T^{-1} \boldsymbol{\eta}_T - \boldsymbol{\alpha}\| > \varepsilon \text{ and } B_T \text{ is non-singular}) + \varepsilon/3 \\ &\quad \leqq P_{\boldsymbol{\alpha}}(\|\boldsymbol{\eta}_T - B_T \boldsymbol{\alpha}\| > \varepsilon \|B_T\|) + \varepsilon/3 \end{aligned}$$

for $T > T_{1\varepsilon}$. Here $\|\cdot\|$ for any vector is the ordinary Euclidean norm and $\|\cdot\|$ for any matrix B is $[Tr(BB')]^{1/2}$. Since $B_T \xrightarrow{p} J$ where J is non-singular as $T \to \infty$, there exists $T_{2\varepsilon} > 0$ such that $\|B_T\| \geqq a > 0$ with probability $1 - \varepsilon/3$ for $T > T_{2\varepsilon}$. Hence, for $T > \max(T_{1\varepsilon}, T_{2\varepsilon})$,

$$P_{\boldsymbol{\alpha}}(\|\hat{\boldsymbol{\alpha}}_T - \boldsymbol{\alpha}\| > \varepsilon) \leqq P_{\boldsymbol{\alpha}}(\|\boldsymbol{\eta}_T - B_T \boldsymbol{\alpha}\| > \varepsilon a) + 2\varepsilon/3.$$

It is sufficient therefore to prove that

$$\boldsymbol{\eta}_T - B_T\boldsymbol{\alpha} \xrightarrow{p} 0 \text{ as } T\to\infty.$$

Equations (60) and (61) together imply that it is sufficient to show that

$$\frac{1}{T}\int_0^T (A^{-1}(t, \xi(t))\boldsymbol{\varphi}_k(t, \xi(t)), d\xi(t) - \mathbf{a}(t, \xi(t))dt)$$

$$-\sum_{j=1}^{n} \alpha_j \frac{1}{T}\int_0^T (A^{-1}(t, \xi(t))\boldsymbol{\varphi}_k(t, \xi(t)), \boldsymbol{\varphi}_j(t, \xi(t)))dt \tag{65}$$

tends to zero in probability as $T\to\infty$ for every $1 \leqq k \leqq n$. But (65) can be written in the form

$$\frac{1}{T}\int_0^T (A^{-1}(t, \xi(t))\boldsymbol{\varphi}_k(t, \xi(t)), d\xi(t) - \mathbf{a}(t, \xi(t))dt - \sum_{j=1}^{n} \alpha_j \boldsymbol{\varphi}_j(t, \xi(t))dt)$$

$$= \frac{1}{T}\int_0^T (A^{-1}(t, \xi(t))\boldsymbol{\varphi}_k(t, \xi(t)), d\xi(t) - \mathbf{a}(t, \xi(t); \boldsymbol{\alpha})dt) \tag{66}$$

$$= \frac{1}{T}\int_0^T (A^{-1}(t, \xi(t))\boldsymbol{\varphi}_k(t, \xi(t)), A^{1/2}(t, \xi(t))d\mathbf{W}(t)).$$

All the above equalities are valid almost surely $\mu_{\boldsymbol{\alpha}}$ and the last equation follows from the fact that $\xi(t)$ is the solution of the stochastic differential equation (53). Clearly the last term tends to zero in probability as $T\to\infty$ from the fact that

$$\frac{1}{\sqrt{T}}\int_0^T (A^{-1/2}(t, \xi(t))\boldsymbol{\varphi}_k(t, \xi(t)), d\mathbf{W}(t)) \xrightarrow{\mathscr{L}} N(0, J)$$

by Theorem 2.1 of Appendix 2 in view of condition (63). Hence $\hat{\boldsymbol{\alpha}}_T$ is a consistent estimator of $\boldsymbol{\alpha}$.

For proving the asymptotic normality of the likelihood equation estimate, consider the limit distribution of $T^{-1/2}\, \partial L/\partial\boldsymbol{\alpha}$. It can be represented in the form

$$T^{-1/2}\frac{\partial L}{\partial \boldsymbol{\alpha}} = \left\{T^{-1/2}\int_0^T (A^{-1/2}(t, \xi(t))\boldsymbol{\varphi}_k(t, \xi(t)), d\mathbf{W}(t)),\ 1 \leqq k \leqq n\right\}$$

and by the condition (63),

$$\frac{1}{T}\int_0^T (A^{-1/2}(t, \xi(t))\boldsymbol{\varphi}_k(t, \xi(t)), A^{-1/2}(t, \xi(t))\boldsymbol{\varphi}_j(t, \xi(t)))dt \xrightarrow{p} b_{kj}$$

as $T\to\infty$. Hence, by Theorem 2.1 of Appendix 2, $T^{-1/2}\, \partial L/\partial\boldsymbol{\alpha}$ is asymptotically normal with mean zero and covariance matrix J. Relation (64) implies that

$$T^{1/2}(\hat{\boldsymbol{\alpha}}_T - \boldsymbol{\alpha}) = B_T^{-1} T^{-1/2} \frac{\partial L}{\partial \boldsymbol{\alpha}}. \tag{67}$$

By the assumptions in the theorem, $B_T^{-1} \xrightarrow{P} J^{-1}$ as $T \to \infty$. Applying Lemma 5.1, we obtain that $T^{1/2}(\hat{\boldsymbol{\alpha}}_T - \boldsymbol{\alpha})$ is asymptotically normal with mean zero and covariance matrix J^{-1}. This completes the proof of the theorem.

Suppose now the drift coefficient and the diffusion matrix in (53) depends only on $\mathbf{x}$:$\mathbf{a}(t, \mathbf{x}) \equiv \mathbf{a}(\mathbf{x})$, $\boldsymbol{\varphi}_k(t, \mathbf{x}) \equiv \boldsymbol{\varphi}_k(\mathbf{x})$, $1 \leq k \leq n$ and $A(t, \mathbf{x}) \equiv A(\mathbf{x})$. It is known that the solution of such an equation is a homogeneous Markov diffusion process (cf. Gikhman and Skorokhod, 1972, p. 105). Let us suppose that it is recurrent and each $\boldsymbol{\alpha} \in K$ has ergodic distribution $G_{\boldsymbol{\alpha}}(x)$. We prove now the Theorem 5.2 which describes the asymptotic behaviour of the likelihood equation estimator $\hat{\boldsymbol{\alpha}}_T$ of $\boldsymbol{\alpha}$.

Theorem 5.2. *If the norm of the matrix $A(\mathbf{x})$ and the norms of the functions $\boldsymbol{\varphi}_k(\mathbf{x})$ are bounded, then for any $\varepsilon > 0$ there is a $T_\varepsilon > 0$ such that for $T > T_\varepsilon$, with probability not less than $1 - \varepsilon$, the likelihood equation has a unique solution $\hat{\alpha}_T$ which is a consistent estimate of the vector α; the distribution of $T^{1/2}(\hat{\boldsymbol{\alpha}}_T - \boldsymbol{\alpha})$ converges to the normal distribution with mean zero and covariance matrix inverse to the matrix*

$$\left(\left(\int (A^{-1}(\mathbf{x})\boldsymbol{\varphi}_k(\mathbf{x}), \boldsymbol{\varphi}_j(\mathbf{x}))\mathrm{d}G_{\boldsymbol{\alpha}}(\mathbf{x})\right)\right)_{n \times n}$$

The conditions of the theorem imply that

$$(A^{-1}(\mathbf{x})\boldsymbol{\varphi}_k(\mathbf{x}), \boldsymbol{\varphi}_j(\mathbf{x})), \; 1 \leq k, j \leq n$$

are bounded and since $\xi(t)$ is ergodic by assumption, it follows that

$$\frac{1}{T} \int_0^T (A^{-1}(\xi(t))\boldsymbol{\varphi}_k(\xi(t)), \boldsymbol{\varphi}_j(\xi(t)))\mathrm{d}t \to \int (A^{-1}(\mathbf{x})\boldsymbol{\varphi}_k(\mathbf{x}), \boldsymbol{\varphi}_j(\mathbf{x}))\mathrm{d}G_{\boldsymbol{\alpha}}(\mathbf{x})$$

with probability one as $T \to \infty$ by Khasminskii (1960). The linear independence of the $\boldsymbol{\varphi}_k(\mathbf{x})$ on $\mathbb{R}^n$ implies the linear independence of $A^{-1/2}(\mathbf{x})\boldsymbol{\varphi}_k(\mathbf{x})$. In fact, if $(c_1, \ldots, c_n) \neq (0, \ldots, 0)$. then the vector

$$A^{-1/2}(\mathbf{x}) \sum_{k=1}^{n} c_k \boldsymbol{\varphi}_k(\mathbf{x})$$

will not be identically zero due to the non-degeneracy of the matrix

$A^{-1/2}(\mathbf{x})$. Hence the matrix

$$\left(\left(\frac{1}{T}\int(A^{-1}(\mathbf{x})\boldsymbol{\varphi}_k(\mathbf{x}),\ \boldsymbol{\varphi}_j(\mathbf{x}))\mathrm{d}G_\alpha(\mathbf{x})\right)\right)$$

will be the Grammian of the system of functions

$$T^{-1/2}A^{-1/2}(x)\boldsymbol{\varphi}_k(x),\ 1\leqq k\leqq n$$

and will consequently be positive definite. Hence all the conditions of Theorem 5.1 hold which proves Theorem 5.2.

Suppose now that $A(t,\ x)\equiv A(t)$ and $\boldsymbol{\varphi}_k(t,\ x)\equiv\boldsymbol{\varphi}_k(t)$ for $1\leqq k\leqq n$. The likelihood equation (64) can be written in the form

$$\boldsymbol{\eta}_T-I_T\boldsymbol{\alpha}=0 \tag{68}$$

since

$$B_T=I_T=\left(\left(\frac{1}{T}\int_0^T(A^{-1}(t)\boldsymbol{\varphi}_k(t),\ \boldsymbol{\varphi}_j(t))\mathrm{d}t\right)\right).$$

The following theorem gives the behaviour of $\hat{\boldsymbol{\alpha}}_T$, the likelihood equation estimator of $\boldsymbol{\alpha}$.

Theorem 5.3. *The solution $\hat{\boldsymbol{\alpha}}_T$ of the likelihood equation* (68) *is an efficient likelihood equation estimator. The vector $T^{1/2}(\hat{\boldsymbol{\alpha}}_T-\boldsymbol{\alpha})$ has normal distribution with mean zero and covariance matrix I_T^{-1}.*

Proof. The matrix I_T is positive definite since it is the Grammian of the linearly independent system of functions $T^{-1/2}\ A^{-1/2}(t)\ \boldsymbol{\varphi}_k(t),\ k=1,\ 2,\ \ldots,\ n$ and hence equation (68) has the unique solution $\hat{\boldsymbol{\alpha}}_T=I_T^{-1}\ \boldsymbol{\eta}_T$. Furthermore the estimate $\hat{\boldsymbol{\alpha}}_T$ is unbiased since

$$\begin{aligned}E_{\boldsymbol{\alpha}}(\hat{\boldsymbol{\alpha}}_T)&=E_{\boldsymbol{\alpha}}(I_T^{-1}\boldsymbol{\eta}_T)=I_T^{-1}E_{\boldsymbol{\alpha}}(\boldsymbol{\eta}_T)\\&=I_T^{-1}E_{\boldsymbol{\alpha}}\left(I_T\boldsymbol{\alpha}+T^{-1}\frac{\partial L}{\partial\boldsymbol{\alpha}}\right)=\boldsymbol{\alpha}+T^{-1}E_{\boldsymbol{\alpha}}\left(\frac{\partial L}{\partial\boldsymbol{\alpha}}\right)=\boldsymbol{\alpha}.\end{aligned}$$

Let us find the distribution of the vector

$$T^{-1/2}\frac{\partial L}{\partial\boldsymbol{\alpha}}=\left\{T^{-1/2}\int_0^T(A^{-1/2}(t)\boldsymbol{\varphi}_k(t),\ \mathrm{d}\mathbf{W}(t)),\ 1\leqq k\leqq n\right\}.$$

Let $\boldsymbol{\lambda}=(\lambda_1,\ \ldots,\ \lambda_n)\in\mathbb{R}^n$ and $\psi(s;\ \boldsymbol{\lambda})$ be the characteristic function of $(\boldsymbol{\lambda},\ T^{-1/2}\ \partial L/\partial\boldsymbol{\alpha})$. Further introduce the notation

$$\boldsymbol{\phi}^{(k)}(t)=\{\phi_1^{(k)}(t),\ \ldots,\ \phi_m^{(k)}(t)\}=A^{-1/2}(t)\boldsymbol{\varphi}_k(t),\ 1\leqq k\leqq n$$

and

$$\boldsymbol{\phi}_i(t) = \{\phi_i^{(1)}(t), \ldots, \phi_i^{(n)}(t)\}, \ 1 \leqq i \leqq m.$$

Then

$$\psi(s; \boldsymbol{\lambda}) = E_{\boldsymbol{\alpha}}\left[\exp\left\{i\, s\left(\boldsymbol{\lambda},\, T^{-1/2}\,\frac{\partial L}{\partial \boldsymbol{\alpha}}\right)\right\}\right].$$

$$= E_{\boldsymbol{\alpha}}\left[\exp\left\{i\, s \sum_{j=1}^{m} T^{-1/2} \int_0^T (\phi_j(t), \boldsymbol{\lambda})\mathrm{d}W_j(t)\right\}\right].$$

The random variables

$$T^{-1/2} \int_0^T (\boldsymbol{\phi}_j(t), \boldsymbol{\lambda})\mathrm{d}W_j(t)$$

are independent and have normal distribution with mean zero and variance $T^{-1} \int_0^T (\boldsymbol{\phi}_j(t), \boldsymbol{\lambda})^2 \mathrm{d}t.$

Hence

$$\psi(s; \boldsymbol{\lambda}) = \exp\left\{-s^2/2 \sum_{j=1}^{m} T^{-1} \int_0^T (\boldsymbol{\phi}_j(t), \boldsymbol{\lambda})^2 \mathrm{d}t\right\} \tag{69}$$

$$= \exp\{-s^2/2(I_T\boldsymbol{\lambda}, \boldsymbol{\lambda})\}.$$

For $s=1$, $\psi(s; \boldsymbol{\lambda})$ is the characteristic function of $T^{-1/2}\, \partial L/\partial \boldsymbol{\alpha}$ and hence (69) implies that $T^{-1/2}\, \partial L/\partial \boldsymbol{\alpha}$ is normally distributed with zero mean and covariance matrix I_T. Since

$$T^{1/2}(\hat{\boldsymbol{\alpha}}_T - \boldsymbol{\alpha}) = I_T^{-1}\, T^{-1/2}\, \frac{\partial L}{\partial \boldsymbol{\alpha}}$$

it follows that $T^{1/2}(\hat{\boldsymbol{\alpha}}_T - \boldsymbol{\alpha})$ is normal with mean zero and covariance matrix I_T^{-1}.

For any unbiased estimator $\boldsymbol{\alpha}_T^*$ of $\boldsymbol{\alpha}$, the covariance matrix B of $T^{1/2}(\boldsymbol{\alpha}_T^* - \boldsymbol{\alpha})$ satisfies $B \geqq I_T^{-1}$ by the Cramer–Rao inequality in the sense that $B - I_T^{-1}$ is non-negative definite. Since covariance matrix of $\hat{\boldsymbol{\alpha}}_T$ is I_T^{-1}, it follows that $\hat{\boldsymbol{\alpha}}_T$ is efficient.

Example 5.1. Let us apply the above results to a geophysical problem. It is known that the instantaneous axis of rotation of the earth is displaced with respect to the minor axis of the terrestrial ellipsoid. This displacement consists of a periodic and a fluctuating part. The latter can be approximately assumed to be a stationary solution of the system of stochastic differential equations of the form

$$\begin{aligned} d\xi(t) &= -\lambda\xi(t)dt - w\eta(t)dt + a^{1/2}dW_1(t), \\ d\eta(t) &= -w\xi(t)dt - \lambda\eta(t)dt + a^{1/2}dW_2(t) \end{aligned} \tag{70}$$

where $W_1(t)$ and $W_2(t)$ are two independent Wiener processes, λ and w are unknown and a is known. This system can be written in the form

$$d\zeta(t) = -r\zeta(t)dt + a^{1/2}dW(t)$$

where

$$\zeta(t) = \xi(t) + i\eta(t),\ W(t) = W_1(t) + iW_2(t),\ r = \lambda - iw.$$

One can construct estimates of λ and w by the methods of Arato *et al.* (1962). Let us choose the dominating measure to be the measure corresponding to the solution of the system with zero parameter and initial value whose distribution coincides with the joint distribution of $(\xi(t), \eta(t))$. In this case, the log-likelihood function can be written as

$$\begin{aligned} L(\lambda; w) = &-\frac{1}{a}\int_0^T \{\lambda\xi(t) + w\eta(t)\}d\xi(t) + \frac{1}{a}\int_0^T \{w\xi(t) - \lambda\eta(t)\}d\eta(t) \\ &-\frac{1}{2a}\int_0^T \{\lambda\xi(t) + w\eta(t)\}^2 dt - \frac{1}{2a}\int_0^T \{w\xi(t) - \lambda\eta(t)\}^2 dt \end{aligned} \tag{71}$$

and it is easily seen from this that the maximum likelihood estimates of λ and w are

$$\hat{\lambda}_T = -\left\{\int_0^T \xi(t)d\xi(t) + \int_0^T \eta(t)d\eta(t)\right\}\left\{\int_0^T |\zeta(t)|^2 dt\right\}^{-1} \tag{72}$$

and

$$\hat{w}_T = \left\{\int_0^T \xi(t)d\eta(t) - \int_0^T \eta(t)d\xi(t)\right\}\left\{\int_0^T |\zeta(t)|^2 dt\right\}^{-1}. \tag{73}$$

For the stationary process $\zeta(t)$, since it is ergodic,

$$\frac{1}{aT}\int_0^T |\zeta(t)|^2 dt \to 1/\lambda \tag{74}$$

with probability one. Equation (74) shows that condition (63) is satisfied. Hence $\hat{\lambda}_T$ and $\hat{w}_T$ are consistent and $\{T^{1/2}(\hat{\lambda}_T - \lambda), T^{1/2}(\hat{w}_T - w)\}$ is asymptotically normal with mean zero and covariance matrix $\left(\begin{smallmatrix}\lambda & 0 \\ 0 & \lambda\end{smallmatrix}\right)$. The asymptotic covariance matrix of $(\hat{\lambda}_T, \hat{w}_T)$ shows that $\hat{\lambda}_T$ and $\hat{w}_T$ are asymptotically independent in view of limiting normality. The above estimate of w coincides with the one found by Arato, Kolmogorov and Sinai (1962) but the estimate of the "damping" parameter here has a more simple form than in their case.

5.2. Testing

We shall assume that all the assumptions stated in Theorem 5.1 hold. Suppose that we wish to test the hypothesis $H_0{:}\boldsymbol{\alpha}=(\alpha_1^0, \ldots, \alpha_n^0)\in K_l$ against alternative hypothesis imposing no restriction other than that $\boldsymbol{\alpha}\in K$ where K_l is an open set of an l-dimensional hyperplane within K and $l<n$. For simplicity, let us assume that the hypothesis has the form $H_0{:}\alpha_{l+1}^0=\ldots=\alpha_n^0=0$ and that the alternative places no restriction on $\alpha_1^0, \ldots, \alpha_1^0, \ldots, \alpha_n^0$.

Let $\hat{\boldsymbol{\alpha}}_T$ be the likelihood equation estimator of $\boldsymbol{\alpha}$ and $\hat{\boldsymbol{\alpha}}_{T0}$ be the likelihood equation estimator of $\boldsymbol{\alpha}$ under H_0. The derivation of $\hat{\boldsymbol{\alpha}}_{T0}$ corresponds exactly to that of $\hat{\boldsymbol{\alpha}}_T$ except that vectors of dimension l instead of n are involved. It can be seen from (57) that

$$\begin{aligned}\frac{1}{T}\log\frac{d\mu_{\boldsymbol{\alpha}}}{d\mu_0}\Big|_{\boldsymbol{\alpha}=\hat{\boldsymbol{\alpha}}_T} &= (\hat{\boldsymbol{\alpha}}_T-\boldsymbol{\alpha}_0)'\boldsymbol{\eta}_T-(\hat{\boldsymbol{\alpha}}_T-\boldsymbol{\alpha}_0)'B_T\boldsymbol{\alpha}_0-\tfrac{1}{2}(\hat{\boldsymbol{\alpha}}_T-\boldsymbol{\alpha}_0)'B_T(\hat{\boldsymbol{\alpha}}_T-\boldsymbol{\alpha}_0)\\ &=(\hat{\boldsymbol{\alpha}}_T-\boldsymbol{\alpha}_0)'(\boldsymbol{\eta}_TB_T\boldsymbol{\alpha}_0)-\tfrac{1}{2}(\hat{\boldsymbol{\alpha}}_T-\boldsymbol{\alpha}_0)'B_T(\hat{\boldsymbol{\alpha}}_T-\boldsymbol{\alpha}_0)\\ &=(\hat{\boldsymbol{\alpha}}_T-\boldsymbol{\alpha}_0)'(B_T\hat{\boldsymbol{\alpha}}_T-B_T\boldsymbol{\alpha}_0)-\tfrac{1}{2}(\hat{\boldsymbol{\alpha}}_T-\boldsymbol{\alpha}_0)'B_T(\hat{\boldsymbol{\alpha}}_T-\boldsymbol{\alpha}_0)\\ &=\tfrac{1}{2}(\hat{\boldsymbol{\alpha}}_T-\boldsymbol{\alpha}_0)'B_T(\hat{\boldsymbol{\alpha}}_T-\boldsymbol{\alpha}_0)\end{aligned}\tag{75}$$

using the definitions of $\boldsymbol{\eta}_T$ and B_T and the fact that $\hat{\boldsymbol{\alpha}}_T=B_T^{-1}\,\boldsymbol{\eta}_T$ is the likelihood equation estimator. Theorem 5.1 implies that

$$T^{1/2}(\hat{\boldsymbol{\alpha}}_T-\boldsymbol{\alpha}_0)\xrightarrow{\mathscr{L}}\mathbf{Z}$$

where $\mathbf{Z}$ has a multivariate normal distribution with mean zero and covariance matrix J^{-1}. But $B_T\xrightarrow{p}J$ as $T\to\infty$. Hence

$$T(\hat{\boldsymbol{\alpha}}_T-\boldsymbol{\alpha}_0)'B_T(\hat{\boldsymbol{\alpha}}_T-\boldsymbol{\alpha}_0)\xrightarrow{\mathscr{L}}\mathbf{Z}'J\ \mathbf{Z}\text{ as }T\to\infty.\tag{76}$$

Under the assumption that H_0 is true and $\alpha_{l+1}=\ldots=\alpha_n=0$, it can be shown that

$$\frac{1}{T}\log\frac{d\mu_{\boldsymbol{\alpha}}}{d\mu_0}\Big|_{\boldsymbol{\alpha}=\hat{\boldsymbol{\alpha}}_{T0}}=\frac{1}{2}(\hat{\boldsymbol{\alpha}}_{T0}-\grave{\boldsymbol{\alpha}}_{00})'B_{T00}(\hat{\boldsymbol{\alpha}}_{T0}-\boldsymbol{\alpha}_{00})\tag{77}$$

where

$$\boldsymbol{\alpha}_0=\begin{pmatrix}\boldsymbol{\alpha}_{00}\\ \hline \boldsymbol{\alpha}_{01}\end{pmatrix}$$

and

$$T(\hat{\boldsymbol{\alpha}}_{T0}-\boldsymbol{\alpha}_{00})'B_{T00}(\hat{\boldsymbol{\alpha}}_{T0}-\boldsymbol{\alpha}_{00})\xrightarrow{\mathscr{L}}\mathbf{Z}_0'\,J_{00}\,\mathbf{Z}_0 \tag{78}$$

where

$$\mathbf{Z}=\begin{pmatrix}\mathbf{Z}_0\\ \hline \mathbf{Z}_1\end{pmatrix} \quad\text{and}\quad J_{00}=p-\lim_{T\to\infty} B_{T00}.$$

The above relations show that

$$2\left\{\log\frac{d\mu_{\boldsymbol{\alpha}}}{d\mu_0}\bigg|_{\boldsymbol{\alpha}=\hat{\boldsymbol{\alpha}}_T}-\log\frac{d\mu_{\boldsymbol{\alpha}}}{d\mu_0}\bigg|_{\boldsymbol{\alpha}=\hat{\boldsymbol{\alpha}}_{T0}}\right\}$$
$$=T\{(\hat{\boldsymbol{\alpha}}_T-\boldsymbol{\alpha}_0)'B_T(\hat{\boldsymbol{\alpha}}_T-\boldsymbol{\alpha}_0)-(\hat{\boldsymbol{\alpha}}_{T0}-\boldsymbol{\alpha}_{00})'B_{T00}\,(\hat{\boldsymbol{\alpha}}_{T0}-\boldsymbol{\alpha}_{00})\}$$

and the term on the right-hand side of the above equality converges in law to $\mathbf{Z}'\,J\,\mathbf{Z}-\mathbf{Z}_0'\,J_{00}\mathbf{Z}_0$, where $\mathbf{Z}$ is multivariate normal with mean zero and covariance matrix J^{-1}, $\mathbf{Z}_0$ is an $(l\times l)$-vector and J_{00} is an $l\times l$-matrix as defined above. From the results in multivariate analysis, it follows that

$$\mathbf{Z}'\,J\,\mathbf{Z}-\mathbf{Z}_0'\,J_{00}\,\mathbf{Z}_0$$

is χ-squared with $n-l$ degrees of freedom. We have proved the following theorem.

Theorem 5.4. *Under the conditions stated above,*

$$2\left\{\log\frac{d\mu_{\boldsymbol{\alpha}}}{d\mu_0}\bigg|_{\boldsymbol{\alpha}=\hat{\boldsymbol{\alpha}}_T}-\log\frac{d\mu_{\boldsymbol{\alpha}}}{d\mu_0}\bigg|_{\boldsymbol{\alpha}=\hat{\boldsymbol{\alpha}}_{T0}}\right\}$$

is asymptotically χ-squared with $n-l$ degree of freedom when H_0 is true.

6. HYPOTHESES TESTING FOR PARAMETERS OF DIFFUSION PROCESSES

6.1. Local Asymptotic Normality (LAN)

Let $\Theta\subset\mathbb{R}^K$ be an open set and θ denote a typical element of Θ. For every $\theta\in\Theta$, $T>0$, consider a probability space $(\Omega,\mathscr{F}^{(T)},P_\theta^{(T)})$ with $\mathscr{F}^{(T)}\subset\mathscr{F}^{(S)}$ if $T\leqq S$ and $P_\theta^{(T)}$ is the restriction of $P_\theta^{(S)}$ to $\mathscr{F}^{(T)}$. Let $\mathbf{X}_T$ be an $\mathscr{F}^{(T)}$-measurable random variable taking values in $(\mathscr{X}^{(T)},\mathscr{U}^{(T)})$. Let $\mathscr{L}\,(\xi|P_\theta^{(T)})$ denote the distribution of the random variable ξ when $P_\theta^{(T)}$ is the probability measure on $(\Omega,\mathscr{F}^{(T)})$. If $\{h_T\}$ is any family of random variables defined on $(\Omega,\mathscr{F}^{(T)},P_\theta^{(T)})$ converging in probability as $T\to\infty$, we shall hereafter denote this by

$$P_\theta - \lim_{T\to\infty} h_T.$$

Definition 6.1. The collection of measures $\{P_\theta^T, \theta\in\Theta\}$ is said to be *locally asymptotically normal* (LAN) in $\theta_0\in\Theta$ as $T\to\infty$ if for some matrices $Q_T\equiv Q(\theta_0, T)$ the following holds:

$$\log \frac{dP^{(T)}_{\theta_0+T^{-1/2}Q_T\mathbf{u}}}{dP^{(T)}_{\theta_0}}(\mathbf{X}_T)=\mathbf{u}' \ \Delta_T(\theta_0, \mathbf{X}_T)-1/2\|\mathbf{u}\|^2+\alpha_T(\mathbf{u}, \mathbf{X}_T, \theta_0) \quad (79)$$

where

$$\mathscr{L}(\Delta_T(\theta_0, \mathbf{X}_T)|P^{(T)}_{\theta 0})\overset{\mathscr{L}}{\to}N(0, I) \text{ as } T\to\infty, \quad (80)$$

I being the identity matrix, and

$$P_{\theta_0}-\lim_{T\to\infty} \alpha_T(\mathbf{u}_T, \theta_0)=0 \quad (81)$$

for every $\mathbf{u}\in\mathbb{R}^K$ and arbitrary bounded $\{\mathbf{u}_T\}$ in $\mathbb{R}^K$.

Let $\mathbf{X}_T=\{\mathbf{X}(t), 0\leqq t\leqq T\}$ and $\mathbf{X}(t)=(X_1(t), \ldots, X_n(t))$ be the solution of the stochastic differential equation

$$d\mathbf{X}(t)=S(\theta, t, \mathbf{X}_t)dt+d\mathbf{W}_t, \mathbf{X}(0)=0, 0\leqq t\leqq T \quad (82)$$

where $S_i(\theta, t, \mathbf{X}_t)$, $1\leqq i\leqq n$, the components of the vector $S(\theta, t, \mathbf{X}_t)$, $\theta\in\Theta$, $t\in[0, T]$ are $\mathscr{F}_t(X)$-measurable and absolutely integrable on $[0, T]$. Here

$$\mathscr{F}_t(X)=\sigma\{\mathbf{X}(s), 0\leqq s\leqq t\}.$$

Let $P_\theta^{(T)}$ be the probability measure induced by $\mathbf{X}_T$ on $\{C[0, T], \mathscr{B}_T\}$ where $C[0, T]$ is the space of continuous functions on $[0, T]$ and $\mathscr{B}_T$ is the σ-algebra of Borel sets of $C[0, T]$ under the supremum norm.

We shall now study the conditions for local asymptotic normality for the family of measures $\{P_\theta^{(T)}\}$ induced by the diffusion process $\mathbf{X}_T$ which is the solution of (82).

Definition 6.2. A collection of random functions $f(v, t)$, $t\in[0, T]$ is said to be differentiable in v in probability in $\mathscr{L}_2[0, T]$ if there exists a function $\partial f(v, t)/\partial t$, $t\in[0, T]$ such that

$$P-\lim_{\Delta\to 0}\int_0^T\left(\frac{1}{\Delta}\{f(v+\Delta)-f(v)\}-\frac{\partial f(v, t)}{\partial v}\right)^2 dt=0. \quad (83)$$

A vector random function with a vector parameter is said to be differentiable in probability with respect to the parameter if each component of the random function is differentiable in probability with respect to each component of the parameter.

Let $\dot{S}(\theta, \mathbf{X}_T)=(\dot{S}(\theta, t, \mathbf{X}_t), 0\leqq t\leqq T)$ be the matrix of order $k\times n$ of derivatives of $S(\theta, t, \mathbf{X}_t)$ with respect to θ when they exist. Let $S(\theta, \mathbf{X}_T)=(S(\theta, t, \mathbf{X}_t), 0\leqq t\leqq T)$ and V_{θ_0} be a neighbourhood of θ_0.

The following theorem gives conditions for LAN of $\{P_\theta^{(T)}, \theta\in\Theta\}$ in θ_0 as $T\to\infty$.

Theorem 6.1. *Suppose the following conditions are satisfied:*

(i) $P_{\tilde{\theta}}^{(T)}\{\|S(\theta, \mathbf{X}_T)\|<\infty\}=1$ *for all* $\theta, \tilde{\theta}$ in V_{θ_0};

(ii) $S(\theta, \mathbf{X}_T)$ *is differentiable in* θ *in a neighbourhood of* θ_0 *in* $P_{\theta_0}^{(T)}$*-measure in* $L_2[0, T]$;

(iii) *there exists a matrix* Q_T *of order* $k\times k$ *such that*

$$P_{\theta_0}-\lim_{T\to\infty}\frac{1}{T}\int_0^T (Q_T\dot{S}(\theta_0, t, \mathbf{X}_T), Q_T\dot{S}(\theta_0, t, \mathbf{X}_t))\mathrm{d}t=I; \tag{84}$$

(iv) *for every bounded* $\{\mathbf{u}_T\}$ *in* $\mathbb{R}^K$,

$$P_{\theta_0}-\lim_{T\to\infty}\int_0^T \|S(\theta_0+T^{-1/2}Q_T, \mathbf{u}_T, t, \mathbf{X}_t)-S(\theta_0, t, \mathbf{X}_t) -T^{-1/2}\mathbf{u}'_T Q_T\dot{S}(\theta_0, t, \mathbf{X}_t)\|^2\mathrm{d}t=0. \tag{85}$$

Then the family of measures $\{P_\theta^{(T)}, \theta\in\Theta\}$ *is LAN in* θ_0 *and furthermore*

$$\Delta_T(\theta_0, \mathbf{X}_T)=T^{-1/2}\int_0^T (Q_T\dot{S}(\theta_0, t, \mathbf{X}_t), (\mathrm{d}\mathbf{X}(t)-S(\theta_0, t, \mathbf{X}_t)\mathrm{d}t)). \tag{86}$$

Proof. Theorem 3.6 implies that the measures $P_{\theta 0}^{(T)}$ and $P_{\theta 0+T^{-1/2}Q_T\mathbf{u}}^{(T)}$ are mutually absolutely continuous for every $\mathbf{u}$ in view of assumption (i). Let

$$Z_T(\mathbf{u}, \mathbf{X}_T)=\frac{\mathrm{d}P_{\theta 0+T^{-1/2}Q_T\mathbf{u}}^{(T)}}{\mathrm{d}P_{\theta 0}^{(T)}}(\mathbf{X}_T) \tag{87}$$

where $\mathbf{X}_T$ is the solution of the stochastic differential equation (82) for $\theta=\theta_0$. Theorem 3.6 again shows that

$$\log Z_T(\mathbf{u}, \mathbf{X}_T)=\int_0^T (S(\theta_0+T^{-1/2}Q_T\mathbf{u}, t, \mathbf{X}_t)-S(\theta_0, t, \mathbf{X}_t), \mathrm{d}\mathbf{W}_t) -\frac{1}{2}\int_0^T \|S(\theta_0+T^{-1/2}Q_T\mathbf{u}, t, \mathbf{X}_t)-S(\theta_0, t, \mathbf{X}_t)\|^2\mathrm{d}t. \tag{88}$$

The stochastic integral on the right-hand side of the above equality can be written in the form

$$T^{-1/2}\mathbf{u}'\int_0^T (Q_T\dot{S}(\theta_0, t, \mathbf{X}_t), \mathrm{d}\mathbf{W}_t)+\int_0^T (\{S(\theta_0+T^{-1/2}Q_T\mathbf{u}, t, \mathbf{X}_t)-S(\theta_0, t, \mathbf{X}_t)$$
$$-T^{-1/2}\mathbf{u}'Q_T\dot{S}(\theta_0, t, \mathbf{X}_t)\}, \mathrm{d}\mathbf{W}_t). \tag{89}$$

The first term in the above expression is asymptotically normal with mean zero and variance $\mathbf{u}'I\mathbf{u}$ by condition (iii) in view of Theorem 2.1 of Appendix 2. Furthermore, Lemma 1.1 of Appendix 2 implies that

$$P_{\theta_0}^{(T)}\Big[\Big|\int_0^T \{S(\theta_0+T^{-1/2}Q_T\mathbf{u}_T, t, \mathbf{X}_t)-S(\theta_0, t, \mathbf{X}_t)$$
$$-T^{-1/2}\mathbf{u}_T'Q_T\dot{S}(\theta_0, t, \mathbf{X}_t)\}, \mathrm{d}\mathbf{W}_t)\Big|>\delta\Big]$$

$$\leqq P_{\theta_0}^{(T)}\Big[\int_0^T \|S(\theta_0+T^{-1/2}Q_T\mathbf{u}_T, t, \mathbf{X}_t)-S(\theta_0, t, \mathbf{X}_t)$$
$$-T^{-1/2}\mathbf{u}'_TQ_T\dot{S}(\theta_0, t, \mathbf{X}_t)\|^2\mathrm{d}t>\alpha\Big]+\frac{\alpha}{\delta^2}$$

for every $\alpha>0$ and $\delta>0$. Condition (iv) implies that the above term tends to zero as $T\to\infty$. Conditions (iii) and (iv) show that

$$P_{\theta_0}-\lim_{T\to\infty}\int_0^T \|S(\theta_0+T^{-1/2}Q_T\mathbf{u}_T, t, \mathbf{X}_t)-S(\theta_0, t, \mathbf{X}_t)\|^2\mathrm{d}t-\|\mathbf{u}_T\|^2=0. \tag{90}$$

Combining these results in (89) and (90) and earlier remarks, we obtain that

$$\log Z_T(\mathbf{u}, \mathbf{X}_T)=\mathbf{u}'\Delta_T(\theta_0, \mathbf{X}_T)-\tfrac{1}{2}\|u\|^2-\alpha_T(\mathbf{u}, \mathbf{X}_T, \theta_0) \tag{91}$$

where $\Delta_T(\theta_0, \mathbf{X}_T)$, $\alpha_T(\mathbf{u}_T, \mathbf{X}_T, \theta_0)$ satisfy (80) and (81) respectively and $\Delta_T(\theta_0, \mathbf{X}_T)$ is given by (86).

6.2. Application to Testing of Hypothesis

Let $\mathbf{X}_T$ be the solution of the stochastic differential equation (82) on $[0, T]$. Suppose that θ is a scalar parameter. Let H_0 be the null hypothesis that $\theta=\theta_0$ and K_0 be the one-sided alternative hypotheses $\theta>\theta_0$. Let $\boldsymbol{\phi}_T$ be a test function for testing H_0 against K_0 of size α based on the process $\mathbf{X}_T$. Hence

$$\int \Phi_T(\mathbf{X}_T)\mathrm{d}P_{\theta_0}^{(T)}=\alpha. \tag{92}$$

Define

$$\beta_T(u, \alpha, \Phi_T) = \int \Phi_T(\mathbf{X}_T) \mathrm{d}P^{(T)}_{\theta_0 + T^{-1/2}Q_{Tu}}. \tag{93}$$

Here we are assuming that the family $\{P_\theta^{(T)}\}$ is LAN in θ_0 as $T \to \infty$ and Q_T is as given in Definition 6.1.

We shall now construct a family of test functions $\Phi^0 = \{\Phi_T^0]$ for testing H_0 against K_0 optimal in some asymptotic sense as $T \to \infty$. If T is sufficiently large and if we use ϕ_T^0 as the test function, then $\beta_T(u, \alpha, \phi_T^0)$ will be sufficiently close to zero.

Definition 6.3. The test-criterion $\Phi^0 = \{\Phi_T^0\}$ is said to be *asymptotically locally most powerful of size* α if for every test-criterion $\Phi = \{\Phi_T\}$ of size α,

$$\varliminf_{T \to \infty} \inf_{0 < u < k} \{\beta_T(u, \alpha, \Phi_T^0) - \beta_T(\mu, \alpha, \Phi_T) \geqq 0 \tag{94}$$

for every $k > 0$.

Consider the stochastic differential equation (82) and suppose that $S(\theta, t, \mathbf{X}_t)$ satisfies conditions (i)–(iv) of Theorem 6.1. Define

$$Q_T^2 \|\dot{S}(\theta_0, t, \mathbf{X}_t)\|^2 = T \quad \text{for} \quad t > T,$$

$$\tau_T = \inf\left\{t: \int_0^t Q_T^2 \|\dot{S}(\theta_0, u, \mathbf{X}_u)\|^2 \mathrm{d}u = T\right\}, \tag{95}$$

and

$$Y_T(\theta_0, \mathbf{X}_T) = T^{-1/2} Q_T \int_0^{\tau_T} (\dot{S}(\theta_0, u, \mathbf{X}_u), (\mathrm{d}\mathbf{X}(u) - S(\theta_0, u, \mathbf{X}_u)\mathrm{d}u))$$
$$= T^{-1/2} Q_T \int_0^{\tau_T} (\dot{S}(\theta_0, u, \mathbf{X}_u), \mathrm{d}\mathbf{W}(u)). \tag{96}$$

Let C_α be the $(1-\alpha)$ percentile of standard normal distribution.

Theorem 6.2. *Under conditions* (i)–(iv) *of Theorem* 6.1, *the test-criterion* $\Phi = \{\Phi_T\}$ *defined by*

$$\Phi_T(\mathbf{X}_T) = \begin{cases} 1 & \text{if } Y_T(\theta_0, \mathbf{X}_T) > C_\alpha \\ 0 & \text{if } Y_T(\Theta_0, \mathbf{X}_T) \leqq C_\alpha \end{cases} \tag{97}$$

is asymptotically locally most powerful of size α.

Proof. Since the family $\{P_\theta^{(T)}\}$ is LAN in θ_0, by Theorem 9.6.1, the criterion "accept H_0 if $\Delta_T(\theta_0, \mathbf{X}_T) \leqq C_\alpha^{(T)}$, otherwise accept K_0" is asymptotically locally most powerful of size α where $C_\alpha^{(T)}$ is selected so that the size of the test is α. This follows by extending the proof given in Roussas (1972) from the discrete case to the continuous case. Furthermore, it can easily be checked that

$$P_{\theta_0} - \lim_{T\to\infty} |\Delta_T(\theta_0, \mathbf{X}_T) - Y_T(\theta_0, \mathbf{X}_T)| = 0.$$

Hence, we can replace $\Delta_T(\theta_0, \mathbf{X}_T)$ by $Y_T(\theta_0, \mathbf{X}_T)$ in the test-criterion. Theorem 1.8 of Appendix 2 implies that $\mathscr{L}(Y_T(\theta_0, \mathbf{X}_T)|P_{\theta 0}^{(T)})$ is N (0, 1). Consequently, the test given by (97) is asymptotically locally most powerful of size α.

Example 6.1. Suppose the one-dimensional diffusion process $X(t)$ is ergodic with stationary distribution $B_\theta(x)$ and satisfies the stochastic differential equation

$$\mathrm{d}X(t) = S(\theta, X(t))\mathrm{d}t + \mathrm{d}W(t), \; X(0) = 0 \tag{98}$$

where $0 \leqq t \leqq T$, $\theta \in \Theta \subset \mathbb{R}$. Suppose further that there exists a neighbourhood V_{θ_0} of $\theta_0 \in \Theta$ such that $\dot{S}(\theta, x) = \partial S(\theta, x)/\partial\theta$ exists in V_{θ_0} and satisfies the Lipschitz condition i.e.

$$|\dot{S}(\theta_1, x) - \dot{S}(\theta_2, x)| \leqq H(x)|\theta_1 - \theta_2|^\alpha, \; \theta_1, \theta_2 \in V_{\theta_0}. \tag{99}$$

Further suppose that

$$\int_{-\infty}^{\infty} H^2(x)\mathrm{d}B_{\theta_0}(x) < \infty, \tag{100}$$

and

$$0 < \gamma = \int_{-\infty}^{\infty} \dot{S}^2(\theta_0, x)\mathrm{d}B_{\theta_0}(x) < \infty. \tag{101}$$

It can now be shown that the family of measures $\{P_\theta^{(T)}, \theta \in \Theta\}$ is locally asymptotically normal in θ_0 as $T \to \infty$ with

$$Q_T = \gamma^{-1/2} \text{ and } \Delta_T(\theta_0, \mathbf{X}_T) = (\gamma T)^{-1/2} \int_0^T \dot{S}(\theta_0, X(t))\mathrm{d}W(t). \tag{102}$$

7. SEQUENTIAL ESTIMATION OF THE PARAMETERS OF A DIFFUSION PROCESS

In Example 4.1, we studied the likelihood equation estimator of the parameter λ of the diffusion process satisfying the stochastic differential equation

$$d\xi(t)=\lambda A(t, \xi)+dW(t). \tag{103}$$

In this section, we shall study sequential estimation of the parameter λ.

Let H be a non-negative number. Define the stopping rule

$$\tau(H)=\inf\left\{t: \int_0^t A^2(s, \xi)ds \geq H\right\}. \tag{104}$$

Assume that $P_\lambda\{\int_0^\infty A^2(s, \xi)ds=\infty\}=1$. Then it is easy to see that $P_\lambda(\tau(H)<\infty)=1$. If the process is observed up to a previously determined non-random moment of time, then Example 4.1 shows that the likelihood equation estimator is

$$\hat{\lambda}_T=\frac{\int_0^T A(s, \xi)d\xi_s}{\int_0^T A^2(s, \xi)ds}. \tag{105}$$

The estimate

$$\hat{\lambda}(H)=\hat{\lambda}_{\tau(H)}=\left\{\int_0^{\tau(H)} A(s, \xi)d\xi_s\right\}\left\{\int_0^{\tau(H)} A^2(s, \xi)ds\right\}^{-1}$$

$$=\frac{1}{H}\int_0^{\tau(H)} A(s, \xi)d\xi_s \tag{106}$$

is called the *sequential* likelihood equation estimator of λ. We shall first study the properties of the estimator $\hat{\lambda}(H)$.

Theorem 7.1. *Let the process $\xi(t)$ satisfy the stochastic differential equation* (103) *where* $|A(t, \xi)|^{1/2}\in H_2[0, T]$ (see Appendix 2 for definition) *for every* $T>0$ *and* $P_\lambda\{\int_0^\infty A^2(t, \xi)dt=\infty\}=1$ *for all* $\lambda\in\Lambda$. *Then the sequential plan* $(\tau(H), \hat{\lambda}(H))$ *has the following properties for all* $\lambda\in\Lambda$.

(i) $\hat{\lambda}(H)=\hat{\lambda}_{\tau(H)}$ *is normally distributed with*

$$E_\lambda\{\hat{\lambda}(H)\}=\lambda,\ \text{var}_\lambda\{\hat{\lambda}(H)\}=1/H; \tag{107}$$

(ii) *the plan is efficient;*

(iii) *the plan is closed i.e.* $P_\lambda(\tau(H)<\infty)=1$.

Proof. Lemma 5.4 implies that the process

$$\int_0^{\tau(H)} A(t, \xi)dW(t)$$

is a Wiener process. Since

$$\hat{\lambda}(H)=1/H\int_0^{\tau(H)}A(t,\xi)\mathrm{d}\xi(t)=1/H\left\{\lambda\int_0^{\tau(H)}A^2(t,\xi)\mathrm{d}t\right.$$

$$\left.+\int_0^{\tau(H)}A(t,\xi)\mathrm{d}W(t)\right\}$$

$$=\lambda+1/H\int_0^{\tau(H)}A(t,\xi)\mathrm{d}W(t),$$

it follows that $\hat{\lambda}(H)$ is normally distributed with mean λ and variance H^{-1}. This proves (i).

We shall now prove a lemma which is an analogue of the Cramer–Rao inequality for sequential plans, for estimating the parameter λ of the process $\xi(t)$. (See Chapter 12, Theorem 2.4 for a related result.)

Lemma 7.1. *Let the sequential plan* $(\tau(\xi), \hat{\lambda}_\tau(\xi))$ *for estimating the parameter* λ *be such that*

$$E_\lambda\{\hat{\lambda}_\tau(\xi)\}=h(\lambda) \tag{108}$$

for all $\lambda\in\Lambda$, *differentiation under the expectation sign on the left-hand side with respect to* λ *being possible. Suppose further that*

$$E_\lambda\left\{\int_0^{\tau(\xi)}A^2(t,\xi)\mathrm{d}t\right\}<\infty \tag{109}$$

for all $\lambda\in\Lambda$. *Then*

$$\mathrm{var}_\lambda\{\hat{\lambda}_\tau(\xi)\}\geqq\{\mathrm{h}'(\lambda)\}^2\left\{E_\lambda\left(\int_0^{\tau(\xi)}A^2(t,\xi)\mathrm{d}t\right)\right\}^{-1} \tag{110}$$

for all $\lambda\in\Lambda$ *where* $h'(\lambda)$ *is the derivative of* $h(\lambda)$.

Proof. Condition (109) ensures the equivalence of measures for the process $\xi(t)$ for all $\lambda\in\Lambda$ and $t\leqq\tau(\xi)$. Let μ_λ be the measure of the process $\xi(t)$, $t\leqq\tau(\xi)$ for $\lambda\in\Lambda$. Then

$$\frac{\mathrm{d}\mu_\lambda}{\mathrm{d}\mu_{\lambda_0}}=\exp\left\{(\lambda-\lambda_0)\int_0^{\tau(\xi)}A(t,\xi)\mathrm{d}\xi(t)-\tfrac{1}{2}(\lambda^2-\lambda_0^2)\right.$$

$$\left.\int_0^{\tau(\xi)}A^2(t,\xi)\mathrm{d}t\right\}\text{ a.s.} \tag{111}$$

Differentiating (108) with respect to λ,

$$E_\lambda\left[\hat{\lambda}_\tau(\xi)\left\{\int_0^{\tau(\xi)} A(t, \xi)\mathrm{d}\xi(t) - \lambda\int_0^{\tau(\xi)} A^2(t, \xi)\mathrm{d}t\right\}\right] = h'(\lambda). \tag{112}$$

In view of condition (109), it follows, from the properties of stochastic integrals and the fact that

$$\int_0^T A(t, \xi)\mathrm{d}\xi(t) = \lambda\int_0^T A^2(t, \xi)\mathrm{d}t + \int_0^T A(t, \xi)\mathrm{d}W(t) \tag{113}$$

that

$$E_\lambda\left\{\int_0^{\tau(\xi)} A(t, \xi)\mathrm{d}\xi(t) - \lambda\int_0^{\tau(\xi)} A^2(t, \xi)\mathrm{d}t\right\} = 0 \tag{114}$$

and

$$E_\lambda\left\{\int_0^{\tau(\xi)} A(t, \xi)\mathrm{d}\xi(t) - \lambda\int_0^{\tau(\xi)} A^2(t, \xi)\mathrm{d}t\right\}^2 = E_\lambda\left\{\int_0^{\tau(\xi)} A^2(t, \xi)\mathrm{d}t\right\}. \tag{115}$$

Applying the Cauchy–Schwartz inequality to the left-hand side of (112), and using (115), we obtain (110).

The plan $(\tau(\xi), \hat{\lambda}(\xi))$ is said to be *efficient* if (110) is an equality for all $\lambda \in \Lambda$ where $h(\lambda) \equiv \lambda$. Since part (i) of the theorem holds for the sequential plan $(\tau(H), \hat{\lambda}_{\tau(H)})$ for all $\lambda \in \Lambda$ and since

$$E_\lambda\left\{\int_0^{\tau(H)} A^2(t, \xi)\mathrm{d}t\right\} = H$$

it follows that the plan is efficient. This proves part (ii). Since

$$P_\lambda(\tau(H) \geq T) = P_\lambda\left\{\int_0^T A^2(t, \xi)\mathrm{d}t < H\right\},$$

from the condition

$$P_\lambda\left\{\int_0^\infty A^2(t, \xi)\mathrm{d}t = \infty\right\} = 1,$$

it follows that $P_\lambda(\tau(H) < \infty) = 1$.

The above properties derived in Theorem 7.1 certainly do credit to the sequential likelihood equation estimator. However a natural question arises whether these advantages are consequences of a rather long mean observation time $E_\lambda(\tau(H))$. The study of this question for general $A(t, \xi)$ is still unsolved. Novikov (1972) examined the quantity $E_{\lambda}(\tau(H))$ for $A(t, \xi) = -\xi_t$ i.e. for a linear stochastic differential equation

$$d\xi(t) = -\lambda\xi(t)dt + dW(t);\ \xi(0) = 0. \tag{116}$$

He obtained the following theorem.

Theorem 7.2.

(i) *For* $-\infty < \lambda < \infty$ *as* $T \to \infty$,

$$P_{\lambda}(\tau(H) \geqq T) = 4\left(\frac{H}{\pi T^2}\right)^{1/2} \exp\left\{-\frac{\lambda^2 H}{2} - \frac{T^2}{8H} + \frac{\lambda T}{2}\right\}\{1 + o(1)\}.$$

(ii) *If* $\lambda^2 H \to \infty$, *then*

$$E_{\lambda}(\tau(H)) = \begin{cases} 2\lambda H\left(1 + \dfrac{3}{4\lambda^2 H} + o\left(\dfrac{1}{\lambda^2 H}\right)^2\right), & \lambda > 0 \\[2ex] \dfrac{\log 8\lambda^2 H}{2|\lambda|}\left(1 + o\left(\dfrac{1}{\lambda^2 H}\right)\right), & \lambda < 0 \end{cases}$$

(iii) *If* $\lambda^2 H \to 0$, *then*

$$E_{\lambda}(\tau(H)) = H^{1/2}[B_1 + B_1^{(1)}\lambda H^{1/2} + o(\lambda^2 H)]$$

where $B_1 \sim 2{\cdot}09$ *and* $B_1^{(1)} \sim 0{\cdot}856$.

We omit the proof of this theorem. Let us compare $E_{\lambda}(\hat{\lambda}_T(\xi) - \lambda\}^2$ and $E_{\lambda}\{\hat{\lambda}(H) - \lambda\}^2$, the mean square errors for the ordinary likelihood equation estimator and the sequential likelihood equation estimator. It is natural to specify that if the time of observation in the ordinary case is equal to T, then for comparison purposes $E_{\lambda}(\tau(H)) \leqq T$, $|H| < \infty$. But the stopping rule τ introduced before does not have this property. Nevertheless let $H = H(\lambda, T)$ be the value for which

$$E_{\lambda}\{\tau(H(\lambda, T))\} \equiv T$$

and define

$$e(\lambda, T) = \frac{E_{\lambda}(\hat{\lambda}_T(\xi) - \lambda)^2}{E_{\lambda}(\hat{\lambda}(H(\lambda, T)) - \lambda)^2}.$$

The quantity $e(\lambda, T)$ characterizes the efficiency of the sequential method in comparison to the ordinary one and it is meaningful if it is known *a priori* that λ is in the neighbourhood of some known point λ_0. In this case, it is natural to choose $H = H(\lambda_0, T)$ as the threshold for H for the construction of $\lambda(H)$. Novikov (1972) studied the efficiency $\theta(\lambda, T)$ and he showed that at fixed T, the gain of using sequential likelihood equation estimation will be especially large if $\lambda \leqq 0$. He proved that

$$e(\lambda, T)=\begin{cases} 1+\dfrac{23}{4\lambda T}+o\left(\dfrac{1}{\lambda T}\right)^2 & \text{if } \lambda T\to\infty \\ C_o - C_1\,\lambda T + o(\lambda T)^2 & \text{if } \lambda T\to 0 \\ \dfrac{\exp(|\lambda|T)}{4\pi^{1/2}(|\lambda|T)^{1/2}}\left(1+o\left(\dfrac{1}{|\lambda|T}\right)\right) & \text{if } \lambda T\to -\infty \end{cases}$$

where $C_0 \sim 3{\cdot}04$ and $C_1 \sim 2{\cdot}13$.

8. SEQUENTIAL TESTS FOR DIFFUSION PROCESSES

In this section, we discuss sequential procedures for unbiased testing of the hypothesis that a diffusion process has a given drift function and diffusion function. Before we study sequential tests for general diffusion processes, we shall first discuss the special case of the Wiener process.

8.1 A Sequential Test for Brownian Motion Drift

Let $\{W(t), t \geqq 0\}$ be a Wiener process with $E(W(t))=\mu t$, $\text{var}(W(t))=\sigma^2 t$ and $W(0)=0$ a.s. We assume that σ^2 is known since

$$\lim_{n\to\infty} \sum_{j=1}^{2^n} [W(j\,t2^{-n})-W(j-1)t2^{-n})]^2=\sigma^2 t \text{ a.s.}$$

by Levy's Theorem, and the variance of the process can be determined by observing the process through a positive length of time. In fact, we shall suppose that $\text{var}(W(t))=t$. Suppose we have to stop observing the process when the fluctuation from the initial position is too large. Let

$$T=\inf\{t>0|W(t)=g \text{ or } -h] \tag{117}$$

where g and h are positive constants. Given $\{W(t), 0\leqq t\leqq T\}$, the problem is to obtain a procedure for unbiased testing of the hypothesis $\mu=0$ i.e. the hypothesis of zero drift against a two-sided alternative.

If the process $W=\{W(t), t\geqq 0\}$ has zero drift, we expect a long time to elapse before $W(T)=g$ or $W(T)=-h$. On the other hand if $\mu\neq 0$, there is a steady drift towards either the upper or lower boundary and we expect a shorter time to elapse before a boundary is reached. In view of this, we can propose the region

$$[T \leqq a,\ W(T) = -h] \cup [T \leqq b,\ W(T) = g], \tag{118}$$

where a and b are to be chosen suitably, as the critical region for the test procedure. a and b have to be chosen so that the test has fixed size $\alpha > 0$ and the test is unbiased i.e. the power function of the test as a function of μ has a minimum value α at $\mu = 0$. Let

$$P_+(t, \mu) = P_\mu(T \leqq t,\ W(T) = g), \tag{119}$$

$$P_-(t, \mu) = P_\mu(T \leqq t,\ W(T) = -h),$$

$$f(t, x, \mu) = \frac{\partial}{\partial x} P_\mu(T > t,\ W(t) \leqq x)$$

and

$$I_r(t, \mu) = \int_{-h}^{g} x^r f(t, x, \mu)\mathrm{d}x,\ r = 0, 1, 2.$$

Further let $P_+(t, 0) = P_+(t)$, $P_-(t, 0) = P_-(t)$, $f(t, x, 0) = f(t, x)$ and $I_r(t, 0) = I_r(t)$, $r = 0, 1, 2$. Here P_μ denotes the probability when μ is the drift parameter. The conditions of the size and unbiasedness of the test imply that a and b satisfy the relations

$$\alpha = P_-(a) + P_+(b)$$

and

$$\frac{\partial}{\partial \mu}\{P_-(a, \mu) + P_+(b, \mu)\}|_{\mu=0} = 0.$$

The probabilities P_+, P_- are the distribution functions of T on the upper and lower boundaries respectively, f is the density function of W before the boundaries are reached and I_r are moments associated with f. It has been shown by Brown (1974b) that α, a and b, g and h are related by the equations

$$1 - \alpha = I_0(a) - g^{-1} I_1(a)$$

and

$$1 - \alpha = I_0(b) + h^{-1} I_1(b)$$

and for most practical purposes, the approximations

$$\frac{g\mu\alpha}{2(g+h)} = \phi(a^{-1/2}(2g+h)) - \phi(a^{-1/2}h),$$

$$\frac{h\alpha}{2(g+h)} = \phi(b^{-1/2}(2g+h)) - \phi(b^{-1/2}g)$$

will be adequate for computing a and b where ϕ is the standard normal distribution function. Observe that a and b are proportional to $(g+h)^2$. Hence, to determine (a, b) for all pairs (g, h) with α fixed, it suffices to find the values of (a, b) corresponding to $g+h=1$, $h=\lambda$, $g=1-\lambda$ for $0<\lambda<1$. If this is done, then it will be enough to find only the values of a, for if we write $a=a(\lambda)$, $b=b(\lambda)$, then the relation $b(\lambda)=a(1-\lambda)$ yields values of b. Brown (1974*b*) gives a table for values of a for which the critical region

$$[T \leqq a,\ W(T)=-\lambda] U [T \leqq b,\ W(T)=1-\lambda]$$

describes an unbiased size α test of the hypothesis that $\mu=0$ against a general alternative.

We shall discuss these results further in the chapter on sequential estimation and testing for stochastic processes.

8.2 Sequential Test for Diffusion Processes

Let $\{X(t),\ t \geqq 0\}$ be a diffusion process with drift function μ and diffusion function σ. We will assume that μ, σ^2 are continuous functions, that $\sigma^2(x) > 0$ for all x in the range of X and that $X(0)=0$ almost surely. X is said to be on the natural scale if $\mu(x)=0$ for all x. It can be shown that there is a strictly increasing function u, called the scale function, such that $\{u(X(t)),\ t \geqq 0\}$ is a diffusion on the natural scale and $u(\cdot)$ is given by the equation.

$$u'(x)=\frac{du(x)}{dx}=\exp\left\{-\int_{x_0}^{x} 2\mu(z)\{\sigma^2(z)\}^{-1} dz\right\}$$

where x_0 is fixed but arbitrary and the diffusion function of $\{u(X(t))\}$ is

$$\sigma_0^2(y)=\sigma^2(x)(u'(x))^2 \text{ where } y=u(x).$$

Diffusion processes on the natural scale can be transformed into Wiener processes by means of a time deformation. Specifically, let $\{Y(t),\ t \geqq 0\}$ be a diffusion process on the natural scale with diffusion function $\sigma_1^2(y)$. If $\{W(t)\}$ is a Wiener process with $\operatorname{var}(W(t))=t$, $\{t, W(t)\}$ is called a Brownian motion path for $t \geqq 0$. Let

$$\tau(t)=\int_0^t \sigma_1^2(Y(y))dy$$

Then $\{\tau(t), Y(t)\}$ is a Brownian motion path for $t \geqq 0$. $\tau(t)$ is called the Brownian time of $Y(t)$ since $\tau(\cdot)$ tells at what rate time should be accumulated in order to transform $\{Y(t)\}$ into a Wiener process. For more details about the scale function for diffusion processes, the reader is referred to

Breiman (1968).

Let

$$T = \inf\{t > 0 | X(t) = g \text{ or } -h\}. \tag{120}$$

where g and h are positive constants. Then

$$T = \inf\{t > 0 | u(X(t)) = u(g) \text{ or } u(-h)\}. \tag{121}$$

$u(X(t))$ is a diffusion process on the natural scale and has Brownian time

$$\begin{aligned} \tau(t) &= \int_0^t \sigma_0^2(u(X(y)))\mathrm{d}y \\ &= \int_0^t \sigma^2(X(y))[u'(X(y))]^2 \mathrm{d}y \end{aligned} \tag{122}$$

and since $\{\tau(t), u(X(t))\}$ is a Brownian motion path, it follows that

$$\tau(T) = \int_0^T \sigma^2(X(y))[u'(X(y))]^2 \mathrm{d}y \tag{123}$$

is the first time a Wiener process starting from $u(X(0)) = u(0)$ at time 0 strikes either of the boundaries $u(g)$ or $u(-h)$.

The procedure for unbiased testing that the diffusion process $\{X(t)\}$ has a prescribed drift function μ and a prescribed diffusion function σ^2 is summarized as follows.

Observe the process $\{X(t)\}$ up to the stopping time T. Compute u' then u, from the hypothesis μ, σ^2 using (117) (choosing $x_0 = X(0) = 0$). Then $\tau(T)$ can be determined from $u';/\sigma^2$ and $\{X(t)\}$, $0 \leqq t \leqq T\}$ by using (123). $\tau(T)$ is the first time a Wiener process starting from $u(X(0)) = u(0)$ strikes either $u(g)$ (if $X(T) = g$) or $u(-h)$ (if $X(T) = -h$). Values of a, b which determine the critical region

$$[\tau(T) \leqq a,\ X(T) = -h] U [\tau(T) \leqq b,\ X(T) = g]$$

can be found by the methods in Section 8.1 for a test of prescribed size α.

9. BAYES ESTIMATION FOR DIFFUSION PROCESSES

Let us consider the stochastic differential equation

$$\mathrm{d}\xi(t) = [a(t, \xi(t)) + \theta b(t, \xi(t))]\mathrm{d}t + \sigma(t, \xi(t))\mathrm{d}W(t),\ \xi(0) = \xi_0 \tag{124}$$

where $a(t, x)$, $b(t, x)$ are real-valued functions and $\sigma(t, x)$ is a positive-valued function such that the equation (124) has a unique solution on $[0, T]$ for every $T>0$ and for every $\theta \in \Theta$ open in $\mathbb{R}$. Let μ_θ be the measure corresponding to the solution of (124) on $[0, T]$ for fixed initial value ξ_0 and we shall assume that $\mu_\theta \sim \mu_{\theta'}$ for all $\theta, \theta' \in \Theta$. The Radon–Nikodyn derivative of μ_θ with respect to μ_θ Θ' is given by

$$\frac{d\mu_\theta}{d\mu_{\theta'}}(\xi_0^T) = \exp\{(\theta - \theta') \int_0^T \frac{b(t, \xi(t))}{\sigma(t, \xi(t))} dW(t) - \frac{1}{2}(\theta - \theta')^2 \int_0^T \left[\frac{b(t, \xi(t))}{\sigma(t, \xi(t))}\right]^2 dt\} \tag{125}$$

where ξ_0^T denotes the sample path of the process ξ on $[0, T]$. This can be seen from Theorem 3.6

Estimators of the form

$$\hat{\theta}_T = \left\{\int \frac{d\mu_v}{d\mu_{\theta'}}(\xi_0^T) p(v) dv\right\}^{-1} \left\{\int u \frac{d\mu_u}{d\mu_{\theta'}}(\xi_0^T) p(u) du\right\}, \tag{126}$$

where θ' is an arbitrary but fixed parameter in Θ and $p(\cdot)$ is a real-valued measurable function defined on Θ, are called *generalized Bayesian estimators*. If p is integrable over Θ, then these estimators are the usual Bayes estimators under quadratic loss function.

We shall now study the behaviour of the estimator $\hat{\theta}_T$ as $T \to \infty$. Let θ_0 denote the true value of θ. Unless otherwise specified P and E will denote probability and expectation when θ_0 is the underlying parameter. Let Q be an arbitrary continuous monotonic increasing function of T such that $Q(T) \to \infty$ as $T \to \infty$. Consider the stochastic process

$$\zeta_T(t) = \frac{d\mu_{\theta_0 + \frac{t}{Q(T)}}}{d\mu_{\theta_0}}(\xi_0^T). \tag{127}$$

It is easy to see that

$$Q(T)(\hat{\theta}_T - \theta_0) = \left\{\int \zeta_T(t) p\left(\theta_0 + \frac{t}{Q(T)}\right) dt\right\}^{-1} \cdot \left\{\int s\zeta_T(s) p\left(\theta_0 + \frac{s}{Q(T)}\right) ds\right\} \tag{128}$$

from (125) to (127). Furthermore,

$$\eta_T(t) = \log \zeta_T(t) = \alpha_T t - \tfrac{1}{2}\beta_T t^2 \tag{129}$$

where

$$\alpha_T = \frac{1}{Q(T)} \int_0^T \frac{b(s, \xi(s))}{\sigma(s, \xi(s))} \mathrm{d}W(s) \tag{130}$$

and

$$\beta_T = \frac{1}{Q^2(T)} \int_0^T \left[\frac{b(s, \xi(s))}{\sigma(s, \xi(s))}\right]^2 \mathrm{d}s. \tag{131}$$

Note that α_T and β_T are well-defined for large T. We shall now state the main theorem of this section.

Theorem 9.1. *Suppose the following conditions are satisfied:*

(i) *p is continuous in a neighbourhood of θ_0, $p(\theta_0) \neq 0$ and*

$$\sup_t |p(t)|(1+|t|^{\lambda})^{-1} < \infty \tag{132}$$

for some $\lambda \geqq 0$.

(ii)
$$\beta_T \xrightarrow{p} \beta > 0 \text{ as } T \to \infty. \tag{133}$$

Then $Q(T)(\hat{\theta}_T - \theta_0)$ is asymptotically normal with mean 0 and variance β^{-1} as $T \to \infty$.

Before we give a proof of this theorem, we shall prove a few lemmas.

Lemma 9.1. *Under condition* (ii) *of Theorem* 9.1, *the finite-dimensional distributions of the process $\{\zeta_T(t)\}$ converge weakly to the finite-dimensional distributions of the process $\{\zeta(t)\}$ as $T \to \infty$ where*

$$\zeta(t) = \exp\{\xi(\sqrt{\beta})t - \tfrac{1}{2}\beta t^2\}, \quad -\infty < t < \infty \tag{134}$$

and ξ is a standard normal variable.

Proof. Define $\psi_T(t) = t^{-1}\eta_T(t) + \frac{1}{2}\beta t$. α_T is asymptotically normal with mean 0 and variance β in view of condition (ii) by a slight modification of Theorem 2.1 of Appendix 2. Hence

$$\eta_T(t) \xrightarrow{\mathcal{L}} \xi(\sqrt{\beta})t - \tfrac{1}{2}\beta t^2 \equiv \eta(t) = \log \zeta(t)$$

as $T \to \infty$ where ξ is a standard normal random variable. Furthermore, for any $t_1 \neq t_2$,

$$\psi_T(t_2) - \psi_T(t_1) = \tfrac{1}{2}(\beta_T - \beta)(t_1 - t_2) \xrightarrow{p} 0$$

as $T \to \infty$ by condition (ii). The lemma now follows from Cramer–Wold technique.

It is clear from the definition of processes $\{\zeta_T(t)\}$ and $\{\zeta(t)\}$ that they have continuous sample paths almost surely. We shall now prove that the processes $\{\zeta_T(t), -A \leqq t \leqq A\}$ converge in distribution to the process $\{\zeta(t), -A \leqq t \leqq A\}$ for any fixed $A > 0$. In other words, the probability measures ν_T generated by the process $\{\zeta_T(t), -A \leqq t \leqq A\}$ converge weakly to the probability measure ν generated by the process $\{\zeta(t), -A \leqq t \leqq A\}$ on $C[-A, A]$. In view of Lemma 9.1, it is sufficient to prove the following lemma (cf. Billingsley, 1968).

Lemma 9.2. *Under condition* (ii) *of Theorem* 9.1,

$$\lim_{h \to 0} \overline{\lim_{T \to \infty}} P\{ \sup_{\substack{|t_1 - t_2| \leqq h \\ t_1, t_2 \in [-A, A]}} |\zeta_T(t_2) - \zeta_T(t_1)| > \varepsilon\} = 0. \tag{135}$$

Proof. We shall suppose that t varies over $[-A, A]$ in the following discussion. Clearly

$$\sup_{|t_2 - t_1| \leqq h} |\zeta_T(t_2) - \zeta_T(t_1)|$$

$$\leqq \sup_{|t_2 - t_1| \leqq h} |\eta_T(t_2) - \eta_T(t_1)| \exp\{\sup_{|t| \leqq A} \eta_T(t)\}$$

and hence for any $N > 0$,

$$P\{ \sup_{|t_1 - t_2| \leqq h} |\zeta_T(t_2) - \zeta_T(t_1)| > \varepsilon\}$$

$$\leqq P\{\sup_{|t| \geqq A} \eta_T(t) > N\}$$

$$+ P\{ \sup_{|t_2 - t_1| \leqq h} |\eta_T(t_2) - \eta_T(t_1)| > \varepsilon e^{-N}\}. \tag{136}$$

In order to prove (135), it is sufficient to prove that

$$\lim_{N \to \infty} \overline{\lim_{T \to \infty}} P\{\sup_{|t| \leqq A} \eta_T(t) > N\} = 0 \tag{137}$$

and that for every $\varepsilon > 0$,

$$\lim_{h \to 0} \overline{\lim_{T \to \infty}} P\{\sup_{|t_2 - t_1| \leqq h} |\eta_T(t_2) - \eta_T(t_1)| > \varepsilon\} = 0. \tag{138}$$

Since

$$\sup_{|t|\leqq A} \eta_T(t) \leqq A|\alpha_T| + \tfrac{1}{2}A^2|\beta_T|,$$

it follows that

$$P\{\sup_{|t|\leqq A} \eta_T(t) > N\}$$

$$\leqq P\{|\alpha_T| > \frac{N}{2A}\} + P\{|\beta_T| > \frac{N}{A}\}.$$

The last two terms tend to zero as $T\to\infty$ and then $N\to\infty$ since α_T is asymptotically normal with mean 0 and variance β and $|\beta_T| \xrightarrow{p} |\beta|$ as $T\to\infty$. Hence (137) holds. It is easy to see that

$$\sup_{\substack{|t_2-t_1|\leqq h \\ t_1,\ t_2\in[-A,\ A]}} |\eta_T(t_2)-\eta_T(t_1)| \leqq h(|\alpha_T| + A|\beta_T|)$$

and therefore

$$P\{\sup_{\substack{|t_2-t_1|\leqq h \\ t_1, t_2\in[-A, A]}} |\eta_T(t_2)-\eta_T(t_1)| > \varepsilon\}$$

$$\leqq P\left\{|\alpha_T| > \frac{\varepsilon}{2h}\right\} + P\left\{|\beta_T| > \frac{\varepsilon}{2Ah}\right\}. \tag{139}$$

Again one can show that the last two terms tend to zero as $T\to\infty$ and then $h\to 0$. Hence (138) holds. This completes the proof of the lemma.

Lemma 9.3. *Under conditions* (i) *and* (ii) *of Theorem* 9.1, *for every* $\varepsilon > 0$

$$\lim_{\substack{A\to\infty \\ T\to\infty}} P\left\{\left|\int_{|t|>A} |t|\zeta_T(t)p\left(\theta_0 + \frac{t}{Q(T)}\right)\mathrm{d}t\right| > \varepsilon\right\} = 0. \tag{140}$$

Proof: In view of condition (i), there exist constants $c > 0$, $\lambda \geqq 0$ such that

$$|p(t)| \leqq c(1 + |t|^{\lambda})$$

for all t. Hence, it is sufficient to prove that

$$\lim_{\substack{A\to\infty \\ T\to\infty}} P\left\{\left|\int_{|t|>A} |t|^{1+\lambda}\zeta_T(t)\mathrm{d}t\right| > \varepsilon\right\} = 0 \tag{141}$$

for every $\varepsilon > 0$. Since

$$\eta_T(t) \leqq |\alpha_T||t| - \tfrac{1}{2}(\beta - |\beta_T - \beta|)t^2$$

it is easy to see that for any $N>0$ and $0<\delta<\beta$,

$$P\left\{\int_{|t|>A}|t|^{1+\lambda}\zeta_T(t)\mathrm{d}t>\varepsilon\right\}$$

$$\leqq P\{|\alpha_T|>N\}+P\{|\beta_T-\beta|>\delta\}$$

$$+P\left(\int_{|t|>A}|t|^{1+\lambda}\exp\{N|t|-\tfrac{1}{2}(\beta-\delta)t^2\}\mathrm{d}t>\varepsilon\right). \tag{142}$$

Since α_T is asymptotically normal with mean zero and $\beta_T \xrightarrow{p} \beta$ as $T\to\infty$, it can now be shown that the three terms on the right-hand side of (142) converge to zero as $A\to\infty$ and $T\to\infty$. We omit the details.

Proof of Theorem 9.1. Relation (128) says that

$$Q(T)(\hat{\theta}_T-\theta_0)=\frac{\int s\zeta_T(s)p\left(\theta_0+\dfrac{s}{Q(T)}\right)\mathrm{d}s}{\int \zeta_T(t)p\left(\theta_0+\dfrac{t}{Q(T)}\right)\mathrm{d}t}. \tag{143}$$

The finite-dimensional distributions of the process $\{\zeta_T(t)\}$ converge weakly to the finite-dimensional distributions of the process $\{\zeta(t)\}$ as $T\to\infty$ by Lemma 9.1 where $\zeta(t)=\exp\{\xi\sqrt{\beta}t-\tfrac{1}{2}\beta t^2\}$, ζ being the standard normal random variable. Passing to the limit formally in (143) as $T\to\infty$, we obtain

$$Q(T)(\hat{\theta}_T-\theta_0)\xrightarrow{\mathscr{L}}\frac{\int s\zeta(s)\mathrm{d}s}{\int \zeta(s)\mathrm{d}s}=\beta^{-1/2}\xi. \tag{145}$$

We shall now justify this limiting argument. Let $A>0$. For any fixed Y_1, Y_2 the functional

$$f_{\gamma_1,\gamma_2}(x)=\gamma_1\int_{-A}^{A}sx(s)\mathrm{d}s+\gamma_2\int_{-A}^{A}x(s)\mathrm{d}s$$

defined on $C[-A, A]$ is continuous. Lemmas 9.1 and 9.2 imply that the processes $\{\zeta_T(t),\ -A\leqq t\leqq A\}$ converge in distribution to the process $\{\zeta(t),\ -A\leqq t\leqq A\}$. Hence the distribution of $f_{\gamma_1,\gamma_2}(\zeta_T)$ converges weakly to the distribution of $f_{\gamma_1,\gamma_2}(\zeta)$ as $T\to\infty$ (cf. Billingsley, 1968). From this fact and the continuity of $p(\theta)$ at θ_0, it follows that the distribution of the random vector

$$\left(\int_{-A}^{A}s\zeta_T(s)p\left(\theta_0+\frac{s}{Q(s)}\right)\mathrm{d}s,\ \int_{-A}^{A}\zeta_T(s)p\left(\theta_0+\frac{s}{Q(s)}\right)\mathrm{d}s\right)$$

converges weakly to the distribution of the random vector

$$\left(p(\theta_0)\int_{-A}^{A} s\zeta(s)\mathrm{d}s,\ p(\theta_0)\int_{-A}^{A}\zeta(s)\mathrm{d}s\right).$$

It can be checked that for any $\varepsilon>0$, there exists $\delta>0$ such that

$$P\left\{\left|\int_{|t|<A} t\zeta(t)p(\theta_0)\mathrm{d}t\right|>\delta\right\}>1-\varepsilon$$

and

$$P\left\{\left|\int_{|t|<A} \zeta(t)p(\theta_0)\mathrm{d}t\right|>\delta\right\}>1-\varepsilon$$

and similar inequalities hold for the processes $\{\zeta_T(t)\}$ for sufficiently large T and large A by the convergence described above and Lemma 9.3. Hence, given any $\varepsilon>0$,

$$Q(T)(\hat{\theta}_T-\theta_0)=\frac{\int_{-A}^{A} s\zeta_T(s)\mathrm{d}s}{\int_{-A}^{A}\zeta_T(s)\mathrm{d}s}\{1+r_n(T)\}$$

where $P\{|r_n(T)|>\varepsilon\}<\varepsilon$ by Lemma 9.3. Therefore one can pass to the limit in (143) in view of the convergence obtained above and (145) holds. Since $\beta^{-1/2}\xi$ is normal with mean 0 and variance β^{-1}, the proof of the theorem is complete.

10. FURTHER TOPICS

In Section 5, we discussed estimation problems for stochastic differential equations of the type

$$\mathrm{d}\xi(t)=a(t,\ \xi(t);\ \alpha)\mathrm{d}t+\sigma(t,\ \xi(t))\mathrm{d}W(t),\ \xi(0)=\xi_0 \tag{146}$$

where $a(t,\ x;\ \alpha)$ and $\sigma(t,\ x)$ are non-random and

$$a(t,\ x;\ \alpha)=a(t,\ x)+\sum_{k=1}^{n}\alpha_k Q_k(t,\ x).$$

Taraskin (1971b) has studied estimation when $a(t,\ x;\ \alpha)$ is a random function of the type

$$a(t,\ \xi(t);\ \alpha)=a(t,\ \xi(t))+\sum_{k=1}^{n}\alpha_k Q_k(t,\ \xi(t);\ \theta(t))$$

where $\{\theta(t)\}$ is a new random process taking values in a measurable space. He has given regularity conditions under which the corresponding likelihood equation estimator $\hat{\alpha}_T$ of α is consistent, asymptotically normal and asymptotically efficient. Confidence regions for parameters of diffusion processess have been discussed in Taraskin (1971a, 1975).

Let us consider the stochastic differential equation

$$d\xi(t)=\theta f(t, \xi)dt+dW(t), \quad \xi(0)=0, \ t\geqq 0 \tag{147}$$

where θ is an $\mathscr{F}_0$-measurable random variable which is normally distributed with mean m and variance r independent of the Wiener process. Here $\{\mathscr{F}_t, t\geqq 0\}$ is a non-decreasing sequence of sub σ-algebras such that $\xi(t)$ and $W(t)$ are measurable with respect to $\mathscr{F}_t$. Musiela (1976) studied the likelihood equation estimation of m and r. We shall now briefly discuss these results.

Let $\mathscr{B}$ be the σ-field of Borel subsets of $C\equiv C[0, \infty)$ relative to the topology of uniform convergence on compact subsets of $\mathbb{R}_+$. Suppose that $f(t, x)$ is defined on $(C, \mathscr{B})$ and measurable with respect to $\mathscr{B}_t=\sigma\{x:x_s(\cdot), s\leqq t\}$. Suppose that $f(t, \xi)\in H_2[0, T]$ for every $T>0$, $m\in\mathbb{R}$, $r\in[0, \infty)$ and

$$P\left(\int_0^\infty f^2(t, \xi)dt=\infty\right)=1.$$

Let $\mu_{m,r}$ be the measure induced by the process ξ when $E\theta=m$ and var $\theta=r$ and μ be the measure induced by the Wiener process on $(C, \mathscr{B})$. For any Markov time τ, let $\mu_{m,r}^\tau$ and μ^τ be the restrictions of $\mu_{m,r}$ and μ respectively to the σ-algebra $\mathscr{B}_\tau$, Further suppose that

$$P\left(\int_0^\tau f^2(t, \xi)dt<\infty\right)=1$$

for all $m\in\mathbb{R}$ and $r\geqq 0$. Musiela (1977a) showed that the Radon–Nikodym derivative of $\mu_{m,r}^\tau$ with respect to μ^τ exists and is given by

$$\frac{d\mu_{m,r}^\tau}{d\mu^\tau}(\tau, \xi)$$
$$=\exp\left\{\frac{1}{2}\left[\frac{r\left(\int_0^\tau f(t, \xi)d\xi_t\right)^2+2m\int_0^\tau f(t, \xi)d\xi_t-m^2\int_0^\tau f^2(t, \xi)dt}{1+r\int_0^\tau f^2(t, \xi)dt}\right.\right. \tag{148}$$
$$\left.\left.-\log\left(1+r\int_0^\tau f^2(t, \xi)dt\right)\right]\right\}.$$

Hence the sequential likelihood equation estimator of m, when the process is stopped at time τ, is

$$\hat{m}_\tau = \frac{\int_0^\tau f(t, \xi) d\xi_t}{\int_0^\tau f^2(t, \xi) dt} \tag{149}$$

and if m is known, the sequential likelihood equation estimator of r is

$$\hat{r}_\tau = \left\{ \frac{\int_0^\tau f(t, \xi) d\xi_t}{\int_0^\tau f^2(t, \xi) dt} - m \right\}^2 - \frac{1}{\int_0^\tau f^2(t, \xi) dt}. \tag{150}$$

Musiela (1977a) obtained the bias and mean square error of m_τ and r_τ. Estimation problems for stochastic differential equations of the type

$$d\xi(t) = f(t, \xi) a\left(\theta, \int_0^t f^2(u, \xi) du, \int_0^t f(u, \xi) d\xi_u\right) dt + dW(t), \ \xi_0 = 0 \tag{151}$$

where $\theta \in \Theta \subset \mathbb{R}^k$, $a(\theta, t, x)$ is defined on $\mathbb{R}^k \times [0, \infty) \times \mathbb{R}$, are studied in Musiela (1977b).

We discussed the asymptotic theory of Bayes estimation in Section 9.9 for stochastic differential equations containing a one-dimensional parameter θ. These results are extended by Linkov (1975b) to the multi-dimensional case. He considered stochastic differential equations of the type

$$\begin{aligned} d\xi(t) = \{a(t, \xi(t)) + (\psi(\theta), b(t, \xi(t))\} dt \\ + \sigma(t, \xi(t)) dW(t), \ \xi_0 = 0 \end{aligned} \tag{152}$$

where $\psi(\theta)$ is a vector-valued function of $\theta \in \Theta \subset \mathbb{R}^k$. He has found sufficient conditions for asymptotic normality of generalized Bayes estimator $\hat{\theta}_T$ both in cases $\psi(\theta) = \theta$ and $\psi(\theta) \neq \theta$. Linkov (1977) has found sufficient conditions for asymptotic normality and convergence of moments of $Q_T(\tilde{\theta}_T - \theta)$ and $Q_T(\hat{\theta}_T - \theta)$ to the corresponding moments of the limiting normal distribution where $\tilde{\theta}_T$ and $\hat{\theta}_T$ are the likehood equation estimator and generalized Bayes estimator respectively for parameter θ of a diffusion process given by the stochastic differential equation

$$d\xi(t) = a(t, \theta, \xi(t)) dt + \sigma(t, \xi(t)) dW(t), \ \xi_0 = 0. \tag{153}$$

These results can also be obtained from the Beinstein-von Mises Theorem for diffusion processes discussed in Prakasa Rao (1981). Kutoyants (1976, 1977a,b) discusses properties of $\hat{\theta}_T$ and $\tilde{\theta}_T$ stochastic differential equations of the type

$$d\xi(t)=a(\theta, \xi(t))dt+\sigma(\xi(t))dW(t), \xi_0=0. \tag{154}$$

Further results of a similar nature are given in Lipcer and Sirjaev (1974). General theory of sequential estimators for diffusion processes is discussed in Ershav (1971).

Two-sided sequential tests for drift μ for the Wiener process were discussed in Section 9.8. Brown (1974) has also studied one-sided tests when the alternative is $\mu>0$. These tests belong to a class of tests introduced by Nadas (1973).

BIBLIOGRAPHICAL NOTES

The theory of diffusion processes has been extensively studied from both theoretical and applied points of view in Doob (1953), Dynkin (1965), Gikhman and Skorokhod (1969, 1972) and Arnold (1974). Treatment of the theory of diffusion processes given in Section 2 is similar to Arnold (1974). Results in Section 3 are from Lipcer and Sirjaev (1972). These have been exhaustively studied in Lipcer and Sirjaev (1974). For processes of diffusion type, the densities have been studied by Duncan (1967) and Kailath (1971). Section 4 contains the work of Le Breton (1974, 1975). Example 4.1 is due to Sirjaev (1972). Asymptotic theory of likelihood equation estimators for parameters of diffusion processes in Section 5 is an improvement over similar results of Taraskin (1974) under stronger conditions. Holevo (1967) has also considered this problem by using methods of stochastic approximation. Example 5.1 is from Taraskin (1974). Subsection 5.3 contains some extensions of results of Taraskin (1974) for problems of tests of hypotheses. Similar work has been done independently by Brown and Hewitt (1975). Section 6 contains results on local asymptotic normality for processes of diffusion type. These are analogues of results of Kutoyants (1975a,b) for "scheme of series" as $T\to\infty$. Kutoyants (1977a) discusses these results further. Example 6.1 is from Kutoyants (1975b). Section 7 contains the work of Novikov (1972) and Sirjaev (1972). Sequential tests for diffusion processes discussed in Section 8 were obtained by Brown (1974a,b). Section 9 contains results of Linkov (1975a) on asymptotic theory of Bayes estimation for parameters of diffusion processes.

COMPLEMENTS

1. Let $\{W(t), 0\leqq t\leqq 1\}$ be the standard Wiener process and $\theta(t)$ be a differentiable function of t whose derivative is of bounded variation in [0, 1]. Let μ_n and μ_W be the

measures generated by processes $\eta(t)=W(t)+\theta(t)$ and $W(t)$ respectively on $C[0,1]$. Show that $\mu_\eta \ll \mu_W$. Find the Radon–Nikodym derivative of μ_η with respect to μ_W.
(Cameron and Martin, 1945)

2. Consider the diffusion branching process $\{X_t\}$ which is a solution of the stochastic differential equation

$$\mathrm{d}X_t=\theta X_t\mathrm{d}t+X_t^{1/2}\,\mathrm{d}W_t,\ X_0=1,\ t\geqq 0.$$

This is the limiting diffusion for a discrete state branching process. Let

$$I_t=\int_0^t X_s\mathrm{d}s.$$

Prove that, if $\hat{\theta}(t)$ is the likelihood equation estimator, then, as $t\to\infty$

$$P(I_t^{1/2}(\hat{\theta}(t)-\theta_0)\leqq x|X_t>0)\to\int_0^x (2\pi)^{-1/2}\,\mathrm{e}^{-t^2/2}\,\mathrm{d}t.$$

(Feigin, 1977)

3. Let $\{Z_t\}$ be the solution of the stochastic differential equation

$$\mathrm{d}Z_t=-(\lambda+\mu)Z_t\mathrm{d}t+\frac{2\lambda\mu}{\lambda+\mu}\,\mathrm{d}W_t.$$

Let $\theta=-(\lambda+\mu)$, $\phi=2\lambda\mu(\lambda+\mu)^{-1}$. Obtain estimates for θ and ϕ.
(Feigin, 1977)

4. Suppose for each $n\geqq 1$, $\{Z_t^{(n)};\ t\geqq 0\}$ is an age dependent branching process with $Z_0^{(n)}=n$ and that it has reproduction generating function f_n satisfying

$$f_n'(1)=1+\alpha/n+o\left(\frac{1}{n}\right).$$

$$f_n''(1)=\beta_n\to\beta,\ \text{and}$$

$$f_n'''(1-)<K.$$

Let the life-length distribution be L with $L(0)=0$ and $\gamma=\int_0^\infty t\ L(t)\mathrm{d}t$. Show that the finite dimensional distributions of $\{n^{-1}\ Z_{nt}^{(n)};\ t\geqq 0\}$ converge as $n\to\infty$ to those of $\{Z_t;\ t\geqq 0\}$ where $Z_0=1$ and

$$\mathrm{d}Z_t=\frac{\alpha}{\gamma}\,Z_t\ \mathrm{d}t+\left(\frac{\beta}{\gamma}\right)^{1/2}\,Z_t^{1/2}\ \mathrm{d}W(t),\ t\geqq 0.$$

$\gamma=(\lambda+\mu)^{-1}$ and $f(s)=(\mu+s^2\lambda)(\lambda+\mu^{-1})$ for linear birth and death process. Obtain likelihood equation estimators of λ and μ.
(Feigin, 1977)

5. Let $\{X_t\}$ be the solution of the stochastic differential equation

$$\mathrm{d}X_t=\theta X_t\mathrm{d}t+G\ \mathrm{d}W_t;\ t\geqq 0; X_0\equiv 0.$$

If $\{X_t\}$ is a one-dimensional process, prove that the likelihood, equation estimator of θ is

$$\hat{\theta}_G(T)=\frac{X_T^2-G^2T}{2\int_0^T X_s^2\,ds}$$

and if $\{X_t\}$ is an n-dimensional process ($n\geqq 2$) and the parameter space Θ consists of symmetric matrices, then $\hat{\theta}_G(T)$ is the solution of the equation,

$$\theta\left(\int_0^T X_sX_s'\,ds\right)+\left(\int_0^T X_sX_s'\,ds\right)\theta=X_TX_T'-TGG' \text{ a.s.}$$

(Le Breton, 1974)

6. For the general age-dependent branching process (Bellman–Harris process) consider the Malthusian parameter δ i.e. δ satisfying the equation

$$m\int_0^\infty e^{-\delta t}\,L(dt)=1$$

where $1<m<f'(1)<\infty$ where $f(s)$ is the generating function of the offspring distribution and $L(t)$ is the life-length distribution. Under suitable assumptions, show that

$$\hat{\delta}_T=\frac{X_T-X_0}{\int_0^T X_s\,ds}$$

is a likelihood equation estimator of δ.

(Athreya and Keiding, 1977)

7. Let $\{X_t, t\geqq 0\}$ be a Wiener process with drift and diffusion parameters μ and σ^2 respectively (μ and σ^2 are both unknown). Let T be a stopping time defined by $T=\min\{t, t\geqq 0: X_t=g\}$ where $g>0$. For testing $H_0:\mu=0$ against $H_1:\mu>0$ based on a random sample of size n from the distribution of T show that the test which rejects H_0 where M/H is small is most powerful unbiased. (Here M and H are the arithmetic and harmonic means respectively of the n observations on T).

(Nadas, 1973)

8. Consider the stochastic differential equation

$$d\xi(t)=[a(\xi(t))+\theta b(\xi(t))]dt+\sigma(\xi(t))dW(t),\ t\geqq 0.$$

Discuss the asymptotic behaviour of the likelihood equation estimate $\hat{\theta}_T^*$ of θ when the process is observed over $[0, T]$ as $T\to\infty$.

(Kulinic, 1975)

9. Consider the Stochastic differential equation

$$d\xi(t)=[\alpha+\alpha_1\xi(t)]dt+(\beta_0+\beta_1\xi(t)+\beta_2\xi^2(t))^{1/2}dW(t),\ t\geqq 0$$

where α_1, β_0, β_1 and β_2 are known real numbers with $\alpha_1<0$, $\beta_2>0$ and $\beta_0+\beta_1x+\beta_2x^2=0$ has complex roots. Discuss the asymptotic behaviour of the likelihood equation estimator $\hat{\alpha}_T$ of α as $T\to\infty$ when the process is observed over $[0, T]$.

(Taraskin, 1970)

10. Consider the stochastic differential equation

$$d\xi(t)=\theta f(t, \xi)dt+dW(t),\ \xi(0)=0,\ t\geqq 0$$

where θ is an $\mathscr{F}_0$-measurable, normally distributed, random variable with mean m and variance r as discussed in Section 10. Suppose the process is observed over $[0, T]$. Show that the likelihood equation estimators $\hat{m}_T$ and $\hat{r}_T$ of m and r (assuming m is known in the latter case) are not consistent but asymptotically unbiased as $T\to\infty$. Find the bias and mean square errors of $\hat{m}_T$ and $\hat{r}_T$.

(Musiela, 1976)

11. Consider the stochastic differential equation

$$d\xi(t)=f(t, \xi)\frac{m+r\displaystyle\int_0^t f(u, \xi)d\xi_u}{1+r\displaystyle\int_0^t f^2(u, \xi)du}dt+dW(t), \quad \xi(0)=0; t\geqq 0$$

where $m\in I\!R$, $r>0$ are unknown and f is a real-valued function. Under suitable assumptions, obtain a sufficient statistic for (m, r) and discuss the behaviour of the likelihood equation estimators $\hat{m}_T$ and $\hat{r}_T$ of m and r (assuming m is known in the latter case) as $T\to\infty$.

(Musiela, 1977b).

12. Consider the stochastic differential equation

$$d\xi(t)=\theta|\xi(t)|^\alpha dt+dW(t),\ \xi(0)=1;\ t\geqq 0$$

where $0\leqq\alpha<1$, $\theta\neq 0$. Obtain the likelihood equation estimator $\hat{\theta}_T$ of θ and discuss its asymptotic behaviour as $T\to\infty$ assuming that α is known.

(Kutoyants, 1977b).

13. Consider the stochastic differential equation

$$d\xi(t)=S(\theta, \xi(t))dt+b(\xi(t))dW(t),\ \xi(0)=0;\ t\geqq 0$$

where $\theta\in\Theta=(\alpha, \beta)\subset I\!R$. Let $f(\theta, x)=S(\theta, x)/b(x)$. Suppose

(i) the process $\{\xi(t)\}$ is ergodic for each $\theta\in\Theta$ with stationary distribution μ_θ,

(ii) $P_\theta^{(T)}\left\{\int_0^T f^2(\theta, \xi(t))dt<\infty\right\}=1,\ \theta\in\Theta,\ T<\infty,$

(iii) $f(x, \theta)$ is differentiable with respect to θ with derivative $\dot{f}(x, \theta)$ satisfying

$$|\dot{f}(\theta_1, x)-\dot{f}(\theta_2, x)|\leqq H(x)|\theta_1-\theta_2|^d$$

where

$$\int H^2(x)d\mu_\theta(x)<\infty,$$

(iv) $$\inf_{x,\theta}|\dot{f}(\theta, x)|>0$$

Show that the family of measures $\{P_\theta^{(T)}, \theta\in\Theta\}$ is LAN. Further, show that the likelihood equation estimator and Bayes estimators for loss functions

$W(\theta, y) = |\theta - y|^a, a \geqq 1$ are consistent, asymptotically normal and asymptotically efficient with limiting variance

$$\left(T \int_0^T \left\{ \frac{\dot{s}(\theta, x)}{b(x)} \right\}^2 \mathrm{d}\mu_\theta(x) \right)^{-1}.$$

(Kutoyants, 1977a)

Chapter 10

Bayesian Inference for Stochastic Processes

1. INTRODUCTION

Let $\{X_j, j=1, 2, \ldots\}$ be independent observations whose individual distribution has a density depending on a real or k-dimensional parameter θ. Assume that some *a priori* distribution of θ has been chosen. It was observed by Laplace and later by von Mises, that under suitable circumstances, the *a posteriori* distribution of θ given the observations $X_1, \ldots, X_n$ for large n is very close to a normal distribution. A related but different result was also proved by Bernstein who considered the posterior distribution of θ given the average $n^{-1}(X_1+\ldots+X_n)$. von Mises extended this result to posterior distributions conditioned by a finite number of differentiable functionals of empirical distribution function.

In this chapter, we deal with the asymptotic theory of Bayes solutions in estimation and testing when the observations are from a discrete parameter stochastic process. We obtain the fundamental theorem in the asymptotic theory of Bayesian inference namely, the approach of the posterior density to the normal for discrete parameter stochastic processes. From the result obtained, we derive the asymptotic efficiency of Bayes estimators obtained from smooth loss functions. Bayesian testing and a limit theorem for the nth root of posterior risk are discussed at the end of the chapter.

2. PRELIMINARIES

Let $\mathbf{Y}^n=(Y_1, \ldots, Y_n)$, $n \geqq 1$ be a sequence of random vectors defined on a probability space $(\Omega, \mathscr{F}, P_\theta)$ and taking values in measure space $(\mathscr{Y}^n, \mathscr{A}^n, \mu^n)$, μ a σ-finite measure on $(\mathscr{Y}, \mathscr{A})$. $\mathscr{Y}$ could be $\mathbb{R}^m$ and $\theta \in \Theta$, a subset of $\mathbb{R}^p$. Let $\|\cdot\|$ be the ordinary Euclidean norm on $\mathbb{R}^p$. Suppose that $\mathbf{Y}^n$ has density $p(\mathbf{y}^n; \theta) \equiv p(y_1, \ldots, y_n; \theta)$ with respect to μ^n. Let $\dot{\theta}_0$ be the true parameter. Let $p(y^0; \theta) \equiv 1$ and suppose the conditional probability density of Y_k given $\mathbf{Y}^{k-1}$ exists for every $\theta \in \Theta$. Let

$$p_k(\theta) = p(\mathbf{y}^k; \theta)/p(\mathbf{y}^{k-1}; \theta)$$

for every $k \geqq 1$. Define $\phi_k(\theta)$, $\dot{\phi}_k(\theta)$ and $\ddot{\phi}_k(\theta)$ as in Chapter 7. Unless otherwise specified, all probability statements are with respect to the P_{θ_0}-measure.

We shall now give a set of regularity conditions which will be assumed in the sequel.

(B1) Θ is an open subset of $\mathbb{R}^p$.

(B2) Λ is a Bayes prior measure on Θ and Λ has a density λ with respect to the p-dimensional Lebesgue measure which is continuous and positive in an open neighbourhood of θ_0.

(B3) $\dot{\phi}_k(\theta)$ and $\ddot{\phi}_k(\theta)$ exist almost surely and $\ddot{\phi}_k(\theta)$ is continuous in θ uniformly in $\mathbf{y}^k \in \mathscr{Y}^k$, $k \geqq 1$ and is a measurable function of $\mathbf{Y}^k$.

(B4) $E_\theta[\dot{\phi}_k(\theta)|\mathbf{Y}^{k-1}]=0$ $a \cdot s$ for all $k \geqq 1$.

(B5) $\Gamma_k(\theta) \equiv E_\theta[\dot{\phi}_k(\theta)\, \dot{\phi}_k'(\theta)|\mathbf{Y}^{k-1}] = -E_\theta[\ddot{\phi}_k(\theta)|\mathbf{Y}^{k-1}]$ $a \cdot s$ for all $k \geqq 1$.

(B6) $n^{-1} \Sigma_{k=1}^n \Gamma_k(\theta)$ converges uniformly to $\Gamma(\theta)$ $a \cdot s$ whenever θ is in a compact set and $\Gamma(\theta)$ is a positive definite non-random matrix.

(B7) For some $\delta > 0$ and for every $\boldsymbol{\alpha} \in \mathbb{R}^p$,

$$n^{-(1+\delta/2)} \sum_{k=l}^{n} E\{|\boldsymbol{\alpha}'\phi_k(\theta_0)|^{2+\delta}|\mathbf{Y}^{k-1}\} \to 0 \text{ a.s.}$$

(B8) There exists $\varepsilon > 0$, $K > 0$ and random variables $B_{k,ij}(\mathbf{Y}^k)$ such that

(i) $\sup\{|\ddot{\phi}_{k,ij}|:\|\mathbf{t}-\theta_0\| \leqq \varepsilon\} \leqq B_{k,ij}$;

(ii) $E\{|B_{k,ij}(\mathbf{Y}^k)|^2|\mathbf{Y}^{k-1}\} \leqq K < \infty$a.s. for all $k \geqq 1$.

(B9) $\hat{\theta}_n \to \theta_0$ a.s. where $\hat{\theta}_n$ is a maximum likelihood estimator.

(B10) Let $\mathbb{R}_k(\theta_0, \varepsilon) = \sup\{\phi_k(\theta) - \phi_k(\theta_0):\|\theta-\theta_0\| > \varepsilon,\ \theta \in \Theta\}$ for any $\varepsilon > 0$. Assume that

(i) $\displaystyle\sum_{k=1}^{\infty} \frac{\operatorname{var}(R_k(\theta_0, \varepsilon))}{k^2} < \infty$ for every $\varepsilon > 0$;

(ii) $\frac{1}{n}\sum_{k=1}^{n} E_{\theta_0}(R_k(\theta_0, \varepsilon)|\mathbf{Y}^{k-1}) \overset{\text{a.s.}}{\to} W(\theta_0, \varepsilon) < 0$

where $W(\theta_0, \varepsilon)$ is a constant almost surely.

(B11) Let $H(\mathbf{t}) = \hat{H}(\|\mathbf{t}\|)$ be a non-negative measurable function and suppose there exists $\varepsilon > 0$ such that $\Gamma(\theta_0) - \varepsilon I$ is positive definite matrix,

$$\int_{\mathbb{R}^p} H(t)\exp\{-\tfrac{1}{2}\mathbf{t}'(\Gamma(\theta^0) - \varepsilon I)\mathbf{t}\}d\mathbf{t} < \infty$$

and for every $h > 0$ and $\delta > 0$,

$$e^{-n\delta}\int_{\|\mathbf{t}\| > h} H(\|\mathbf{t}\| n^{1/2})\lambda(\hat{\theta}_n + \mathbf{t})d\mathbf{t} \to 0 \text{ a.s.}$$

We remark that even though the conditions stated above are somewhat stringent (B1) and (B3)–(B9) are mainly used to obtain strong consistency and asymptotic normality of the maximum likelihood estimator (MLE). Some of these conditions can be relaxed. We shall now state a theorem which will be useful later in the study of Bayes estimators.

Theorem 2.1 *Under assumptions* (B1) *and* (B3)–(B9), *if* $\hat{\theta}_n$ *is a MLE of* θ_0, *then*

(i) $\hat{\theta}_n \to \theta_0$ *a.s.*;

(ii) $\sum_{k=1}^{n} \dot{\phi}_k(\hat{\theta}_n) = 0$ *for* $n \geq N$ *where N might depend on the sample observed;*

(iii) $n^{1/2}(\hat{\theta}_n - \theta_0) \overset{\mathcal{L}}{\to} N(0, \Gamma^{-1}(\theta_0))$.

The proof of this theorem is omitted; it can be obtained from results in Chapter 7 or it can be found in Prakasa Rao (1974). The regularity conditions can be simplified further when the type of dependence among the observations is known.

3. THE BERNSTEIN–VON MISES THEOREM

We shall now prove the Bernstein–von Mises theorem for stochastic processes. Let $p_n(\theta|\mathbf{y}^n) = p_n(\theta| y_1, \ldots, y_n)$ be the posterior density of θ given the observations $y_1, \ldots, y_n$ corresponding to the prior probability density λ. Let

$$p_n^*(\mathbf{t}| y_1, \ldots, y_n) = n^{-1/2} p_n(\theta| y_1, \ldots, y_n) \tag{1}$$

i.e. $p_n^*(\mathbf{t}| y_1, \ldots, y_n)$ is the posterior density of $\mathbf{t} = n^{1/2}(\theta - \hat{\theta}_n)$.

Theorem 3.1. *Under assumptions* (B1)–(B11),

$$\lim_{n\to\infty} \int_{\mathbb{R}^p} H(\mathbf{t})|p_n^*(\mathbf{t}|y_1, \ldots, y_n) - \varphi(\Gamma^{-1}(\theta_0), \mathbf{t})|d\mathbf{t} = 0 \text{ a.s.} \tag{2}$$

where $\varphi(\mathrm{B}, \mathbf{t})$ *is the density of the multivariate normal distribution with mean* $\mathbf{0}$ *and covariance matrix* B.

The proof of this theorem will be based on the following lemmas. Define

$$v_n(\mathbf{t}) = \exp\left\{\sum_{k=1}^{n} [\phi_k(\hat{\theta}_n + \mathbf{t}n^{-1/2}) - \phi_k(\hat{\theta}_n)]\right\} \tag{3}$$

and

$$C_n = \int_{IR^p} v_n(\mathbf{t})\lambda(\hat{\theta}_n + \mathbf{t}n^{-1/2})d\mathbf{t}. \tag{4}$$

It is easy to see that

$$p_n^*(\mathbf{t}|\, y_1, \ldots, y_n) = C_n^{-1} v_n(\mathbf{t})\lambda(\hat{\theta}_n + \mathbf{t}n^{-1/2}). \tag{5}$$

Lemma 3.1. *Let assumptions* (B1)–(B10) *be satisfied. Then the following conclusions hold.*

(i) *For every* $\delta > 0$, *there exists* $\varepsilon^*(\delta) > 0$ *and an integer* N *such that for every* $n \geqq N$,

$$\sup\{v_n(\mathbf{t}): \|\mathbf{t}\| \geqq \delta\varepsilon^*(\delta)^{1/2}\} e^{-n\varepsilon^*(\delta)}; \tag{6}$$

(ii) *for every* $\varepsilon > 0$ *sufficiently small, there exists* $\delta_0 > 0$ *an integer* N *such that*

$$\log\, v_n(\mathbf{t}) \leqq \tfrac{1}{2}\mathbf{t}'(-\Gamma(\theta_0) + \varepsilon I)\mathbf{t} \tag{7}$$

whenever $\|\mathbf{t}\| \leqq \delta_0 n^{1/2}$;

(iii) *for every fixed* $\mathbf{t}$

$$\lim_{n\to\infty} v_n(\mathbf{t}) = \varphi(\Gamma^{-1}(\theta_0), \mathbf{t})(2\pi)^{p/2} \det(\Gamma(\theta_0))^{-1/2}. \tag{8}$$

Here det (A) *denotes the determinant of matrix* A.

Proof. Observe that

$$\begin{aligned} \log\, v_n(\mathbf{t}) &= \sum_{k=1}^{n} \{\phi_k(\hat{\theta}_n + \mathbf{t}n^{-1/2}) - \phi_k(\hat{\theta}_n)\} \\ &= n^{-1/2}\mathbf{t}' \sum_{k=1}^{n} \dot{\phi}_k(\hat{\theta}_n) + \frac{n^{-1}}{2}\mathbf{t}'\left\{\sum_{k=1}^{n} \ddot{\phi}_k(\theta_n')\right\}\mathbf{t} \end{aligned} \tag{9}$$

where $\|\hat{\theta}_n-\theta_n'\|\leqq\|\mathbf{t}\|n^{-1/2}$. The first term on the right-hand side of the relation (9) is zero almost surely for $n\geqq N.$ (say). For the second term, we have

$$\begin{aligned}&\frac{n^{-1}}{2}\mathbf{t}'\left\{\sum_{k=1}^{n}\ddot{\phi}_k(\theta_n')\right\}\mathbf{t}\\&=\frac{n^{-1}}{2}\mathbf{t}'\left\{\sum_{k=1}^{n}\ddot{\phi}_k(\theta_0)\right\}\mathbf{t}+\frac{n^{-1}}{2}\mathbf{t}'\left\{\sum_{k=1}^{n}[\ddot{\phi}_k(\theta_n')-\ddot{\phi}_k(\theta_0)]\right\}\mathbf{t}.\end{aligned}\tag{10}$$

Since

$$\sum_{k=1}^{\infty}\frac{\operatorname{var}[\ddot{\phi}_{k,ij}(\theta_0)]}{k^2}\leqq\sum_{k=1}^{\infty}\frac{E(B_{k,ij})^2}{k^2}\leqq K\sum_{k=1}^{\infty}\frac{1}{k^2}<\infty$$

by (B8),

$$\frac{1}{n}\sum_{k=1}^{n}\{\ddot{\phi}_{k,ij}(\theta_0)-E_{\theta_0}[\ddot{\phi}_{k,ij}(\theta_0)|\mathbf{Y}^{k-1}]\}\to 0\text{ a.s.}$$

by the strong law of large numbers for martingale differences given in Loeve (1963), p. 381. But

$$E_{\theta_0}[\ddot{\phi}_{k,ij}(\theta_0)|\mathbf{Y}^{k-1}]=-\Gamma_{k,ij}(\theta_0)\text{ a.s.}$$

and

$$\frac{1}{n}\sum_{k=1}^{n}\Gamma_{k,ij}(\theta_0)\to\Gamma_{ij}(\theta_0)\text{ a.s.}$$

by (B6). Hence

$$\frac{1}{n}\sum_{k=1}^{n}\ddot{\phi}_{k,ij}(\theta_0)\to-\Gamma_{ij}(\theta_0)\text{ a.s.}\tag{11}$$

Choose $\varepsilon>0$ such that $-\Gamma(\theta_0)+\varepsilon I$ is negative definite. Relation (11) implies that there exists an integer $N_2\geqq N_1$ such that for all $\mathbf{t}$

$$\frac{1}{n}\mathbf{t}'\left\{\frac{1}{n}\sum_{k=1}^{n}\ddot{\phi}_k(\theta_0)\right\}\mathbf{t}<\frac{1}{2}\mathbf{t}'\left\{-\Gamma(\theta_0)+\frac{\varepsilon}{2}I\right\}\mathbf{t}.\tag{12}$$

Let us now estimate the second term on the right-hand side of (10). Let $\delta_0>0$ to be chosen later. There exists an integer $N_3\geqq N_2$ such that $\|\hat{\theta}_n-\theta_0\|<\delta_0$ and $\|\theta_n'-\hat{\theta}\|\leqq\|\mathbf{t}\|n^{-1/2}\leqq\delta_0$ for $n\leqq N_3$ by strong consistency of $\hat{\theta}_n$. Hence, if $n\geqq N_3$, $\|\theta_n'-\theta_0\|\leqq 2\delta_0$,

and

$$\begin{aligned} &\frac{1}{n}\sum_{k=1}^{n}|\ddot{\phi}_{k,ij}(\theta'_n)-\ddot{\phi}_{k,ij}(\theta_0)| \\ &\leqq\frac{1}{n}\sum_{k=1}^{n}\sup\{|\ddot{\phi}_{k,ij}(\theta)-\ddot{\phi}_{k,ij}(\theta_0)|:\|\theta-\theta_0\|\leqq 2\delta_0\}. \end{aligned} \tag{13}$$

Since $\ddot{\phi}_k(\theta)$ is continuous in θ uniformly in $\mathbf{y}^k\in\mathscr{S}^k$, $k\geqq 1$, it follows by (B3) that the term on the right-hand side of relation (13) tends to zero as $\delta\to 0$ uniformly in n. Hence there exists a $\delta_0>0$ such that whenever $\|\mathbf{t}\|\leqq\delta_0 n^{1/2}$,

$$\log v_n(\mathbf{t})<\tfrac{1}{2}\mathbf{t}'\{-\Gamma(\theta_0)+\varepsilon I\}\mathbf{t} \tag{14}$$

for $n>\max(N_1, N_2, N_3)$. This proves (7). For $\mathbf{t}$ such that $\|\mathbf{t}n^{-1/2}\|\geqq\delta>0$,

$$\begin{aligned} n^{-1}\log v_n(\mathbf{t}) &= n^{-1}\sum_{k=1}^{n}\{\phi_k(\hat{\theta}_n+\mathbf{t}n^{-1/2})-\phi_k(\hat{\theta}_n)\} \\ &= n^{-1}\sum_{k=1}^{n}\{\phi_k(\hat{\theta}_n+\mathbf{t}n^{-1/2})-\phi_k(\theta_n)\} \\ &\quad + n^{-1}\sum_{k=1}^{n}\{\phi_k(\theta_0)-\phi_k(\hat{\theta}_n)\} \\ &= I_{ln}+I_{2n}(\text{say}). \end{aligned} \tag{15}$$

If $\|\hat{\theta}_n-\theta_0\|<\frac{\delta}{2}$ for $n\geqq N_4$, then $\|\mathbf{t}n^{-1/2}\|\geqq\delta$ implies that

$$\|\hat{\theta}_n+\mathbf{t}n^{-1/2}-\theta_0\|>\delta/2.$$

Hence, for $n\geqq N_4$,

$$\begin{aligned} I_{ln} &\leqq n^{-1}\sum_{k=1}^{n}\sup\{\phi_k(\theta)-\phi_k(\theta_0):\|\theta-\theta_0\|>\delta/2\} \\ &= n^{-1}\sum_{k=1}^{n}R_k(\theta_0, \delta/2). \end{aligned} \tag{16}$$

Assumption (B10) implies that

$$n^{-1/2}\sum_{k=1}^{n}R_k(\theta_0, \delta/2)\to W(\theta_0, \delta/2)\text{ a.s.}$$

by Loeve (1963), p. 301. Moreover

$$\begin{aligned} I_{2n} &= n^{-1}\left\{\sum_{k=1}^{n}\dot{\phi}'_k(\hat{\theta}_n)\right\}\cdot(\theta_0-\hat{\theta}_n)+ \\ &(\theta_0-\hat{\theta}_n)'\left\{\frac{1}{2n}\sum_{k=1}^{n}\ddot{\phi}_k(\theta'_n)\right\}(\theta_0-\hat{\theta}_n) \end{aligned} \tag{17}$$

where $\|\hat{\theta}_n - \theta_n'\| \leqq \|\hat{\theta}_n - \theta_0\|$. Observe that the first term on the right-hand side of relation (17) is zero almost surely and the second term converges to zero almost surely by the method given in (10)–(13) and the fact that $\hat{\theta}_n \to \theta_0$ a.s. Now choose $\varepsilon(\delta)$ such that

$$0 < \varepsilon(\delta) < -W(\theta_0, \delta/2).$$

It is clear that if $n \geqq N = \max(N_i, 1 \leqq i \leqq 4)$ and $\|\mathbf{t}\| \geqq n^{1/2}\delta$, then

$$\log v_n(\mathbf{t}) \leqq -n\frac{\varepsilon(\delta)}{4} = -n\varepsilon^*(\delta).$$

This proves (6). For any fixed $\mathbf{t}$ and $\varepsilon > 0$, choose ε_1 such that

$$0 < \tfrac{1}{2}\|\mathbf{t}\|^2 \varepsilon_1 < \varepsilon.$$

It is now clear from relations (9)–(14) that

$$|\log v_n(\mathbf{t}) - \tfrac{1}{2}\mathbf{t}'\Gamma(\theta_0)\mathbf{t}| \leqq \tfrac{1}{2}\|\mathbf{t}\|^2 \varepsilon_1 < \varepsilon$$

for $n \geqq \max\,(N, \|\mathbf{t}\|^2/\delta^2)$. Hence,

$$\log v_n(\mathbf{t}) \to -\tfrac{1}{2}\mathbf{t}'\Gamma(\theta_0)\mathbf{t}$$

as $n \to \infty$, which implies relation (8).

Lemma 3.2. *Under assumptions* (B1)–(B11), *there exists a* $\delta_0 > 0$ *such that*

$$\lim_{n\to\infty} \int_{\|\mathbf{t}\| \leqq \delta_0 n^{1/2}} H(\mathbf{t}) |v_n(\mathbf{t})\lambda(\hat{\theta}_n + \mathbf{t}n^{-1/2}) - \lambda(\theta_0) \exp\{-\tfrac{1}{2}\mathbf{t}'\Gamma(\theta_0)\mathbf{t}\}| d\mathbf{t} = 0 \text{ a.s.} \tag{18}$$

Proof. It is easy to see that

$$\begin{aligned} &\int_{\|\mathbf{t}\| \leqq \delta_0 n^{1/2}} H(\mathbf{t}) |v_n(\mathbf{t})\lambda(\hat{\theta}_n + \mathbf{t}n^{-1/2}) - \lambda(\theta_0) \exp\{-\tfrac{1}{2}\mathbf{t}'\Gamma(\theta_0)\mathbf{t}\}| d\mathbf{t} \\ &\leqq \int_{\|\mathbf{t}\| \leqq \delta_0 n^{1/2}} H(\mathbf{t})\lambda(\theta_0) |v_n(\mathbf{t}) - \exp\{-\tfrac{1}{2}\,\mathbf{t}'\Gamma(\theta_0)\mathbf{t}\}| d\mathbf{t} \\ &+ \int_{\|\mathbf{t}\| \leqq \delta_0 n^{1/2}} H(\mathbf{t}) v_n(\mathbf{t}) |\lambda(\theta_0) - \lambda(\hat{\theta}_n + \mathbf{t}n^{-1/2})| d\mathbf{t} \\ &= I_{3n} + I_{4n} \text{(say)}. \end{aligned} \tag{19}$$

Choose an $\varepsilon > 0$ such that

$$\int_{\mathbb{R}^p} H(\mathbf{t}) \exp\{-\tfrac{1}{2}\mathbf{t}'(\Gamma(\theta_0) - \varepsilon I)\mathbf{t}\} d\mathbf{t} < \infty. \tag{20}$$

This is possible by assumption (B11). Lemma 3.1 implies that there exists $\delta_1 > 0$ and a positive integer N such that

$$v_n(\mathbf{t}) \leqq \exp\{-\tfrac{1}{2}\mathbf{t}'(\Gamma(\theta_0)\mathbf{t} - \varepsilon I)\mathbf{t}\} \tag{21}$$

whenever $\|\mathbf{t}\| \leqq \delta_1 n^{1/2}$ and $n \geqq N$. Since

$$v_n(\mathbf{t}) \to \exp\{-\tfrac{1}{2}\mathbf{t}'\Gamma(\theta_0)\mathbf{t}\}$$

as $n \to \infty$ for every $\mathbf{t}$ by relation (8) and the integrand in I_{3n} is dominated by the function

$$H(\mathbf{t})\lambda(\theta_0)[\exp\{-\tfrac{1}{2}\mathbf{t}'\Gamma(\theta_0)\mathbf{t}\} + \exp\{-\tfrac{1}{2}\mathbf{t}'(\Gamma(\theta_0) - \varepsilon I)\mathbf{t}\}]$$

which is integrable by relation (20), it follows that

$$\int_{\|\mathbf{t}\| \leqq \delta_1 n^{1/2}} H(\mathbf{t})\lambda(\theta_0)|v_n(\mathbf{t}) - \exp\{-\tfrac{1}{2}\mathbf{t}'\Gamma(\theta_0)\mathbf{t}\}|\,d\mathbf{t} \to 0 \text{ a.s.} \tag{22}$$

by the dominated convergence theorem. Furthermore

$$\int_{\|\mathbf{t}\| \leqq \delta_1 n^{1/2}} H(\mathbf{t})v_n(\mathbf{t})|\lambda(\theta_0) - \lambda(\hat{\theta}_n + \mathbf{t}n^{-1/2})|\,d\mathbf{t}$$

$$\leqq \sup_{\|\theta - \theta_0\| \leqq \delta_1} |\lambda(\theta) - \lambda(\theta_0)| \int_{\|\mathbf{t}\| \leqq \delta_1 n^{1/2}} H(\mathbf{t}) \exp\{-\tfrac{1}{2}\mathbf{t}'(\Gamma(\theta_0) - \varepsilon I)\mathbf{t}\}\,d\mathbf{t} \tag{23}$$

by relation (14) whenever $\delta_1 \leqq \delta$. For a given η, choose $\delta_0 \leqq \delta_1$ so that

$$\sup_{\|\theta - \theta_0\| \leqq \delta_0} |\lambda(\theta) - \lambda(\theta_0)| \int_{\|\mathbf{t}\| \leqq \delta_0 n^{1/2}} H(\mathbf{t}) \exp\{-\tfrac{1}{2}\mathbf{t}'(\Gamma(\theta_0) - \varepsilon I)\mathbf{t}\}\,d\mathbf{t} < \eta. \tag{24}$$

Relations (22) and (24) together prove Lemma 3.2.

Lemma 3.3. *Under the assumptions* (B1)–(B11), *for every* $\delta > 0$,

$$\lim_{n \to \infty} \int_{\|\mathbf{t}\| > \delta n 1/2} H(\mathbf{t})|v_n(\mathbf{t})\lambda(\hat{\theta}_n + \mathbf{t}n^{-1/2}) - \lambda(\theta_0)\exp\{-\tfrac{1}{2}\mathbf{t}'\Gamma(\theta_0)\mathbf{t}\}|\,d\mathbf{t} = 0 \text{ a.s.}$$

Proof. It is clear that

$$\int_{\|\mathbf{t}\| > \delta n^{1/2}} H(\mathbf{t})|v_n(\mathbf{t})\lambda(\hat{\theta}_n + \mathbf{t}n^{-1/2}) - \lambda(\theta_0)\exp\{-\tfrac{1}{2}\mathbf{t}'\Gamma(\theta_0)\mathbf{t}\}|\,d\mathbf{t}$$

$$\leqq \int_{\|\mathbf{t}\| > \delta n 1/2} H(\mathbf{t})v_n(\mathbf{t})\lambda(\hat{\theta}_n + \mathbf{t}n^{-1/2})\,d\mathbf{t}$$

$$+\int_{\|\mathbf{t}\|>\delta n^{1/2}} H(\mathbf{t})\lambda(\theta_0)\exp\{-\tfrac{1}{2}\mathbf{t}'\Gamma(\theta_0)\mathbf{t}\}d\mathbf{t}$$

$$\leqq e^{-n\varepsilon^*(\delta)}\int_{\|\mathbf{t}\|>\delta n^{1/2}} H(\mathbf{t})\lambda(\hat{\theta}_n+\mathbf{t}n^{-1/2})d\mathbf{t}$$

$$+\lambda(\theta_0)\int_{\|\mathbf{t}\|>\delta n^{1/2}} H(\mathbf{t})\exp\{-\tfrac{1}{2}\mathbf{t}'\Gamma(\theta_0)\mathbf{t}\}d\mathbf{t}$$

by (6) and the last two terms tend to zero almost surely by (B11).

Proof of the Theorem. We shall now prove the main theorem. Lemmas 3.2 and 3.3. imply that

$$\lim_{n\to\infty}\int_{\mathbb{R}^p} H(\mathbf{t})|v_n(\mathbf{t})\lambda(\hat{\theta}_n+\mathbf{t}n^{-1/2})-\lambda(\theta_0)\exp\{-\tfrac{1}{2}\mathbf{t}'\Gamma(\theta_0)\mathbf{t}\}|d\mathbf{t}=0 \text{ a.s.}$$

Putting $H(t)\equiv 1$, which satisfies (B11) trivially, we get

$$C_n=\int_{\mathbb{R}^p} v_n(\mathbf{t})\lambda(\hat{\theta}_n+\mathbf{t}n^{-1/2})d\mathbf{t}\to\lambda(\theta_0)\int_{\mathbb{R}^p}\exp\{-\tfrac{1}{2}\mathbf{t}'\Gamma(\theta_0)\mathbf{t}\}d\mathbf{t}. \tag{26}$$

Hence

$$\int_{\mathbb{R}^p} H(\mathbf{t})|p_n^*(\mathbf{t}|\mathbf{y}^n)-\varphi(\Gamma^{-1}(\theta_0),\mathbf{t})|d\mathbf{t}$$

$$\leqq\int_{\mathbb{R}^p} H(\mathbf{t})|C_n^{-1}\lambda(\hat{\theta}_n+\mathbf{t}n^{-1/2})v_n(\mathbf{t})-C_n^{-1}\lambda(\theta_0)\exp\{-\tfrac{1}{2}\mathbf{t}'\Gamma(\theta_0)\mathbf{t}\}|d\mathbf{t} \tag{27}$$

$$+\int_{\mathbb{R}^p} H(\mathbf{t})|C_n^{-1}\lambda(\theta_0)-\Delta(\theta_0)|\exp\{-\tfrac{1}{2}\mathbf{t}'\Gamma(\theta_0)\mathbf{t}\}d\mathbf{t}$$

where

$$\Delta(\theta_0)=(\det\Gamma(\theta_0))^{1/2}(2\pi)^{-p/2}. \tag{28}$$

The last two terms on the right-hand side of relation (27) tend to zero almost surely by (23) and (26).

As a corollary to Theorem 3.1, we have the following result which includes the Bernstein–von Mises Theorem in the independent and identically distributed case.

Theorem 3.4. *Let assumptions* (B1)–(B10) *be satisfied and*

$$\int_{\mathbb{R}^p}\|\theta\|^m\lambda(\theta)d\theta<\infty \tag{29}$$

for some integer $m \geqq 0$. *Then*

$$\lim_{n\to\infty} \int ||\mathbf{t}||^m |p_n^*(\mathbf{t}|\mathbf{y}^n) - \Delta(\theta_0) \exp\{-\tfrac{1}{2}\mathbf{t}'\Gamma(\theta_0)\mathbf{t}\}| d\mathbf{t} = 0 \text{ a.s.} \tag{30}$$

Proof. For $m=0$, the verification of assumption (B11) is easy and relation (30) follows from Theorem 3.1. Suppose $m \geqq 1$. Let $H(\mathbf{t}) = ||\mathbf{t}||^m$, $h>0$ and $\delta>0$. It is easy to see that

$$\begin{aligned}
&e^{-\delta n} \int_{||\mathbf{t}||>h} H(n^{1/2}\mathbf{t})\lambda(\hat{\theta}_n + \mathbf{t}) d\mathbf{t} \\
&= n^{m/2} e^{-\delta n} \int_{||\mathbf{t}-\hat{\theta}_n||>h} \lambda(\mathbf{t}) ||\mathbf{t} - \hat{\theta}_n||^m d\mathbf{t} \\
&\pm n^{m/2} e^{-\delta n} \int_{||\mathbf{t}-\hat{\theta}_n||>h} \lambda(\mathbf{t}) \{||\mathbf{t}|| \mid ||\hat{\theta}_n||\}^m d\mathbf{t} \\
&\leqq 2^{m-1} e^{-\delta n} n^{m/2} \left\{ \int_{||\mathbf{t}-\hat{\theta}_n||>h} \lambda(\mathbf{t}) ||\mathbf{t}||^m d\mathbf{t} + \int_{||\mathbf{t}\theta_n||>h} \lambda(\mathbf{t}) ||\hat{\theta}_n||^m d\mathbf{t} \right\} \\
&\leqq 2^{m-1} e^{-\delta n} n^{m/2} \left\{ \int_p ||\mathbf{t}||^m \lambda(\mathbf{t}) d\mathbf{t} + ||\hat{\theta}_n||^m \right\}
\end{aligned}$$

and the last term tends to zero almost surely from the finiteness condition given by relation (29) and the fact that $\hat{\theta}_n$ is strongly consistent.

4. ASYMPTOTIC BEHAVIOUR OF BAYES ESTIMATORS

In this section, we shall study the asymptotic behaviour of Bayes estimators for discrete time stochastic processes.

4.1 Asymptotic Properties of Bayes Estimators

In this section, we shall give an application of Theorem 3.1 to the asymptotic theory of Bayesian inference for stochastic processes. We define a *Bayes estimator* $T_n = T_n(Y_1, \ldots, Y_n)$ as an estimator which minimizes

$$B_n(\varphi) = \int \hat{W}(\theta, \varphi) p_n(\theta | y_1, \ldots, y_n) d\theta$$

for all sequences $(y_1, y_2, \ldots)$ where $\tilde{W}(\theta, \varphi)$ is a loss function defined on $\Theta \times \Theta$. We shall assume in the sequel that such a measurable Bayes estimator T_n of θ exists.

Theorem 4.1. *Suppose that assumptions* (B1)–(B11) *are satisfied. Let* T_n *be a Bayes estimator of* θ *with respect to a loss function* $\hat{W}(\theta, \varphi)$ *satisfying the* *following conditions.*

(i) $\tilde{W}(\theta, \varphi) = -W(\|\theta - \varphi\|) \geqq 0;$ (31)

(ii) $W(x_1) > W(x_2)$ if $x_1 > x_2 \geqq 0$. (32)

Further suppose there exists a sequence $\{a_n\}$ *and functions* $H(\mathbf{t})$ *on* R^p *and* $G(t)$ *on* R *such that*

(iii) $a_n \geqq 0;$ (33)

(iv) $G(\|\mathbf{t}\|)$ *satisfies assumption* (B11), *and for every* $n \geqq 1$,

$$a_n W(\|\mathbf{t}\| n^{-1/2}) \leqq G(\|\mathbf{t}\|); \tag{34}$$

(v) $a_n W(\|\mathbf{t}\| n^{-1/2}) \to \tilde{H}(\|\mathbf{t}\|) = H(\mathbf{t})$ as $n \to \infty$

uniformly on compact sets of $\mathbf{t}$;

(vi) $\Delta(\theta_0) \int \hat{H}(\|\mathbf{t}\| + m) \exp\{-\frac{1}{2}\mathbf{t}'\Gamma(\theta_0)\mathbf{t}\} d\mathbf{t}$

has a strict minimum at $m = 0$ *where* $\Delta(\theta_0) = (\det(\Gamma(\theta_0)))^{1/2}(2\pi)^{-p/2}$. *Then as* $n \to \infty$

(a) $n^{1/2}(T_n - \hat{\theta}_n) \to 0$ a.s., (35)

(b) $T_n \to \theta_0$ *a.s.*, (36)

(c) $n^{1/2}(T_n - \theta_0) \xrightarrow{\mathscr{L}} N(0, \Gamma^{-1}(\theta_0))$, (37)

(d) $a_n B_n(T_n) \to \Delta(\theta_0 \int_{\mathbb{R}^p} \tilde{H}(\|\mathbf{t}\|) \exp\{-\frac{1}{2}\mathbf{t}'\Gamma(\theta_0)\mathbf{t}\} d\mathbf{t}$ a.s. (38)

Note that relations (36) and (37) ensure the strong consistency and asymptotic normality of the Bayes estimator T_n. (35) proves that the Bayes estimator and the maximum likelihood estimator are asymptotically equivalent. (38) gives an estimate for the Bayesian risk of T_n. An example of a loss function described in the theorem in $\tilde{W}(\theta, \varphi) = |\theta - \varphi|$.

Proof. It is sufficient to prove relations (36) and (38). The rest follow from (35) in view of Theorem 2.1. Let us first consider

$$\begin{aligned}
\limsup a_n B_n(T_n) &\leqq \limsup a_n B_n(\hat{\theta}_n) \\
&= \limsup \int a_n W(\|\theta - \hat{\theta}_n\|) p_n(\theta|\mathbf{y}^n) d\theta \\
&= \limsup \int a_n W(\|\mathbf{t}\| n^{-1/2}) p_n^*(\mathbf{t}|\mathbf{y}^n) d\mathbf{t} \\
&\leqq \limsup \int |a_n W(\|\mathbf{t}\| n^{-1/2}) - H(\mathbf{t})| . \qquad (39)\\
&\qquad |p_n^*(\mathbf{t}|\mathbf{y}^n) - \Delta(\theta_0) \exp\{-\tfrac{1}{2}\mathbf{t}'\Gamma(\theta_0)\mathbf{t}\}| d\mathbf{t} \\
&\quad + \limsup \Delta(\theta_0) \int |a_n W(\|\mathbf{t}\| n^{-1/2}) - H(\mathbf{t})| \exp\{-\tfrac{1}{2}\mathbf{t}'\Gamma(\theta_0)\mathbf{t}\} d\mathbf{t} \\
&\quad + \lim \int \tilde{H}(\|\mathbf{t}\|) p_n^*(\mathbf{t}|\mathbf{y}^n) d\mathbf{t}.
\end{aligned}$$

Let I_1, I_2 and I_3 be the terms on the right-hand side of inequality (39). Observe that

$$0 \leqq I_1 \leqq \limsup \int 2G(\|\mathbf{t}\|)|p_n^*(\mathbf{t}|\mathbf{y}^n) - \Delta(\theta_0)\exp\{-\tfrac{1}{2}\mathbf{t}'\Gamma(\theta_0)\mathbf{t}\}d\mathbf{t} = 0$$

by Theorem 3.1. I_2 is equal to zero by the dominated convergence theorem and

$$I_3 \leqq \int \tilde{H}(\|\mathbf{t}\|)\Delta(\theta_0)\exp\{-\tfrac{1}{2}\mathbf{t}'\Gamma(\theta_0)\mathbf{t}\}d\mathbf{t}.$$

Hence

$$\limsup a_n B_n(T_n) \leqq \Delta(\theta_0)\int \tilde{H}(\|\mathbf{t}\|)\exp\{-\tfrac{1}{2}\mathbf{t}'\Gamma(\theta_0)\mathbf{t}\}d\mathbf{t}. \tag{40}$$

Before we prove (38), we shall first show that

$$n^{1/2}(\hat{\theta}_n - T_n) = V_n < \infty \text{ a.s.}$$

For, if not, for every $M > 0$, there exists a set A_M with positive probability such that $\|V_n\| > M$ infinitely often. Let $\{n_j\}$ be the subsequence for which $\|V_{n_j}\| > M$. Then

$$\begin{aligned} a_{n_j}B_{n_j}(T_{n_j}) &= \int_{a_{n_j}} a_{n_j} W(\|\mathbf{t} + V_{n_j}\| n_j^{-1/2}) p_{n_j}^*(\mathbf{t}|\mathbf{y}^{n_j})d\mathbf{t} \\ &\geqq \int_{\|\mathbf{t}\| \leqq M/4} a_{n_j} W(\|\mathbf{t} + V_{n_j}\| n_j^{-1/2}) p_{n_j}^*(\mathbf{t}|\mathbf{y}^{n_j})d\mathbf{t} \\ &\geqq \int_{\|\mathbf{t}\| \leqq M/4} a_{n_j} W\left(\left\{\|\mathbf{t}\| + \frac{M}{4}\right\} n_j^{-1/2}\right) p_{n_j}^*(\mathbf{t}|\mathbf{y}^{n_j})d\mathbf{t} \end{aligned}$$

since $\|\mathbf{t}\| \leqq M/4$, $\|V_{n_j}\| > M$ implies $\|\mathbf{t} + V_{n_j}\| > \|\mathbf{t}\| + M/4$ and the last term in the above inequality tends to

$$\Delta(\theta_0)\int_{\|\mathbf{t}\| \leqq M/4} \tilde{H}\left(\|\mathbf{t}\| + \frac{M}{4}\right)\exp\{-\tfrac{1}{2}\mathbf{t}'\Gamma(\theta_0)\mathbf{t}\}d\mathbf{t}$$

by Theorem 3.1. Furthermore $\tilde{H}(\|\mathbf{t}\| + M/4)\, I[\|\mathbf{t}\| \leqq M/4]$ is a non-decreasing function of M for each fixed $\mathbf{t}$. Hence

$$\begin{aligned} &\lim_{M\to\infty} \int_{\|\mathbf{t}\| \leqq M/4} \tilde{H}\left(\|\mathbf{t}\| + \frac{M}{4}\right)\Delta(\theta_0)\exp\{-\tfrac{1}{2}\mathbf{t}'\Gamma(\theta_0)\mathbf{t}\}d\mathbf{t} \\ &= \lim_{M\to\infty} \tilde{H}(M) \\ &> \Delta(\theta_0)\int H(\mathbf{t})\exp\{-\tfrac{1}{2}\mathbf{t}'\Gamma(\theta_0)\mathbf{t}\}d\mathbf{t} \end{aligned}$$

by assumption (vi). Hence, for a large enough M, for a set of positive probability

$$\liminf a_n B_n(T_n) > \Delta(\theta_0) \int H(\mathbf{t}) \exp\{-\tfrac{1}{2}\mathbf{t}'\Gamma(\theta_0)\mathbf{t}\} d\mathbf{t}$$

$$\geqq \limsup a_n B_n(\hat{\theta}_n) \tag{41}$$

which contradicts the definition of T_n. Hence $\limsup \|V_n\| < \infty$ a.s.

Let $\varepsilon > 0$ and B_M be the set such that $\|V_n\| < M$ for every n and $P(B_M) > 1 - \varepsilon$. Then V_n is bounded in $\mathbb{R}^p$ whenever B_M occurs. For any fixed $w \in B_M$, let $\mathbf{m}$ be a vector in the limit set of $V_n(w)$. Suppose $\mathbf{m} \neq \mathbf{0}$. Let $V_{n_j}(w) \to \mathbf{m}$ as $n \to \infty$. Then

$$\liminf a_{n_j} B_{n_j}(T_{n_j}) \geqq \liminf \int_{\|\mathbf{t}\| \leqq v_0} a_{n_j} W(\|\mathbf{t} + V_{n_j}\| n_j^{-1/2}) p^*_{nj}(\mathbf{t}|\mathbf{y}^{n_j}) d\mathbf{t}$$

$$\geqq \int_{\|\mathbf{t}\| \leqq v_0} \liminf a_{n_j} W(\|\mathbf{t} + V_{n_j}\| n_j^{-1/2}) p^*_{nj}(\mathbf{t}|\mathbf{y}^{n_j}) d\mathbf{t}$$

$$\geqq \Delta(\theta_0) \int_{\|\mathbf{t}\| \leqq v_0} \tilde{H}(\|\mathbf{t} + \mathbf{m}\|) \exp\{-\tfrac{1}{2}\mathbf{t}'\Gamma(\theta_0)\mathbf{t}\} d\mathbf{t}$$

by assumption (v). By choosing v_0 large, we get that, for any $\eta > 0$,

$$\liminf a_{n_j} B_{n_j}(T_{n_j}) > \Delta(\theta_0) \int \tilde{H}(\|\mathbf{t}\|) \exp\{-\tfrac{1}{2}\mathbf{t}'\Gamma(\theta_0)\mathbf{t}\} d\mathbf{t} - \eta \tag{42}$$

by assumption (vi). Since relation (42) holds for every $\eta > 0$,

$$\liminf a_{n_j} B_{n_j}(T_{n_j}) > \Delta(\theta_0) \int \tilde{H}(\|\mathbf{t}\|) \exp\{-\tfrac{1}{2}\mathbf{t}'\Gamma(\theta_0)\mathbf{t}\} d\mathbf{t}.$$

This is impossible by relation (40). Hence $\mathbf{m} = 0$ and $n^{1/2}(T_n - \hat{\theta}_n) \to 0$ a.s. Moreover

$$\lim a_n B_n(T_n) = \lim a_n B_n(\hat{\theta}_n)$$

$$= \Delta(\theta_0) \int_{\mathbb{R}^p} \tilde{H}(\|\mathbf{t}\|) \exp\{-\tfrac{1}{2}\mathbf{t}'\Gamma(\theta_0)\mathbf{t}\} d\mathbf{t}.$$

This proves (38).

4.2 Weak Convergence of Bayes Estimators

Sufficient conditions for the strong consistency of Bayes estimators for general loss functions have already been given in the previous subsection. Here we shall briefly discuss some recent results for loss functions of quadratic nature for a sequence of statistical problems which have the Bayesian form.

Consider a sequence of measurable spaces $(\mathscr{X}_n, \mathscr{A}_n)_{n=1}^{\infty}$ together with a probability space $(\Theta, \mathscr{B}, \mu)$. Let $\{P_{n\theta}, \theta \in \Theta, n \geqq 1\}$ be a family of probability measures such that $P_{n\theta}$ is defined on the σ-algebra $\mathscr{A}_n$ for $n \geqq 1$ and $P_{n\theta}(A): \Theta \to \mathbb{R}$ is $\mathscr{B}$-measurable for each $A \in \mathscr{A}_n$ where $n \geqq 1$.

From a statistical point of view, we have a sequence of statistical problems which have a Bayesian form. The space Θ can be presented as the set of "possible states of nature" and the probability measure μ as our preliminary description of the "true state of nature" which is to be estimated on the basis of observations taking their values in the space $\mathscr{X}_n$. Observe that the system $\{(\mathscr{X}_n, \mathscr{A}_n)\}_{n=1}^{\infty}$ can be thought of as a system corresponding to a stochastic process whose joint densities depend on a parameter θ.

Definition 4.1. Let $f: \Theta \to \mathbb{R}$ be a measurable function. A sequence $h_n: \mathscr{X}_n \to \mathbb{R}$ of measurable functions is said to be *μ-consistent* for f if for every $\varepsilon > 0$

$$P_{n\theta}(|h_n - f(\theta)| > \varepsilon) \to 0 \text{ in measure } \mu.$$

Definition 4.2. Let $f \in L_1(\mu)$, the space of μ-integrable functions. A sequence $h_n: \mathscr{X}_n \to \mathbb{R}$ of measurable functions is said to be *L_1-consistent* for f if

$$\int_{\mathscr{X}_n \times \Theta} |h_n(x) - f(\theta)| Q_n(dx, d\theta) \to 0 \text{ as } n \to \infty$$

where Q_n is defined on $(\mathscr{X}_n \times \Theta, \mathscr{A}_n \times \mathscr{B})$ by the relation

$$Q_n(C \times D) = \int_C P_{n\theta}(D) \mu(d\theta). \tag{43}$$

Definition 4.3. Consider a measurable function $f: \Theta \to \mathbb{R}$. A sequence $h_n: \mathscr{X}_n \to \mathbb{R}$ of measurable function is said to be *weakly consistent* for f if

$$h_n \xrightarrow{\mathscr{L}} f$$

(h_n and f could be considered as random variables on the probability space $(\mathscr{X}_n \times \Theta, \mathscr{A}_n \times \mathscr{B}, Q_n)$).

From standard results in probability, it follows that L_1-consistency implies μ-consistency which in turn implies weak consistency. Let

$$b_n = E_{Q_n}(f | \tilde{\mathscr{A}}_n) \tag{44}$$

where $\hat{\mathscr{A}}_n$ is the σ-algebra defined by

$$\hat{\mathscr{A}}_n = \{\hat{\mathscr{A}} \in \mathscr{A}_n \times \mathscr{B}: \hat{A} = A \times \Theta \text{ where } A \in \mathscr{A}_n\}.$$

The conditional expectation can be looked upon as the Bayes estimator corresponding to the loss functions which are quadratic. The following theorem can be proved.

Theorem 4.2. *If $f \in L_1(\mu)$ and b_n defined as in (44) then $\{b_n\}$ is μ-consistent for f if and only if it is L_1-consistent and if and only if it is weakly consistent.*

We omit the proof of this theorem as it is of technical interest only. The reader is referred to Stepan (1973) for a detailed proof.

4.3 Bayes Estimation for Markov Chains

In this section, we shall discuss the behaviour of generalized Bayes estimators for Markov chains including, in particular, transient Markov chains. For discrete time stationary Markov processes satisfying Deeblin's condition (cf. Doob, 1953), these results have been discussed in Borwanker *et al.* (1971) and they follow from those of Section 4.1. The principal peculiarity of the problem is the fact that weighted probability laws arise as limit distributions of the estimators θ_n.

We shall first describe the class of Markov chains considered in this section. It is assumed that the state space E of the Markov chain X is countable. Further, we assume that for any $x, y \in E$, the average number of transitions from x to y is finite i.e.

$$g(x, y) = \sum_{n=1}^{\infty} p^{(n)}(x, y) < \infty \tag{45}$$

where $p^{(n)}(x, y)$ is the transition probability in n steps. We shall assume that the chain $X - \{X_0, X_1, \ldots\}$ satisfies condition (45) for any $\theta \in \Theta$, the parameter space. We shall first state the conditions the chain should satisfy using the terminology of boundary theory for chains satisfying condition (45) (cf. Dynkin, 1965). Let P_x and E_x be the probability and expectation respectively corresponding to the initial distribution concentrated at the point x. For convenience of reference, we shall state some results from the boundary theory for Markov chains satisfying condition (45).

Lemma 4.1. *The space E can be extended to a topological space $\tilde{E}$ such that there exists a subset Γ contained in $\tilde{E} - E$ called the exit space with the property that for any $x \in E$,*

$$\lim_{n \to \infty} X_n = X_\infty \in \Gamma$$

exists with probability $P_x = 1$.

Lemma 4.2. *For each $\gamma \in \Gamma$, there corresponds a non-negative function $f_\gamma(x)$ such that for every $x \in E$,*

$$\sum_y p(x, y) f_\gamma(y) = f_\gamma(x) \tag{46}$$

and such that, with respect to the conditional probability distribution $P_\gamma(\cdot) = P_x(\cdot|X_\infty = \gamma)$, the sequence $X_0 = x, X_1, \ldots, X_n, \ldots$ is also a homogeneous Markov chain on the set $\{x: f_\gamma(x) > 0\}$ with transition probabilities

$$p_\gamma(x, y) = p(x, y) f_\gamma(y)/f_\gamma(x). \tag{47}$$

Observe that the n-step transition probability $p_\gamma^{(n)}(x, y)$ corresponding to $p_\gamma(x, y)$ is given by the relation

$$p_\gamma^{(n)}(x, y) = p^{(n)}(x, y) f_\gamma(y)/f_\gamma(x).$$

We shall restrict to the case when Γ is finite. Then it can be shown that $f_\gamma(x)$ is the probability of hitting γ starting from x i.e.

$$f_\gamma(x) = P_x\left(\lim_{n\to\infty} X_n = \gamma\right)$$

and $f_\gamma(x)$ satisfies the relation

$$f_\gamma(x) \to 1 \quad \text{as} \quad x \to \gamma.$$

The following regularity conditions will be assumed in the sequel.

(L1) $\theta \in \Theta$, non-degenerate interval (a, b) of the real line.

(L2) The transition probabilities $p(x, y; \theta)$ are absolutely continuous with respect to each other for any fixed x and different θ.

(L3) $\pi(\theta)$ is a non-negative continuous function defined on Θ which grows not faster than a certain power of θ.

(L4) For each $\gamma \in \Gamma$, there exist sequences of neighbourhoods A_n, B_n, $A_n \downarrow \gamma$, $B_n \downarrow \gamma$ such that

$$p(x, y; \theta) = 0 \text{ for all } x \in A_n,\ y \notin B_n. \tag{48}$$

(L5) There exists a constant $c > 0$ and for any $\theta \in \Theta$, a constant $C > 0$ such that for all sufficiently small h

$$ch^2 \leqq \sum_y [(p(x, y; \theta + h))^{1/2} - (p(x, y; \theta))^{1/2}]^2 \leqq Ch^2 \tag{49}$$

for any $x \in E$, where $C \leqq \tilde{c}(1 + |\theta|^p)$ for some $\tilde{c} > 0$ and $p > 0$.

(L6) As $x \to \gamma$ and $h_1, h_2 \to 0$, there exists the non-zero limit

$$\lim_{\substack{x \to \gamma \\ h_1 \to 0 \\ h_2 \to 0}} (h_1 - h_2)^{-2} \sum_y [(p(x, y; \theta + h_1))^{1/2} - (p(x, y; \theta + h_2))^{1/2}]^2 = \tfrac{1}{4} I_\gamma(\theta). \quad (50)$$

(L7) Let

$$V(x, h; \delta) = \left\{ y : \left| \left\{ \frac{p(x, y; \theta + h)}{p(x, y; \theta)} \right\}^{1/2} - 1 \right| > \delta \right\}. \quad (51)$$

for any $\delta > 0$ and $\gamma \in \Gamma$,

$$\lim_{\substack{x \to \gamma \\ h \to 0}} h^{-2} \sum_{y \in V(x, h, \delta)} [(p(x, y; \theta + h))^{1/2} - (p(x, y; \theta))^{1/2}]^2 = 0. \quad (52)$$

(L8) There exists a continuous function $g(h) > 0$ such that $g(h) < 1$ for $h > 0$ and for some $\delta > 0$

$$\limsup_{h \to \infty} g(h) h^\delta < \infty \quad (53)$$

and such that, for all $x \in E$,

$$\sum_y \{p(x, y; \theta + h)\, p(x, y; \theta)\}^{1/2} \leqq g(|h|). \quad (54)$$

We shall consider properties of generalized Bayes estimators of the form

$$\theta_n = \frac{\int \theta p(x_1, \ldots, x_n; \theta) \pi(\theta) \mathrm{d}\theta}{\int p(x_1, \ldots, x_n; \theta) \pi(\theta) \mathrm{d}\theta} \quad (55)$$

where $p(X_1, \ldots, X_n; \theta)$ is the joint density of $(X_1, \ldots, X_n)$ when θ is the true parameter. Note that if $\pi(\theta)$ is the density of θ corresponding to a prior measure on Θ and if the loss function is of quadratic nature, then θ_n is the usual Bayes estimator. Furthermore, the study of generalized Bayes estimators is connected, in particular, with the fact that in a certain sense they are the best (for $\pi(\theta) \equiv 1$) if θ is a shift parameter. We shall come back to this case later. Behaviour of generalized Bayes estimators for diffusion processes is given in Section 9, Chapter 9.

Condition (L4) is a restriction on the behaviour of the chain X. In concrete cases, this condition is often fulfilled. Conditions (L5)–(L7) are of local character. We shall give later some simpler conditions which ensure (L5)–(L7). It can be shown that conditions (L5)–(L7) alone are not sufficient even to ensure the existence of consistent estimators for θ.

Theorem 4.3. *Suppose the initial distribution is concentrated at the point x and that conditions (L1)–(L7) are satisfied. Then the limit distribution of the variable* $n^{1/2}(\theta_n-\theta)$ *is the weighted normal distribution*

$$\sum_{\gamma\in\Gamma} f_\gamma(x,\theta)\, N(0, I_\gamma^{-1}(\theta)).$$

This theorem is an obvious consequence of the following lemma.

Lemma 4.3. *With respect to the condition distribution* $P_\gamma=P_x(\cdot/X_\infty=\gamma)$, *the variable* $n^{1/2}(\theta_n-\theta)$ *is asymptotically normal* $N(0, I_\gamma^{-1}(\theta))$.

Proof. For the proof of this lemma, we shall assume that $\pi(\cdot)\equiv 1$ on Θ. The general case of $\pi(\cdot)$, $\pi(\theta)>0$ and continuous can be dealt with as in Ibragimov and Khasiminskii (1972, 1973). We can assume that $\Theta=\mathbb{R}$ by putting $\pi(\theta)=0$ for $\theta\in\mathbb{R}-\Theta$. It is clear that

$$n^{1/2}(\theta_n-\theta)=\frac{\int z \prod_{i=0}^{n-1} p(X_i, X_{i+1};\theta+n^{-1/2}z)\mathrm{d}z}{\int \prod_{i=0}^{n-1} p(X_i, X_{i+1};\theta+n^{-1/2}z)\mathrm{d}z}$$

$$=\frac{\int z\, Z_n(z)\mathrm{d}z}{\int Z_n(z)\mathrm{d}z} \tag{56}$$

where

$$Z_n(z)=\prod_{i=0}^{n-1}\frac{p(X_i, X_{i+1};\theta+n^{-1/2}z)}{p(X_i, X_{i+1};\theta)}=\exp Y_n(z) \tag{57}$$

and

$$Y_n(z)=\sum_{i=0}^{n-1}\log\frac{p(X_i, X_{i+1};\theta+n^{-1/2}z)}{p(X_i, X_{i+1};\theta)}. \tag{58}$$

We shall find the limit distribution of $Y_n(z)$.

Lemma 4.4. *The finite-dimensional distributions of the process* $\{Y_n(z), -\infty<z<\infty\}$ *corresponding to the measure* P_γ *converge to the corresponding distributions of the process*

$$Y(z)=I_\gamma^{1/2}(\theta)\xi z-I_\gamma(\theta)z^2/2, -\infty<z<\infty \tag{59}$$

where ξ *is standard normal random variable.*

Proof. Let us write $Y_n(z)$ in the form

$$Y_n(z)=\sum_{i=0}^{l_n-1}\log\frac{p(X_i, X_{i+1}; \theta+zn^{-1/2})}{p(X_i, X_{i+1}; \theta)}$$
$$+\sum_{i=l_n}^{n-1}\log\frac{p(X_i, X_{i+1}; \theta+n^{-1/2}z)}{p(X_i, X_{i+1}; \theta)} \tag{60}$$

where $l_n=0(n^{1/3})$ and $l_n\to\infty$. Observe that

$$P_\gamma\left\{|\log\frac{p(X_i, X_{i+1}; \theta+h)}{p(X_i, X_{i+1}; \theta)}|>\delta\right\}\leqq c\, E_\gamma|\left[\frac{p(X_i, X_{i+1}; \theta+h)}{p(X_i, X_{i+1}; \theta)}\right]^{1/2}-1|$$
$$\leqq c\, E_\gamma\left\{\sum_y[(p(X_i, y; \theta+h))^{1/2}-(p(X_i, y; \theta))^{1/2}]^2 f_\gamma(y)/f_\gamma(X_i)\right\}. \tag{61}$$

The first inequality follows from the observation that $|\log x|>\delta$ for sufficiently small δ implies that $|\sqrt{x}-1|>\delta/4$ and from Tchebyshev's inequality and the second inequality follows by Schwartz's inequality. Relation (61) implies that the first sum on the right-hand side of relation (60) converges to zero in probability by condition (*L*5).

Let us consider the second sum in (60). We assume that a sequence of neighbourhoods A_n of the point γ, $A_n\downarrow\gamma$ is chosen so that

$$P_\gamma\{X_i\in A_n,\ i=l_n, \dots, n\}\to 1 \text{ as } n\to\infty.$$

Such a sequence A_n exists since $P_\gamma\{X_\infty=\gamma\}=1$. Moreover, by condition (L4), there is a sequence B_n, $B_n\downarrow\gamma$ such that $p(x, y; \theta)=0$ for any $x\in A_n$, $y\notin B_n$. By the definition of the set A_n,

$$P_\gamma\left\{\sum_{l_n}^{n}\log\frac{p(X_i, X_{i+1}; \theta+zn^{-1/2})}{p(X_i, X_{i+1}; \theta)}\right.$$
$$\left.\neq\sum_{l_n}^{n}\log\frac{p(X_i, X_{i+1}; \theta+zn^{-1/2})}{p(X_i, X_{i+1}; \theta)}\chi_{A_n}(X_i)\right\}=o(1)$$

as $n\to\infty$ where χ_B denotes the indicator function of the set B. Let δ be an arbitrary small positive number. Set

$$\alpha_{in}=\log\frac{p(X_i, X_{i+1}; zn^{-1/2})}{p(X_i, X_{i+1}; \theta)}\chi_{A_n}(X_i).$$

By condition (L6),

$$P_\gamma\{|\alpha_{in}|>\delta\}\leqq\frac{16}{\delta^2}E_\gamma E_{\gamma,X_i}|\left\{\frac{p(X_i, X_{i+1}; \theta+zn^{-1/2})}{p(X_i, X_{i+1}; \theta)}\right\}^{1/2}-1|^2\chi_{A_n}(X_i)\chi^{(X_{i+1})}_{[|\alpha_{in}|>\delta]}$$
$$\leqq\frac{16}{\delta^2}E_\gamma\chi_{A_n}(X_i)\frac{1}{f_\gamma(X_i)}.$$

$$\sum_{y\in V(X_i, zn^{-1/2}\delta/4)} \{(p(X_i, y; \theta+zn^{-1/2}))^{1/2}-(p(X_i, y; \theta)^{1/2}\}^2 f_\gamma(y) \tag{62}$$

$$\leq \frac{1}{n}o(1). \tag{63}$$

Relation (63) implies that

$$P_\gamma\left\{\sum_{l_n}^{n} \log \frac{p(X_i, X_{i+1}; \theta+n^{-1/2}z)}{p(X_i, X_{i+1}; \theta)} \chi_{A_n}(X_i)\right.$$

$$\left.\neq \sum_{l_n}^{n} \log^{(\delta)} \frac{p(X_i, X_{i+1}; \theta+n^{-1/2}z)}{p(X_i, X_{i+1}; \theta)} \chi_{A_n}(X_i)\right\} = o(1). \tag{64}$$

Here, for any function f, $f^{(\delta)}=f$ if $|f|\leq\delta$ and is zero if $|f|>\delta$. Since δ is arbitrarily small, one can set, without violating condition (64), $\delta=\delta_n$ sufficiently slowly tending to zero. Assuming $\delta=\delta_n$ is sufficiently small, it can be seen, by using expansions for $\log x$, that

$$\begin{aligned}\alpha_{in}^{(\delta)} &= \beta_{in}+\gamma_{in}\\ &= 2\left(\left\{\frac{p(X_i, X_{i+1}; \theta+n^{-1/2}z)}{p(X_i, X_{i+1}; \theta)}\right\}^{1/2}-1\right)\chi_{A_n}(X_i)\\ &\quad -\left(\left\{\frac{p(X_i, X_{i+1}; \theta+n^{-1/2}z)}{p(X_i, X_{i+1}; \theta)}\right\}^{1/2}-1\right)^3+\gamma_{in}\end{aligned} \tag{65}$$

where

$$\gamma_{in}=C_{in}\left(\left\{\frac{p(X_i, X_{i+1}; \theta+n^{-1/2}z)}{p(X_i, X_{i+1}; \theta)}\right\}^{1/2}-1\right)^3-\beta_{in}\chi_{[|\alpha_{in}|>\delta]}(X_{i+1})$$

and $|C_{in}|<c<\infty$. Relation (63) implies that

$$\sum_{l_n}^{n}\gamma_{in}\to 0 \text{ in } P_\gamma\text{-probability.}$$

It is enough to study the limit distribution of the sum $\Sigma_{l_n}^{n}\beta_{in}$ as $n\to\infty$. Let $\sigma_{in}^2=\operatorname{var}(\beta_{in})$ and $\sigma_{in}^{2\prime}=\operatorname{var}(\beta_{in}|X_0, \ldots, X_n)$ making index γ implicit in the definition. We have

$$E_{\gamma, X_i}\left(\left(\left(\frac{p(X_i, X_{i+1}; \theta+h)}{p(X_i, X_{i+1}; \theta)}\right)^{1/2}-1\right)^2\chi_{A_n}(X_i)\right.$$

$$=\chi_{A_n}(X_i)\sum_{y}[(p(X_i, y; \theta+h))^{1/2}-(p(X_i, y; \theta))^{1/2}]^2\frac{f_\gamma(y)}{f_\gamma(X_i)}. \tag{66}$$

But it can be shown that

$$E_{\gamma, X_i}\left(\left(\left(\frac{p(X_i, X_{i+1}; \theta+h)}{p(X_i, X_{i+1}; \theta)}\right)^{1/2} - 1\right)^2 \chi_{A_n}(X_i)\right.$$
$$= \chi_{A_n}(X_i)\frac{1}{4} h^2 I_\gamma(\theta)(1+o(1)) \tag{67}$$

which implies the relation

$$E_\gamma\left(\left(\left(\frac{p(X_i, X_{i+1}; \theta+h)}{p(X_i, X_{i+1}; \theta)}\right)^{1/2} - 1\right)^2 \chi_{A_n}(X_i)\right.$$
$$= \frac{1}{4} h^2 I_\gamma(\theta)\,(1+o(1)), \tag{68}$$

Relation (68) shows that

$$E_\gamma\left(\left(\left(\frac{p(X_i, X_{i+1}; \theta+h)}{p(X_i, X_{i+1}; \theta)}\right)^{1/2} - 1\right)\chi_{A_n}(X_i)\right.$$

$$= E_\gamma \chi_{A_n}(X_i) \sum_{y \in B_n} ((p(X_i, y; \theta+h)\, p(X_i, y; \theta))^{1/2} - p(X_i, y; \theta))\, \frac{f_\gamma(y)}{f_\gamma(X_i)}$$

$$= (1+o(1))E_\gamma \chi_{A_n}(X_i)\left\{\left(\sum_y [p(X_i, y; \theta+h)\, p(X_i, y; \theta)]^{1/2}\right) - 1\right\}$$

$$= -\frac{1}{2}(1+o(1))E_\gamma \chi_{A_n}(X_i) \sum_y [(p(X_i, y; \theta+h))^{1/2} - (p(X_i, y; \theta))^{1/2}]^2$$

$$= -\frac{1}{8} I_\gamma(\theta) h^2 (1+o(1)). \tag{69}$$

Relations (67)–(69) shows that

$$E_\gamma \beta_{in} = -\frac{1}{2} I_\gamma(\theta)\, \frac{z^2}{n}\, (1+o(1)), \tag{70}$$

$$\sigma_{in}'^2 = \chi_{A_n}(X_i) I_\gamma(\theta) \frac{z^2}{n}(1+o(1)), \tag{71}$$

and

$$\sigma_{in}^2 = I_\gamma(\theta) \frac{z^2}{n}(1+o(1)). \tag{72}$$

Note that, from relations (70)–(72), we can conclude that

$$\sum_{i=l_n}^{n} \sigma_{in}^2 \leqq c < \infty, \; E_\gamma \sum_{i=l_n}^{n} |\sigma_{in}^2 - \sigma_{in}'^2| \to 0 \tag{73}$$

as $n \to \infty$. Conditions (73) imply in turn that $\Sigma_{l_n}^n \beta_{in}$ is asymptotically normal by a central limit theorem for dependent summands in Loeve (1963), p. 377. Furthermore

$$\sum_{i=l_n}^{n} \beta_{in} \xrightarrow{\mathscr{L}} N(-\tfrac{1}{2} I_y(\theta), I_y(\theta) z^2).$$

Hence, by (63), $Y_n(z)$ has the same distribution asymptotically.

In order to establish the convergence of the finite-dimensional distributions of $Y_n(z)$, we use the standard Cramer–Wold technique. Here we make use of the fact

$$\operatorname{cov}(\alpha_{in}(z_1), \alpha_{in}(z_2)) = I_y(\theta) \frac{z_1 z_2}{n} (1 + o(1)) \tag{74}$$

which can again be obtained by expansions of logarithms. This proves Lemma 4.4.

To complete the proof of Lemma 4.3, we shall first prove that the processes $\{Z_n^{1/2}(z), -\infty < z < \infty\}$ satisfy a "tightness" condition. This will ensure the convergence in distribution of the stochastic processes involved. (cf. Billingsley, 1968, Parthasarathy, 1967).

Let z_1 and z_2 belong to the interval $(l, l+1)$. Then there is an n_0 not depending on l, such that for all $n > n_0$ and any sufficiently close z_1, $z_2 \in (l, l+1)$,

$$E_y |Z_n^{1/2}(z_1) - Z_n^{1/2}(z_2)|^2 \leqq c |z_1 - z_2|^2 |l|^p \tag{75}$$

where $c > 0$. Moreover, for any $N > 0$, there exists a constant C_N such that for $n > n_0$

$$E_y Z_n^{1/2}(z) \leqq \frac{c_N}{|z|^N} \tag{76}$$

for all z. This can be shown as follows

$$\begin{aligned} E_y |Z_n^{1/2}(z_1) - Z_n^{1/2}(z_2)|^2 &\leqq \frac{1}{f_y(x, \theta)} E |Z_n^{1/2}(z_1) - Z_n^{1/2}(z_2)|^2 \\ &= \frac{2}{f_y(x, \theta)} (1 - E Z_n^{1/2}(z_1) Z_n^{1/2}(z_2)). \end{aligned}$$

Applying the Markov property and using successively the following inequality, (see condition (L5)),

$$E_{X_i} \left\{ \left(\frac{p(X_i, X_{i+1}; \theta + n^{-1/2} z_1)}{p(X_i, X_{i+1}; \theta)} \right)^{1/2} \left(\frac{p(X_i, X_{i+1}; \theta + n^{-1/2} z_2)}{p(X_i, X_{i+1}; \theta)} \right)^{1/2} \right\}$$

$$=1-\frac{1}{2}\sum_{y}[(p(X_i, y; \theta+n^{-1/2}z_1))^{1/2}-(p(X_i, y; \theta+n^{-1/2}z_2))^{1/2}]^2$$

$$\geqq 1-\frac{C}{2}\frac{|z_1-z_2|^2}{n}$$

$$\geqq 1-\tilde{c}\frac{|z_1-z_2|^2}{n}\ |l|^p$$

we obtain that for sufficiently close z_1, z_2 in $(l, l+1)$,

$$1-E_\gamma Z_n^{1/2}(z_1)Z_n^{1/2}(z_2)\leqq 1-\left(1-\tilde{\tilde{c}}|l|^p\frac{|z_1-z_2|^2}{n}\right)^n$$

$$\leqq c|z_1-z_2|^2|l|^p$$

for suitably chosen constant $c>0$. Here $\tilde{c}$, $\tilde{\tilde{c}}$ are suitably chosen non-negative constants. Similarly, for $|z|\ n^{-1/2}<k$, where k is sufficiently small, we have by condition (L5),

$$E_\gamma Z_n^{1/2}(z_1)\leqq\left(1-c\frac{z_1^2}{n}\right)^n\leqq e^{-c'z_1^2} \tag{77}$$

and for $|z|\ n^{-1/2}>K$, where K is sufficiently large, and for all sufficiently large n, we obtain, using condition (L7),

$$E_\gamma Z_n^{1/2}(z)\leqq\frac{1}{f_\gamma(x)}EZ_n^{1/2}(z)\leqq c^n\left(\frac{\sqrt{n}}{z}\right)^{n\delta}\leqq\frac{c_N}{|z|^N} \tag{78}$$

for any integer N, c_N not depending on n and in the intermediate region, by condition (L7),

$$E_\gamma Z_n^{1/2}(z)\leqq\frac{1}{f_\gamma(x)}\Delta^n\leqq\frac{c}{|z|^N}\text{ where } 0<\Delta<1 \tag{79}$$

for all sufficiently large n. (76) is a consequence of (77)–(79). It can now be shown that the processes $\{Z_n(z), -\infty<z<\infty\}$ form a "tight" family in view of the inequalities (75) and (76). This can be proved as in Ibragimov and Khasminskii (1973). We defer the proof to the end of the chapter.

The property of "tightness" of the family of measures generated by the processes $\{Z_n\}$ and the fact that the finite-dimensional distributions of $\{Z_n\}$ converge prove that the measures generated by the processes on $C_0(-\infty, \infty)$ (the space of continuous functions f for which $\lim_{|x|\to\infty} f(x)=0$ endowed with the uniform metric) converge weakly. Hence if f is any continuous functional on C_0 then

$$\lim_{n\to\infty} P_\gamma\{f(Z_n)\leqq x\}=P_\gamma\{f(Z)\leqq x\}.$$

In particular, it follows that

$$R_n = \frac{\int z \exp\{Y_n(z)\} dz}{\int \exp\{Y_n(z)\} dz} \xrightarrow{\mathscr{L}} \frac{\int z \exp\{I_\gamma^{1/2}(\theta)\xi z - \frac{1}{2} I_\gamma(\theta) z^2\} dz}{\int \exp\{I_\gamma^{1/2}(\theta)\xi z - \frac{1}{2} I_\gamma(\theta) z^2\} dz}$$

$$= \frac{\xi}{I_\gamma^{1/2}(\theta)}$$

with respect to P_γ-measure. Note that $R_n = n^{1/2}(\theta_n - \theta)$. Hence

$$n^{1/2}(\theta_n - \theta) \xrightarrow{\mathscr{L}} N(0, I_\gamma^{-1}(\theta))$$

with respect to P_γ-measure. This proves Lemma 4.3.

In view of the convergence of measures, it can be shown under stronger conditions that

$$E_\gamma R_n^a \to \frac{1}{(I_\gamma(\theta))^{a/2}} E|\xi|^a \text{ as } n \to \infty$$

for any $a > 0$ where ξ is a normal random variable with mean 0 and variance 1. The results of Theorem 4.3 can be generalized to the case of an arbitrary initial distribution concentrated on a finite subset of the space E.

We shall now give some simpler conditions for Theorem 4.3 to hold. Let us suppose that, for any $x, y \in E$, $p(x, y; \theta)$ is absolutely continuous in θ and that

$$I(x, \theta) = \sum_y \left[\frac{\partial p(x, y; \theta)}{\partial \theta}\right]^2 / p(x, y; \theta) < \infty.$$

Suppose that

(i) $I(x, \theta)$ is continuous in θ uniformly in x;

(ii) there exist constants $c > 0$ and $p > 0$ such that $I(x, \theta) \leqq c(1 + |\theta|^p)$;

(iii) uniformly in x,

$$\lim_{\substack{h_1 \to 0 \\ h_2 \to 0}} (h_1 - h_2)^{-2} \sum_y [(p(x, y; \theta + h_1))^{1/2} - (p(x, y; \theta + h_2))^{1/2}]^2 = \frac{I(x, \theta)}{4},$$

(iv) $\lim_{x \to \gamma} I(x, \theta) = I(\gamma, \theta) > 0, \ \gamma \in \Gamma$.

It is not hard to show that conditions (L5) and (L6) are consequences of (i)–(iv).

Example 4.1. Let the set of states E of the chain X be finite and decomposed into classes of states $A_0, A_1, \ldots, A_i$ such that for any $\theta \in \Theta$, A_0 is the set of transient states and the classes $A_1, \ldots, A_i$ are closed persistent states. Let $X_0 = x \in A_0$ and $f_\gamma(x, \theta)$ be the probability of hitting A_γ starting from $x \in A_0$. Thus $f_\gamma(x, \theta)$ is the probability that the chain, starting from x, is absorbed in the class A_γ so that, in particular $\Sigma_{\gamma=1}^{i} f_\gamma(x, \theta) = 1$.

The following theorem can be proved along lines similar to Theorem 4.3. Suppose, for any $x \in E$, the probability distributions defined by the transition density function $p(x, y; \theta)$ are absolutely continuous with respect to each other and the distribution of X_0 is concentrated on the set A_γ and coincides with a stationary distribution μ_γ on A_γ. Further suppose that the following regularity conditions are satisfied:

(i) $p(x, y, \theta)$ is continuously differentiable with respect to θ for $x, y \in E$;

(ii) there exists constants $c > 0$ and $p > 0$ such that, for all $x, y \in E$,

$$\left\{\frac{\partial p(x, y; \theta)}{\partial \theta}\right\}^2 \leqq c\, p(x, y; \theta)\,(1 + |\theta|^p);$$

(iii) there exist $c > 0$ and $\delta > 0$ such that, for any $x, y \in E$ and all $h \neq 0$, $\theta, \theta + h \in \Theta$,

$$p(x, y; \theta + h)\, p(x, y; \theta) < \frac{c}{|h|^\delta}.$$

Then the limit distribution of the variable $n^{1/2}(\theta_n - \theta)$ is the weighted normal distribution

$$\sum_{\gamma=1}^{i} f_\gamma(x, \theta)\, N(0, I_\gamma^{-1}(\theta))$$

where

$$I_\gamma(\theta) = \sum_{A_\gamma} \left[\mu_\gamma(x) \sum_{y \in A_\gamma} \frac{\{\delta p(x, y; \theta)/\partial \theta\}^2}{p(x, y; \theta)}\right].$$

We omit the proof as it is similar to that of Theorem 4.3.

5. BAYESIAN TESTING

In this section, we deal with the behaviour of the Bayes posterior risk in testing disjoint hypotheses, or hypotheses which are separated by an indifference region in which the losses due to taking the wrong decision are zero. We shall prove that the posterior risk goes to zero exponentially under some regularity conditions. The regularity conditions assumed here are weaker than those used in Section 2 to discuss asymptotic behaviour of Bayes estimators.

Under the set-up given in Section 2, assume that Θ is an open subset of $\mathbb{R}^p$ and Λ is a prior measure on Θ which assigns positive mass to every non-empty open subset of Θ.

Suppose the decision space D contains two members d_0 and d_1 and that for every θ at least one of $L(\theta, d_0)$ and $L(\theta, d_1)$ equals zero and both loss functions are non-negative and measurable in θ. Let

$$L(\theta)=\max\{L(\theta, d_1), L(\theta, d_0)\}. \tag{80}$$

We shall assume that

$$\int_\Theta L(\theta)\Lambda(\mathrm{d}\theta)<\infty. \tag{81}$$

Define

$$\Lambda^*(B)=\int_B L(\theta)\Lambda(\mathrm{d}\theta) \tag{82}$$

for every $B\subset\Theta$. Let the hypothesis be given by the set

$$H=\{\theta: L(\theta, d_1)=L(\theta)\} \tag{83}$$

and H' be the complement of H in Θ. A Bayes procedure for testing H against H' is

$$\begin{aligned}\delta(y_1,\ldots, y_n)&=d_0 \quad \text{if} \quad \int_H p(y_1,\ldots, y_n;\ \theta)\Lambda^*(\mathrm{d}\theta)\geqq\int_{H'} p(y_1,\ldots, y_n;\ \theta)\Lambda^*(\mathrm{d}\theta)\\ &=d_1 \quad \text{otherwise.}\end{aligned} \tag{84}$$

The Bayes posterior risk is

$$L_n=\min\left\{\frac{\int_H p(y_1,\ldots, y_n;\ \theta)\Lambda^*(\mathrm{d}\theta)}{\int_\Theta p(y_1,\ldots, y_n;\ \theta)\Lambda(\mathrm{d}\theta)}, \frac{\int_{H'} p(y_1,\ldots, y_n;\ \theta)\Lambda^*(\mathrm{d}\theta)}{\int_\Theta p(y_1,\ldots, y_n;\ \theta)\Lambda(\mathrm{d}\theta)}\right\}. \tag{85}$$

We shall assume that the following regularity conditions are satisfied.

(T1) $\Lambda(B)>0$ for all open sets $B\subset\Theta$ and

$$0<\Lambda^*(H)<\Lambda^*(\Theta). \tag{86}$$

(T2) Define

$$h_k(\theta_0, \delta, \mathbf{s})=E_{\theta_0}[\sup\{|\phi_k(\mathbf{s})-\phi_k(\mathbf{t})|:\|\mathbf{s}-\mathbf{t}\|\leqq\delta\}|\mathbf{Y}^{k-1}]$$

where $\phi_k(\mathbf{s})$ is as defined in Chapter 7. Suppose that, for δ sufficiently small,

$$h_k(\theta_0, \delta, \mathbf{s}) < \infty \text{ a.s.}$$

and

$$n^{-1} \sum_{k=1}^{n} h_k(\theta_0, \delta, \mathbf{s}) \to h(\theta_0, \delta, \mathbf{s}) < \infty \text{ a.s.} \tag{87}$$

Further suppose that

$$\lim_{\delta \to 0} h(\theta_0, \delta, \mathbf{s}) = 0. \tag{88}$$

Let

$$R_k(\mathbf{s}, \delta) \equiv \sup\{|\phi_k(\mathbf{s}) - \phi_k(\mathbf{t})| : \|\mathbf{s} - \mathbf{t}\| \leqq \delta\} \tag{89}$$

and assume that

$$\sum_{k=1}^{\infty} \frac{\operatorname{var}_{\theta_0}(R_k(\mathbf{s}, \delta))}{k^2} < \infty. \tag{90}$$

(T3) Define

$$V_k(r) = \sup\{[\phi_k(\theta) - \phi_k(\theta_0)] : \|\theta\| > r\}.$$

Assume that there exists an $r > 0$ such that

$$\sum_{k=1}^{\infty} \frac{\operatorname{var}_{\theta_0}[V_k(r)]}{k^2} < \infty, \tag{91}$$

and

$$\frac{1}{n} \sum_{k=1}^{n} E_{\theta_0}[V_k(r) | \mathbf{Y}^{k-1}] \xrightarrow{\text{a.s.}} V(\theta_0) \tag{92}$$

where

$$P_{\theta_0}[V(\theta_0) \leqq B(\theta_0) \leqq 0] = 1 \tag{93}$$

and $B(\theta_0)$ is as defined below.

(T4) Assume that

$$n^{-1} \sum_{k=1}^{n} E_{\theta_0}[\phi_k(\theta) - \phi_k(\theta_0) | \mathbf{Y}^{k-1}] \xrightarrow{\text{a.s.}} J(\theta, \theta_0) \tag{94}$$

uniformly on compact sets of θ.

Theorem 5.1. *Under assumptions* (T1)–(T4),

$$n^{-1} \log L_n \to B(\theta_0) \text{ a.s.} \tag{95}$$

where

$$B(\theta_0)=\min((\Lambda^*)\ \text{ess}\cdot\sup\{J(\theta,\ \theta_0):\theta\in H\},\\ (\Lambda^*)\ \text{ess}\cdot\sup\{J(\theta,\ \theta_0):\theta\in H'\}). \tag{96}$$

Proof. Let

$$J_n(\theta,\ \theta_0)=n^{-1}\sum_{k=1}^{n}\{\phi_k(\theta)-\phi_k(\theta_0)\}. \tag{97}$$

It is easy to see from (85) that

$$n^{-1}\log L_n=\min\left(n^{-1}\log\left\{\int_H \exp\,(nJ_n(\theta,\ \theta_0))\Lambda^*(\mathrm{d}\theta)\right\},\right.\\ \left.n^{-1}\log\left\{\int_{H'} \exp\,(nJ_n(\theta,\ \theta_0))\Lambda^*(\mathrm{d}\theta)\right\}\right)\\ -n^{-1}\log\left\{\int_\Theta \exp\,(nJ_n(\theta,\ \theta_0))\Lambda(\mathrm{d}\theta)\right\}. \tag{98}$$

Define $\bar{S}=\{\theta:\|\theta-\theta_0\|\leqq \mathrm{d}(\theta_0)\}$ where $\mathrm{d}(\theta_0)=\|\theta_0\|+r$. Without loss of generality, we shall suppose that $\bar{S}$ is disjoint from the boundary of Θ and hence is compact. We shall first prove that

$$n^{-1}\log\left\{\int_\Theta \exp\,(nJ_n(\theta,\ \theta_0))\Lambda(\mathrm{d}\theta)\right\}\to 0 \text{ a.s.} \tag{99}$$

It is sufficient to prove that

$$n^{-1}\log\left\{\int_{\bar{S}} \exp\,(nJ_n(\theta,\ \theta_0))\Lambda(\mathrm{d}\theta)\right\}\to 0 \text{ a.s.} \tag{100}$$

and

$$\limsup_n n^{-1}\log\left\{\int_{\bar{S}^c} \exp\,(nJ_n(\theta,\ \theta_0))\Lambda(\mathrm{d}\theta)\right\}\leqq 0 \text{ a.s.} \tag{101}$$

where $\bar{S}^c$ is the complement of the set $\bar{S}$ in Θ. Note that

$$n^{-1}\log\left\{\int_{\bar{S}^c} \exp\,(nJ_n(\theta,\ \theta_0))\Lambda(\mathrm{d}\theta)\right\}\\ \leqq n^{-1}\log\Lambda(\bar{S}^c)+n^{-1}\sum_{k=1}^{n}\sup\{[\phi_k(\theta)-\phi_k(\theta_0)]:\|\theta-\theta_0\|\geqq \mathrm{d}(\theta_0)\}\\ \leqq n^{-1}\sum_{k=1}^{n}V_k(r). \tag{102}$$

Assumption (T3) implies that

$$n^{-1}\sum_{k=1}^{n} V_k(r)\rightarrow V(\theta_0) \text{ a.s.}$$

and $P_{\theta_0}[V(\theta_0)\leqq 0]=1$ and hence (101) holds. Let

$$Q_n(\theta,\,\theta_0)=J_n(\theta,\,\theta_0)-J(\theta,\,\theta_0). \tag{103}$$

Let $\varepsilon>0$. By (T2), there exists $\delta(\varepsilon,\,\theta,\,\theta_0)>0$ such that

$$h(\theta_0,\,\delta,\,\theta)<\varepsilon. \tag{104}$$

Cover $\bar{S}$ with spheres of radius $\delta(\varepsilon,\theta,\theta_0)$ about each θ. One can extract a finite subcover centred at θ_j, $1\leqq j\leqq n$ (say) so that:

$$\begin{aligned}
&\sup\,\{|Q_n(\theta,\,\theta_0)-Q_n(\theta_j,\,\theta_0)|:\|\theta_j-\theta\|\leqq\delta(\varepsilon,\,\theta_j,\,\theta_0)\}\\
&\quad\leqq \sup\,\{|J_n(\theta,\,\theta_0)-J_n(\theta_j,\,\theta_0)|:\|\theta_j-\theta\|\leqq\delta(\varepsilon,\,\theta_j,\,\theta_0)\}\\
&\quad+\sup\,\{|J(\theta,\,\theta_0)-J(\theta_j,\,\theta_0)|:\|\theta_j-\theta\|\leqq\delta(\varepsilon,\,\theta_j,\,\theta_0)\}\\
&\quad\leqq n^{-1}\sum_{k=1}^{n}\sup\,\{|\phi_k(\theta)-\phi_k(\theta_j)|:\|\theta_j-\theta\|\leqq\delta(\varepsilon,\,\theta_j,\,\theta_0)\}\\
&\quad+\sup\,\{|J(\theta,\,\theta_0)-n^{-1}\sum_{k=1}^{n}E_{\theta_0}\{\phi_k(\theta)-\phi_k(\theta_0)|\mathbf{Y}^{k-1}\}|:\|\theta_j-\theta\|\leqq\delta(\varepsilon,\,\theta_j,\,\theta_0)\}\\
&\quad+|J(\theta_j,\,\theta_0)-n^{-1}\sum_{k=1}^{n}E_{\theta_0}\{\phi_k(\theta_j)-\phi_k(\theta_0)|\mathbf{Y}^{k-1}\}|\\
&\quad+n^{-1}\sum_{k=1}^{n}E_{\theta_0}[\sup\,\{|\phi_k(\theta_j)\quad\phi_k(\theta)|:\|\theta_j-\theta\|\leqq\delta(\varepsilon,\,\theta_j,\,\theta_0)\}|\mathbf{Y}^{k-1}]\\
&\quad=R_{1n}+R_{2n}+R_{3n}+R_{4n}\text{ (say).}
\end{aligned} \tag{105}$$

Condition (T2) implies that

$$\limsup R_{1n}\leqq\varepsilon \text{ a.s.} \tag{106}$$

in view of relation (104). Condition (T4) shows that

$$\limsup R_{2n}=\limsup R_{3n}=0 \text{ a.s.} \tag{107}$$

and condition (T2) again gives that

$$\limsup R_{4n}\leqq\varepsilon \text{ a.s.} \tag{108}$$

Hence, for $1\leqq j\leqq r$,

$$\limsup_{n}\,\{|Q_n(\theta,\,\theta_0)-Q_n(\theta_j,\,\theta_0)|:\|\theta_j-\theta\|\leqq\delta(\varepsilon,\,\theta_j,\,\theta_0)\}\leqq 2\varepsilon \text{ a.s.} \tag{109}$$

It is clear that

$$\limsup_n \sup \{|Q_n(\theta_j, \theta_0)|: \theta \in \bar{S}\}$$

$$\leqq \limsup_n \max_{1 \leqq j \leqq r} |Q_n(\theta_j, \theta_0)|$$

$$+ \limsup_n \max_{1 \leqq j \leqq r} \sup \{|Q_n(\theta, \theta_0) - Q_n(\theta_j, \theta_0)|: \|\theta_j - \theta\| \leqq \delta(\varepsilon, \theta_j, \theta_0)\}$$

$$= R_5 + R_6 \text{ (say).}$$

In view of conditions (T2) and (T4), it can be shown that $Q_n(\theta_j, \theta_0) \to 0$ a.s. for $1 \leqq j \leqq r$. Hence $R_5 = 0$ a.s. and $R_6 \leqq 2\varepsilon$ a.s. by relation (109). Hence

$$\sup \{|Q_n(\theta, \theta_0)|: \theta \in \bar{S}\} \to 0 \text{ a.s.} \tag{110}$$

Note that

$$(\Lambda) \text{ ess sup } J(\theta, \theta_0) = \sup_\theta J(\theta, \theta_0) = J(\theta_0, \theta_0) = 0 \tag{111}$$

by assumptions (T1), (T2) and (T4). An application of Lemma 4.2 of Bickel and Yahav (1967) yields relations (99). Similar arguments show that

$$n^{-1} \log \left\{ \int_H \exp(nJ_n(\theta, \theta_0)) \Lambda^*(d\theta) \right\} \to (\Lambda^*) \text{ ess sup } \{J(\theta, \theta_0): \theta \in H\}$$

and

$$n^{-1} \log \left\{ \int_{H'} \exp(nJ_n(\theta, \theta_0)) \Lambda^*(d\theta) \right\} \to (\Lambda^*) \text{ ess sup } \{J(\theta, \theta_0): \theta \in H'\}.$$

Hence the theorem is proved in view of (98).

The results obtained in this section can be used to study asymptotic sequential analysis in the dependent case.

6. FURTHER TOPICS

In Section 2, we discussed convergence of posterior density to normal in mean. It is of theoretical interest and practical use to find out the rate of approximation of posterior distribution to normal distribution. If the observations are independent and identically distributed, Hipp and Michel

(1976) proved that if $R_{n,x}$ denotes the posterior distribution of θ given $\mathbf{x}=(x_1, \ldots, x_n)$ and $Q_{n,x}$ denotes the normal distribution with mean θ_n, the maximum likelihood estimator based on $\mathbf{x}$, and the covariance matrix

$$\left(\left(\sum_{v=1}^{n} - \frac{\partial^2 \log p(x_v, \theta)}{\partial\theta_j \partial\theta_j}\bigg|_{\theta=\theta_n}\right)\right)_{p\times p}$$

($p(x, \theta)$ denotes the density with respect to a dominating measure), then, under some regularity conditions depending on an integer $s \geqq 2$, for every compact subset K of θ there exists a constant $c_K > 0$ such that

$$\sup_{\theta\in K} p_\theta\{d(R_{n,x}, Q_{n,x}) \geqq c_K n^{-1/2}\} = 0(n^{-s/2})$$

where d is the variational distance on $\mathbb{R}^p$ between measures defined by

$$d(\mu, v) = \sup\{|\mu(B) - v(B)| : B \in \mathscr{B}_p\}$$

$\mathscr{B}_p$ being the σ-field of Borel Subsets of $\mathbb{R}^p$. This result is an improvement over some earlier results of Strasser (1975). Results in the case of discrete time stationary Markov processes have been proved in Prakasa Rao (1977a, b) extending results of Hipp and Michel (1976) and Strasser (1977) in the i.i.d. case.

Doob (1949) obtained a fundamental result regarding consistency of Bayes estimators. He showed that under very weak measurability assumptions, for every prior distribtuion Λ on the parameter space Θ, the Bayes estimators are consistent except possibly for a set of values θ in Θ having Λ-measure zero. Schwarz (1965) proved consistency of Bayes estimators under conditions which are weaker than those given in Le Cam (1953) or Wald (1949) for consistency of Bayes and maximum likelihood estimators. These results can be extended to arbitrary discrete time stochastic processes. Freedman (1963) showed that if the sample space is discrete, then in the case of independent and identically distributed random variables taking on only finitely many values, the Bayes estimators are consistent and asymptotically normal. Majone (1966) extended Freedman's results to random elements taking values in Borel sets of a complete separable metric space. We shall not go into details of these results. Strasser (1973) discussed general theorems on the existence of Bayes estimators. Foutz (1974) discussed degenerate convergence of posterior distributions in the independent not identically distributed case generalizing results of Le Cam (1953) in the i.i.d. case. See the recent paper of Yamada (1976) in this connection.

7. PROOF OF TIGHTNESS OF PROCESSES

Before we discuss the tightness property of the processes $\{Z_n(z), -\infty < z < \infty\}$, we shall prove some lemmas.

Lemma 7.1. *There exist positive numbers k and $c_1 > 0$ such that for all z, $|z| < k\sqrt{n}$,*

$$P_\gamma\{Z_n(z) > \exp(-c_1 z^2)\} \leqq \exp(-c_1 z^2). \tag{112}$$

Proof. This inequality follows from relation (77) and Chebyshev's inequality.

Lemma 7.2. *For any positive integer N, there are numbers n_0 and K such that*

$$P_\gamma\left\{Z_n(z) > \frac{1}{|z|^N}\right\} \leqq C_N |z|^{-N} \tag{113}$$

for all $n > n_0$ where C_N is a constant not depending on n and z.

Proof. This again follows from relation (77) and Chebyshev's inequality in view of Lemma 7.1.

Lemma 7.3. *Given any positive integer N, there exists an integer n_0 and a constant C_N depending only on N such that, for $n > n_0$,*

$$P_\gamma\left\{\sup_{l \leqq z \leqq l+1} Z_n(z) \geqq \frac{1}{l^N}\right\} \leqq \frac{C_N}{l^N}, \quad l \geqq 1. \tag{114}$$

Proof. In view of the estimate obtained in relation (75), it follows from results in Prohorov (1956), p. 180 (cf. Gikhman and Skorokhod, 1969) that

$$P_\gamma\left\{\sup_{\substack{|z_2 - z_1| \leqq h \\ z_1, z_2 \in [l, l+1]}} |Z_n^{1/2}(z_2) - Z_n^{1/2}(z_1)| > h^{1/4}\right\} \leqq C|l|^p h^{1/4} \tag{115}$$

(see also Lemma 2.10 of Ibragimov and Khasminskii, 1972). Let us split the interval $[l, l+1]$ into r equal parts. In view of Lemma 7.2,

$$P_\gamma\left\{\max_{0 \leqq k \leqq r} Z_n^{1/2}(l + kr^{-1}) > \frac{1}{l^{6N}}\right\} \leqq (r+1)\frac{C_N}{l^{6N}}.$$

Thus, setting $r=l^{5N}$, we obtain

$$P_\gamma\left\{\max_{l\leq z\leq l+1} Z_n^{1/2}(z)>\frac{1}{l^N}\right\}\leqq\frac{(r+1)C_N}{l^{6N}}$$

$$+P_\gamma\left\{\max_{|z_1-z_2|\leq l^{-5N}}|Z_n^{1/2}(z_2)-Z_n^{1/2}(z_1)|>\frac{1}{l^N}-\frac{1}{l^{6N}}\right\}$$

$$\leqq\frac{C_N}{l^N}$$

provided N is sufficiently large (say) $N\geqq N_0$ by Lemma 7.2 and relation (115). However, it is clear that if (114) holds for $N>N_0$, by suitable choice of C_N, it holds for $N<N_0$.

Lemma 7.4 *Given any positive integer N, there exists an integer n_0 and a constant C_N depending only on N such that, for $n>n_0$,*

$$P_\gamma\left\{\sup_{|z|>M} Z_n(z)>\frac{1}{M^N}\right\}\leqq\frac{C_N}{M^N} \tag{116}$$

for any $M\geqq 1$.

Proof. In view of the inequality

$$P_\gamma\left\{\sup_{|\theta|>a} Z_n(z)>2\sum_{k=0}^{\infty}\frac{1}{(M+k)^N}\right\}$$

$$\leqq\sum_{k-0}^{\infty}P_\gamma\left\{\sup_{0\leqq|z|-M-k\leq 1} Z_n(z)>\frac{1}{(M+k)^N}\right\} \tag{117}$$

relation (116) is a consequence of relation (114).

Define the new random processes $\bar{Z}_n(z)$ as follows:

$$\bar{Z}_n(z)=\begin{cases} Z_n(z) \text{ if } \sqrt{n}(a-\theta)<z<\sqrt{n}(b-\theta) \\ 0 \qquad \text{if } z\leqq\sqrt{n}(a-\theta)-1 \text{ or } z\geqq\sqrt{n}(b-\theta)+1 \\ \text{linear and continuous in all other intervals.}\end{cases}$$

Theorem 7.1. *Under the assumptions stated earlier, the sample paths of $\bar{Z}_n(z)$, $-\infty<z<\infty$ and of the limiting process $Z(z)$ belong to $C_0(\mathbb{R})$ almost surely where $C_0(\mathbb{R})$ is the space of continuous functions f on $\mathbb{R}$) such that $\lim_{|x|\to\infty} f(x)=0$ endowed with the uniform metric.*

Proof. This theorem follows from Lemmas 7.3 and 7.4 by the Borel–Canteli lemma.

Theorem 7.2. *Under the assumptions stated earlier, for any* $\varepsilon>0$,

$$\lim_{d\to 0}\overline{\lim_{n\to\infty}}\, P_\gamma\left\{\sup_{|z_1-z_2|<d}|Z_n(z_1)-Z_n(z_2)|>\varepsilon\right\}=0.$$

Proof. It is enough to prove that if L is a constant such that $(\theta-Ln^{-1/2}, \theta+Ln^{-1/2})\subset\Theta$, then there exists a positive δ such that

$$P_\gamma\left\{\sup_{|z_2-z_1|\leqq h}|Z_n(z_1)-Z_n(z_2)|>h^\delta\right\}\leqq ch^\delta \tag{119}$$

for some $c>0$, c and δ not depending on L and n. By relation (115), if $L_1=h^{[8(p+1)]^{-1}}$, then

$$P_\gamma\left\{\sup_{\substack{|z_2-z_1|\leqq h\\ |z_i|<L_1}}|Z_n^{1/2}(z_1)-Z_n^{1/2}(z_2)|>h^{1/4}\right\}\leqq ch^{1/8}. \tag{120}$$

Again, by Lemma 7.4,

$$P_\gamma\left\{\sup_{\substack{|z_2-z_1|\leqq h\\ |z_i|>L_1}}|Z_n^{1/2}(z_1)-Z_n^{1/2}(z_2)|>h^{1/4}\right\}$$

$$\leqq P_\gamma\left\{\sup_{|z|>L_1}Z_n^{1/2}(z)>L_1^{-2(p+1)}\right\}\leqq ch^{1/4}. \tag{121}$$

Relations (120) and (121) prove (119). Thus an inequality of the form (119) holds for the process $Z_n^{1/2}(z)$. It is also clear that a similar inequality holds for the process $Z_n(z)$ whenever $|z|>L_1$. In a similar fashion, it can be shown that there exists $\delta_1>0$ such that

$$P_\gamma\left\{\sup_{|z|\leqq L_1}Z_n^{1/2}(z)>L_1^{\delta_1}\right\}\leqq L_1^{-\delta_1}<h^\delta.$$

Relation (119) now follows from the relation

$$Z_n(z_2)-Z_n(z_1)=(Z_n^{1/2}(z_2)-Z_n^{1/2}(z_1))(Z_n^{1/2}(z_2)+Z_n^{1/2}(z_1))$$

which in turn proves (118).

Sample paths of $\{\overline{Z}_n(z)\}$ can be considered as bounded continuous functions on the compactification $(\overline{R}, \delta)$ with probability one where for any z_1, z_2 in $\overline{\mathbb{R}}$ the distance $\delta(z_1, z_2)$ can be defined by

$$\delta(z_1, z_2)=\frac{|z_1-z_2|}{1+|z_1-z_2|}.$$

Further $\overline{Z}_n(0)=1$. Then, it follows from Theorem 5.6 of Straf (1970),

Prakasa Rao (1975) or from results of Billingsley (1968) that the tightness of the measures generated by $\{\bar{Z}_n\}$ is equivalent to the assertion made in Theorem 7.2.

Theorem 7.3. *Under the assumptions stated, the measures in* $C_0(\mathbb{R})$ *of the processes* $\{Z_n(z)\}$ *converge weakly to the measure generated by* $\{Z(z)\}$ *where* $Z(z)=Y(z)$, $Y(z)$ *as defined in* (59). *In particular, for measurable functionals on* $C_0(\mathbb{R})$, $\{\psi_n\}$ *which continuously converge to* ψ,

$$\lim_{n\to\infty} P_\gamma\{\psi_n(\bar{Z}_n)\leqq x\}=P_\gamma\{\psi(Z)\leqq x\}.$$

BIBLIOGRAPHICAL NOTES

Le Cam (1953, 1958) gave a rigorous proof of the convergence of posterior distribution to normal for independent and identically distributed observations under some conditions. His result is an improvement over earlier results of Laplace, von Mises and Bernstein of similar type. Various modifications and generalizations of this result have been made recently by Bickel and Yahav (1969), Ghao (1970), David (1970), Walker (1969), Borwanker *et al.* (1971), Prasad (1973) and Marc Moore (1977). Le Cam (1970) gave an extensive discussion of conditions underlying the Bernstein–von Mises Theorem. Methods used in this chapter are similar to those in Bickel and Yahav (1969) and Borwanker *et al.* (1971). An alternative approach to these problems through the likelihood ratio process is due to Ibragimov and Khasminskii (1973) and Levit (1974). Johnson (1967) has obtained the asymptotic series expansion for posterior distribution with the leading term of the expansion as normal distribution for observations from exponential families.

Sections 2 and 3 and Theorem 4.1 are from Prakasa Rao (1974). Section 4.2 contains results from Stephan (1973). Bayes estimation for Markov chains discussed in Section 4.3 is due to Levit (1974). Results on Bayes testing in Section 5 are due to Prakasa Rao (1974) extending the work of Bickel and Yahav (1969) in the i.i.d. case.

COMPLEMENTS

1. Let the observations X_n have the form: $X_n=q^n\theta+\xi_n$ where $q>1$, θ is unknown, $\xi_1, \xi_2, \ldots$ are independent identically distributed "noises" with density p with respect

to the Lebesgue measure. Consider the generalized Bayes estimator

$$\theta_n = \frac{\displaystyle\int z \prod_{i=1}^{n} p(X_i - q^i z)\mathrm{d}z}{\displaystyle\int \prod_{i=1}^{n} p(X_i - q^i z)\mathrm{d}z}.$$

Under suitable conditions, show that $q^n(\theta_n - \theta)$ converges to the distribution of ξ_n as $q \to \infty$. Thus as $q \to \infty$, the estimators θ_n have asymptotically the same distribution as the estimator $\tilde{\theta}_n = X_n q^{-n}$ constructed over the single last observation X_n. (This is an example where the information I_n in the nth observation X_n increases exponentially and only the latter terms of the sequence $X_1, \ldots, X_n$ render an essential influence on the distribution of θ_n).

(Levit, 1972, 1974)

2. Suppose that observations are independent and that X_n is distributed with density $p_n(x, \theta)$. Further assume that for some $\theta \neq \theta'$, the following two conditions hold:

(i) $$\sum_{i=1}^{\infty} \int |(p_i(x, \theta'))^{1/2} - (p_i(x, \theta))^{1/2}|^2 \mathrm{d}x < \infty;$$

(ii) $$\int (p_i(x, \theta) p_i(x, \theta'))^{1/2} \mathrm{d}x > 0 \text{ for all } i \geqq 1.$$

Prove that there exists no consistent estimator of θ. (Levit, 1974)

3. Prove the result state in Example 4.1. (Levit, 1974)

4. Show that conditions (L5) and (L6) are consequences of (i)–(iv) given before Example 4.1. (Levit, 1974)

5. Suppose the observed random variables $X_1, \ldots, X_n$ can be represented in the form

$$X_i = f_i(X'_i), \ i = 1, \ldots, n$$

where the f_i are one-to-one, continuously differentiable functions and the random variables X'_i have joint density $p'(x'_1, \ldots, x'_n; \theta)$ with respect to the Lebesgue measure satisfying $p'(x'_1, \ldots, x'_n; \theta) = p'(x'_1 - \theta, \ldots, x'_n - \theta)$ and having finite second moments. Then, among the class of all estimators θ_n of θ such that

$$\theta_n(f_1(X'_1 + \theta), \ldots, f_n(X'_n + \theta)) = \theta_n(f_1(X'_1) + \ldots + f_n(X'_n)) + \theta$$

show that the estimator

$$\theta_n = \frac{\displaystyle\int z p(X_1, \ldots, X_n; z)\mathrm{d}z}{\displaystyle\int p(X_1, \ldots, X_n; z)\mathrm{d}z}$$

has the smallest mean square error i.e. $E(\theta_n - \theta)^2$ is the smallest.

(Levit, 1974)

6. Let $\eta_1, \eta_2, \ldots$ be independent noise functions in the memoryless channel with noiseless feedback with a common density function p and let $\zeta_1, \ldots, \zeta_n, \ldots$ be observations at the channel output. Suppose

$$\zeta_n = X_n + \eta_n;\ X_n = q^n(\xi - \xi_{n-1})$$

where ξ is the message to be transmitted,

$$\xi_n = \frac{\int y p(\zeta_1 - qy) p(\zeta_2 - q^2(y - \xi_1)) \ldots p(\zeta_n - q^n(y - \xi_{n-1}))\mathrm{d}y}{\int p(\zeta_1 - qy) p(\zeta_2 - q^2(y - \xi_1)) \ldots p(\zeta_n - q^n(y - \xi_{n-1}))\mathrm{d}y}$$

is the generalized Bayes estimator of ξ constructed from the preceding observations, $q > 1$. The estimate ξ_n becomes known at the channel input due to the presence of noiseless feedback. Suppose that there exist constants $c > 0$, $C > 0$, $\alpha > 0$, $\beta > 0$, such that for all h

$$1 - c|h|^\alpha \leqq \int (p(x-h)p(x))^{1/2}\mathrm{d}x \leqq C|h|^{-\beta}.$$

Show that X_n converges in law to a limit distribution and

$$\lim_n EX_n^2 = S(q) < \infty.$$

$S(q)$ is called the average transmitted power. Let $R = \log q$ be the attainable transmission rate. Suppose p is absolutely continuous and

$$I = \int_{-\infty}^{\infty} \frac{|p'(x)|^2}{p(x)}\mathrm{d}x < \infty.$$

Show that $S(q) \sim (q^2 - 1)I^{-1}$ as $q \to 1$.

(Levit, 1972)

7. Suppose that for each $n \geqq 1$ there exists a probability measure λ_n on $\mathbb{R}^2$ with marginals λ_{1n} and λ_{2n} on $\mathbb{R}$ such that

$$\int_C x\lambda_{1n}(\mathrm{d}x) = \int_{\mathbb{R}_2} y\chi_C(x)\lambda_n(\mathrm{d}x, \mathrm{d}y)$$

for all Borel sets C in $\mathbb{R}$ and for all $n \geqq 1$. Further suppose that there exists a measure α on $\mathbb{R}$ such that $\lambda_{1n} \to \alpha$ weakly, $\lambda_{2n} = \alpha$ for all $n \geqq 1$, and $\int_{\mathbb{R}} |y|\alpha(\mathrm{d}y) < \infty$. Show that

$$\int_{\mathbb{R}^2} |x - y|\lambda_n(\mathrm{d}x, \mathrm{d}y) \to 0 \text{ as } n \to \infty.$$

Deduce Theorem 4.2 from the above result.

(Stepan, 1971)

8. Let Θ be the real line and $Z_1, Z_2, \ldots, Z_n, \ldots$ be independent and identically distributed observations each $N(\delta(\theta), 1)$ where $\delta(\theta)=1-\theta$ if θ belongs to the Cantor set on $[0, 1]$ and $\delta(\theta)=\theta$ otherwise. Show that the Bayes estimator namely $E(\theta|Z_1, \ldots, Z_n)$ of θ converges to $\delta(\theta)$ a.e. P_θ for every $\theta \in \Theta$ which shows that Bayes estimator of θ is not consistent. Show that θ has consistent estimators.

(Schwartz, 1965)

9. Let $\Theta=[1, 2)$ and λ be the Lebesgue measure on the Borel sets of Θ. Let Z_i, $i \geqq 1$ have uniform distribution on $[0, 1)$ if $\theta=1$ or on $[0, 2/\theta)$ if $1<\theta<2$. Obtain the maximum likelihood estimator δ_n and the Bayes estimator β_n of θ corresponding to the sample $Z_1, \ldots, Z_n$ given that the loss function is quadratic. Show that δ_n is consistent and β_n is not.

(Schwartz, 1965)

10. Simplify the regularity conditions given in Section 4.1 for stationary Markov processes.

(Borwanker *et al.*, 1971)

11. Let $X_0, X_1, \ldots$ be a Markov chain for which the state space consists of the non-negative integers, the transitional probabilities are

$$p(0, 1; \theta)=1, \ p(i, i+1; \theta)=\theta, \ p(i, i-1; \theta)=1-\theta, \ i=1, 2, \ldots$$

and the initial distribution is

$$f(0, \theta)=(1-2\theta)/(2-2\theta), \ f(i; \theta)=\frac{1}{\theta}\left(\frac{\theta}{1-\theta}\right)^i f(0; \theta), \ i=1, 2, \ldots$$

where $\theta \in \Theta=(0, \frac{1}{2})$. Let

$$\lambda(\theta)=\begin{cases} k(\alpha, \beta)\theta^\alpha(\frac{1}{2}-\theta)^\beta & \text{if } 0<\theta<\frac{1}{2} \\ 0 & \text{otherwise,} \end{cases}$$

where $\alpha>0$, $\beta>0$ and $k(\alpha, \beta)$ is a normalizing factor, be a prior density on $(0, \frac{1}{2})$. Suppose $L(\theta, d)=B(\theta)l(\|\theta-d\|^2)$ is the loss function where

(i) $l: \mathbb{R}^+ \cup \{0\} \to \mathbb{R}^+ \cup \{0\}$, $l(0)=0$, $l(t)>0$ if $t>0$;

(ii) $l'(t)$ exists and is non-negative for $t>0$;

(iii) there exist $\gamma, \delta \in \mathbb{R}^+$ and $s \geqq 1$ such that

$$l(t)=\gamma t^s/s \text{ for } 0 \leqq t<\delta;$$

(iv) $\limsup_{t\to\infty} l'(t)t^{-r}<\infty$ for some $r \geqq 0$;

(v) B is a bounded continuous function and $B(\theta)>0$ for each $\theta \in \Theta$.

Under the above assumptions, show that the Bayes estimator θ_n' corresponding to the prior density λ and loss function L is strongly consistent i.e.

$$\theta_n' \to \theta_0 \text{ a.s. } P_{\theta_0}.$$

(Marc Moore, 1977)

12. Let $X_0, X_1, \ldots$ be a Markov chain with state space $\mathscr{X} = \{0, 1\}$, transition matrix

	0	1
0	$(1-p)+\pi p$	$(1-\pi)p$
1	$(1-\pi)(1-p)$	$\pi+(1-\pi)p$

and initial distribution $f(1; p, \pi) = 1 - f(0; p, \pi) = p$ where $(p, \pi) \in \Theta = [0, 1] \times [0, 1]$. Consider the prior density $\lambda(p, \pi)$ on Θ defined by

$$\lambda(p, \pi) = \begin{cases} \dfrac{\Gamma(\alpha_1+\alpha_2)}{\Gamma(\alpha_1)\Gamma(\alpha_2)} \dfrac{\Gamma(\beta_1+\beta_2)}{\Gamma(\beta_1)\Gamma(\beta_2)} p^{\alpha_1-1}(1-p)^{\alpha_2-1}\pi^{\beta_1-1}(1-\pi)^{\beta_2-1} & \\ & \text{for } 0 < p, \pi < 1 \\ 0 & \text{otherwise.} \end{cases}$$

For prior density λ given and loss function L satisfying the conditions in the previous example, show that

$$\theta_n' \to \theta_0 \text{ a.s. } P_{\theta_0}$$

for every $\theta_0 = (p_0, \pi_0) \in (0, 1) \times (0, 1)$.

(Marc Moore, 1977)

Chapter 11

Nonparametric Inference for Stochastic Processes

1. INTRODUCTION

In the preceding chapters, we studied statistical inference for stochastic processes under the assumption that measures generated by the processes are completely known except for certain parameters which have to be estimated from the observed realizations. As in the case of independent and identically distributed observations, this might not always be possible and we are forced to use non-parameteric techniques for inference. Here we study non-parametric estimation for stationary processes and non-parametric tests for stochastic processes.

2. NONPARAMETRIC ESTIMATION FOR STOCHASTIC PROCESSES

Estimates of the density function of a population based on a sample of independent observations have been considered by several authors. Questions of bias, variance and asymptotic distribution of the estimates have also been discussed. Our object in this section is to look at such estimates when the observations are dependent. In particular, we shall discuss some results on estimation of marginal density for stationary Markov processes and for stationary mixing processes. It should be noted that the asymptotic results for density estimation in general for dependent observations are essentially the same as those in the case of independent observations as long as some type of asymptotic independence is present in the process.

We shall also study the behaviour of robust nonparametric estimators for dependent data. Specifically, we shall discuss the asymptotic behaviour of the Hodges–Lehmann estimator for location parameters when the observations are from a stationary mixing process.

2.1 Non-existence of Unbiased Estimates of Density

Let $\{X_n, n \geqq 1\}$ be a strictly stationary stochastic process with a continuous one-dimensional marginal probability density f. The following theorem can be proved when the observations are independent and identically distributed.

Theorem 2.1. *Let $\{X_n, n \geqq 1\}$ be as stated above. Then, for any n, there does not exist a function $h_n(y{:}x_1, \ldots, x_n)$ which is measurable in $(x_1, \ldots, x_n)$ for each y, and such that*

$$\int h_n(y{:}x_1, \ldots, x_n) P\mathrm{d}(x_1, \ldots, x_n) = f(y)$$

for all continuous f.

The theorem states that an unbiased estimate of f under reasonable conditions does not exist. The proof of the above theorem, in the i.i.d. case, given by Rosenblatt (1956) is simple and interesting. The following proof uses the notion of generalized functions. We refer the reader to Gelfond and Shilov (1964) and Gelfond and Vilenkin (1964) for a discussion on generalized functions. A short summary of these concepts is given in Appendix 4.

Let $K_m = C_0^\infty(\mathbb{R}^m)$ be the linear space of all real-valued infinitely differentiable functions φ defined on $\mathbb{R}^m$ such that φ vanishes outside a compact interval. We shall denote K_m by $\mathscr{D}(\mathbb{R}^m)$ if the following topology is imposed on K_m: $\varphi_h \to \varphi$ in $\mathscr{D}(\mathbb{R}^m)$ if

(i) there exists a compact set $K \subset \mathbb{R}^m$ such that support $(\varphi_h) c K$, $h = 1, 2, \ldots$

(ii) for any $s \geqq 0$, $\varphi_h^{(s)}(x) \to \varphi^s(x)$ uniformly on K where $\varphi_h^{(s)}(x)$ denotes any sth partial derivative of $\varphi_h(x)$ for $s \geqq 1$ and $\varphi_h^{(0)}(x) \equiv \varphi_h(x)$.

We shall first state two lemmas. The first lemma is from Yosida (1968) p. 65 and the second is from Gelfond and Vilenkin (1964) p. 20. The proofs are omitted.

Lemma 2.1. *Let* $\mathbf{x} \in \mathbb{R}^n$, $\mathbf{y} \in \mathbb{R}^m$ *and* $\mathbf{z} = (\mathbf{x}, \mathbf{y}) \in \mathbb{R}^{n+m}$. *Then, for any function* $\varphi(\mathbf{z}) = \varphi(\mathbf{x}, \mathbf{y}) \in C_0^\infty(\mathbb{R}^{n+m})$, *we can choose functions* $u_{ij} \in C_0^\infty(\mathbb{R}^n)$ *and* $v_{ij} \in C_0^\infty(\mathbb{R}^m)$ *such that the sequence of functions*

$$\varphi_i(\mathbf{z}) = \varphi_i(\mathbf{x}, \mathbf{y}) = \sum_{j=1}^{k_i} u_{ij}(\mathbf{x}) v_{ij}(\mathbf{y}) \tag{2}$$

tend to $\varphi(\mathbf{z}) = \varphi(\mathbf{x}, \mathbf{y})$ *as* $i \to \infty$ *in the topology of* $\mathscr{D}(\mathbb{R}^{n+m})$.

Lemma 2.2 *Let* $B(\varphi_1, \varphi_2, \ldots \varphi_m)$ *be a poly-linear functional i.e. linear in each* φ_i, $1 \leqq i \leqq m$ *on the space* $K = K_1 = C_0^\infty(\mathbb{R})$. *Then there is a linear functional* F *on* K_m *such that*

$$B(\varphi_1, \varphi_2, \ldots, \varphi_m) = \langle F, \varphi_1 \varphi_2 \ldots \varphi_m \rangle \tag{3}$$

where

$$\langle F, \varphi_1 \varphi_2 \ldots \varphi_m \rangle = \int \varphi_1(x_1)\, \varphi_2(x_2) \ldots \varphi_m(x_m)\, \mathrm{d}F(x_1, \ldots, x_m). \tag{4}$$

Proof of Theorem 2.1. Let us assume that for some n there exists a measurable function h satisfying (1). We can assume without loss of generality that $h(y; x_1, \ldots, x_n)$ is a symmetric function of $x_1, \ldots, x_n$. Otherwise, we can take

$$h^*(y; x_1, \ldots, x_n) = \frac{1}{n!} \sum h(y; x_{i_1}, \ldots, x_{i_n}) \tag{5}$$

where the summation is over all permutations $x_{i_1}, \ldots, x_{i_n}$ of $x_1, \ldots, x_n$. Relation (1) implies that if φ is any non-negative continuous integrable function with

$$I(\varphi) = \int \varphi(x) \mathrm{d}x > 0 \tag{6}$$

then

$$\int h(y; x_1, \ldots, x_n) \frac{\varphi(x_1)}{I(\varphi)} \cdots \frac{\varphi(x_n)}{I(\varphi)} \mathrm{d}x_1 \ldots \mathrm{d}x_n = \frac{\varphi(y)}{I(\varphi)}. \tag{7}$$

Therefore

$$\int h(y; x_1, \ldots, x_n) \varphi(x_1) \ldots \varphi(x_n) \mathrm{d}x_1 \ldots \mathrm{d}x_n = \varphi(y)[I(\varphi)]^{n-1}. \tag{8}$$

Let

$$G_y(\varphi_1, \ldots, \varphi_n) = \int h(y; x_1, \ldots, x_n)\varphi_1(x_1) \ldots \varphi_n(x_n)\mathrm{d}x_1 \ldots \mathrm{d}x_n \tag{9}$$

for all continuous integrable functions $\varphi_1, \ldots, \varphi_n$. We shall first show that, for non-negative continuous integrable $\varphi_1, \varphi_2, \ldots, \varphi_n$,

$$G_y(\varphi_1, \ldots, \varphi_n) = \frac{1}{n} \sum_{i=1}^{n} \left\{\varphi_i(y) \prod_{j \neq i} I(\varphi_j)\right\}. \tag{10}$$

Let us first suppose that $n=3$. Then

$$G_y(\varphi_1+\varphi_2, \varphi_1+\varphi_2, \varphi_1+\varphi_2) = (\varphi_1+\varphi_2)(y)[I(\varphi_1+\varphi_2)]^2$$

from (8) and hence

$$3G_y(\varphi_1, \varphi_2, \varphi_2) + 3G_y(\varphi_1, \varphi_1, \varphi_2) = \varphi_1(y)[I(\varphi_2)]^2 + \varphi_2(y)[I(\varphi_1)]^2 + 2[\varphi_1(y)+\varphi_2(y)]I(\varphi_1)I(\varphi_2). \tag{11}$$

Again, from (8),

$$G_y(\varphi_1+\varphi_2+\varphi_3, \varphi_1+\varphi_2+\varphi_3, \varphi_1+\varphi_2+\varphi_3) = (\varphi_1+\varphi_2+\varphi_3)(y)[I(\varphi_1+\varphi_2+\varphi_3)]^2. \tag{12}$$

Expanding and using (8) and (12), we get that

$$G_y(\varphi_1, \varphi_2, \varphi_3) = \frac{1}{3} \sum_{i=1}^{3} \left[\varphi_i(y) \prod_{j \neq i} I(\varphi_j)\right]. \tag{13}$$

The proof for an arbitrary positive integer n proceeds on similar lines by considering $G_y(\varphi_1+\varphi_2, \ldots, \varphi_1+\varphi_2)$, $G_y(\Sigma_{i=1}^3 \varphi_i, \ldots, \Sigma_{i=1}^3 \varphi_i), \ldots, G_y(\Sigma_{i=1}^n \varphi_i, \ldots, \Sigma_{i=1}^n \varphi_i)$.

Suppose now that φ is any arbitrary continuous integrable function. Write $\varphi = \varphi^+ - \varphi^-$ where $\varphi^+ = \varphi$ if $\varphi \geqq 0$, $\varphi^+ = 0$ if $\varphi < 0$ and $\varphi^- = -\varphi$ if $\varphi \leqq 0$, $\varphi^- = 0$ if $\varphi > 0$. Then, if $n=3$, we get

$$\begin{aligned}
G_y(\varphi, \varphi, \varphi) &= G_y(\varphi^+ - \varphi^-, \varphi^+ - \varphi^-, \varphi^+ - \varphi^-) \\
&= G_y(\varphi^+, \varphi^+, \varphi^+) - G_y(\varphi^-, \varphi^-, \varphi^-) \\
&\quad + 3G_y(\varphi^+, \varphi^-, \varphi^-) - 3G_y(\varphi^+, \varphi^+, \varphi^-) \\
&= \varphi^+(y)I^2(\varphi^+) - \varphi^-(y)I^2(\varphi^-) + \varphi^+(y)I^2(\varphi^-) \\
&\quad + 2\varphi^-(y)I(\varphi^+)I(\varphi^-) - \varphi^-(y)I^2(\varphi^+) - 2\varphi^+(y)I(\varphi^+)I(\varphi^-)
\end{aligned}$$

$$=(\varphi^+ - \varphi^-)(y)[I(\varphi^+ - \varphi^-)]^2$$
$$=\varphi(y)I^2(\varphi).$$

In general, for arbitrary n,

$$G_y(\varphi, \varphi, \ldots, \varphi) = \varphi(y)[I(\varphi)]^{n-1}. \tag{14}$$

Proceeding exactly as in the case of non-negative functions, it can be shown that

$$G_y(\varphi_1, \ldots, \varphi_n) = \frac{1}{n}\sum_{i=1}^{n}\left[\varphi_i(y)\prod_{j\neq i} I(\varphi_j)\right] \tag{15}$$

for arbitrary continuous integrable functions $\varphi_1, \varphi_2, \ldots, \varphi_n$. In particular, (15) holds for $\varphi_i \in K_1$, so that, by Lemma 2.2 there exists a linear functional H on K_n given by

$$\langle H, \theta\rangle = \int h(y; x_1, \ldots, x_n)\, \theta(x_1, \ldots, x_n)\, dx_1 \ldots dx_n \tag{16}$$

for all $\theta \in K_n$ and such that for φ_i in K_1, $1 \leqq i \leqq n$,

$$\langle H, \varphi_1\varphi_2 \ldots, \varphi_n\rangle = G_y(\varphi_1, \ldots, \varphi_n) = \frac{1}{n}\sum_{i=1}^{n} \langle \delta_y^*(x_i), \varphi_i(x_1) \ldots \varphi_n(x_n)\rangle$$

by (15) where $\delta_y^*(x_1)$ is defined by

$$\langle \delta_y^*(x_1), \varphi\rangle = \int \varphi(y, x_2, \ldots, x_n) dx_2 \ldots dx_n$$

for all $\varphi \in K_n$. According to Lemma 2.1, finite sums of products of φ in K_1 like

$$\sum_{i=1}^{r} \varphi_{1,i}(x_1) \ldots \varphi_{n,i}(x_n)$$

are dense in K_n. We get that

$$\langle H, \theta\rangle = \frac{1}{n}\sum_{i=1}^{n} \langle \delta_y^*(x_i), \theta\rangle \tag{17}$$

for all $\theta \in K_n$ which implies that $h(y; x_1, \ldots, x_n)$ cannot be a measurable function of $y, x_1, \ldots, x_n$ in view of (16). This contradicts the assumption that there exists h measurable satisfying (1).

This theorem does not require the condition assumed by Rosenblatt (1956). But the proof of the theorem in the i.i.d. case, given in Rosenblatt

(1956), uses the fact that order statistic $(X_{(1)}, \ldots, X_{(n)})$ is a complete sufficient statistic. Note that the theorem proved above implies that there are no unbiased estimates for any finite-dimensional densities.

In view of Theorem 2.1, it is natural to ask whether there are any estimators which are asymptotically unbiased. In the next two sections, we shall discuss the kernel method of density estimation, the method using orthogonal expansions, and the method of delta sequences.

2.2 Kernel Type Density Estimation for Stationary Markov Processes

Let us consider a strictly stationary Markov process defined on a probability space $(\Omega, \mathscr{J}, P)$ and taking values in $(\mathbb{R}, \mathscr{B})$ where $\mathbb{R}$ is the real line and $\mathscr{B}$ is the Borel subsets of $\mathbb{R}$. Let p and q be the initial and two-dimensional joint density of the process with respect to the appropriate Lebesgue measure. For $p(x)>0$, transition density of the process is then defined by $t(x'|x)=q(x, x')/p(x)$, $(x, x')=y$.

We shall assume that the underlying process satisfies the Doeblin's condition (D_0) (cf. Doob, 1953, p. 221) Namely, there is a finite-valued measure λ on $\mathscr{B}$ with $\lambda(\mathbb{R})>0$, an integer $v \geqq 1$ and $\varepsilon>0$ such that

$$T^{(v)}(\xi, A) \leqq 1-\varepsilon \quad \text{if } \lambda(A) \leqq \varepsilon$$

and there is only a single ergodic set and this set contains no cyclically moving subsets. (Here $T^{(n)}(.,.)$ is the n-step transition probability measure). Under (D_0) it can be shown that there exists positive constants $\alpha \geqq 1$ and $0<\rho<1$ such that

$$|T^{(n)}(\xi, E)-\pi(E)| \leqq \alpha\rho^n, \ n \geqq 1, \ E \in \mathscr{B}$$

where π is the stationary initial distribution of the process. It is clear that $p(\cdot)$ is the density of π with respect to the Lebesgue measure on $\mathbb{R}$. Let K be a probability density function defined on $\mathbb{R}$ which satisfies the following conditions.

(K1) $K(x) \leqq M < \infty$, $x \in \mathbb{R}$; $|x| \, K(x) \to 0$ as $|x| \to \infty$;
(K2) $K(-x)=K(x)$, $x \in \mathbb{R}$; $\int x^2 \, K(x) dx < \infty$.
(K3) If $\hat{K}$ is the Fourier transform of K i.e. $\hat{K}(u)=\int e^{iux} K(x) dx$, then $\hat{K}$ is absolutely integrable.

Let $\{h_n\}$ be a sequence of positive constants tending to zero as $n \to \infty$ such that

(H1) $n \, h_n \to \infty$ as $n \to \infty$,

(H2) $n \, h_n^2 \to \infty$ as $n \to \infty$.

The following additional conditions are imposed on the process
(P1) Both probability densities p and q are continuous.
(P2) p is uniformly continuous.
(P3) q is uniformly continuous.

Some examples of probability density functions K satisfying (K1)–(K3) are given below:

(i) $K_1(x)=1/2 \quad |x|\leqq 1$
$\qquad =0 \quad |x|>1$,

(ii) $K_2(x)=1-|y| \quad$ for $|y|\leqq 1$
$\qquad =0 \quad$ for $|y|>1$,

(iii) $K_3(x)=(2\pi)^{-1/2}\,e^{-(1/2)x^2}$,

(iv) $K_4(x)=\dfrac{1}{\pi(1+x^2)}$,

(v) $K_5(x)=\dfrac{1}{2\pi}\left(\dfrac{\sin x/2}{x/2}\right)^2$,

(vi) $K_6(x)=\frac{3}{4}\lambda^{-3}(\lambda^2-x^2) \quad$ if $x^2\leqq\lambda^2$
$\qquad =0 \quad$ if $x^2>\lambda^2$, $\lambda>0$.

We will need the following result which is a variation of a theorem of Bochner (1955) in the sequel.

Theorem 2.2. *Let* $(\mathbb{R}^m, \mathscr{B}_m)$ *be the m-dimensional Euclidean space with the corresponding Borel σ-field and* $(\mathbb{R}, \mathscr{B})$ *be the real line with* $\mathscr{B}=\mathscr{B}_1$. *Let* $K:(\mathbb{R}, \mathscr{B}_m)\to(\mathbb{R}, \mathscr{B})$ *be measurable and such that*

$$|K(\mathbf{z})|\leqq M,\ z\in\mathbb{R}^m;\ \int_{\mathbb{R}^m}|K(\mathbf{z})|d\mathbf{z}<\infty, \tag{18}$$

$$\|\mathbf{z}\|^m|K(\mathbf{z})\to 0 \text{ as } \|\mathbf{z}\|\to\infty, \tag{19}$$

where $\|\cdot\|$ *is the ordinary Euclidean norm. Furthermore let* $g:(\mathbb{R}^m, \mathscr{B}_m)\to(\mathbb{R}, \mathscr{B})$ *be measurable such that*

$$\int|g(\mathbf{z})|d\mathbf{z}<\infty. \tag{20}$$

Define

$$g_n(\mathbf{x})=\frac{1}{h_n^m}\int K\left(\frac{\mathbf{z}}{h_n}\right)g(\mathbf{x}-\mathbf{z})\,d\mathbf{z} \tag{21}$$

where $h_n>0$, $n=1, 2, \ldots$ *and*

$$h_n\to 0 \text{ as } n\to\infty. \tag{22}$$

Then, if g is continuous, we have

$$\lim g_n(\mathbf{x}) = g(\mathbf{x}) \int K(\mathbf{z})\mathrm{d}\mathbf{z} \tag{23}$$

and if g is uniformly continuous, then the convergence in (23) *is uniform.*

Proof. Let us consider

$$\begin{aligned}
&|g_n(\mathbf{x}) - g(\mathbf{x}) \textstyle\int K(\mathbf{z})\mathrm{d}\mathbf{z}| \\
&\leqq \sup\{|g(\mathbf{x}-\mathbf{z}) - g(\mathbf{x})| : \|\mathbf{z}\| \leqq \delta\} \textstyle\int |K(\mathbf{z})|\mathrm{d}\mathbf{z} \\
&\quad + \sup\{\|\mathbf{z}\|^m |K(\mathbf{z})| : \|\mathbf{z}\| > \delta h_n^{-1}\}\{\textstyle\int |g(\mathbf{z})|\mathrm{d}\mathbf{z}\}\delta^{-m} \\
&\quad + |g(\mathbf{x})| \int_{\|\mathbf{z}\| > \delta h_n^{-1}} |K(\mathbf{z})|\mathrm{d}\mathbf{z}
\end{aligned}$$

where $\delta > 0$. Condition (20) together with the uniform continuity assumption about g implies its boundedness. Therefore, for a fixed $\delta > 0$, as $n \to \infty$, the last term above tends to zero uniformly in $\mathbf{x}$. The same is true of the second term by condition (19). Finally, letting $\delta \to 0$; the first term tends to zero uniformly in $\mathbf{x}$ by the uniform continuity of g. This completes the proof. If g is continuous only at $\mathbf{x}$, then (23) follows from the above inequality.

Lemma 2.3 *For $n \geqq 1$, let g, G, g_n, G_n, f and f_n be measurable functions defined on* $\mathbb{R}$ *into itself and such that*

$$g_n(x) \leqq f_n(x) \leqq G_n(x), \ x \in \mathbb{R}$$

and as $n \to \infty$,

$$g_n \to g, \ G_n \to G, \ f_n \to f$$

in the Lebesgue measure on $\mathbb{R}$*. Further suppose that*

$$\int g_n(x)\mathrm{d}x \to \int g(x)\mathrm{d}x < \infty$$

and

$$\int G_n(x)\mathrm{d}x \to \int G(x)\mathrm{d}x < \infty.$$

Then

$$\int f_n(x)\mathrm{d}x \to \int f(x)\mathrm{d}x, \text{ as } n \to \infty.$$

This is a generalized version of the dominated convergence theorem. We refer the reader to Pratt (1960) for a proof of the above lemma. In analogy with density estimators on the real line, given that the process is observed upto time $n+1$, we estimate $p(x)$ by

$$p_n(x)=\frac{1}{nh_n}\sum_{j=1}^{n} K\left(\frac{x-X_j}{h_n}\right) \tag{24}$$

and $q(y)$ by

$$q_n(y)=\frac{1}{nh_n}\sum_{j=1}^{n} K\left(\frac{x-X_j}{h_n^{1/2}}\right) K\left(\frac{x'-X_{j+1}}{h_n^{1/2}}\right) \tag{25}$$

where $y=(x, x')$ and K and h_n satisfy conditions (K1)–(K3) and (H1)–(H2) respectively.

We shall now study some properties of the estimators $p_n(x)$.

Theorem 2.3 *Let the assumptions* (K1), (H1) *and* (P1) *be satisfied. Then*

$$E\int|p_n(x)-p(x)|\mathrm{d}x\to 0 \text{ as } n\to\infty$$

and

$$\int|Ep_n(x)-p(x)|\mathrm{d}x\to 0 \text{ as } n\to\infty. \tag{26}$$

Proof. Fubini's Theorem implies that

$$E\int|p_n(x)-p(x)|\mathrm{d}x=\int E|p_n(x)-p(x)|\mathrm{d}x. \tag{27}$$

But

$$\{E|p_n(x)-p(x)|\}^2\leqq E|p_n(x)-p(x)|^2. \tag{28}$$

Now

$$E(p_n(x)-p(x))^2=\operatorname{var}(p_n(x))+[Ep_n(x)-p(x)]^2. \tag{29}$$

Since $p(x)$ is continuous in x by assumption,

$$Ep_n(x)\to p(x)$$

by Theorem 2.2. Furthermore

$$\begin{aligned}\operatorname{var}(p_n(x))&=n^{-1}h_n^{-2}\operatorname{var}[K((x-X_1)h_n^{-1})]\\&+2(nh_n)^{-2}\sum_{i<j}\operatorname{cov}[K((x-X_i)h_n^{-1}), K((x-X_j)h_n^{-1})].\end{aligned}$$

But

$$h_n^{-1} \operatorname{var}[K((x-X_1)h_n^{-1})]=h_n^{-1}\int K^2((x-z)h_n^{-1})p(z)\mathrm{d}z$$

$$-h_n[h_n^{-1}\int K((x-z)h_n^{-1})p(z)\mathrm{d}z]^2$$

and this tends to

$$p(x)\int K^2(z)\mathrm{d}z$$

by Theorem 2.2 since

$$\int K^2(z)\mathrm{d}z<\infty.$$

As for the covariance term, we have the following. Under Doeblin's condition (Do), Lemma 7.1 of Doob, 1953, p. 222 applies and

$$|\operatorname{cov}[K((x-X_1)h_n^{-1}),\ K((x-X_{j+1})h_n^{-1})]|$$
$$\leqq 2\gamma^{1/2}\rho^{j/2}\{E/K((x-X_1)h_n^{-1})|\}$$

for some γ, ρ such that $\gamma>0$, $0<\rho<1$. Therefore

$$|(nh_n)^{-1}\sum_{i<j}\operatorname{cov}[K((x-X_i)h_n^{-1}),\ K((x-X_j)h_n^{-1})]|$$
$$\leqq (nh_n)^{-1}\sum_{j=1}^{n-1}(n-j)2\gamma^{1/2}\rho^{j/2}E|K((x-X_1)h_n^{-1})|^2$$
$$\leqq (nh_n)^{-1}n\rho^{1/2}(1-\rho^{1/2})^{-1}2\gamma^{1/2}E|K((x-X_1)h_n^{-1})|^2$$
$$=2\gamma^{1/2}\rho^{1/2}(1-\rho^{1/2})^{-1}h_n^{-1}E|K((x-X_1)h_n^{-1})|^2$$

and the last expression converges as $n\to\infty$ to

$$2\gamma^{1/2}\rho^{1/2}(1-\rho^{1/2})^{-1}p(x)\int K^2(z)\mathrm{d}z$$

by Theorem 2.2. Since $nh_n\to\infty$ as $n\to\infty$ it follows that

$$\operatorname{var}(p_n(x))\to 0.$$

Hence

$$E|p_n(x)-p(x)|\to 0 \tag{30}$$

for every x by relation (29). Furthermore

$$0\leqq E|p_n(x)-p(x)|\leqq Ep_n(x)+p(x)$$

and

$$Ep_n(x)+p(x)\to 2p(x) \text{ as } n\to\infty$$

by (30). In addition

$$\int[E(p_n(x))+p(x)]\mathrm{d}x=\int[h_n^{-1}\,EK((x-X_1)h_n^{-1})]\mathrm{d}x+1$$
$$=E[h_n^{-1}\int K((x-X_1)h_n^{-1})\mathrm{d}x]+1=2.$$

Hence, by Lemma 2.3, it follows that

$$E\int|p_n(x)-p(x)|\mathrm{d}x\to 0.$$

Again

$$\int Ep_n(x)\mathrm{d}x=\int[h_n^{-1}EK((x-X_1)h_n^{-1})]\mathrm{d}x$$
$$=E[h_n^{-1}\int K((x-X_1)h_n^{-1})\mathrm{d}x]=1.$$

But

$$Ep_n(x)\to p(x).$$

Hence, by Scheffe's Theorem (cf. Rao, 1973, p. 124), we get the desired result.

Another set of interesting properties of the estimator $p_n(x)$ is given in the following theorem.

Theorem 2.4 *Let assumptions* (K1)–(K3), (H2) *and* (P2) *be satisfied. Then we have*

(i) $\sup\{Ep_n(x)-p(x):x\in\mathbb{R}\}\to 0,$ (31)

(ii) $E[\sup\{|p_n(x)-p(x)|^2:x\in\mathbb{R}\}]\to 0,$ (32)

(iii) $\sup\{|p_n(x)-p(x)|:x\in\mathbb{R}\}\xrightarrow{p}0.$ (33)

Proof. (i) is a consequence of Theorem 2.2 since p is uniformly continuous by hypothesis. Now

$$E^{1/2}[\sup\{|p_n(x)-p(x)|^2:x\in\mathbb{R}\}]$$
$$\le E^{1/2}[\sup\{|p_n(x)-Ep_n(x)|^2:x\in\mathbb{R}\}]+ \tag{34}$$
$$[\sup\{|Ep_n(x)-p(x)|:x\in\mathbb{R}\}]$$

by the Minkowski inequality and by (31), it is sufficient to prove that

$$E^{1/2}[\sup\{|p_n(x)-Ep_n(x)|^2 : x\in \mathbb{R}\}]\to 0. \tag{35}$$

But the left-hand side of (35) is bounded by

$$(n^{-1/2}h_n)^{-1}\int|\hat{K}(u)|du$$

where $\hat{K}(u)=\int e^{iux}K(x)dx$. Assumption $(K3)$ and $(H2)$ complete the proof of (35) and hence that of (32). (33) is an immediate consequence of (32) by Tchebyshev's inequality.

We shall now prove that the estimators are asymptotically normal under some further restrictions on the process.

Before we study the asymptotic normality, we shall state a central limit theorem, the proof of which is given in Roussas (1969).

Theorem 2.5. *Suppose X_n is a Markov process satisfying Doeblin's condition. Set*

$$Y_j=(X_j,\ \dots,\ X_{j+r-1}),\ j\geqq 1$$

where r is an integer greater than or equal to 1. Let $0<k_n\to\infty$ as $n\to\infty$ and $\{L_n\}$ be a sequence of uniformly bounded real-valued functions on $(\mathbb{R}^r,\ \mathscr{B}_r)$ such that

(i) $E|L_n(Y_1)|^2=0(k_n n^{-1})$.
(ii) $E|f_n(Y_1)f_n(Y_j)|=0(k_n^2 n^{-2})$ *uniformly in j, $1\leqq j\leqq n$,*
(iii) $E|f_n(Y_1)f_n(Y_i)f_n(Y_j)|=0(k_n^3 n^{-3})$ *uniformly in i and j, $1\leqq i<j\leqq n$, $n=2$, $3,\dots$*
(iv) $k_n^{-1}n \operatorname{var} L_n(Y_1)\to\sigma^2<\infty$ *as $n\to\infty$ $\sigma^2>0$*

where $$f_n(Y_j)=L_n(Y_j)-E\,L_n(Y_j).$$

Further suppose that there exist sequences $\{\alpha_n\}$, $\{\beta_n\}$, $\{\mu_n\}$ tending to infinity such that

$$\beta_n\mu_n\alpha_n^{-1}\to 0 \quad \textit{and} \quad \alpha_n k_n n^{-1}\to 0$$

as $n\to\infty$. Then

$$k_n^{-1/2}\sum_{j=1}^{n}[L_n(Y_j)-E(L_n(Y_j))]\overset{\mathscr{L}}{\to}N(0,\ \sigma^2).$$

In addition to the assumptions stated in $(K1)$–$(K2)$, $(H1)$ and $(P1)$, suppose that the joint densities of X_1, X_i and X_1, X_i and X_j are bounded by a constant for all i, j such that $1<i\leqq n$ and $1<i<j\leqq n$, $n=2,\ 3,\dots$ Let

$k_n = nh_n$ and set $r=1$ and

$$L_n(z) = K((x-z)h_n^{-1})$$

in the above theorem. Suppose there exist sequences $\{\alpha_n\}$, $\{\beta_n\}$, $\{\mu_n\}$ of integers tending to infinity as $n \to \infty$ such that

$$\beta_n \mu_n \alpha_n^{-1} \to 0 \text{ and } \alpha_n \mathrm{h}_n \to 0$$

as $n \to \infty$. We have the following theorem from Theorem 2.5.

Theorem 2.6. *For any x,*

$$(nh_n)^{1/2}(p_n(x) - E(p_n(x)) \xrightarrow{\mathscr{L}} N(0, \sigma^2(x))$$

where

$$\sigma^2(x) = p(x) \int K^2(z) \mathrm{d}z.$$

Asymptotic properties of the estimator $q_n(y)$ of $q(y)$ can be derived in an analogous manner. Let

$$F_n(x) = \int_{-\infty}^{x} p_n(z) \mathrm{d}z$$

where $p_n(x)$ is as defined above. Roussas (1969) discussed asymptotic properties of the estimator $F_n(x)$ of the initial distribution function. He proved that

$$EF_n(x) \to F(x) \qquad \text{for all } x \in \mathbb{R},$$

$$F_n(x) \to F(x) \qquad \text{in quadratic mean,}$$

and

$$\sup_{x \in \mathbb{R}} \{E|F_n(x) - F(x)|\} \to 0$$

as $n \to \infty$ provided ($K1$), ($H1$) and ($P1$) are satisfied.

Since the estimator $p_n(x)$ depends on the choice of K, a natural question is how to choose K. Suppose we are in the i.i.d. case. It can be shown that, as $n \to \infty$,

$$nh_n E|p_n(x) - p(x)|^2 \to p(x) \int K^2(z) \mathrm{d}z +$$
$$+ \tfrac{1}{4} c[p''(x) \int z^2 K(z) \mathrm{d}z]^2$$

provided $\{h_n\}$ is chosen so that $nh_n^5 \to$a finite limit c under some regularity conditions. Suppose we are interested in choosing K so that the mean integrated squared error

$$J_n = E\int |p_n(x) - p(x)|^2 dx$$

is minimum. It is easy to see that, under appropriate conditions,

$$nh_n J_n \to \int K^2(z)dz + \tfrac{1}{4}c\left(\int (p''(x))^2 dx\right)\left(\int z^2 K(z)dz\right)^2. \tag{36}$$

Since the variance of the limiting normal distributions is $p(x)\int K^2(z)dz$ in the i.i.d. case, an optimal choice of K is the one which minimizes $\int K^2(z)dz$ and $\int z^2 K(z)dz$.

Consider the class of densities which are square integrable and have finite second moments. Without loss of generality, let us assume that their first moments are zero. Within this class, an optimal choice of K can be shown to be

$$K(x) = \begin{cases} B(A^2 - x^2) & \text{if } x^2 \leqq A^2 \\ 0 & \text{if } x^2 > A^2 \end{cases}$$

where A and B are determined by the requirements that K is a density and

$$\int z^2 K(z)dz = m^2.$$

By setting $A = \lambda$, we see that

$$K(x) = \begin{cases} \frac{3}{4}\lambda^{-3}(\lambda^2 - x^2) & \text{if } x^2 \leqq \lambda^2 \\ 0 & \text{if } x^2 > \lambda^2 \end{cases} \tag{37}$$

and

$$\int_{-\lambda}^{\lambda} K^2(x)dx = \frac{3}{5\lambda}, \quad \int_{-\lambda}^{\lambda} x^2 K(x)dx = \frac{1}{5}\lambda^2. \tag{38}$$

By suitable choice of λ, asymptotic variance can be made as small as one pleases. Note that if we fix the second moment of K, K is completely determined.

Theorem 2.6 proves the asymptotic normality of the estimator. Prakasa Rao (1977) obtained the rate of convergence. It was shown that, under some regularity conditions,

$$\sup_{y \in \mathbb{R}} \left| P\left(\frac{p_n(x) - p(x)}{\operatorname{var}(p_n(x))^{1/2}} \leqq y\right) - \phi(y)\right| = 0(n^{1/3-\tau}) \tag{39}$$

for $0<\tau<\frac{1}{3}$, provided $h_n=n^{-1/3}$, uniformly for x over compact sets. Similar results in the i.i.d. case were given by Wertz (1971). Here $\phi(\cdot)$ is the standard normal distribution function.

2.3 Density Estimation using Delta-sequences for Stationary Markov Processes

Let us again consider a strictly stationary Markov process $\{X_n, n\geqq 1\}$ defined on a probability space $(\Omega, \mathscr{F}, P)$ and taking values in $(\mathbb{R}, \mathscr{B})$ where $\mathbb{R}$ is the real line and $\mathscr{B}$ is the σ-algebra of Borel sets of $\mathbb{R}$. Let p be the one-dimensional marginal density of the process with respect to Lebesgue measure on the real line. We shall assume that the process satisfies Doeblin's condition as stated in the previous section with constants $r\geqq 1$ and $0<\rho<1$.

A sequence $\{\delta_m(x)\}$ of functions in $L_\infty(\mathbb{R})$ is called a *delta-sequence of positive type* $\alpha>0$ if

(i) $\delta_m(x)\geqq 0,\ m\geqq 1,\ x\in\mathbb{R}$,

(ii) there exist constants $A>0$, $B>0$ such that

$$|1-\int_{-A}^{B}\delta_m(x)dx|=0(m^{-\alpha}), \tag{40}$$

(iii) $\sup\{|\delta_m(x)\}:|x|\geqq m^{-\alpha}\}=0(m^{-\alpha})$, (41)

(iv) ess $\sup\{|\delta_m(x)\}:x\in\mathbb{R}\}\sim m$. (42)

An example of a delta-sequence of type 1 is

$$\delta_m(x)=m\chi_{[0,\frac{1}{m}]}(x),\ m=1, 2, \ldots$$

when χ_E denotes the indicator function of the set E. Let $K(x)$ be any bounded probability density function such that $K(x)=0(|x|^{-1-\beta})$ as $|x|\to\infty$ for some $\beta>0$. It is not difficult to check that the sequence

$$\delta_m(x)=mK(mx)$$

is a delta-sequence of type $\beta/(\beta+2)$.

Suppose that we have observed the process $\{X_n, n\geqq 1\}$ up to time n. Let $\{\delta_m\}$ be a delta-sequence of postive type α. Define

$$p_{nm}(x)=\frac{1}{n}\sum_{i=1}^{n}\delta_m(x-X_i),\ n,\ m=1, 2, \ldots. \tag{43}$$

We choose $p_{nm}(x)$ as an estimator for $p(x)$. Since $p_{nm}(x)$ are random variables with finite variance, mean square error can be used as an indicator of the rate of convergence of these estimators. In fact

$$E(p_{nm}(x)-p(x))^2 = \frac{1}{n} E(\delta_m^2(x-X_1))$$
$$+\frac{2}{n^2}\sum_{j=1}^{n}\sum_{i=1}^{j-1} E[\delta_m(x-X_i)\delta_m(x-X_j)]$$
$$+p^2(x)-2p(x)E[\delta_m(x-X_1)]$$

by stationarity of the process. Since

$$|\mathrm{cov}(\delta_m(x-X_i),\ \delta_m(x-X_j))| \leqq 2r^{1/2}\rho^{(j-i)/2}E[\delta_m(x-X_1)]^2$$

by Doob (1953), p. 222, it follows that

$$E(p_{nm}(x)-p(x))^2$$
$$\leqq \frac{1}{n} E(\delta_m(x-X_1))^2 + \{E(\delta_m(x-X_1))-p(x)\}^2$$
$$+4\frac{r^{1/2}\rho^{1/2}}{n(1-\rho^{1/2})}E(\delta_m(x-X_1))^2 \tag{44}$$

by simple computation. As a consequence of the above estimate, we have the following theorem which gives a uniform bound for the mean square error of the estimator $p_{nm}(x)$ of $p(x)$.

Theorem 2.7. *Let $\{\delta_m(x)\}$ be a delta-sequence of positive type α. If $p(x)$ satisfies the Lipschitz condition of order λ, $0<\lambda\leqq 1$ i.e.*

$$|p(x)-p(y)|\leq c|x-y|^{\lambda}$$

for some constant $c>0$, then

$$\sup_{-\infty<x<\infty} E(p_{nm}(x)-p(x))^2 \leqq c_0 mn^{-1} + c_1{}^{m-2\alpha\lambda} \tag{45}$$

where c_0 and c_1 are constants independent of x and n.

Proof. In view of (44), it is enough to estimate

$E(\delta_m(x-X_1))^2$ and $\{E(\delta_m(x-X_1))-p(x)\}$. Let us now consider

$$E(\delta_m(x-X_1)-p(x)) = \int_{-\infty}^{\infty} \delta_m(x-t)p(t)\mathrm{d}t - p(x)$$
$$= \int_{-\infty}^{\infty} \delta_m(t)p(x-t)\mathrm{d}t - p(x)$$

$$= \int_{-m^{-\alpha}}^{m} \delta_m(t)[p(x-t)-p(x)]\mathrm{d}t + p(x)\left\{\int_{-m^{-\alpha}}^{m^{-\alpha}} \delta_m(t)\mathrm{d}t - 1\right\}$$

$$+ \int_{-\infty}^{-m^{-\alpha}} \delta_m(t)p(x-t)\mathrm{d}t + \int_{m^{-\alpha}}^{\infty} \delta_m(t)p(x-t)\mathrm{d}t.$$

Choose m large so that $m^{-\alpha} < \min(A, B)$. The sum of the last two terms is $0(m^{-\alpha})$ since p integrates to one and relation (41) holds. Since p satisfies the Lipschitz condition, p is uniformly bounded on $\mathbb{R}$. Furthermore, the second term is bounded by

$$\left|\int_{-A}^{B} \delta_m(t)\mathrm{d}t - 1\right| + \left|\int_{m^{-\alpha}}^{B} \delta_m(t)\mathrm{d}t + \int_{-A}^{-m^{-\alpha}} \delta_m(t)\mathrm{d}t\right|$$

which in turn is bounded by

$$c'm^{-\alpha} + (B+A)c''m^{-\alpha}$$

for large m for some constants c' and c''. Since p is Lipschitz of order λ,

$$\left|\int_{-m^{-\alpha}}^{m^{-\alpha}} \delta_m(t)[p(x-t)-p(x)]\mathrm{d}t\right| \leqq c'''m^{-\alpha\lambda}$$

for some constant c'''. Combining these bounds, we obtain (45).

In particular, if we consider estimators

$$p_n^*(x) = p_{n,m_n}(x) = \frac{1}{n}\sum_{i=1}^{n} \delta_{n,m_n}(x - X_i)$$

where $m_n = [n^{(1+2\alpha\lambda)^{-1}}]$, then

$$\sup_{-\infty < x < \infty} E(p_n^*(x) - p(x))^2 = 0(n^{-1+(1+2\alpha\lambda)^{-1}}).$$

The advantage of the method of density estimation described in this section is that it unifies several methods, like the kernel type of density estimation described in Section 2.2 and the density estimation by orthogonal series which will be discussed in Section 2.4. Large sample properties of the estimators $p_{nm}(x)$ have not been studied as yet even in the i.i.d. case.

2.4 Density Estimation by the Orthogonal Series Method for Stationary ϕ-mixing Processes

We now propose to discuss a different type of method for density estimation. This method has been extensively discussed in the i.i.d. case.

Let $\{\xi_n, n \geqq 0\}$ be a sequence of random variables defined on a probability space $(\Omega, \mathscr{A}, p)$ with values in a measurable space $(E, \mathscr{B})$. Suppose that $\{\xi_n\}$ is a strictly stationary process and that it is φ-mixing.

Let $\mathscr{M}_0^t$ and $\mathscr{M}_{t+n}^{\infty}$ represent the σ-fields generated by $\xi_0, \xi_1, \ldots, \xi_t$ and $\xi_{t+n}, \xi_{t+n+1}, \ldots$ respectively. We shall define $\varphi(0)=1$.

We shall first state two lemmas which will be used in the sequal. Proofs of these lemmas can be found in Billingsley (1968).

Lemma 2.4. *Suppose X and Y are random variables measurable with respect to σ-algebras $\mathscr{M}_0^t$ and $\mathscr{M}_{t+k}^{\infty}(k \geqq 0)$ respectively. Further suppose that*

$$E\{|X|^r\}<\infty, \ E\{|Y|^s\}<\infty$$

for some $r>1$, $s>1$, $\frac{1}{r}+\frac{1}{s}=1$. Then

$$|E(XY)-E(X)E(Y)| \leqq 2(\varphi(k))^{1/r}(E|X|^r)^{1/r}(E|Y|^s)^{1/s}. \tag{46}$$

In particular

(i) *if X and Y are square integrable, then*

$$|E(XY)-E(X)E(Y)| \leqq 2\varphi^{1/2}(k)E^{1/2}(X^2)E^{1/2}(Y^2); \tag{47}$$

(ii) if $E(|X|)<\infty$ *and* $P(|Y| \leqq C)=1$ *where C is a constant, then*

$$|E(XY)-E(X)E(Y)| \leqq 2C\varphi(k)E(|X|); \tag{48}$$

(iii) *if $|X| \leqq C_1$ and $|Y| \leqq C_2$, then*

$$|E(XY)-E(X)E(Y)| \leqq 2C_1C_2\varphi(k). \tag{49}$$

Lemma 2.5. *Suppose that $E(\xi_0)=0$ and $|\xi_0|<C$ with*

$$\sum_{n=0}^{\infty} \varphi^{1/2}(n)<\infty. \tag{50}$$

Then

$$E(S_n^4) \leqq 768 \cdot C^4 \left\{ \sum_{i=0}^{\infty} \varphi^{1/2}(i) \right\}^2 n^2 \tag{51}$$

where

$$S_n=\sum_{i=0}^{n} \xi_i.$$

Suppose that the common distribution P_0 of ξ_n admits *a* Radon–Nikodym derivative $dP_0/d\mu$ with respect to a σ-finite measure μ on $(E, \mathscr{B})$ and $dP_0/d\mu \in L^2(\mu)$ where $L^2(\mu)$ is the space of μ-squared integrable functions on E. We suppose that $L^2(\mu)$ is separable. Let $\{e_i, i \geqq 1\}$ be an orthonormal basis of $L^2(\mu)$. Let

$$\frac{dP_0}{d\mu} = \sum_{i=1}^{\infty} a_i e_i \tag{52}$$

where

$$a_i = \int e_i \frac{dP_0}{d\mu} d\mu \tag{53}$$

where equality in (52) is in the sense of convergence in $L^2(\mu)$. Let $\{e_i^0, i \geqq 1\}$ be an orthonormal fixed basis of $L^2(\mu)$ and for any version $\{e_i^*, i \geqq 1\}$ of $\{e_i, i \geqq 1\}$, define

$$\hat{f}_n^*(x) = \sum_{i=0}^{q(n)} \hat{a}_{in} e_i^*(x), \; x \in E \tag{54}$$

where

$$\hat{a}_{in} = \frac{1}{n} \sum_{j=0}^{n-1} e_i^0(\xi_j), \; 0 \leqq i \leqq q(n) \tag{55}$$

where $\{q(n)\}$ is a sequence of integers tending to infinity as $n \to \infty$. $\hat{f}_n^*$ is a version of estimator $\hat{f}_n$ of $dP_0/d\mu$ based on the process $\{\xi_0, \ldots, \xi_{n-1}\}$.

We shall now discuss some asymptotic properties of estimators as regards mean integrated sequared error and asymptotic normality etc.

Theorem 2.8. *If $\Sigma_{i=0}^{\infty} \varphi^{1/2}(i) < \infty$, then the condition*

$$\text{(B)} \qquad \lim_{n \to \infty} \int_E \frac{1}{n} \left(\sum_{i=0}^{q(n)} e_i^2 \right) \frac{dP}{d\mu} = 0 \tag{56}$$

is sufficient for $\hat{f}_n$ to converge to f in mean integrated squared error (MISE) i.e.

$$\lim_{n \to \infty} E\left(\int_E \left(\hat{f}_n - \frac{dP_0}{d\mu} \right)^2 d\mu \right) = 0. \tag{57}$$

Proof. Note that

$$\int_E \left(\hat{f}_n - \frac{dP_0}{d\mu} \right)^2 d\mu = \int_E \left\{ \sum_{i=0}^{q(n)} (\hat{a}_{in} - a_i) e_i^*(x) + \sum_{i > q(n)} a_i e_i^*(x) \right\}^2 d\mu(x)$$

$$= \sum_{i=0}^{q(n)} (\hat{a}_{in} - a_i)^2 + \sum_{i > q(n)} a_i^2 \tag{58}$$

since $\{e_i^*\}$ is an orthonormal system. The last term on the right-hand side of (58) tends to zero as $n \to \infty$ since

$$\int_E f^2 \mathrm{d}\mu = \sum_{i=0}^{\infty} a_i^2 < \infty$$

and $q(n) \to \infty$. In order to prove (57) it is sufficient to show that

$$\sum_{i=0}^{q(n)} E(\hat{a}_{in} - a_i)_i^2 \to 0 \tag{59}$$

as $n \to \infty$. But

$$\hat{a}_{in} - a_i = \left(\sum_{j=0}^{n-1} \frac{1}{n} e_i^0(\xi_j) - a_i \right) = \frac{1}{n} \sum_{j=0}^{n-1} (e_i^0(\xi_j) - a_i) \tag{60}$$

and

$$E(e_1^0(\xi_j)) = \int_E e_i(x) \frac{\mathrm{d}P_0}{\mathrm{d}\mu}(x) \mathrm{d}\mu(x) = a_i, \tag{61}$$

and hence

$$\begin{aligned} E(\hat{a}_{in} - a_i)^2 &= E\left\{ \frac{1}{n} \sum_{j=0}^{n-1} (e_i(\xi_j) - a_i) \right\}^2 \\ &= \frac{1}{n^2} \sum_{j=0}^{n-1} \operatorname{var}(e_i(\xi_j)) + \frac{1}{n^2} \sum_{j \neq j'} \operatorname{cov}(e_i(\xi_j), e_i(\xi_{j'})). \end{aligned} \tag{62}$$

This implies in turn that

$$\begin{aligned} \sum_{i=0}^{q(n)} E(\hat{a}_{in} - a_i)^2 = \frac{1}{n^2} \sum_{j=0}^{n-1} \left(\sum_{i=0}^{q(n)} \operatorname{var}(e_i(\xi_j)) \right) \\ + \frac{1}{n^2} \sum_{j \neq j'} \left(\sum_{i=0}^{q(n)} \operatorname{cov}(e_i(\xi_j), e_i(\xi_{j'})) \right). \end{aligned} \tag{63}$$

Since the ξ_i have the same distribution, it is easy to see that the first term on the right-hand side of (63) is given by

$$\frac{1}{n} \sum_{i=0}^{q(n)} E(e_i^2(\xi_0)) - \frac{\sum_{i=0}^{q(n)} a_i^2}{n}. \tag{64}$$

Since $\mathrm{d}P_0/\mathrm{d}\mu \in L^2(\mu)$,

$$\frac{\sum_{i=0}^{q(n)} a_i^2}{n} \leqq \frac{\int_E \left(\frac{\mathrm{d}P_0}{\mathrm{d}\mu} \right)^2 \mathrm{d}\mu}{n} \leqq \frac{K}{n}$$

and hence

$$\frac{\sum_{i=0}^{q(n)} a_i^2}{n} \to 0 \text{ as } n \to \infty.$$

Let

$$J = \sum_{j \neq j'} \left(\sum_{i=0}^{q(n)} \operatorname{cov}(e_i(\xi_j), e_i(\xi_{j'})) \right). \tag{64}$$

It now follows that

$$\begin{aligned}
|J| &\leqq 2 \sum_{i=0}^{q(n)} \left(\sum_{j=0}^{n-1} \sum_{j'=j+1}^{n-1} |\operatorname{cov}(e_i(\xi_j), e_i(\xi_{j'}))| \right) \\
&\leqq 4 \sum_{i=0}^{q(n)} \left(\sum_{j=0}^{n-1} \sum_{j'=j+1}^{n-1} \varphi^{1/2}(j'-j) E^{1/2}(e_i^2(\xi_j)) E^{1/2}(e_i^2(\xi_{j'})) \right) \\
&\qquad \text{(by Lemma 2.4)} \\
&= 4 \sum_{i=0}^{q(n)} \left(\sum_{j=0}^{n-1} \sum_{j'=j+1}^{n-1} \varphi^{1/2}(j'-j) E(e_i^2(\xi_0)) \right) \qquad (65) \\
&\qquad \text{(by stationarity of } \{\xi_n\}) \\
&\leqq 4 \sum_{i=0}^{q(n)} \left(\sum_{j=0}^{n-1} \left(\sum_{k=1}^{\infty} \varphi^{1/2}(k) E(e_i^2(\xi_0)) \right) \right) \\
&\leqq 4n \left\{ \sum_{k=1}^{\infty} \varphi^{1/2}(k) \right\} \sum_{i=0}^{q(n)} E(e_i^2(\xi_0)).
\end{aligned}$$

In view of the fact that $\Sigma_{k=1}^{\infty} \varphi^{1/2}(k) < \infty$, (59) holds provided

$$\lim_{n \to \infty} \frac{1}{n} \sum_{i=0}^{q(n)} E(e_i^2(\xi_0)) = 0$$

by earlier remarks. But

$$\begin{aligned}
\frac{1}{n} \sum_{i=0}^{q(n)} E(e_i^2(\xi_0)) &= \frac{1}{n} \sum_{i=0}^{q(n)} \int_E e_i^2 \frac{dP_0}{d\mu} d\mu \\
&= \int_E \frac{1}{n} \left(\sum_{i=0}^{q(n)} e_i^2 \right) \frac{dP_0}{d\mu} d\mu
\end{aligned}$$

and the last term tends to zero by the hypothesis.

In addition to the hypothesis stated earlier, suppose $(E, \mathscr{B})$ is a topological space with $\mathscr{B}$ as the σ-field of Borel sets of E, $dP_0/d\mu$ admits a unique continuous version denoted by f. Further suppose that there exists a continuous version $\{e_i^*\}$ of $\{e_i\}$. Let

$$a_i = \int f\, e_i^* \, \mathrm{d}\mu$$

and

$$f_n = \sum_{i=0}^{q(n)} a_i\, e_i^*$$

as before. We shall now prove the following theorem which gives almost sure uniform convergence of f_n to f.

Theorem 2.9. *In addition to the assumptions stated above, suppose that*

$$\text{(i)} \quad \sum_{i=0}^{\infty} \varphi^{1/2}(\mathrm{i}) < \infty, \tag{66}$$

$$\text{(ii)} \quad M = \sup_{i \geqq 1} \sup_{x \in E} |e_i^*(x)| < \infty, \tag{67}$$

(iii) *the series* $\sum_{i=0}^{\infty} e_i^*$ *converges uniformly to* f, *and*

$$\text{(iv)} \quad \sum_{n=1}^{\infty} n^{-2}\, q(n)^5 < \infty. \tag{68}$$

Then

$$\sup_{x \in E} |\hat{f}_n^*(x) - f(x)| \to 0 \text{ a.s. } as\ n \to \infty.$$

Proof. From the definitions of $\hat{f}_n^*$ and f, it follows that

$$\sup_{x \in E} |\hat{f}_n^*(x) - f(x)| \leqq \sup_{x \in E} \Big| \sum_{i=0}^{q(n)} (\hat{a}_{in} - a_i) e_i^*(x)\Big| + \sup_{x \in E} \Big| \sum_{i > q(n)} a_i e_i^*(x)\Big|$$

and the last term tends to zero in view of (iii). As a consequence of (67), we have

$$\sup_{x \in E} \Big|\sum_{i=0}^{q(n)} (\hat{a}_{in} - a_i)\, e_i^*(x)\Big| \leqq M \sum_{i=0}^{q(n)} |\hat{a}_{in} - a_i|$$

and hence, for any $\varepsilon > 0$,

$$P\left\{\sup_{x \in E} \Big| \sum_{i=0}^{q(n)} (\hat{a}_{in} - a_i) e_i^*(x)\Big| > \varepsilon\right\}$$

$$\leqq P\left\{\sum_{i=0}^{q(n)} |\hat{a}_{in}-a_i|>\frac{\varepsilon}{M}\right\}$$

$$\leqq \sum_{i=0}^{q(n)} P\left\{|\hat{a}_{in}-a_i|>\frac{\varepsilon}{M(q(n)+1)}\right\}$$

$$\leqq \sum_{i=0}^{q(n)} \frac{E|\hat{a}_{in}-a_i|^4}{\varepsilon^4 M^4(q(n)+1)^4}. \tag{69}$$

Note that

$$\hat{a}_{in}-a_i=\sum_{j=0}^{n-1}\frac{(e_i(\xi_j)-a_i)}{n}$$

and

$$\left|\frac{e_i(\xi_0)-a_i}{n}\right|\leqq\frac{2M}{n}$$

since

$$|a_i|\leqq\left|\int e_i^* f\,\mathrm{d}\mu\right|\leqq M\int f\,\mathrm{d}\mu\leqq M$$

and hence

$$E|\hat{a}_{in}-a_i|^4\leqq 768\left(\frac{2M}{n}\right)^4\left\{\sum_{i=0}^{\infty}\varphi^{1/2}(i)\right\}n^2$$

by Lemma 2.5. Therefore,

$$P\left\{|\hat{a}_{in}-a_i|>\frac{\varepsilon}{M(q(n)+1)}\right\}=0\left(n^2\left(\frac{2M}{n}\right)^4\right)|\varepsilon^4(1+q(n))^4$$

$$=0\left(\frac{q(n)^4}{n^2}\right). \tag{70}$$

Relations (69) and (70) prove that

$$P\left(\sup_{x\in E}\left|\sum_{i=0}^{q(n)}(\hat{a}_{in}-a_i)e_i^*(x)\right|>\varepsilon\right)=0\left(\frac{q(n)^5}{n^2}\right). \tag{71}$$

An application of the Borel–Cantelli Lemma gives the desired result in view of condition (iv).

We now briefly discuss asymptotic normality of the estimator $\hat{f}_n^*(x)$. Delecroix (1975) has proved that

$$S_n(x)\equiv\frac{\hat{f}_n^*(x)-f_n(x)}{\operatorname{var}(f_n^*(x))^{1/2}}\xrightarrow{\mathscr{L}}N(0,\ 1) \tag{72}$$

provided $n^{-1}\, q(n)^4 \to 0$ as $n \to \infty$ under some regularity conditions where

$$\hat{f}_n^*(x) = \sum_{i=0}^{q(n)} \hat{a}_{in}\, e_i^*(x)$$

and

$$f_n(x) = \sum_{i=0}^{q(n)} a_i e_i^*(x).$$

He uses central limit theorems for mixing sequence due to Phillip (1969a). He has further obtained rates of convergence by using some theorems also due to Phillip (1969b). We shall state the result.

Theorem 2.10. *Let F_n be the distribution function of $S_n(x)$ defined above. Suppose that*

(i) $\sum_{j=1}^{\infty} \varphi^{1/2}(j) < 1/4,$

(ii) $f_n(x) \to f(x)$ *for all* $x \in \mathbb{R}^p$,

(iii) $E\left(\sum_{i=0}^{q(n)} e_i^*(\xi_0) e_i^*(x)\right)^2 \geqq Bq(n)$, $B > 0$ for n large, and

(vi) $n^{-1} q(n)^6 \to 0$ *as* $n \to \infty$.

Then

$$\sup_{u \in \mathbb{R}} |F_n(u) - \Phi(u)| = 0\left[\left(\frac{q(n)^6}{n}\right)^{1/12}\right] + 0(n^{3/4}\varphi(n^{1/4}))$$

where $\Phi(u)$ is the standard normal distribution function.

2.5 Robustness of Hodges–Lehmann Estimators for Stationary ϕ-mixing Processes

Let $X_1, X_2, \ldots$ be a sequence of i.i.d. random variables with a common distribution F with location parameter θ. Hodges and Lehmann (1963) proposed a class of estimators $\hat{\theta}_n$ for estimation of θ based on signed rank statistics of $\{X_i - \theta, 1 \leqq i \leqq n\}$. They proved that their estimators are asymptotically normal. In this section, we shall study the behaviour of these estimators when the observations are from a stationary ϕ-mixing process.

Let $\{X_n, n \geqq 1\}$ be a strictly stationary ϕ-mixing process such that the joint distribution of $(X_1, X_2, \ldots, X_n)$ is continuous for every n. Suppose the one-dimensional marginal distribution F of X_1 is symmetric and absolutely continuous with bounded uniformly continuous density f. Let θ be the location parameter of F. Since the problem is to obtain asymptotic distribution of the Hodges–Lehmann (H–L) estimator of θ and since signed rank statistics are invariant under location change, we shall assume that the true location parameter is zero and that $\theta \in [-a, a]$. Define

$$\begin{aligned} H(x) &= F(x) - F(-x) = 2F(x) - 1 \quad & 0 \leqq x < \infty \\ &= 0 & -\infty < x < 0, \end{aligned} \tag{73}$$

$$\mu_1(t, \theta) = F(H^{-1}(t) + \theta) - F(\theta) \qquad 0 \leqq t \leqq 1, \tag{74}$$

$$\mu_2(t, \theta) = F(\theta) - F(-H^{-1}(t) + \theta) \qquad 0 \leqq t \leqq 1, \tag{75}$$

and

$$\mu(t, \theta) = \mu_1(t, \theta) - \mu_2(t, \theta) \qquad 0 \leqq t \leqq 1. \tag{76}$$

(Here $G^{-1}(t) = \inf\{x : G(x) > t\}$ for any function G.) Let $R_i(\theta)$ be the rank of $|X_i - \theta|$ in $\{|X_i - \theta|, 1 \leqq i \leqq n\}$ for any fixed θ. For any bounded non-decreasing score function ψ on $[0, 1]$ such that $\psi(1/2) = 0$, define

$$S_n(J, \theta) = n^{-1} \sum_{i=1}^{n} J\left(\frac{R_i(\theta)}{n}\right) s(X_i - \theta) \tag{77}$$

where $J(u) = \psi(u + 1)/2)$, $0 \leqq u \leqq 1$, $s(x) = I(x \geqq 0) - I(x \leqq 0)$ and $I(A)$ denotes the indicator function of set A. $S_n(J, \theta)$ is called the signed rank statistic based on the score function ψ. $S_n(J, \theta)$ can also be written in the form of a stochastic integral namely

$$S_n(J, \theta) = \int_0^1 J(t) dS_n(t, \theta) \tag{78}$$

where

$$S_n(t, \theta) = n^{-1} \sum_{i=1}^{n} I(R_i(\theta) \leqq nt) s(X_i - \theta),\ 0 \leqq t \leqq 1. \tag{79}$$

Since ψ is non-decreasing and $\psi(1/2) = 0$, it follows that $J(\cdot) \geqq 0$ and non-decreasing and hence $S_n(J, \theta)$ is a non-increasing function of θ. Let $\hat{\theta}_n$ be the average of the infimum of θ for which $S_n(J, \theta) \geqq 0$ and the supremum of θ for which $S_n(J, \theta) \leqq 0$. We shall call $\hat{\theta}_n$ a Hodges–Lehmann estimator of θ. If there exists $\hat{\theta}_n$ such that $S_n(J, \hat{\theta}_n) = 0$, clearly $\hat{\theta}_n$ is an H-L estimator.

We shall now introduce some additional notation. Let

$$\mu(\theta) = \int_0^1 J(t)\ d[\mu_1(t, \theta) - \mu_2(t, \theta)], \tag{80}$$

$$T_n(t, \theta) = n^{1/2}(S_n(t, \theta) - \mu(t, \theta)), \tag{81}$$

and

$$T_n(J, \theta) = n^{1/2}(S_n(J, \theta) - \mu(\theta)) \tag{82}$$

For $|\theta| \leqq an^{-1/2}$, $0 \leqq t \leqq 1$, define

$$W_1(t, \theta) = n^{-1/2} \sum_{i=1}^{n} [I(0 < X_i - \theta \leqq H^{-1}(t)) - \mu(t, \theta)]. \tag{83}$$

$$W_2(t, \theta) = n^{-1/2} \sum_{i=1}^{n} [I(-H^{-1}(t) < X_i - \theta \leqq 0) - \mu_2(t, \theta)], \tag{84}$$

$$V(t, \theta) = W_1(t, \theta) + W_2(t, \theta) \tag{85}$$

and

$$Q(t, \theta) = W_1(t, \theta) - W_2(t, \theta). \tag{86}$$

Let

$$H_{n\theta}(x) = F_n(x + \theta n^{-1/2}) - F_n(-x + \theta n^{-1/2}),\ x \geqq 0 \tag{87}$$

where F_n is the empirical distribution function of $X_1, X_2, \ldots, X_n$. It is not difficult to check that another representation for $T_n(t, \theta)$ defined by (81) is

$$\begin{aligned} T_n(t, \theta n^{-1/2}) &= V(H(H_{n\theta}^{-1}(t)), \theta n^{-1/2}) \\ &+ n^{1/2}[\mu((H_{n\theta}^{-1}(t)), \theta n^{-1/2}) - \mu(t, \theta n^{-1/2})] \text{ a.s.} \end{aligned} \tag{88}$$

for all $0 \leqq t \leqq 1$, $|\theta| \leqq a$.

We shall first prove the weak convergence (cf. Billingsley, 1968) of the processes $\{T_n(t, \theta n^{-1/2}), 0 \leq t \leq 1\}$ to a continuous Gaussian process and then prove asymptotic normality of $T_n(J, \theta n^{-1/2})$ by using representation (78).

In addition to the assumptions stated above, we shall suppose that the process $\{X_n, n \geqq 1\}$ is ϕ-mixing with

$$\sum_{n=1}^{\infty} n\phi^{1/2}(n) < \infty. \tag{89}$$

Let

$$Z_n(t) = n^{1/2}[F_n(F^{-1}(t)) - t],\ 0 \leqq t \leqq 1. \tag{90}$$

Lemma 2.6. *Under condition* (89),

$$Z_n \xrightarrow{\mathscr{L}} Z$$

where $\{Z(t), 0 \leqq t \leqq 1\}$ *is a continuous Gaussian process with mean function zero and covariance kernel*

$$K(t, s)=\min (t, s)+\sum_{k=1}^{\infty} \operatorname{cov} [Y_1(t), Y_{k+1}(s)] + \sum_{k=1}^{\infty} \operatorname{cov}[Y_1(s), Y_{k+1}(t)],\ 0\leqq t,\ s\leqq 1 \tag{91}$$

where

$$Y_j(t)=[I(F(X_j)\leqq t)-t],\ j\geqq 1,\ 0\leqq t\leqq 1.$$

As a consequence of this lemma, for every $\varepsilon>0$,

$$\lim_{\delta\to 0}\ \overline{\lim_{n\to\infty}}\ P\left\{\sup_{|t-s|\leqq\delta} |Z_n(t)-Z_n(s)|>\varepsilon\right\}=0. \tag{92}$$

We omit the proof of Lemma 2.6. It is given in Sen (1971). Hereafter we shall use $\|\cdot\|$ to denote the supremum norm over $0\leqq t\leqq 1,\ -a\leqq\theta\leqq a$.

Lemma 2.7. *Under condition* (89),

$$\|H\|(H_{n\theta}^{-1}(t))-t\|=o_p(1) \tag{93}$$

Proof. Note that

$$\|H(H_{n\theta}^{-1}(t))-t\|\leqq \sup_{x,\theta}|H_{n\theta}(x)-H_\theta(x)| + \sup_{x,\theta}|H_\theta(x)-H(x)|+o(n^{-1}) \tag{94}$$

where $H_\theta(x)=H(x+\theta n^{-1/2})$. Since F is continuous with bounded derivative, it is easy to see that

$$\sup_{x,\theta}|H_\theta(x)-H(x)|=o(1). \tag{95}$$

Furthermore, one can check that

$$F_n(x+\theta n^{-1/2})-F(x+\theta n^{-1/2})=o_p(1) \tag{96}$$

for $-\infty<x<\infty$ and $-a\leqq\theta\leqq a$ by showing that

$$\operatorname{var}[F_n(x+\theta n^{-1/2})-F(x+\theta n^{-1/2})] \leqq n^{-1}\left\{\frac{1}{4}+\sum_{j=1}^{\infty} \phi^{1/2}(n)\right\}.$$

The last inequality follows by Lemma 1, p. 170 of Billingsley (1968). The convergence in (96) will be uniform in x and θ in view of the fact that $H_\theta(x)$

and $H_{n\theta}(x)$ are bounded and non-decreasing. Therefore

$$\sup_{x,\theta}|H_{n\theta}(x)-H_{\theta}(x)|=o_p(1). \tag{97}$$

Relations (94), (95) and (97) prove the lemma.

For any process $X(\cdot,\theta)$, let $w(X,\delta)$ be defined by

$$w(X,\delta)=\sup_{|\theta|\leqq a,|t-s|\leqq\delta}\{|X(t,\theta n^{-1/2})-X(s,\theta n^{-1/2})|\}. \tag{98}$$

Lemma 2.8. *Under condition* (89),

$$\|W_j(t,\theta n^{-1/2})-W_j(t,0)\|=o_p(1).\ j=1,2. \tag{99}$$

Proof. This lemma follows from (92) and the fact that

$$W_1(t,\theta)=Z_n(F(H^{-1}(t)+\theta))-Z_n(F(\theta))$$

and

$$W_2(t,\theta)=Z_n(F(\theta))-Z_n(F(-H^{-1}(t)+\theta)).$$

Lemma 2.9. *Under condition* (89),

$$n^{1/2}[H(H_{n\theta}^{-1}(t))-t]=-Q(t,0)+o_p(1) \tag{100}$$

uniformly in $|\theta|\leqq a,\ 0\leqq t\leqq 1$.

Proof. This lemma follows from the fact that f is bounded, even, uniformly continuous and from Lemmas 2.6 and 2.8.

Theorem 2.11. *Under the conditions stated above*

$$\|T_n(t,\theta n^{-1/2})-T_n(t,0)\|=o_p(1) \tag{101}$$

and

$$\{T_n(t,\theta n^{-1/2}),\ 0\leqq t\leqq 1,\ |\theta|\leqq a\}\overset{\mathscr{L}}{\to}\{Y(t),\ 0\leqq t\leqq 1\} \tag{102}$$

where $\{Y(t),\ 0\leqq t\leqq 1\}$ *is a Gaussian process with* $E[Y(t)]=0$ *for all* t *and*

$$\operatorname{cov}(Y(t),Y(s))=\lim_{n\to\infty}\operatorname{cov}[V(t),0),V(s,0)].$$

Proof. In view of Lemmas 2.8 and 2.9,

$$V(H(H_{n\theta}^{-1}(t)),\theta n^{-1/2})=V(t,\theta n^{-1/2})+o_p(1)=V(t,0)+o_p(1) \tag{103}$$

and

$$n^{1/2}[\mu(H(H_{n_\theta}^{-1}(t)), \theta n^{-1/2}) - \mu(t, \theta n^{-1/2})] = o_p(1) \tag{104}$$

uniformly for $0 \leqq t \leqq 1$ and $|\theta| \leqq a$. Representation (88) gives the relation

$$\|T_n(t, \theta n^{-1/2}) - V(t, 0)\| = o_p(1). \tag{105}$$

Putting $\theta = 0$, we obtain that

$$\|T_n(t, 0) - V(t, 0)\| = o_p(1). \tag{106}$$

Relations (105) and (106) together prove (101). Note that

$$\begin{aligned} V(t, 0) &= W_1(t, 0) - W_2(t, 0) \\ &= Z_n(F(H^{-1}(t))) + Z_n(F(-H^{-1}(t))) - 2Z_n(F(0)) \end{aligned}$$

and the last term converges in distribution to

$$Y(t) = Z(F(H^{-1}(t))) + Z(F(-H^{-1}(t))) - 2Z(F(0))$$

where $\{Z(t), 0 \leqq t \leqq 1\}$ is a continuous Gaussian process as defined in Lemma 2.6.

As a consequence of this theorem, we obtain the asymptotic linearity of signed rank statistics as given in the following theorem.

Theorem 2.12. *Under the conditions of Theorem* 2.11,

$$n^{1/2}\|S_n(t, \theta n^{-1/2}) - S_n(t, 0) - \theta n^{-1/2} q(t)\| = o_p(1) \tag{107}$$

where

$$q(t) = 2[f(H^{-1}(t)) - f(0)], \quad 0 \leqq t \leqq 1$$

and

$$\sup_{|\theta| \leqq a} n^{1/2} |S_n(J, \theta n^{-1/2}) - S_n(J, 0) \ + n^{-1/2} q(J)| = o_p(1) \tag{108}$$

where

$$q(J) = 2\int_0^1 f(H^{-1}(t)) \mathrm{d}J(t) = \int_0^1 q(t) \mathrm{d}J(t) + 2f(0)\, J(1).$$

Proof. Relation (107) follows from (101) and the fact that

$$n^{1/2}\|\mu(t, \theta n^{-1/2}) - \mu(t, 0) - \theta n^{-1/2} q(t)\| \to 0.$$

It is clear from (78) that

$$S_n(J, \theta) = S_n(1, \theta) J(1) - \int_0^1 S_n(t, \theta) \mathrm{d}J(t)$$

for $|\theta| \leqq a$ with probability one. Since $J(0)=0$ and $S_n(1, \theta)=1-2F_n(\theta)$,

$$\begin{aligned}
&n^{1/2}[S_n(J, \theta n^{-1/2})-S_m(J, 0)+n^{-1/2}\theta q(J)] \\
&=2J(1)n^{1/2}[F_n(0)-F_n(\theta n^{-1/2})] \\
&\quad -n^{1/2}\int_0^1 [S_n(t, \theta n^{-1/2})-S_n(t, 0)-n^{-1/2}\,\theta q(t)]\mathrm{d}J(t) \\
&\quad +2\theta f(0)J(1) \\
&=o_p(1)
\end{aligned}$$

by (107), (92) and the fact that f and J are bounded.

Theorem 2.13. *Under the conditions stated above*

$$\frac{n^{1/2}S_n(J, 0)}{\sigma} \xrightarrow{\mathscr{L}} N(0, 1) \tag{109}$$

where

$$\sigma^2=\sigma_J^2+2\sum_{k=1}^{\infty} \operatorname{cov}\,[Y_1, Y_{k+1}],\ \sigma_J^2=\int_0^1 J^2(u)\mathrm{d}u, \tag{110}$$

$$Y_k=J(H|X_k|)s(X_k),\ k\geqq 1 \tag{111}$$

Proof. Observe that

$$\begin{aligned}
n^{1/2}S_n(J, 0)=T_n(J, 0)&=\int_0^1 J(t)\mathrm{d}T_n(t, 0) \\
&=\int_0^1 J(t)\mathrm{d}V(t, 0)+o_p(1).
\end{aligned} \tag{112}$$

The last relation follows from the fact that J is bounded and by (105). But

$$\int_0^1 J(t)\mathrm{d}V(t, 0)=n^{-1/2}\sum_{i=1}^{n} J(H(|X_i|))s(X_i)$$

and the last term is asymptotically normal, with mean 0 and variance σ^2 given by (110), by Theorem 20.1 of Billingsley (1968). The convergence of the series defining σ^2 follows from the fact that

$$\sum_{n=1}^{\infty} \phi^{1/2}(n)<\infty.$$

Theorem 2.14. *Under the conditions stated above,*

$$n^{1/2}(\hat{\theta}_n-\theta)\xrightarrow{\mathscr{L}} N(0, \tau^2) \tag{113}$$

where $q(J)\tau=\sigma$.

Proof. This theorem now follows from Theorems 2.12 and 2.13.

3. NONPARAMETRIC TESTS FOR STOCHASTIC PROCESSES

In this section, we shall briefly discuss non-parametric test theory for discrete parameter stochastic processes. Results obtained here mainly apply to processes with "independent structure" i.e. those processes which are either mixtures of i.i.d. processes or from which one can generate i.i.d. random variables. The study of distribution-free or non-parametric tests is essentially a study of similar tests and test-functions. Distribution-free (DF) methods can be used whenever the underlying probability distributions are either unknown or not completely specified. Wilcoxon two-sample statistics for ϕ-mixing processes is discussed at the end of this Section.

3.1. Preliminaries

We shall first state a few definitions.

Definition 3.1. A statistic T is *Distribution Free* (DF) with respect to a family of distributions $\Omega(H_0)$ if there exists a single distribution G such that

$$P(T \leqq t|F) = G_T(t)$$

for all real t and all F in $\Omega(\mathrm{H}_0)$.

Definition 3.2. A set A is *non-trivial similar* with respect to $\Omega(H_0)$ if there exists an $\alpha(0<\alpha<1)$ such that $P(A|F)=\alpha$ for all F in $\Omega(H_0)$.

Definition 3.3. A test function $\varphi(z)$ is *similar* with respect to $\Omega(H_0)$ if there exists an $\alpha(0<\alpha<1)$ such that $E_F(\varphi)=\alpha$ for all F in $\Omega(H_0)$.

Let T be a sufficient statistic for Ω and let Ω_T be the induced family of distributions. Any test satisfying

$$E(\varphi(Z)|T=t) = \alpha \text{ a.s. } \Omega_T \tag{114}$$

is similar with respect to Ω since

$$E_F(\varphi(Z)) = E_F\{E(\varphi(Z)|T)\} = \alpha$$

for all $F \in \Omega$. A test which satisfies (114) is said to have *Neyman structure* with respect to T. If Ω is complete (cf. Lehmann, 1959), then it is easy to find a most powerful test among all similar tests possessing Neyman structure.

Construction of similar sets and test functions consists of choosing a fixed proportion of the points of almost every orbit. This is done by means of Basic-Pitman functions.

Definition 3.4. h is called a *B-Pitman function* with respect to groups S if for all $G \in \Omega(H_0)$ and for all non-identity elements γ in S,

$$P\{h(Z)=h(\gamma(Z))|G\}=0.$$

The *B-Pitman statistic* induced by the B-Pitman function h is defined by

$$\tilde{R}(h(z))=\sum_{\gamma \in S} \varepsilon(h(z)-h(\gamma(z)))$$

where $\varepsilon(u)=1$ or 0 according to whether $u \geqq 0$ or not.

Definition 3.5. $\hat{h}$ is called a *basic non-sequential Pitman function* (BNS-Pitman function) with respect to group S and its subgroup S_q if $\hat{h}(Z)=h(\gamma(Z))$ for all $\gamma \in S_q$ and

$$P\{\hat{h}(Z)=h(\gamma(Z))|G\}=0$$

for all $G \in \Omega(H_0)$ and for all γ not in S_q. The *BNS-Pitman statistic* induced by BNS-Pitman function $\hat{h}$ is defined by

$$\tilde{R}(\hat{h}(z))=\frac{1}{C(S_q)} \sum_{\gamma \in S} \varepsilon(h(z)-h(\gamma(z)))$$

where $C(S_q)$ is the cardinality or order of the group S_q.

Note that $\tilde{R}(h)$ assumes the values 1, 2, ..., $C(S)$ and $\tilde{R}(\hat{h})$ takes the values 1, 2, ..., $C(S)/C(S_q)$ on almost every orbit. Furthermore each B-Pitman function is a BNS-Pitman function for $S_q=\{e\}$ where e is the identity element.

The basic theorem in terms of functions and groups is as follows.

Theorem 3.1

(i) *For any test function* $\varphi(z)$,

$$\int \varphi(z)\ G(dz)=\alpha$$

for all $G \in \Omega$ *if and only if* $\Sigma_{\gamma \in S}\ \varphi(\gamma(z))=C(S)\alpha$ for almost all z.

(ii) *A statistic T is DF, with respect to Ω if and only if there exists a B-Pitman function h with respect to S and a function ψ such that $T=\psi(\tilde{R}(h))$. Furthermore, if T is a distribution-free statistic, then the* *null distribution is discrete with probabilities which are integral multiples of* $(C(S))^{-1}$ *and conversely.*

(iii) *The following three conditions are equivalent:*
   (a) *A set A is similar of size* α *with respect to* Ω;
   (b) *A contains* $C(S)\alpha$ *elements of almost every orbit* $\{\gamma(z):\gamma \in S\}$;
   (c) *There exists a B-Pitman function h with respect to S such that* $A = \{z : \tilde{R}(h(z)) \leqq C(S)\alpha\}$.

(iv) *The most powerful distribution-free (MPDF) test against a simple alternative likelihood function* L_1 *is of the form*

$$\varphi(z) = \begin{cases} 1 & \text{if } \tilde{R}(h_1(z)) > K(\alpha, S) \\ \delta & \text{if } \tilde{R}(h_1(z)) = K(\alpha, S) \\ 0 & \text{if } \tilde{R}(h_1(z)) < K(\alpha, S) \end{cases}$$

where h_1 *is a B-Pitman function with respect to S whose ordering on almost every orbit is consistent with that of* L_1. *The power of this test is greater than* α *if and only if* L_1 *is not invariant under S.*

(v) *Consider the family*

$$\Omega(H_1) = \{L_1 : L_1(z) = \exp\,(Q(\theta)h_1(z) + q(z, \theta))\}$$

where $Q(\theta) > 0$ *for all* θ *and* h_1 *is a B-Pitman function with respect to S.*
   (a) *The test in* (iv) *is uniformly most powerful distribution free (UMPDF) with respect to* $\Omega(H_1)$ *if* $q(\cdot, \cdot)$ *is invariant under S.*
   (b) *Assume that when* $\theta = \theta_0$, L_1 *is in* Ω *and that there is sufficient regularity that* $q_2(z, \theta) = o(Q(\theta))$ *in some neighbourhood of* θ_0 *where* $q = q_1 + q_2$ *and* q_1 *is invariant under S. Then the test in* (iv) *above is locally MPDF against* $\Omega(H_1)$.

This theorem characterizes the family of all similar tests and distribution-free statistics in terms of orbits and *BP* statistics. This is a consequence of the complete sufficiency of the orbit, symmetric completeness of the null hypothesis class and an extension of a result due to Lehmann and Stein (1949). Thus, similar *DF* tests are conditional tests given the orbit.

We shall now sketch a proof of the theorem in the case of the group $S^* = S^*(n_1, \ldots, n_k)$ where S^* is the subgroup of the symmetric group S_N of $N!$ permutations of $(1, 2, \ldots, N)$ where $N = \Sigma_{i=1}^k n_i$ which leaves k sets of order statistics of random samples of size n_i, $1 \leqq i \leqq k$ invariant. This group corresponds to the k-sample problem of testing equality of distributions. For proofs in other cases, we refer the reader to the papers cited in Bell, Woodroofe and Avadhani (1970).

Proof. Let

$$B_0 = \{(y_1, \ldots, y_N) : y_1 < \ldots < y_N\}$$

and

$$D = \prod_1^n \{u_i(1) < \ldots < v_i(n_i)\}.$$

where $v_i(1), \ldots, v_i(n_i)$ is the ordered sequence corresponding to ith group of size n_i. Then

$$D = U\gamma(B_0)$$

where the union is over all γ in $S^* = S^*(n_1, \ldots, n_k)$. Further, the joint distribution of $Z = (V_1(1), \ldots, V_k(n_k))$ is given by

$$\mathrm{d}G^*(z) = \begin{cases} \pi_{i=1}^{k} (n_i!) \; \pi_{i=1}^{k} \; \pi_{j=1}^{n_i} \; \mathrm{d}G(V_i(j)) & \text{if } z \in D \\ 0 & \text{otherwise.} \end{cases}$$

If

$$\alpha = \int_D \varphi \mathrm{d}G^*, \text{ then}$$

$$\alpha = \sum_{\gamma \in S^*} \int_{\gamma^{-1}(B_0)} \varphi(z) \mathrm{d}G^*(z)$$

$$= \sum_{\gamma \in S^*} \int_{B_0} \varphi(\gamma(z)) \mathrm{d}G^*(\gamma(z))$$

$$= \int_{B_0} \left[\sum_{\gamma \in S^*} \varphi(\gamma(z)) \right] \mathrm{d}G^*(z).$$

Let $\psi(z) = \Sigma_{\gamma \in S^*} \varphi(\gamma(z))$. Then, for all continuous G on $\mathbb{R}$, we have

$$\alpha = \int_{B_0} \psi(z) \mathrm{d}G^*(z) = (N!)^{-1} \prod_{i=1}^{k} (n_i!) \int_{B_0} \psi(y_1, \ldots, y_N)(N! \prod_1^N \mathrm{d}G(y_i)).$$

But the order statistics are complete. Hence (i) follows.

(iii) follows from (i) if φ is taken as the indicator function of A where A is a similar set.

The B-Pitman function is used to select points (see Bell and Doksum, 1967).

(ii) follows from (iii) in that each $T^{-1}(B) = A$ must be similar. (iv) is the result of applying the Neyman–Pearson Lemma to each orbit. (v) (a) follows from the fact that the orderings on the orbit of h_1, $Q_1(\theta)h_1(z)$,

$Q(\theta)h_1(z)+q(z,\theta)$ are identical if $q(z,\theta)$ is invariant under S. (v) (b) can be seen by observing that $h_1(z)+q(z,\theta)\ (Q(\theta))^{-1}$ is negligible for θ sufficiently close to θ_0.

3.2 Tests for processes with stationary independent increments (SII)

Suppose $\{X_t,\ t\geqq 0\}$ is an SII process. Then

$$Y_j = X_{jt} - X_{(j-1)t},\ 1\leqq j\leqq n$$

are independent and identically distributed for each j. The common distribution function F of Y_j is uniquely determined by the probability measure P of the process and the time interval $(0, t]$. Conversely F and t determine P if $X_0=0$ (Doob, 1953). Since the Y's are i.i.d. and univariate, one can make use of known results for the univariate goodness-of-fit two-sample and randomness hypotheses.

Goodness-of-fit. The hypothesis $H_0{:}P=P_0$ becomes $H_0{:}F=F_0$ in terms of $Y_1,\ldots,\ Y_n$. Since this is a simple hypothesis, the usual parametric theory holds. On the other hand, a large family of DF statistics is obtained by making the probability transformation $U=F_0(Y)$. Let $\Omega(H_0)$ denote the family of probability measures satisfying the hypothesis H_0.

The following theorem follows from results in Birnbaum and Rubin (1954) and Bell (1964).

Theorem 3.2.

(i) *Each statistic* $T=\varphi[F_0(Y_1),\ldots,\ F_0(Y_n)]$ *is* DF *with respect to* $\Omega(H_0)$ *in the sense that there exists* Q *such that*

$$P\{\psi[F_0(Y_1),\ldots,\ F_0(Y_n)]\leqq t|F_0\}=Q(t)$$

for all F *in* $\Omega(H_0)$ *and for all* t.

(ii) $Q(t)=P(\psi(U_1,\ldots,\ U_n)\leqq t|U_0]$ *for all* t *where* U_0 *is the standard uniform distribution.*

(iii) *There exists a* DF *statistic which is not of the form in* (i) *above.*

For example $\overline{U}=n^{-1}\ \Sigma_{i=1}^{n}\ F_0(Y_i)$ is DF with respect to $\Omega(H_0)$ and $\overline{Z}=n^{-1/2}$ $\Sigma_{i=1}^{n}\ \phi^{-1}\ (F_0(Y_i))$ is DF with Q exactly $N(0,\ 1)$. If F_0 is $N(\mu,\ \sigma^2)$, then $Z=(X-\mu/\sigma)\sqrt{n}$.

Two-sample case. The hypothesis $H_0{:}P_1=P_2$ reduces to $H_0{:}F_1=F_2$ where the data is $X'_1,\ldots,X'_m$ and $Y'_1,\ldots,Y'_n$ with $X'_i=X_{it}-X_{(i-1)t}$ and Y'_j, $1\leqq j\leqq m$ defined similarly. The following theorems can be obtained from the results of Bell and Smith (1969).

Theorem 3.3. *A statistic T is DF with respect to $\Omega(H_0)$ if and only if there exists a B-Pitman function h with respect to S_N, $N=m+n$ and a function W such that $T=W(\tilde{R}(h))$.*

For a simple alternative and a Koopman–Pitman type class related to it, one can obtain a *UMPDF* test.

Theorem 3.4. *Let $Z=(X'_1, \ldots, Y'_n)$ and let h_1 be a B-Pitman function. Then the UMPDF test against the alternative family*

$$\{(P_1, P_2){:}L(z)=\exp\{ah_1(z)+h(z)\} \text{ with } a>0 \text{ and } h \text{ invariant under } S_N\}$$

is of the form

$$\psi(z)=\begin{cases} 1 & \text{if } \tilde{R}(h_1(z))>K(\alpha, m, n) \\ \delta & \quad = \\ 0 & \quad < \end{cases}$$

Example 3.1. (a) Let $W_1, \ldots, W_N$, $N=n+m$ be standard normal sample independent of the two stochastic processes. Let $W(RX'_i))$ be that W whose rank in the W-sample is equal to that of X'_i in the combined sample

$$\{Z_1, \ldots, Z_N\}=\{X'_1, \ldots, X'_m, Y'_1, \ldots, Y'_n\}$$

and let $W(R(Y'_j))$ be analogously defined. Form

$$T=\frac{1}{m}\sum_1^m W(R(X'_i))-\frac{1}{n}\sum_1^n W(R(Y'_j)).$$

For each fixed set of W's, T is a DF statistic i.e. T is conditionally a DF statistic. When the W's constitute a random normal sample, T is a randomized DF statistic (cf. Bell and Doksum, 1965). The null distribution of T is $N(0, 1/m+1/n)$.

(b)

$$T_1=\int (F_m(z)-G_n(z))\,\mathrm{d}\left(\frac{mF_m+nG_n}{N}\right)$$

$$=N^{-1}m^{-2}n^{-2}\sum_{r=1}^{N}[NnG_n(Z_r)-rn]^2$$

is a non-randomized DF statistic.

Tests for the Model. One wishes to determine whether or not the process being observed is an SII process. On restricting to the continuous case, as usual, one finds for the increments

$$X_i' = X_{it} - X_{(i-1)t} \, (1 \leqq i \leqq n)$$

the null hypothesis class is that of randomness. Hence, one has the following theorem from Bell and Donoghue (1969).

Theorem 3.5. *If P is such that $L(x_1, \ldots, x_n) = \pi_1^n f(x_i)$ or is otherwise invariant under S_N, then*

(i) *the MPDF test of size α has power α.*

Further, consider the family of processes with measures P yielding

$$\{L{:}L(z) = \exp[a(\theta)h(z) + b(z, \theta)]\}$$

with $a(\theta) > 0$ and h a fixed B-Pitman function. For this family

(ii) *$\tilde{R}(h)$ is UMPDF if $b(\cdot, \theta)$ is invariant under S_N for all θ;*

(iii) *$\tilde{R}(h)$ is locally MPDF if $b = b_1 + b_2$ where b_1 is invariant under S_N for all θ and $b_2(z, \theta) = o(a(\theta))$ where $a(\theta)$ is continuous and $a\,(\theta_0) = 0$.*

Example 3.2. Let the process be such that its increments $X_1', \ldots, X_n'$ form a first-order Markov process with the following properties

(i) $X_1' = aX_0' + Y_1, X_2' = aX_1' + Y_2, \ldots, X_n' = aX_{n-1}' + Y_n$;

(ii) $Y_1, \ldots, Y_n$ are i.i.d. $N(0, \sigma^2)$;

(iii) X_0' is arbitrarily distributed.

Then the Pitman function which generates *MPDF* test is

$$h = \sum_{i=1}^{n} x_i x_{i-1}.$$

An exponential family with respect to which $\tilde{R}(h)$ is UMPDF is

$\{L{:}L = \exp[a(\theta) \sum_{i=1}^{n} x_i x_{i-1} + b(\theta, z) + \sum_{i=1}^{n} g(x_i)]$ where $b(\theta, z)$ is invariant under S_n for all $\theta\}$.

Further $\tilde{R}(h)$ generates the locally MPDF test against any family of processes whose likelihood function are of the form

$$\left\{\exp\left[a(\theta) \sum_{i=1}^{n} x_i x_{i-1} + b(\theta, z) + \sum_{i=1}^{n} g(x_i) + Q(z, \theta, i)\right]\right\}$$

where $Q(z, \theta, i) = o(a(\theta))$ for all i and for almost all z.

3.3 Tests for Exchangeable Processes

We will now consider inference for processes which are exchangeable in the following sense. A discrete parameter process $\{X_n, n \geqq 1\}$ is called *exchangeable* if and only if for each fixed $k \geqq 1$, the joint distribution of $(X_{i_1}, \ldots, X_{i_k})$ is the same for all choices of distinct positive integers $i_1, \ldots, i_k$. Note that any process of i.i.d. random variables is exchangeable. Suppose Z has an exponential distribution and conditionally given $Z = z > 0$, let $X_1, X_2, \ldots,$ be i.i.d. $N(0, z)$. Then $\{X_n, n \geqq 1\}$ is an exchangeable process. The following theorem due to De Fennetti (Loeve, 1963), characterizes the exchangeable processes.

Theorem 3.6. *Let $\{X_n, n \geqq 1\}$ be an exchangeable process on a probability space $(\Omega, \mathscr{F}, P)$; then there is a sub-σ-algebra $\mathscr{F}_0 \subset \mathscr{F}$ such that $X_1, X_2, \ldots$ are conditionally independent and identically distributed given $\mathscr{F}_0$.*

Let $\mathbb{R}^\infty$ denote the countably infinite product of real lines with elements denoted by $x = (x_1, x_2, \ldots)$ and let $\mathscr{B}_\infty$ be the Borel sets of $\mathbb{R}^\infty$. Then any discrete parameter stochastic process $\{X_n, n \geqq 1\}$ induces a measure P^* on $\mathscr{B}_\infty$ by

$$P^*(B) = P[(X_1, X_2, \ldots) \in B], \quad B \in \mathscr{B}_\infty.$$

Obviously, the process is exchangeable if and only if P^* is symmetric i.e. invariant under permutations of any finite number of coordinates of $\mathbb{R}^\infty$. The following theorem is a special case of a result due to Hewitt and Savage (1955).

Theorem 3.7. *Let $\mathscr{M}$ denote the class of all univariate distributions endowed with its weak topology. For each $F \in \mathscr{M}$, let π_F denote the product measure induced on $\mathscr{B}_\infty$ by F. If P^* is any symmetric probability measure on $\mathscr{B}_\infty$, then there is a uniquely determined probability measure λ on the Borel sets of $\mathscr{M}$ for which*

$$P^*(B) = \int_{\mathscr{M}} \pi_F(B)\lambda(\mathrm{d}F). \tag{115}$$

In particular, (115) holds if P^* is the measure of an exchangeable process. In this case we will write $P^* = P_\lambda$ and λ is called the *mixing measure* of the process.

As a consequence of the above theorem, we get that exchangeable processes and i.i.d. processes have the same similar sets subject to certain continuity conditions. More precisely, the following theorem holds.

Theorem 3.8. *Let Λ_0 be the set of all λ for which $P_\lambda([X : X_1 = X_2]) = 0$. If*

$B \in \mathscr{B}_{\infty}$, then $P_{\lambda}(B) = \alpha$ for all $\lambda \in \Lambda_0$ if and only if $\pi_F(B) = \alpha$ for all continuous $F \in \mathscr{M}$.

Proof. If F is continuous, then π_F is a P_{λ} with $\lambda \in \Lambda_0$ so that the "only if" part is trivial. To prove the "if" assertion, it will suffice to say that for $\lambda \in \Lambda_0$ almost everywhere, F is continuous (λ). This follows from

$$\begin{aligned} 0 &= P_{\lambda}([X : X_1 = X_2]) \\ &= \int_{\mathscr{M}} \left[\int_{-\infty}^{\infty} (F(x+0) - F(x-0))\, \mathrm{d}F(x) \right] \lambda(\mathrm{d}F) \end{aligned}$$

which implies that for almost every $F[\lambda]$, $F(x+0) = F(x-0)$ for almost every $x(F)$.

Hereafter we shall assume that $\lambda \in \Lambda_0$. Theorem 3.8 has the following application. Suppose $X_1, \ldots, X_k$ is a sample from an exchangeable process. Then the distribution of any rank or permutation statistic will be the same as if $X_1, \ldots, X_k$ were a random sample from a continuous distribution function.

Example 3.4. To test the null hypothesis that $(X_1, \ldots, X_k)$ is a sample from an exchangeable process with unspecified mixing measure $\lambda \in \Lambda_0$ against alternatives of increasing trend, an appropriate test would be to reject the hypothesis for large values of the Spearman Rank correlation coefficient

$$\rho = \frac{R}{k^2 - k} \sum_{i=1}^{k} i \left(R_i - \frac{k+1}{2} \right)$$

where R_i denotes the rank of X_i, $i = 1, \ldots, k$.

Ahmad (1974) deals with structure and characterisation of the exchangeable processes. He shows that, for exchangeability hypothesis, all distribution free statistics are based on permutation-rank statistics.

The hypothesis H_0:*λ is degenerate* is equivalent to the randomness hypothesis i.e. X's are i.i.d. Since any exchangeable sample function contains conditionally i.i.d. random variables, one needs cross-section data $X_{11}, \ldots, X_{1k}, \ldots, X_{n1}, \ldots, X_{nk}$. The following theorem can be derived from Theorem 3.1.

Theorem 3.9. (i) *A statistic T is DF with respect to $\Omega(H_0)$ if and only if there exists a B-Pitman function h with respect to S_{nk} and a function ψ such that $T = \psi(\tilde{R}(h))$.*

(ii) *A test function φ is similar of size α if and only if*

$$\sum \varphi(\gamma(z)) = (nk)!\, \alpha$$

for almost all z where the sum is over all γ in S_{nk}.

(iii) *The MPDF test against a simple alternative with likelihood function L is of the form*

$$\psi(z)=\begin{cases}1 & \text{if } \tilde{R}(h(z))>C(\alpha, nk)\\ \delta & \qquad = \\ 0 & \qquad < \end{cases}$$

where h is a B-Pitman function which in almost every orbit gives an ordering consistent with that of L.

As before, one can derive UMPDF test, locally MPDF tests for Koopman-Pitman type exponential families.

Example 3.5. Consider a normal regression alternative i.e. $X_{i1},\ldots, X_{ik}$ are independent normal with mean $\mu_i=\mu+ib$ and variance σ^2, $1\leqq i\leqq n$. Then the MPDF test is based on the likelihood function

$$L(z)=(2\pi\sigma^2)^{\frac{-nk}{2}}\exp\left(-\frac{1}{2}\sigma^{-2}\sum_{i,j}(x_{ij}-\mu_i)^2\right).$$

Two *BNS*-Pitman functions giving the same orderings on the orbits are

$$h_i=\sum_1^n\mu_i\bar{x}_i \text{ and } h_i^*=\sum_1^n i\bar{x}_i.$$

A *DF* statistic simpler to compute and directed against the same alternatives is $T=\Sigma_i\, iR(\bar{x}_i)$.

Tests for the model. Let $X_{i1}, X_{i2},\ldots, X_{ik}$, $i=1,\ldots, n$ denote n independent observations on an initial segment of a stochastic process $X_1, X_2,\ldots$. For testing the hypothesis H_0 that the process is exchangeable with an unspecified mixing measure $\lambda\in\Lambda_0$, one can show that a statistic T is *DF* with respect to $\Omega(\mathrm{H}_0)$ if and only if there exists a *B*-Pitman function h with respect to $S_k WrS_n$ and a function ψ such that $\psi(\tilde{R}(h))\equiv T$. Furthermore φ is a test function similar of size α with respect to $\Omega(H_0)$ if $\Sigma\varphi(\gamma(x_{11},\ldots, x_{1k}; x_{21};\ldots x_{2k}; x_{n1},\ldots, x_{nk})=(k!)^n n!\,\alpha$ for almost all $(x_{11},\ldots, x_{nk})$ where the sum is over $S_k WrS_n$. Here $S_k WrS_n$ is the Wreath product of S_k with S_n.

These results follow from Theorems 3.1 and 3.8.

Example 3.6. Consider a Gaussian alternative with mean function μ and covariance function Γ. Let $\Sigma=\Gamma(s, t)$, where $\Gamma(s, t)$ is the covariance between $X_s$and X_t and $\mu=(\mu(1),\ldots, \mu(k))$ and the cross-section data $X_{i1},\ldots, X_{ik}$,

$i=1,\ldots, n$ be as usual. The likelihood function is

$$L(z)=K(n, k)\exp\left\{-\frac{1}{2}\sum_{i=1}^{n}(\mathbf{X}^{(i)}-\mu)'\sum^{-1}(\mathbf{X}^{(i)}-\mu)\right\}$$

where $\mathbf{X}^{(i)}=(X_{i1},\ldots, X_{ik})$. A *BNS*-Pitman function giving the same ordering on the orbits is

$$h(z)=\sum_{i=1}^{n}(\mathbf{x}^{(i)})'\sum^{-1}\mu.$$

The MPDF test is based on $R^*(h(z))$ which is a generalized B-Pitman statistics (cf. Bell, Woodwork, Avadhani (1970) p. 247). Note that if $\mu(t)\equiv 0$, then the resulting Gaussian process is exchangeable.

Goodness-of-fit tests. We shall now consider briefly tests drawing inferences about the mixing measure λ. Suppose, for example, that we are interested in testing the hypothesis that μ_λ is of the form

$$\mu_\lambda(B)=\int_0^\infty \pi_\theta(B)\lambda^*(d\theta),\ B\in\mathscr{B}_\infty \tag{116}$$

where, for each $\theta>0$, π_θ is the measure on $\mathscr{B}_\infty$ making the coordinate functions independent *exponential* random variables with mean θ and λ is a specified or unspecified measure on the Borel sets of $(0, \infty)$. To test the hypothesis H_1 that μ_λ is of the form (116) with λ^* unspecified on the basis of cross-sectional data $X_{i1},\ldots, X_{ik}$, $i=1,\ldots, n$, we simply test that each of the samples $X_{i1},\ldots, X_{ik}$ are random samples from an exponential with unspecified mean and reject H_1 otherwise. This can be done by using results of Darling (1955) if k is large or by using the fact that

$$\left\{\frac{X_{i1}+\ldots+X_{im}}{m}\right\}\left\{\frac{X_{im+1}+\ldots+X_{ik}}{k-m}\right\}^{-1}$$

has an F-distribution if H_1 is true. If we accept H_1, then we may test H_0 that $\lambda^*=\lambda_0^*$ (specified) by testing that the common distribution function of

$$\theta_i=\frac{X_{i1}+\ldots+X_{ik}}{k}$$

is

$$G(x, \lambda_0^*)=\int_0^\infty F_k\left(\frac{kx}{\theta}\right)\lambda_0^*(d\theta)$$

where F_k is the gamma distribution with parameter k. This can be done by standard methods.

3.4 Tests for Spherically Exchangeable (SE) Processes

These processes represent generalizations of i.i.d. Gaussian processes with mean function $\mu(k) \equiv 0$ and covariance function $\Gamma(r, k) = 0$ if $r \neq k$ and $\Gamma(r, k) = \sigma^2$ if $r = k$. Formally, a stochastic process $\{X_n, n \geqq 1\}$ is called SE (*spherically exchangeable*) if there exists a function ψ on $\mathbb{R}_+$ such that for each finite set $(i_1, \ldots, i_k)$ of natural numbers, the joint characteristic function of $(X_{i_1}, \ldots, X_{i_k})$ satisfies

$$\tilde{\psi}(t_1, \ldots, t_k) = \psi\left(\sum_{i=1}^{k} t_i^2\right). \tag{117}$$

If $\{X_n, n \geqq 1\}$ are i.i.d. $N(0, \sigma^2)$, then the process is SE. Let η be a random variable with distribution λ on $[0, \infty)$ and $\{Y_n, n \geqq 1\}$ be a process such that conditionally given $\eta = \sigma_0$, the Ys are i.i.d. $N(0, \sigma_0^2)$. Then, the unconditional process is SE. In fact

$$\tilde{\psi}(t_1, \ldots, t_k) = \int_0^\infty \exp\left(-\frac{1}{2}\left(\sum_1^k t_i^2\right)\sigma^2\right)\lambda(d\sigma).$$

Kelker (1968) and Smith (1969) have treated finite families of spherically exchangeable random variables. Many of the properties developed by them carry over to stochastic processes. Ahmad (1975) also discussed some characterization theorems for spherically exchangeable random variables.

SE processes are mixtures of zero mean i.i.d. Gaussian processes in the same sense that exchangeable processes are mixtures of i.i.d. processes. The following lemma follows from a theorem of Schonberg (1938) and the fact that a completely monotone function is the Laplace transform of a measure on $(0, \infty)$.

Lemma 3.1. *If $\{X_n, n \geqq 1\}$ is an SE process, then the function ψ in (117) is completely monotone and hence can be written as*

$$\psi(t) = \int_0^\infty e^{-ts}\lambda_1(ds) \tag{118}$$

where λ_1 is a probability measure on $[0, \infty)$.

Theorem 3.11. *Let $\{X_n; n \geqq 1\}$ be an SE process and for each $\sigma \geqq 0$, let $\pi'_\sigma = \pi_{F_\sigma}$, where F_σ is the $N(0, \sigma^2)$ distribution function, (π_F for any distribution function F is as defined in Theorem 3.8). There is a uniquely determined measure λ on the Borel sets of $[0, \infty)$ for which*

$$P(B) = \int_0^\infty \pi'_\sigma(B)\lambda(d\sigma) \tag{119}$$

for all $B \in \mathscr{B}_\infty$. λ *is the mixing measure of the process and the right side of* (119) *will be denoted by* $P_\lambda(B)$.

Proof. Applying the transformation $s=\frac{1}{2}\sigma^2$, the relation (118) can be written as

$$\psi(t)=\int_0^\infty \exp(-\tfrac{1}{2}t\sigma^2)\lambda(dt)$$

where λ is a measure on the Borel sets of $[0, \infty)$. Since mixtures of characteristic functions correspond uniquely to mixtures of distribution, it now follows that (119) holds whenever B is a Borel cylinder set. An application of the monotone class theorem extends the validity of (119) to all Borel sets. Finally, the uniqueness follows from the uniqueness of λ in Theorem 3.8.

Note that λ can be viewed as a parameter for the family of spherically exchangeable processes. For an initial segment $(X_1, \ldots, X_k)$ of an SE process, $S_k^2=1/k\ \Sigma_{i=1}^k X_i^2$ is a sufficient statistic for λ and the conditional distribution of $(X_1, \ldots, X_k)$ given $S_k^2=s^2$ is the uniform distribution over the hypersphere of radius ks^2 centred at the origin. Let $G_k(\cdot; \lambda)$ be the distribution function of S_k^2 when λ is the mixing measure. It is not difficult to see that if $\{X_n, n \geqq 1\}$ and $\{X'_n, n \geqq 1\}$ are spherically exchangeable, then X_1 and X'_1 have identical distributions if and only if the two processes have identical laws.

H_0:$P=P_0$ **(Goodness-of-fit)**. For this, one needs cross-sectional data $X_{11}, \ldots, X_{1k}; \ldots; X_{n1}, \ldots, X_{nk}$. This can be reduced to $S_1^2, \ldots, S_n^2$ where $S_i^2=1/k\ \Sigma_{j=1}^k X_{ij}^2 \cdot S_1^2, \ldots, S_n^2$ are distributed as a random sample of size n from $G_k(\cdot; \lambda_0)$ under H_0. Each statistic of the form $T=\psi[G_k(S_1^2; \lambda_0), \ldots, G_k(S_n^2; \lambda_0)]$ is *DF* with respect to the goodness-of-fit hypothesis. For example

$$T=n^{-1}\sum_{i=1}^n J^{-1}(G_k(S_i^2; \lambda_0)),$$

where J, a strictly monotone distribution function on $\mathbb{R}$, is *DF* and the null distribution of T is that of the mean of a random sample of size n from a population with distribution J.

H_1:$P_1=P_2$**(Two-sample case).** Here one needs cross-sectional data $X_{11}, \ldots, X_{mk}; Y_{11}, \ldots, Y_{nk}$ which is reduced to $S_1^2, \ldots, S_m^2$ where $S_i^2=k^{-1}\Sigma_1^k X_{ij}^2$ and $V_1^2, \ldots, V_n^2$ where $V_i^2=k^{-1}\Sigma_1^k Y_{ij}^2$. The hypothesis is then H_0:$G_k(\cdot; \lambda_1)=G_k(\cdot; \lambda_2)$.

Let F_m and J_n be the empirical distribution functions of S^2's and V^2's respectively and

$$[Z_1,\ldots, Z_N]=[S_1^2,\ldots, S_m^2; V_1^2,\ldots, V_n^2];\ 0=q_0<q_1<\ldots<q_{s+1}<1$$

be such that Nq_i is a positive integer. Let

$$\eta_{1i}=m[F_m(X(Nq_i))-F_m(Z(Nq_{i-1}))],$$

$$\eta_{2i}=n[J_n(Z(Nq_i))-J_n(Z(Nq_{i-1}))],$$

and

$$\hat{T}=2-2(nm)^{-1/2}\sum_1^{s+1}(\eta_{1i}\eta_{2i})^{1/2}$$

$$=\sum_{i=1}^{s+1}[(\eta_{1i}m^{-1})^{1/2}-(\eta_{2i}n^{-1})^{1/2}]^2.$$

$\hat{T}$ is DF with respect to $\Omega(H_0)$ and it is asymptotically χ-squared with s degrees of freedom (cf. Matusita, 1955).

Bell *et al.* (1970) have also given DF tests for testing the hypothesis that the SE process is degenerate in the sense that there exists some unknown $\sigma>0$ such that the variables are i.i.d. $N(0,\sigma^2)$. We will not go into the details here.

3.5 Wilcoxon Test on φ-mixing Processes

Suppose $\{X_1, \ldots, X_m\}$ and $\{Y_1, \ldots, Y_n\}$ are independent realizations of stationary φ-mixing processes observed up to times m and n respectively with marginals F and G. We shall now discuss a property of Wilcoxon two-sample statistic for testing the hypothesis $F=G$. This statistic is widely used if the observations are independent and identically distributed.

Among the desirable features of the Wilcoxon two-sample statistic, when defined on independent random samples, is its asymptotic normality, which for large samples facilitates tests of the hypothesis that $F=G$ and a calculation of the power for any alternative (F, G). We shall now prove that these aspects are true also when the samples arise from stochastic processes belonging to a wide class, including strictly stationary φ-mixing processes.

The hypothesis $H{:}F=G$ may be tested (conservatively) by testing the hypothesis $H_0{:}\gamma=0$ where $\gamma=2P(Y>X)-1$. A representation of the Wilcoxon two-sample statistic is the U-statistic with sign function as kernel namely,

$$U=(mn)^{-1}\sum_{i=1}^{m}\sum_{j=1}^{n}\varepsilon(Y_j-X_i) \tag{120}$$

where $\varepsilon(u) = -1, 0, 1$ according as $u < 0, = 0, > 0$. Since $E[\varepsilon(Y - X)] = \gamma$, the statistic U provides a test statistic for testing H_0. Serfling (1968) obtained the following theorems as a consequence of a central limit theorem for φ-mixing processes due to Ibragimov (1962). We will not give the proof here.

Theorem 3.11. *Let $\{X_n, n \geqq 1\}$ and $\{Y_n, n \geqq 1\}$ be independent strictly stationary φ-mixing processes with $\varphi(n) = 0(n^{-\lambda})$, $\lambda > 1$. Let F and G be the respective continuous marginal distributions. Then, as m and $n \to \infty$ such that m/n has a limit $c \neq 0$, the statistic*

$$Z = m^{1/2}(U - \gamma) \tag{21}$$

is asymptotically normal with mean 0 and variance

$$A^2 = 4 \lim_{k \to \infty} k^{-1} \operatorname{var}\left(\sum_1^k G(X_i)\right) + 4c \lim_{k \to \infty} k^{-1} \operatorname{Var}\left(\sum_1^k F(Y_i)\right). \tag{22}$$

The proof of the theorem consists in showing that $Z - W \xrightarrow{p} 0$ where

$$W = m^{-1/2} \sum_{i=1}^{m} [f_{10}(X_i) - \gamma] + m^{1/2}\, n^{-1} \sum_{j=1}^{n} [f_{01}(Y_j) - \gamma] \tag{22}$$

and

$$f_{10}(t) = E[\varepsilon(Y - t)] = 1 - 2G(t);\ f_{01}(t) = E[\varepsilon(t - X)] = 2F(t) - 1 \tag{23}$$

and then applying the central limit theorem for φ-mixing processes due to Ibragimov (1962).

As an application, we may test for "shift" in an otherwise stationary process $\{\xi_n, n \geq 1\}$ by taking two portions $\{X_i\}_1^m$ and $\{Y_j\}_1^n$ sufficiently for apart to be considered independent and treating these as samples from continuous distributions $F(x)$ and $G(y) = F(y - \Delta)$ respectively. The hypothesis $\Delta = \Delta_0$ may be tested by applying the Wilcoxon two-sample procedure to the observations $\{X_i\}_1^m$ and $\{Y_j'\}_1^n$ where $Y_j' = Y_j - \Delta_0$.

The theorem can be applied to find the robustness of Wilcoxon two-sample statistic under departures from standard assumption of random samples (i.i.d. observations).

4. FURTHER TOPICS

Results in Section 2.3 on density estimation can be easily extended to stationary ϕ-mixing processes as was done in Ahmad, I (1977a). One can also use kernel type of density estimation for this class of processes. This was discussed by Ahmad, I (1977b). A general method of density estimation is given in Foldes and Revesz (1974) for i.i.d. random variables and in Foldes (1974) for dependent sample. They obtain exponential bounds for

$$P\left\{\sup_{a+\delta\leqq x\leqq b-\delta}|p_n(x)-p(x)|\geqq\varepsilon\right\}$$

where

$$p_n(x)=\frac{1}{n}\sum_{k=1}^{n}\Phi_n(x, X_k)$$

estimates density $p(x)$ and $\Phi_n(x, y)$, $n\geqq 1$ is a sequence of Borel-measurable functions on $\mathbb{R}^2$. Nonparametric density estimators for continuous time Markov processes have been studied recently in Banon (1978) and Prakasa Rao (1979). They have applied these results to nonparametric estimation of the drift coefficient of a diffusion process.

Gastwirth and Rubin (1975b) investigated the effect of serial dependence in the data on the efficiency of robust estimators. They showed that if the observations are from a stationary process satisfying certain mixing condition, then linear combinations of order statistics and the H–L estimator are asymptotically normal. The processes need not be ϕ-mixing in general. They have studied Gaussian processes in detail and showed that if the serial correlations $\{\rho_n\}$ are greater than or equal to zero, then the efficiency of the robust estimator relative to the mean is greater than in the case of i.i.d. observations. Related work on these problems is in Gastwirth and Rubin (1971, 1975a) and Gastwirth *et al.* (1967). Koul (1977) proved the asymptotic normality of some robust estimators of the regression parameter when the errors in the regression model are stationary and strong mixing. It is worthwhile noting that a Chernoff–Savage representation for a general class of rank order statistics for stationary ϕ-mixing processes has been studied by Sen and Ghosh (1973). Sen (1977) has also discussed almost sure representation of linear combinations of order statistics with smooth weight functions from stationary ϕ-mixing processes as sums of i.i.d. random variables plus a remainder term which is $0(n^{-1/2-\eta})$, $\eta>0$. Portnoy (1977) obtained estimators of location parameter which are approximately asymptotically optimal in the sense of Huber (1964, 1972) for models which are similar to moving average models.

Albers (1977) has studied one-sample linear rank tests when the observations are from an autoregressive process. He showed that these tests applied to certain transformed observations have asymptotically the same properties as under independence both under the hypothesis and under contiguous alternatives. In particular, the tests are distribution-free asymptotically. Similar work has also been discussed in Modestino (1969). In addition to the various classes of stochastic processes discussed in Section 3,

Bell, *et al.* (1970) have also studied tests for Poisson processes and processes with monotone marginals. These processes can also be obtained from i.i.d. random variables by suitable transformations. They have studied distribution-free tests of goodness-of-fit, two sample tests and tests of the model for these processes.

BIBLIOGRAPHICAL NOTES

Rosenblatt (1956), Parzen (1962) and Bartlett (1963) have studied density estimation for independent and identically distributed observations among others. These results have been extended to stationary Markov processes by Rosenblatt (1970), Roussas (1968, 1969), and Prakasa Rao (1977, 1978b) etc. and for stationary ϕ-mixing processes by Bosq (1973, 1975), Delecroix (1975) and Ahmad, I. (1977a,b). Robustness of non-parametric estimators for dependent data was first studied by Gastwirth and Rubin (1975a,b) and later by Koul (1975, 1977). Most of the work on non-parametric tests for stochastic processes is surveyed in Bell *et al.* (1970). The basic distribution-free test theory can be found in Bell and Smith (1969a,b) and Bell and Donoghue (1969). The fundamental ideas of distribution free tests for i.i.d. observations are due to Pitman (1939), Scheffē (1949) and Lehmann and Stein (1949).

Theorem 2.1 is due to Rosenblatt (1956) in the i.i.d. case. It can also be obtained from general results in Bickel and Lehmann (1969) for the i.i.d. case. Proof in the case of stochastic processes is due to Borwanker (1967). Section 2.2 contains results from Roussas (1968, 1969). Section 2.3 discusses work of Prakasa Rao (1978b). These results are extensions of some recent work of Walter and Blum (1976) for i.i.d. observations. Section 2.4 follows the results of Delecroix (1975). Robustness of the H–L estimator studied in Section 2.5 is due to Koul (1975). Results in Sections 3.1 to 3.3 are taken from Bell, *et al.* (1970) and Section 3.4 contains the work of Serfling (1968).

COMPLEMENTS

1. Let $\{X_n, n \geqq 1\}$ be a strictly stationary φ-mixing process such that $\Sigma_{n=1}^{\infty}\varphi(n) < \infty$ and let F be the distribution of X_i. Let μ_n be the measure on $(\mathscr{K}, \mathscr{B})$ generated by $T_n = n^{1/2}(F_n - F)$ where F_n is the empirical distribution function. Show that μ_n converges weakly to μ where μ is a Gaussian measure with characteristic functional

$$C(\varphi) = \exp(-1/2 G(\varphi, \varphi))$$

where

$$G(\varphi, \psi) = \int (\varphi(x) - E\varphi)(\psi(x) - E\psi) dF(x)$$

$$+\sum_{i=2}^{\infty}\int(\varphi(x)-E\varphi)(\psi(y)-E\psi)dF_i(x,y)$$

$$+\sum_{i=2}^{\infty}\int(\psi(x)-E\psi)(\varphi(y)-E\varphi)dF_i(x,y).$$

Here $E\varphi=\int\varphi(x)dF(x)$ and F_i is the joint distribution of X_1 and X_i.

(Borwanker, 1971)

2. Let $\{X_n, n\geqq 1\}$ be as above and $\{g_n\}$ be a δ-sequence vanishing outside a compact support such that $\int|g_n(x)|dx\leqq M<\infty$ for all n. Let $H_n(x)=n^{-1}\Sigma_1^n g_n(x-X_i)$ and let μ_n^* be the generalized stochastic process generated by $n^{1/2}(H_n-DF)$. Show that $EH_n(x)\to f(x)$ for all x *and* $\mu^{*n}\xrightarrow{\mathscr{L}}D\mu$ where $D\mu$ is Gaussian with $C_{D\mu}(\varphi)=\exp(-1/2G(\varphi,\varphi))$ for all $\varphi\in\mathscr{K}$.

(Borwanker, 1971)

3. Suppose that the transition density $t(x'|x)$ of a stationary Markov process is estimated by $t_n(x'|x)=q_n(x,x')/p_n(x)$ where $p_n(x)$ and $q_n(x,x')$ are as defined in relations (24) and (25). Show that, under assumptions (K1), (H1) and (P1),

$$\int|t_n(x'|x)-t(x'|x)|dx'\xrightarrow{p}0$$

for every $x\in\mathbb{R}$ and under assumptions (K1)–(K3), (H2), (P1) and (P2),

$$\sup_{x\in\mathbb{R}}|t_n(x'|x)-t(x'|x)|\xrightarrow{p}0.$$

(Roussas, 1968)

4. (Bernstein's inequality) Let $\{X_n, n\geqq 1\}$ be a stationary φ-mixing process with $|X_1|\leqq d$, $EX_1=0$ and $EX_1^2>0$. For $k=k(n)$, $n\geqq 2$ with $2k\leqq n$, consider the sequences

$$Y_j=X_{2(j-1)k+1}+\ldots+X_{(2j-1)k},\quad 1\leqq j\leqq l,$$

$$Z_j=X_{(2j-1)k+1}+\ldots+X_{2jk},\quad 1\leqq j\leqq l,$$

$$Z_{l+1}=\begin{cases}0 & \text{if } 2lk=n\\ X_{2lk+1}+\ldots+X_n & \text{otherwise}\end{cases}$$

where l is determined by the inequalities $2lk\leqq n<2(l+1)k$. Let $S_n=X_1+\ldots+X_n$ and $\sigma_n=[2\Sigma_{j=1}^l E(Y_j^2)]^{1/2}$, $n\geqq 2$. Show that for any $\varepsilon>0$ and $0<t<\sigma_n/2kd$,

$$P(S_n>\varepsilon\sigma_n)\leqq e^{1-t}\varepsilon[e^{3/2t^2}+\varphi(k)e^{n/k}].$$

(Bosq, 1975)

5. Suppose $\{\xi_n, n\geqq 0\}$ is stationary φ-mixing with $\varphi(n)=a\rho^n$, $a>0$, $0<\rho<1$, $n\geqq 1$. Further suppose that

(i) $M=\sup_{i\geqq 1}\sup_{x\in E}|e_i^*(x)|<\infty$,

(ii) $\Sigma_{i=1}^{\infty}a_ie_i(x)$ converges uniformly to $f(x)$, and

(iii) $\Sigma_{n=1}^{\infty}q(n)\exp[-\gamma(\sqrt{n}/q(n))]<\infty$ for every $\gamma>0$.

Show that for $\delta>0$ and $\alpha>(-\log\rho)^{-1/2}$,

$$P\left\{\sup_{x\in E}|\hat{f}_n(x)-f(x)|>\delta\right\}=0\left[q(n)\exp\left(-\frac{\delta}{4M^2\alpha}\right)\left(\frac{\sqrt{n}}{q(n)}\right)\right]$$

and hence $\sup_{x\in E}|\hat{f}_n(x)-f(x)|\to 0$ a.s. Here $\hat{f}_n(x)$ is as defined in Section 2.4.

(Bosq, 1975)

6. Let $\{N(t),\ t\geqq 0\}$ be a Poisson process with mean function μ strictly monotone, continuous with $\mu(0)=0$ and $\mu(\infty)=\infty$. If the process is observed over $[0, t]$, obtain a most powerful test for testing $H_0:\mu=\mu_0$ against $H_1:\mu=\mu_1$ assuming that μ_1 and μ_2 are differentiable.

(Bell, *et al.*, 1970)

7. Let $\{N(t)\},\ t\geqq 0\}$ be a Poisson process with mean function μ having the same properties as in the previous problem and let $\{W_n,\ n=1, 2, \ldots,\}$ be the dual waiting time process corresponding to the Poisson process $\{N(t),\ t\geqq 0\}$. Let

$$T_1=\frac{1}{12n}+\sum_{i=1}^{n}\left\{\frac{2i-1}{2n}-\left(\mu_0(W_i)/u_0(t)\right)\right\}^2$$

where $[0, t]$ is the interval of time over which the process is observed. Show that T_1 is *DF* for testing the goodness-of-fit hypothesis $H_0:\mu(t)=\mu_0(t)$ and further show that T_1 has the distribution of Cramer von Mises statistic under H_0.

(Bell *et al.*, 1970)

Chapter 12

Sequential Inference for Stochastic Processes

1. INTRODUCTION

In this chapter, we discuss inference procedures for stochastic processes through sequential procedures. Sequential procedure is a method of statistical inference whose characteristic feature is that the number of observations required or the time required for observation of the process is not determined in advance. The decision to terminate the observation on the process depends, at each stage, on the results of the observations previously made. A merit of the sequential method is that test procedures and estimation procedures can be derived which require on the average, a substantially smaller time period of observation or smaller number of observations than equally reliable procedures based on a predetermined time of observation or number of observations respectively.

Large sample theory of sequential estimation for parameters of stochastic processes will be studied in the next section. We will show that asymptotic properties valid for fixed number of observations are also valid for estimators obtained under sequential sampling. Sequential estimation for processes with continuous parameters is also discussed in this section. We will study efficient sequential estimation for stationary processes with independent increments. Some results on Wald's equation in continuous time and sequential probability ratios tests for stationary processes with independent increments and for continuous time Markov processes are given. Finally, sequential tests for the increase in the drift as well as the sign of the drift of Wiener process are discussed at the end of the chapter. Similar results were discussed earlier in Chapter 9 on sequential problems for diffusion processes.

2. SEQUENTIAL ESTIMATION FOR STOCHASTIC PROCESSES

2.1 Large Sample Theory of Sequential Estimation

Large sample theory of sequential estimation is well known for independent and identically distributed observations. It has been shown that results valid for fixed sample size are asymptotically valid for sequential sampling if the sample size is large. But the advantage of the sequential procedure is that one can obtain estimators with preassigned accuracy. These results have been recently extended to stationary Markov processes. Here we discuss the case of arbitrary discrete parameter stochastic processes.

Let $\mathbf{Y}^n=(Y_1, \ldots, Y_n)$, $n \geqq 1$ be a sequence of random vectors defined on a probability space $(\Omega, \mathscr{F}, P_\theta)$ and taking values in measure spaces $(\mathscr{Y}^n, \mathscr{A}^n, \mu^n)$, μ σ-finite on $(\mathscr{Y}, \mathscr{A})$. $\mathscr{Y}$ could be $\mathbb{R}^m$ and $\theta \in \Theta$, an open subset of $\mathbb{R}^p$. Let $\|\cdot\|$ denote the Euclidean norm of $\mathbb{R}^p$. Suppose that $\mathbf{Y}^n$ has density $p(\mathbf{y}^n; \theta)=p(y_1, y_2, \ldots, y_n; \theta)$ with respect to σ-finite measure μ^n. Let θ_0 be the true parameter and $\hat{\theta}_n$ be a MLE of θ based on the process observed up to time n. We shall assume that $\hat{\theta}_n$ is strongly consistent. For any estimator θ_n of θ, the generalized variance of θ_n is defined to be the determinant of the covariance matrix of θ_n. In this section, we shall prove that the method of maximum likelihood can be used with a suitable stopping rule to obtain estimators of θ with generalized variance smaller than any preassigned positive number. Let

$$Z_n(\theta_0)=\sum_{k=1}^{n} \dot{\phi}_k(\theta_0) \tag{1}$$

where ϕ is as defined in Chapter 7. Let

$$\Gamma_k(\theta)=E_\theta[\dot{\phi}_k(\theta)\dot{\phi}'_k(\theta)|\mathbf{Y}^{k-1}] \tag{2}$$

and assume that

$$n^{-1} \sum_{k=1}^{n} \Gamma_k(\theta) \rightarrow \Gamma(\theta) \text{ a.s.} \tag{3}$$

uniformly for θ over compact sets and $\Gamma(\theta)$ is a positive definite constant matrix. Let us write

$$n^{-1/2}Z_n(\theta_0)=I_n n^{1/2}(\hat{\theta}_n-\theta_0). \tag{4}$$

Suppose that the regularity conditions assumed are such as to ensure that

$$I_n \xrightarrow{p} \Gamma(\theta_0), \tag{5}$$

and

$$n^{1/2}(\hat{\theta}_n-\theta_0)\xrightarrow{\mathscr{L}} N(0,\ \Gamma^{-1}(\theta_0)). \tag{6}$$

For example, conditions (N1)–(N9) of Section 3.4 of Prakasa Rao (1974) will yield relations (5) and (6). Let

$$\Gamma_n^*(\theta)=\sum_{k=1}^{n}\Gamma_k(\theta).$$

It is easy to see that

$$\{\Gamma_n^*(\theta_0)\}^{-1/2}Z_n(\theta_0)\xrightarrow{\mathscr{L}} N(0,\ I) \tag{7}$$

where I is the identity matrix by relations (4)–(6). Let $\{N_n\}$ be a sequence of positive integer valued random variables such that

$$n^{-1}N_n \xrightarrow{p} c>0 \tag{8}$$

where c is a constant. Further suppose that, for every $\boldsymbol{\lambda}\neq 0$ in $\mathbb{R}^p$,

$$\lim_{\varepsilon\to\pm 0}\overline{\lim_{n}}\ \frac{E_{\theta_0}\left(\sum_{k=1}^{[nc]}\boldsymbol{\lambda}'\dot{\phi}_k(\theta_0)\right)^2-E_{\theta_0}\left(\sum_{k=1}^{[nc(1\pm\varepsilon)]}\boldsymbol{\lambda}'\dot{\phi}_k(\theta_0)\right)^2}{n\lambda'\Gamma(\theta_0)\lambda}=c. \tag{9}$$

Theorem 2.1 *Under the assumptions stated above,*

$$\{\Gamma_{N_n}^*(\theta_0)\}^{-1/2}Z_{N_n}(\theta_0)\xrightarrow{\mathscr{L}} N(0,\ I). \tag{10}$$

This result follows from the random central limit theorem for martingales proved in Appendix 1. As a consequence of this theorem and the remarks made above, it follows that

$$\{\Gamma_{N_n}^*(\theta_0)\}^{1/2}(\hat{\theta}_{N_n}-\theta_0)\xrightarrow{\mathscr{L}} N(0,\ I). \tag{11}$$

Suppose we want to estimate θ with a given generalized variance δ^2. Let $V_n(\hat{\theta}_n)$ be an estimate of the covariance matrix of $\hat{\theta}_n$. Suppose that this matrix is non-singular for all n. Let $\delta_r \downarrow 0$ and $\{N_r\}$ be a sequence of positive integer valued random variables defined by the condition that N_r is the smallest integer n such that

$$\det\,[V_n(\hat{\theta}_n)] \leqq \delta_r^2. \tag{12}$$

Let n_r be a sequence of positive integers defined by the condition

$$\det\,[n\Gamma(\theta_0)]^{-1} \leqq \delta_r^2. \tag{13}$$

Further suppose that $N_r/n_r \xrightarrow{p} c > 0$ as $r \to \infty$. It follows from relation (11) that $\hat{\theta}_{N_r}$ is asymptotically normal with mean θ_0 and generalized variance less than δ_r^2 for sufficiently large r.

One can obtain asymptotic confidence ellipsoids for θ_0 of given volume and given confidence coefficient by sequential procedures given above,.

Suppose $\{T_n\}$ is a sequence of estimators of θ_0 satisfying

$$\|\{\Gamma_n^*(\theta_0)\}^{1/2}(T_n - \theta_0) - \boldsymbol{\alpha} - \beta\{\Gamma_n^*(\theta_0)\}^{-1/2}Z_n(\theta_0)\| \xrightarrow{p} 0 \tag{14}$$

for some $\boldsymbol{\alpha} \in \mathbb{R}^p$, $\beta \in \mathbb{R}^1$ not depening on the sample and the following multivariate extension of Anscombe's condition: given $\varepsilon > 0$, $\eta > 0$, there exists a positive integer n_0 and a positive number c such that for every $n \geqq n_0$,

$$P\left\{\max_{|n-n'|<cn} \|\{\Gamma_n^*(\theta_0)\}^{1/2}(T_n - T_{n'})\| < \varepsilon\right\} > 1 - \eta. \tag{15}$$

Gleser (1969) proved that if $\{N_n\}$ is a sequence of positive integer valued random variables such that $n^{-1}N_n \xrightarrow{p} 1$, then the asymptotic distribution of

$$\{\Gamma_n^*(\theta_0)\}^{1/2}(T_{N_n} - \theta_0)$$

is the same as that of

$$\{\Gamma_n^*(\theta_0)\}^{1/2}(T_n - \theta_0)$$

which in turn is the same as the asymptotic distribution of

$$\boldsymbol{\alpha} + \beta\{\Gamma_n^*(\theta_0)\}^{-\frac{1}{2}}Z_n(\theta_0)$$

by relation (14). Thus, if it is required to find an estimator of θ_0 with prescribed generalized variance δ^2, one can use the same type of sequential rule as described above.

Under fairly general conditions, it can be shown that the LEE satisfies (14). Hence one can use the method of maximum likelihood to obtain sequential estimates with a given generalized variance.

2.2 Sequential Estimation for Processes with a Continuous Time Parameter

Let $\{X(t, \theta), t \geqq 0\}$, $\theta \in \Theta$ be a family of stochastic processes. Let $c(t)$ be a cost function for observing the process upto time t and $W(\theta, \theta')$ be the loss involved in giving the estimate θ' when θ is the correct value of the parameter. Let T be a Markov time corresponding to the process $\{X(t), t \geqq 0\}$ and δ be a decision rule depending on $X(t)$ only through its values for $0 \leqq t \leqq T$ if $T < \infty$. The rule corresponding to T and δ is as follows: observe the process up to time T and then adopt the estimate δ. Such a procedure is called a *sequential plan* (T, δ). If T is a constant independent of the sample function $x(t)$, then the procedure is not truly sequential.

We assume here that the processes are continuous from the right. For a given θ, the risk associated with such a procedure is

$$R_\theta(T, \delta) = E_\theta[c(T) + W(\theta, \delta)] \tag{16}$$

provided the expectation exists. Assuming that the expected value exists for every $\theta \in \Theta$, define the maximum risk associated with T and δ by

$$R(T, \delta) = \sup_{\theta \in \Theta} R_\theta(T, \delta). \tag{17}$$

Definition 2.1. A sequential plan $(\hat{T}, \hat{\delta})$ is called a *minimax* rule if

$$R(\hat{T}, \hat{\delta}) \leqq R(T, \delta) \tag{18}$$

for any T and δ for which relation (17) is defined.

Let ζ be a σ-field of subscts of Θ and $R_\theta(\delta, T)$ be ζ-measurable in θ and F be a probability distribution on (Θ, ζ). Define

$$R_F(T, \delta) = \int R_\theta(T, \delta) F(\mathrm{d}\theta) \tag{19}$$

assuming the integral (19) to exist. Let D be a class of decision rules δ described above.

Definition 2.2. A sequential plan (T_F, δ_F) is called a *Bayes rule* for F if

$$R_F(T_F, \delta_F) = \inf_{\substack{\delta \in D \\ 0 \leqq T \leqq \infty}} R_F(T, \delta). \tag{20}$$

Let δ^T denote fixed time procedure with constant observation time T and in this case write δ^T instead of (T, δ). Define

$$R_\theta(\delta^T) = E_\theta[W(\theta, \delta^T)] \tag{21}$$

and

$$R_F(\delta^T) = \int R_\theta(\delta^T) F(\mathrm{d}\theta). \tag{22}$$

Definition 2.3. A procedure δ_F^T is called a *T-Bayes rule* for F if relation (22) assumes its minimum value for $\delta^T = \delta_F^T$.

Let A be any set of sample paths $x(t)$ for which $P[X(t) \in A]$ is a well-defined ζ-measurable function. For every A for which

$$P(A) \equiv \int P[X(t) \in A] F(\mathrm{d}\theta)$$

is positive, define the *a posteriori* probability distribution $F(\theta|A)$ by assigning for every Borel set $S \in \zeta$, the probability

$$\{P(A)\}^{-1} \int_S P[X(t) \in A] F(\mathrm{d}\theta).$$

The *a posteriori risk* corresponding to F and δ^T is

$$R_F(\delta^T|A) = \int R_\theta(\delta^T) F(\mathrm{d}\theta|A).$$

One can give precise measurability conditions for sets A for which $R_F(\delta^T|A)$ exists, but we will not discuss them here. We shall now state a theorem which is familiar from results in decision theory (see Ferguson, 1967). Le Cam (1955) gives an extension of Wald's theory of statistical decision functions.

Theorem 2.2. *Suppose that, for every $T \geqq 0$, there exists a sequence of probability distributions F_n, $n \geqq 1$ for which there are corresponding T-Bayes solutions δ_n^T with the property that the a posteriori risk associated with F_n and δ_n^T is independent of the sample function $x(t)$ and suppose there exists $\hat{\delta}^T$ for which*

$$R(T) = \sup_\theta R_\theta(\hat{\delta}^T) = \lim_{n\to\infty} R_{F_n}(\delta_n^T).$$

If there exists a T_0, $0 \leqq T_0 < \infty$ for which

$$c(T_0) + R(T_0) = \min_{T \geqq 0} c(T) + R(T) \tag{23}$$

holds, then the fixed time procedure $\hat{\delta}^{T_0}$ is a minimax procedure.

We shall now illustrate the concepts discussed in this section by means of some examples. Here $R(T)$ is a non-negative continuous function. We assume that the cost function $c(T)$ is non-negative, lower semi-continuous and tends to infinity as $T \to \infty$. These conditions will guarantee the existence of T_0 which satisfies (23).

Example 2.1. (Poisson process) Consider the family of Poisson processes $\{X_\lambda(t), t \geqq 0\}$ with stationary independent increments satisfying

$$P\{X_\lambda(t) = x\} = e^{-\lambda t} \frac{(\lambda t)^x}{x!}, \; x = 0, 1, 2, \ldots$$

We let $\Theta = \{\lambda : 0 < \lambda < \infty\}$ and ζ be the σ-field of Borel sets of $\Theta \cup \{0\}$. The problem is to estimate λ. If $X(t)$ is observed for $0 \leqq t \leqq T$, then $X(T)$ is a sufficient statistic for λ. Let the loss function be of the form

$$W(\lambda, \delta) = \lambda^{-1}(\delta - \lambda)^2. \tag{24}$$

Let $F_n(\lambda)$, $n \geqq 1$ be the probability distribution on the half line $(0, \infty)$ with density

$$\begin{aligned} f_n(\lambda) &= \frac{1}{n} e^{-\lambda/n} \quad && 0 < \lambda < \infty \\ &= 0 && \text{otherwise.} \end{aligned}$$

Let the process be observed during the time $0 \leqq t \leqq T$. The *a posteriori* probability distribution, when $X(T) = x$, is well-defined and its density is

$$\begin{aligned} f_n(\lambda|x) &= \frac{\lambda^x}{x!}\left(T + \frac{1}{n}\right)^{x+1} \exp(-\lambda(T + 1/n)), \quad && 0 < \lambda < \infty \\ &= 0 && \text{otherwise} \end{aligned}$$

and the *a posteriori* T-risk is

$$R_n(\delta^T|x) = \int_0^\infty \frac{1}{\lambda}(\delta^T - \lambda)^2 f_n(\lambda|x) d\lambda.$$

It can be seen that this risk is minimized by taking

$$\delta^T(X(T) = x) = 1 \Big/ \int_0^\infty \frac{1}{\lambda} f_n(\lambda|x) d\lambda = \frac{x}{T + \frac{1}{n}}.$$

Hence the T-Bayes solution corresponding to $F_n(\lambda)$ is

$$\hat{\delta}_n^T = \frac{X(T)}{T + 1/n}.$$

The corresponding *a posteriori* risk is

$$R_n(\hat{\delta}_n^T|x) = \frac{1}{T+1/n}$$

which is independent of $X(T)$ and hence of the sample since $X(T)$ is a sufficient statistic. On the other hand, by taking

$$\hat{\delta}^T = \frac{X(t)}{T} \tag{25}$$

we see that

$$R_\lambda(\hat{\delta}^T) = \frac{1}{T} \text{ for all } \lambda > 0.$$

Hence the following holds: for a Poisson process with $0 < \lambda < \infty$ and loss function given by relation (24), the fixed time estimate $\hat{\delta}_T$ defined by (25) with $T = T_0$ given by

$$c(T_0) + \frac{1}{T_0} = \min_{T>0} \{c(T) + 1/T\}$$

is minimax.

Example 2.2. (Wiener process) Consider the Wiener process $\{W(t), t \geqq 0\}$ with $E[W(t)] = \mu t$, $\text{var}[W(t)] = t$, $W(0) = 0$, $-\infty < \mu < \infty$. Let the loss function be of the form

$$W(\mu, \delta) = w(|\mu - \delta|)$$

with $w(x)$ non-negative and non-decreasing for $x \geqq 0$. Wolfowitz (1950) proved that if $X_1, \ldots, X_n$ is a random sample from a normal distribution with mean μ and variance unity, then $\overline{X}$, the sample average, is a minimax estimate provided the loss function is a non-decreasing function (cf. Ferguson, 1967, p. 176). The arguments of Wolfowitz (1950) carry over to the present case also. The fixed time estimator

$$\hat{\delta}^T = \frac{X(T)}{T}$$

gives for $T > 0$ and any real μ, the T-risk

$$R_\mu(\hat{\delta}^T) = (2T/\pi)^{1/2} \int_0^\infty w(x) \exp(-x^2 T/2) \mathrm{d}x.$$

Note that $R_\mu(\hat{\delta}^T)$ is non-increasing and continuous from the right as a function of T, $T>0$ and in fact

$$\lim_{T\downarrow 0} R_\mu(\hat{\delta}^T) = \lim_{x\to\infty} w(x).$$

We shall suppose that $w(x)\neq 0$ and

$$\int_0^\infty w(x)e^{-x^{2h}}dx < \infty$$

for some $h\geqq 0$. Then the estimate $\hat{\delta}^T$ define above with $T=T_0$ given by

$$c(T_0)+(2T_0/\pi)^{1/2}\int_0^\infty w(x)\exp(-x^2T_0/2)dx$$

$$=\min_{T\geqq 0}[c(T)+(2T/\pi)^{1/2}\int_0^\infty w(x)\exp(-x^2T/2)dx] \tag{26}$$

is minimax, where the terms following $c(T)$ and $c(T_0)$ are replaced by $\sup_x w(x)$ for $T=0$ or $T_0=0$ and δ_T may be replaced by any estimator if the minimum in (26) is attained at $T=0$ (which can only occur if w is bounded).

Kiefer (1975) discussed invariant, minimax sequential estimation for continuous time processes when the invariance is in the decision problem or the invariance has to do with the time parameter. We shall not pursue them here.

2.3 Efficient Sequential Estimation

In the previous section, we studied Bayes and minimax estimators. We shall now discuss some optimality properties of a restricted class of unbiased sequential plans by obtaining an analogue of the Cramer–Rao inequality.

Let $\xi_\theta(t)$ be a stochastic process with continuous time parameter defined on a probability space $(\Omega, \mathscr{F}, P)$ and with probability distribution depending on a scalar, a parameter $\theta\in\Theta$. Suppose the process is observed up to a Markov time τ. Knowing the values τ and $\xi_\theta(\tau)$, we wish to estimate the value of a function $g(\theta)$ of the parameter θ. Thus the problem is to find Markov time τ and estimator $f(\tau, \xi_\theta(\tau))$ having some optimal properties. An unbiased sequential plan will be defined by a Markov time τ and an unbiased estimator f of the function $g(\theta)$. We assume that the loss function is quadratic.

Before we obtain the analogue of the Cramer-Rao inequality for unbiased sequential plans, we shall state Sudakov's Lemma. Let

$$\{\xi_\theta\}=\{\xi_\theta(t)=x(t, w),\ w\in(\Omega.\ \mu_\theta),\ \theta\in\Theta,\ t\in T\}$$

be a family of stochastically continuous processes whose probability distributions are entirely defined by a given parameter θ. Let $\mathscr{X} \subset \mathbb{R}^n$ be the space of values of the processes $\{\xi_\theta\}$ and $T = [0, \infty)$. Let $\mathscr{F}_t$ be the σ-algebra generated by the functions $x(s, w)$, $s \leqq t$. We suppose that, for any $t < \infty$, all measures μ_θ are absolutely continuous with respect to a certain measure μ_{θ_0} on $\mathscr{F}_t$. Let τ be a Markov time. If the process $\xi_\theta(t) = x(t, w)$ is measurable, then the function $x(\tau(w), w)$ is measurable on Ω. The functions $\tau(w)$ and $x(\tau(w), w)$, both measurable on Ω, generate, for every θ, the measure m on the product space $T \times \mathscr{X}$ according to the mapping

$$w \to (\tau(w),\ x(\tau(w),\ w)) = z(w)$$

in the following manner: for any Borel set $C \subset T \times \mathscr{X}$,

$$m(c) = \mu(Z^{-1}(C)).$$

Let $p_{\theta_0}(t, x; \theta)$ be the density of the process $\xi_\theta(t)$ relative to $\xi_{\theta_0}(t)$ at time t.

Theorem 2.3. (*Sudakov's Lemma*). *Let the functions* $x(t, w)$ *of the variable* t *be right continuous for almost every* $w \in \Omega$ *(with respect to* μ_{θ_0}*). Suppose that the function* $p_{\theta_0}(t, x; \theta)$ *is continuous with respect to both arguments* t *and* x *on* $T \times \mathscr{X}$. *If the family* $\{\xi_\theta\}$ *is such that the end time* t *determines a sufficient statistic, then the measures* m_θ *are absolutely continuous with respect to the measure* m_{θ_0} *and*

$$dm_\theta / dm_{\theta_0} = p_{\theta_0}(t,\ x;\ \theta).$$

Proof of this theorem is given in Appendix 3. Taking into account the Sudakov result of the determination of the Radon–Nikodym derivative dm_θ/dm_{θ_0}, we shall now give the lower bound for the variance for the estimators of θ in a class of processes for which the Sudakov lemma is valid.

Let $U = T \times \mathscr{X}$ and $\mathscr{B}$ be the σ-algebra of Borel subsets of U. By $t = t(u)$ and $x = x(u)$, we mean the components of the point $u \in U$. Hereafter, we assume that $\theta \in [a, b]$. Any real function f defined on U and $\mathscr{B}$-measurable is called an *estimator* of the parameter $Q = g(\theta)$ where g is a function of θ.

Definition 2.4. By an *unbiased sequential plan*, we mean a triplet (τ, g, f) consisting of a Markov time τ, the function $g(\theta)$ and its estimator f, if for every θ in $[a, b]$,

$$P(\{w : 0 < \tau(w) < \infty\}) = 1$$

and f is an unbiased estimator of $g(\theta)$ i.e.

$$E_\theta(f) = \int_U f(u) p_{\theta_0}(t,\ x;\ \theta) m_{\theta_0}(du) = g(\theta)$$

for every θ in $[a, b]$.

We consider only such estimators for which $E_\theta(f^2)$ exists and is finite i.e.

$$E_\theta(f^2) = \int_U f^2(u) p_{\theta_0}(t, x; \theta) m_{\theta_0}(du) < \infty$$

for all $\theta \in [a, b]$. We further assume that $g(\theta)$ and $p_{\theta_0}(t, x; \theta)$ satisfy certain regularity conditions namely

(M1) $g(\theta)$ is differentiable in $[a, b]$ with derivative $g'(\theta) \neq 0$ for all θ in $[a, b]$;

(M2) For any fixed θ, there exists a function $G(t, x)$ and an $\varepsilon > 0$ such that

$$\left| \frac{p_{\theta_0}(t, x; \theta') - p_{\theta_0}(t, x; \theta)}{(\theta' - \theta) p_{\theta_0}(t, x; \theta)} \right| < G(t, x), \ |\theta' - \theta| < \varepsilon \tag{27}$$

where

$$\int_U G^2(t, x) p_{\theta_0}(t, x; \theta) m_{\theta_0}(du) < \infty. \tag{28}$$

Theorem 2.4. *Under the above assumptions, for each unbiased sequential plan* (τ, g, f),

$$\operatorname{var}_\theta(f) \geqq \frac{[g'(\theta)]^2}{\displaystyle\int_U (\partial \log p_{\theta_0}(t, x; \theta)/\partial\theta)^2 p_{\theta_0}(t, x; \theta) m_{\theta_0}(du)} \tag{29}$$

equality holding at a particular parameter θ *if and only if*

$$f(u) - g(\theta) = h(\theta) \frac{\partial \log p_{\theta_0}(t, x; \theta)}{\partial \theta}$$

almost everywhere with respect to measure m_{θ_0} *where* $h(\theta) \neq 0$.

Proof. Since f is an unbiased estimator of $g(\theta)$, it follows that

$$\int_U f(u) \frac{p_{\theta_0}(t, x; \theta') - p_{\theta_0}(t, x; \theta)}{(\theta' - \theta) p_{\theta_0}(t, x; \theta)} p_{\theta_0}(t, x; \theta) m_{\theta_0}(du) = \frac{g(\theta') - g(\theta)}{\theta' - \theta} \tag{30}$$

for any $\theta \neq \theta'$. Inequality (27) implies that

$$|f(u) \frac{p_{\theta_0}(t, x; \theta') - p_{\theta_0}(t, x; \theta)}{(\theta' - \theta) p_{\theta_0}(t, x; \theta)}| < |f(u) G(t, x)|$$

where $E_\theta(|fG|) < \infty$ since $E_\theta(f^2) < \infty$ and $E_\theta(G^2) < \infty$ by assumptions. Then, from the Lebesgue Dominated Convergence Theorem, one can evaluate limits on both sides of relation (30) as $\theta' \to \theta$. Hence

$$\int_U f(u) \frac{\partial \log p_{\theta_0}(t, x; \theta)}{\partial \theta} p_{\theta_0}(t, x; \theta) m_{\theta_0}(du) = g'(\theta). \tag{31}$$

In a similar way, relations (27) and (28) and the condition

$$\int_U p_{\theta_0}(t, x; \theta) m_{\theta_0}(du) = 1$$

for each $\theta \in [a, b]$ show that

$$\int_U \frac{\partial \log p_{\theta_0}(t, x; \theta)}{\partial \theta} p_{\theta_0}(t, x; \theta) m_{\theta_0}(du) = 0. \tag{32}$$

Hence

$$\begin{aligned} &\int_U f(u) \frac{\partial \log p_{\theta_0}(t, x; \theta)}{\partial \theta} p_{\theta_0}(t, x; \theta) m_{\theta_0}(du) \\ &= \int_U [f(u) - g(\theta)] \frac{\partial \log p_{\theta_0}(t, x; \theta)}{\partial \theta} p_{\theta_0}(t, x; \theta) m_{\theta_0}(du). \end{aligned} \tag{33}$$

Relations (31) and (33) imply that

$$\begin{aligned} [g'(\theta)]^2 \leqq &\int_U [f(u) - g(\theta)]^2 p_{\theta_0}(t, x; \theta) m_{\theta_0}(du). \\ &\int_U \left\{\frac{\partial \log p_{\theta_0}(t, x; \theta)}{\partial \theta}\right\}^2 p_{\theta_0}(t, x; \theta) m_{\theta_0}(du) \end{aligned} \tag{34}$$

by the Cauchy–Schwartz inequality. Since f is an unbiased estimator of $g(\theta)$, the above inequality proves that

$$[g'(\theta)]^2 \leqq \operatorname{var}_\theta(f) \int_U \left\{\frac{\partial \log p_{\theta_0}(t, x; \theta)}{\partial \theta}\right\}^2 p_{\theta_0}(t, x; \theta) m_{\theta_0}(du) \tag{35}$$

which yields (29). Inequality (29) is an equality if and only if there exist constants c_1 and c_2, possibly depending on θ, not both zero such that

$$c_1(\theta)[f(u) - g(\theta)] = c_2(\theta) \frac{\partial \log p_{\theta_0}(t, x; \theta)}{\partial \theta} \quad \text{a.s. } m_\theta. \tag{36}$$

Since the density $p_{\theta_0}(t, x; \theta)$ should be strictly positive, the equality holds almost everywhere m_{θ_0}. Further, since $g'(\theta) \neq 0$ for θ in $[a, b]$, inequality (35) implies that neither $f(u) - g(\theta)$ nor $\partial \log p_{\theta_0}(t, x; \theta)/\partial\theta$ can be equal to zero almost everywhere m_{θ_0}. Hence $c_1(\theta)$ and $c_2(\theta)$ cannot be zero for any θ in

$[a, b]$. Therefore, equality (36) holds if and only if equality

$$f(u) - g(\theta) = h(\theta) \frac{\partial \log p_{\theta_0}(t, x; \theta)}{\partial \theta} \text{ a.s. } m_{\theta_0} \tag{37}$$

holds where $h(\theta) \neq 0$.

Definition 2.5. An unbiased sequential plan (τ, g, f) is said to be *efficient for a given* θ if inequality (29) becomes equality at θ. The estimator f is then called *efficient* at this value of θ and the function g is *efficiently estimable* at the point θ.

An estimator f is efficient at a given θ if and only if it is of the form

$$f(u) = h(\theta) \frac{\partial \log p_{\theta_0}(t, x; \theta)}{\partial \theta} + g(\theta) \text{ a.s. } m_{\theta_0}. \tag{38}$$

This follows from the theorem proved above.

Definition 2.6. An unbiased sequential plan (τ, g, f) is said to be *efficient* if it is efficient for each θ in $[a, b]$. The estimator f is then called *efficient* and the function g is said to be *efficiently estimable.*

We shall now study some classes of processes for which the above results can be applied.

Definition 2.7. A class of homogeneous processes $\{X_t\}$ with independent increments satisfying $P(\{w: X_0(w) = 0\}) = 1$ is said to be of *exponential type* if the density of X_t, with respect to the Lebesgue or counting measure, belongs to the exponential family of distributions i.e.

$$p(t, x; \theta) = r(t, x) \exp\{w_1(\theta)t + w_2(\theta)x\}, \; t > 0, \; x \in X \subset \mathbb{R}, \; \theta \in [a, b] \tag{39}$$

where the function $r(t, x)$ is non-negative.

We shall assume that these processes are right continuous. For this class of processes, $X_t(w)$ is a sufficient statistic for the parameter θ. We shall assume further that $w_1(\theta)$, $w_2(\theta)$ are differentiable in $[a, b]$, $w_2'(\theta) > 0$ for θ in $[a, b]$ and the function $w_1'(\theta)/w_2'(\theta)$ is strictly decreasing in $[a, b]$.

Let us remark that Poisson, negative-binomial, gamma and Wiener process with linear drift are of exponential type.

Relation (39) gives that

$$p_{\theta_0}(t, x; \theta) = \exp\{w_1(\theta)t + w_2(\theta)x\} \exp\{-w_1(\theta_0)t - w_2(\theta_0)x\} \tag{40}$$

and from Sudakov's Lemma, there exists a coutably additive measure ρ_τ on

$\mathscr{B}$ independent of θ such that, for each θ in $[a, b]$

$$P(\{w:(\tau(w), X_\tau(w))\in C\})$$

$$= \int_C \exp\{w_1(\theta)t + w_2(\theta)x\}\exp\{-w_1(\theta_0)t - w_2(\theta_0)x\}m_{\theta_0}(du)$$

$$= \int_C \exp\{w_1(\theta)t + w_2(\theta)x\}\rho_\tau(du)$$

for all $C\in\mathscr{B}$ where the last equality defines the measure ρ_τ. It is easy to see now that

$$\text{var}_\theta(f) = \frac{[g'(\theta)]^2}{E_\theta[w_1'(\theta)\tau + w_2'(\theta)X_\tau]^2} \tag{41}$$

for any unbiased estimator f of $g(\theta)$ since

$$\frac{\partial \log p_{\theta_0}(t, x; \theta)}{\partial\theta} = w_1'(\theta)t + w_2'(\theta)x.$$

Hence an unbiased estimator f of $g(\theta)$ is efficient at θ if and only if it is of the form

$$f(u) = h(\theta)[w_1'(\theta)t(u) + w_2'(\theta)x(u)] + g(\theta) \text{ a.e. } \rho_\tau$$

3. SEQUENTIAL TESTS OF HYPOTHESES FOR STOCHASTIC PROCESSES

3.1 Wald's Equations in Continuous Time

Suppose $\{Z(t), t\geqq 0\}$ is a separable homogeneous process with independent increments. Further suppose that $Z(0)=0$. Let $(\Omega, \mathscr{F}, P)$ be the probability space on which the process is defined. Let $\{\mathscr{F}_t; t\geqq 0\}$ denote an increasing family of sub-σ-fields of $\mathscr{F}$, $\mathscr{F}_t$ containing the σ-field generated by $\{Z(s):0\leqq s\leqq t\}$. Let T denote a Markov time corresponding to the family $\{\mathscr{F}_t, t\geqq 0\}$.

Let $Z=Z(1)$, $M(\theta)=E(e^{-\theta Z})$. Let δ and σ^2 denote the drift and variance per unit time of the process $Z(\cdot)$. Assume that $\delta\geqq 0$ and $0<\sigma^2<\infty$. Note that $E(e^{-\theta Z(t)})=M(\theta)^t$. Let

$$\Theta = \{\theta | M(\theta) < \infty\}$$

and

$$\Theta_0 = \{\theta | M(\theta) \leqq 1\}.$$

Θ and Θ_0 could be degenerate or infinite intervals with 0 as the left end point of Θ_0. Let $K(\theta)=\log E(e^{i\theta Z})$.

Lemma 3.1.

(i) $\{\exp(i\theta Z(t)\to tK(\theta)), \mathscr{F}_t; t\geqq 0\}$ *is a unit mean martingale for all* θ.

(ii) $\{\exp(-\theta Z(t))M(\theta)^{-t}, \mathscr{F}_t; t\geqq 0\}$ *is a unit mean martingale for each* $\theta\in\Theta$.

(iii) $\{\exp(-\theta Z(t), \mathscr{F}_t; t\geqq 0\}$ *is a positive super martingale for each* $\theta\in\Theta_0$.

Proof of this lemma is left as an exercise for the reader. Wald's identity in continuous time states that the martingale in (ii) has unit mean when t is replaced by a Markov time T satisfying certain conditions. We now present the following theorem.

Theorem 3.1. *Suppose* Θ_0 *is non-degenerate. Then for any Markov Time T,*

$$E[\exp(-\theta Z(T))]\leqq 1 \tag{42}$$

for every $\theta\in\Theta_0$. *and* $E[Z(T)^r.]$ *exists* $(\leqq\infty)$ *for every positive integer* $r\geqq 1$.

Proof. Note that

$$E[e^{-\theta Z(T)}]=E\left[\lim_{n\to\infty}\chi_{[T\leqq n]}e^{-\theta Z(T)}\right]$$

where χ_A denotes the indicator function of the set A. "E" and "lim" may be interchanged by the monotone convergence theorem and T may then be replaced by $T_n=\min(n, T)$. Hence

$$E[e^{-\theta Z(T)}]\leq\limsup_n E[e^{-\theta Z(T_n)}]. \tag{43}$$

The supermartingale $\{e^{-\theta Z(t)}, t\geqq 0\}$, $\theta\in\Theta_0$ remains a supermartingale when stopped at the bounded stopping time T_n (c.f. Doob, 1953, Theorem 11.6, Chapter VII), so that

$$E[\exp(-\theta Z(T_n))]\leqq E[\exp(-\theta Z(0))]=1 \tag{44}$$

for all $\theta\in\Theta_0$ and for all n. Relations (43) and (44) together prove that

$$E[e^{-\theta Z(T)}]\leqq 1$$

for all $\theta\in\Theta_0$. (42) implies that $Z(T)^-$ has a moment generating function so that moments for all orders of $Z(T)^-$ are finite and hence moments of all orders of $Z(T)$ exist.

Theorem 3.2. *If either* (i) Θ_0 *is non-degenerate or* (ii) $ET<\infty$ *and* $EZ=0$, *then* $E[Z(T)]$ *exists and*

$$EZ(T)=EZET. \tag{45}$$

This result follows as a corollary to the following continuous time version of a result due to Robbins and Samuel (1966). For any random Y, define the generalized expectation E^*Y by $E^*Y=a$ if $Y_1+\ldots+Y_n/n\to a$ a.s. where Y_i are i.i.d. versions of Y. Robbins and Samuel (1966) proved a discrete version of the following theorem.

Theorem 3.3. *If* $Z(t)/t\to a$ a.s. *as* $t\to\infty$ $(0\leqq a\leqq\infty)$, *then* $E^*Z(t)$ *exists and equals* $a\,E(T)$ *unless both* $a=0$ *and* $E(T)=\infty$.

Note that, if $E[Z]<\infty$, then a theorem due to Doob (1953), p. 364 asserts that $Z(t)/t\to E[Z]$ a.s. We leave the proof of Theorem 3.3. as an exercise for the reader.

An alternative method of proving a continuous version of Wald's equation is through martingales. We shall now sketch a proof due to Doob (1953), p. 380. Suppose the sample paths of the process are right continuous and $E|Z(t)|<\infty$ for all $t\geqq 0$. Then $E[Z(t)]=E[Z]t$, $t\geqq 0$ by the stationarity of the process. Let $X(t)=Z(t)-tE(Z)$. Then the process $\{X(t),\ t\geqq 0\}$ is a martingale with independent increments. Further, if T is a stopping time, then the stopped process $\{X(t),\ 0\leqq t\leqq T\}$ is also a martingale and hence

$$E[X(T)]=E[X(0)]$$

i.e.

$$E[Z(T)-TE(Z)]=0$$

which gives Wald's equation. This method can also be used to derive Wald's fundamental identity

$$E[\exp(-\theta Z(T))M(\theta)^{-T}]=1$$

for any stopping time T by observing that $\{\exp(-\theta Z(t))M(\theta)^{-t},\ t\geqq 0\}$ is a martingale and applying Theorem 11.8 of Chapter VII of Doob (1953). Note that the above result holds in particular for stopping time of the form $T=t$ if $Z(t)\geqq a$ or $Z(t)\leqq b$ while $b<Z(\tau)<a$ for $\tau<t$, a, b being constants, $a>0$, $b<0$.

Hall (1969) derived moment identities of higher order. We will not pursue them here. We shall now give another method for deriving Wald's identity for the Wiener process.

Theorem 3.4. *Let* $\{W(t),\ t\geqq 0\}$ *be the standard Wiener process. For any bounded Markov time* T,

$$E[\exp(\lambda W(T)-\lambda^2 T/2)]=1 \tag{46}$$

for every $\lambda > 0$.

Proof. Let us suppose that T is a bounded Markov time and that $T \leqq M$ a.s. The process

$$Y(t) = \exp(\lambda W(t) - \lambda^2 t/2), \ t \geqq 0 \tag{47}$$

satisfies

$$E[Y(t)] = 1 \tag{48}$$

for each t. Let χ_A denote the indicator of a set A and $E_A g = E[\chi_A g]$ for any integrable function g and set A. For any stopping time T, the properties of the conditional expectation give, for fixed $t > 0$,

$$\begin{aligned} E_{[T<t]}\{Y(t)\} &= E\{E(\chi_{[T<t]} Y(t) | T, Y(T))\} \\ &= E\{\chi_{[T<t]} E[Y(t) | T, Y(T)]\}. \end{aligned} \tag{49}$$

The strong Markov property (cf. Ito and McKean, 1965) gives

$$E[Y(t) | T, Y(T)] = Y(T) \quad \text{a.s.}$$

and hence

$$E_{[T<t]}\{Y(t)\} = E_{[T<t]}\{Y(T)\}. \tag{50}$$

Choosing $t > M$, it follows from (49) and (50) that

$$E(\exp(\lambda W(T) - \lambda^2 T/2)) = 1.$$

3.2 Wald's Sequential Probability Ratio Test for Continuous Time Stochastic Processes

Let $\{X_1(t), t \geqq 0\}$ and $\{X_2(t), t \geqq 0\}$ be two different stochastic processes. Suppose the statistician observes continuously, beginning at $t = 0$, a process $\{X(t), t \geqq 0\}$ which is either $\{X_1(t)\}$ or $\{X_2(t)\}$, and has to decide, as soon as possible, whether $\{X(t)\}$ is $\{X_1(t)\}$ or $\{X_2(t)\}$. "As soon as possible" means the following. Let T be the stopping time when the statistician reaches a decision and let $E_i T$ denote the expected value when $\{X(t)\} = \{X_i(t)\}$, $i = 1, 2$. Let α_1, α_2 be two positive constants, $\alpha_1 + \alpha_2 < 1$. Subject to the requirement that the probability of an incorrect decision when $\{X(t)\} = \{X_i(t)\}$ be at most α_i, the problem is to give a procedure for deciding between $\{X_1(t)\}$ and $\{X_2(t)\}$ such that $E_i(T)$ is a minimum for $i = 1, 2$.

We shall assume that $X(t)$ is a sufficient statistic for the process i.e. the conditional distribution of $X(\tau)$, $0 \leqq \tau \leqq t$ given $X(t)$ is, with probability one for every t, the same for the processes $\{X_1(t)\}$ and $\{X_2(t)\}$. Let $f_1(x, t)$

and $f_2(x, t)$ be the density functions of $X_1(t)$ and $X_2(t)$ respectively with respect to a σ-finite measure. Let

$$L(t)=\log\frac{f_2(X(t), t)}{f_1(X(t), t)};\ L(0)=0.$$

We shall suppose that $\{L(t), t\geqq 0\}$ is a process with stationary independent increments.

Let $b<0<a$ be given numbers. We shall use Wald's sequential probability ratio test (SPRT) as follows: As long as $L(t)$ lies between b and a, continue observing $X(t)$. As soon as $L(t)\leqq b$, stop observing $\{X(t)\}$ and decide $\{X(t)\}=\{X_1(t)\}$. As soon as $L(t)\geqq a$, stop observing $\{X(t)\}$ and decide $\{X(t)\}=\{X_2(t)\}$.

We shall now apply the above test for the Wiener and Poisson processes.

Example 3.1. Let $\{X_1(t), t\geqq 0\}$ and $\{X_2(t), t\geqq 0\}$ be Wiener processes with $X_1(0)\equiv X_2(0)=0$, $E[X_1(t)]=m_1 t$, $E[X_2(t)]=m_2 t$ and $\text{var}[X_1(t)=\text{var}[X_2(t)]=t$. It is easy to check that

$$L(t)=X(t)(m_2-m_1)-\frac{t}{2}(m_2^2-m_1^2). \tag{51}$$

The sample functions of the $\{X(t)\}$ process are continuous with probability one. We choose a and b, $b<0<a$, such that the statistician will continue to observe $\{X(t)\}$ until $L(t)=a$ or $L(t)=b$. In view of results in Wald (1947) which are approximations in the discrete case since excess of $L(t)$ over a or b is neglected, Wald's formulae cease to be mere approximations and become exact here. Thus, for example,

$$\alpha_1(a, b)=\frac{1-e^b}{e^a-e^b}, \tag{52}$$

and

$$\alpha_2(a, b)=\frac{e^b(e^a-1)}{e^a-e^b} \tag{53}$$

where $\alpha_i(a, b)$ is the probability of error when $\{X(t)\}=\{X_i(t)\}$.

For any Wiener process $\{X(t)\}$ with mean m and variance one per unit time, not necessarily either $\{X_1(t)\}$ or $\{X_2(t)$, the probability that $L(t)$ will reach b before reaching a is given exactly by

$$\frac{A^{h(m)}-1}{A^{h(m)}-B^{h(m)}} \tag{54}$$

where $A=\log a$, $B=\log b$ and for any m, $h(m)$ is the non-zero solution of the equation

$$E_m[\exp(L(1)h(m))]=1.$$

It is easy to see that

$$h(m)=\frac{m_1+m_2-2m}{m_1-m_2}. \tag{55}$$

Let H be the quantity defined in (54). Then

$$ET=\frac{Hb+(1-H)a}{E_m(L(1))}. \tag{56}$$

ET is the expected time for the termination of the test. One can also obtain the exact distribution of T as in the discrete case. All these results are exact as the excess of $L(T)$ over the boundaries a and b is zero with probability one. These relations can also be derived directly from the Wald's identity and Wald's equation discussed in the previous section.

Example 3.2. (Poisson process). In this example we consider the problem of deciding which of two values given in advance represent the correct mean occurence time of a Poisson process with stationary independent increments.

The probability that a Poisson process with mean occurence time $1/\lambda$ will result in exactly k occurences in time $[0, T]$ is

$$\frac{(\lambda T)^k e^{-\lambda T}}{k!}, \quad k=0, 1, 2, \ldots .$$

Let H_i, $i=1, 2$ denote the hypothesis that $\lambda=\lambda_i$, $\lambda_i>0$. It is clear that the two processes satisfy the conditions imposed above. Hence, given two positive numbers α_1, $\alpha_2(\alpha_1+\alpha_2<1)$, a test procedure for deciding between H_1 and H_2, which satisfies the condition that the probability of a wrong decision when H_i is true does not exceed α_i, $i=1, 2$ is a Wald sequential probability ratio test. This is also optimal in some sense. Observe that

$$L(t)=X(t)\log\frac{\lambda_2}{\lambda_1}-(\lambda_2-\lambda_1)t. \tag{57}$$

Thus, assuming $\lambda_2>\lambda_1$, the best decision rule is specified by two numbers a, $b(b<0<a)$ in the manner described earlier.

Suppose that α_1 and α_2 are the actual probabilities of error. It is easily seen that

$$\frac{1-\alpha_2}{\alpha_1}=\frac{P_2(H_2)}{P_1(H_2)}, \frac{\alpha_2}{1-\alpha_1}=\frac{P_2(H_1)}{P_1(H_1)} \tag{58}$$

where $P_i(H_j)$ is the probability that hypothesis H_j is accepted when H_i is true. Furthermore

$$E_1^*(e^{L(T)}|S) = \frac{P_2(H_1)}{P_1(H_1)}$$

and

$$E_1^{**}(e^{L(T)}|S) = \frac{P_2(H_2)}{P_1(H_2)}$$

where $E_1^*(u|S)$ denotes the conditional expected value of u under the hypothesis H_1 and under the restriction that H_1 is accepted and $E_1^{**}(u|S)$ is the conditional expectation of u under the hypothesis H_1 and under the restriction that H_2 is accepted. Hence

$$\exp\left(\inf_i L(T)\right) = \inf_i e^{L(T)} \leqq \frac{P_2(H_i)}{P_1(H_i)} \leqq \sup_i e^{L(T)} \leqq e^{\sup_i L(T)} \tag{59}$$

where the $\sup_i$ and $\inf_i$ being taken over all values of $L(T)$ where the observation is stopped at time T with decision to adopt H_i. If the decision to accept H_2 is adopted at time T, then $L(T) \geqq a$ and $L(t) < a$ for $t < T$. Since $L(t+0) - L(t) \leqq \log \lambda_2/\lambda_1$ with probability one, relations (58) and (59) imply that

$$e^a \leqq \frac{1-\alpha_2}{\alpha_1} \leqq \frac{\lambda_2}{\lambda_1} e^a. \tag{60}$$

Similarly, if at time T, we decide to terminate observation and adopt H_1, we must have $L(T) \leqq b$ and $L(t) > b$ for $t < T$. Since $L(t) \geqq L(t-0)$ almost surely, $L(T) = b$ with probability one. Hence

$$\frac{\alpha_2}{1-\alpha_1} = e^b.$$

Thus $b = \log (\alpha_2/1-\alpha_1)$ and

$$\log\frac{\lambda_1}{\lambda_2} + \log\frac{1-\alpha_2}{\alpha_1} \leqq a \leqq \log\frac{1-\alpha_2}{\alpha_1}.$$

Dvoretzky, *et al.* (1953) give a method for exact computation of a and the power function of the test.

3.3. Generalized Sequential Probability Ratio Tests

Suppose $\{Z_n\}$ is a sequence of random vectors (perhaps of varying dimensions) on a measure space $(\Omega, \mathscr{F})$, Z_n representing the data to be collected at stage n of the experiment. Let $\mathscr{F}_n$ represent the sub σ-field of $\mathscr{F}$ generated by

$$\mathbf{Z}^{(n)} = (Z_1, \ldots, Z_n)$$

and let $\mathscr{F}_0 = \mathscr{F}$; let P_0 and P_1 represent two alternative probability measures on $(\Omega, \mathscr{F})$ and E_i represent expectation with respect to P_i. Let H_i represent the hypothesis that P_i is the correct measure. We consider testing H_0 against H_1 sequentially.

Let p_{in} denote the density of $\mathbf{Z}^{(n)}$ under H_i (with respect to a mutually dominating measure) and let R_n denote the likelihood ratio $R_n = p_{1n}/p_{0n}$ (well defined unless $p_{1n} = p_{0n} = 0$). We define the log-likelihood ratio $L_n = \log R_n$, with increments $X_n = L_n - L_{n-1}$ $(L_0 = 0)$ which is well defined, though possibly infinite, at least under H_0 and H_1.

Note that $\{R_n\}$ is a supermartingale when H_0 is true (cf. Doob, 1953, p. 348). For

$$E_0(R_n|\mathscr{F}_{n-1}) = E_0\left(\frac{p_{1n}(Z_n|Z_1, \ldots, Z_{n-1})}{p_{0n}(Z_n|Z_1, \ldots, Z_{n-1})} \frac{p_{1,n-1}(\mathbf{Z}^{(n-1)})}{p_{0,n-1}(\mathbf{Z}^{(n-1)})} |\mathscr{F}_{n-1}\right)$$

$$= \int_{[p_{0n}>0]} \frac{p_{1n}(z_n|\mathbf{Z}^{(n-1)})}{p_{0n}(z_n|\mathbf{Z}^{(n-1)})} \frac{p_{1,n-1}(\mathbf{Z}^{(n-1)})}{p_{0,n-1}(\mathbf{Z}^{(n-1)})} p_{0,n}(z_n|\mathbf{Z}^{(n-1)} dP_0(z_n)$$

$$\leqq \frac{p_{1,n-1}}{p_{0,n-1}} = R_{n-1}. \tag{61}$$

If $P_1 \ll P_0$, the equality will hold above and $\{R_n\}$ will be a martingale when H_0 is true.

Let $\{W_\delta(t), t \geqq 0\}$, $W_\delta(0) = 0$ be the Wiener process with drift δ and variance one per unit time. From the results on embeddable super martingale given in the Appendix 1, one can prove the following theorem.

Theorem 3.5. *There exist successive stopping times $\{T_n\}$ on $W_\delta(\cdot)$, with $\delta = -\frac{1}{2}(+\frac{1}{2}$ respectively) for which $\{L_n\}$ and $\{W_\delta(T_n)\}$ have the same finite-dimensional distributions under $H_0(H_1$ respectively). If $P_1 \ll P_0(P_0 \ll P_1$ respectively), then $\{\exp(-2\delta L_n), \mathscr{F}_n\}$ is a martingale.*

Wald's SPRT terminates the experiment at stage n if $L_n \leqq b$ (in favour of H_0) or $L_n \geqq a$ (in favour of H_1) while $b < L_m < a$ for all $m < n$. Wald proposed using $a = \log A$, $b = \log B$ with $A = (1-\beta)/\alpha$ and $B = \beta/1-\alpha$ where (α, β) are the desired error probabilities $(\alpha + \beta < 1)$. Let $P_i(H_j)$ be the probability that SPRT decides in favour of H_j when H_i is true. Let $\alpha' - P_0(H_1)$ and $\beta' - P_1(H_0)$.

The relevance of embedding theory in this context is that instead of watching the discrete time process $\{L_n\}$, we can equivalently (for probability calculations under H_0) watch a corresponding continuous time Wiener process. As an approximation to SPRT, we can terminate in favour of H_0 and H_1 according to whether $W_\delta(\cdot)$ first crosses the levels b and a respectively. The process will terminate, at time T, say, not later than $T_{N'}$ where N' is the termination time for the discrete time problem. Wald's approximation formulae can be derived through this analogy. It is clear that

$$\begin{aligned}\alpha' &= P_0(L_n \text{ crosses } a \text{ before } b)\\ &\leqq P_0\left(\sup_n L_n \geqq a\right)\\ &= Pr\left(\sup_n W_{-1/2}(T_n) \geqq a\right) \text{(by the embedding Theorem 3.5)} \qquad (62)\\ &\leqq Pr\left(\sup_t W_{-1/2}(t) \geqq a\right) = \mathrm{e}^{-a} = \frac{\alpha}{1-\beta}.\end{aligned}$$

Similarly, one can prove that $\beta' \leqq \beta/(1-\alpha)$. We have used the following result for Wiener processes with drift $\delta \neq 0$ at the last step.

Lemma 3.2. *For $\delta \neq 0$, $-\infty \leqq b < c < a \leqq \infty$,*

$$p \equiv Pr[W_\delta(\cdot) \textit{ hits } a \textit{ before } b \textit{ starting from } c].$$

$$= \frac{\mathrm{e}^{-2\delta b} - \mathrm{e}^{-2\delta c}}{\mathrm{e}^{-2\delta b} - \mathrm{e}^{-2\delta a}}. \qquad (63)$$

Proof. Let $Z_\delta(\cdot) = \mathrm{e}^{-2\delta W_\delta(\cdot)}$. Then $Y = Z_\delta(T)$ is a stopped martingale where T is the hitting time. Y can only equal $\mathrm{e}^{-2\delta a}$ or $\mathrm{e}^{-2\delta b}$ and $E(Y) = \mathrm{e}^{-2\delta c}$ from martingale theory. Since $E(Y) = p\mathrm{e}^{-2\delta a} + (1-p)\mathrm{e}^{-2\delta b}$ and $E(Y) = \mathrm{e}^{-2\delta c}$, it follows that

$$p = \frac{\mathrm{e}^{-2\delta b} - \mathrm{e}^{-2\delta c}}{\mathrm{e}^{-2\delta b} - \mathrm{e}^{-2\delta a}}$$

as was to be proved.

Let P and Q be fixed probability measures on a measurable space $(\Omega, \mathscr{F})$. Let $\mathscr{F}_1 \subset \mathscr{F}_2 \subset \ldots \subset \mathscr{F}_\infty \subset \mathscr{F}$ be an increasing sequence of sub-σ-fields of $\mathscr{F}$ where $\mathscr{F}_n$ represents the "information" available at the nth stage of the experiment. The likelihood ratio λ_n associated with the nth stage is defined directly in terms of $\mathscr{F}_n$. An extended non-negative random variable λ will be called a *likelihood ratio* for a σ-field $\mathscr{E}$ if it is $\mathscr{E}$-measurable and

$$Q(E, \lambda \neq \infty) = \int_E \lambda \mathrm{d}P, \; P(E, \lambda \neq 0) = \int_E \lambda^{-1} \mathrm{d}Q \tag{64}$$

for all $E \in \mathscr{E}$ where we define $\lambda^{-1} = 0$ if $\lambda = \infty$. Note that λ is unique up to a P and Q equivalence. If $\mathscr{E}$ is generated by a random mapping X, then λ becomes the likelihood ratio of X in the usual ratio of densities sense.

Definition 3.1. A *stopping variable* N is an extended random variable taking values in $1, 2, \ldots, \infty$ such that $[N=n] \in \mathscr{F}_n$, $n \geqq 1, 2, \ldots, \infty$.

Let $\mathscr{F}_N$ be the σ-field of sets E such that $E \cap [N=n] \in \mathscr{F}_n$ for every $n \geqq 1$. It is easy to see that λ_N is the likelihood ratio of $\mathscr{F}_N$. Here $\lambda_N = \lambda_n$ on $[N=n]$.

Definition 3.2. A *decision* D is a random mapping taking values in the "action space" $\{P, Q, \Delta\}$ such that the event $[D=a] \in \mathscr{F}_N$ for each action a and such that $[D=\Delta] = [N=\infty]$.

Definition 3.3. A *test* T is a pair (N, D). The *error probabilities* for T are defined by $\alpha = P(D=Q)$, $\beta = Q(D=P)$. Two tests $T = (N, D)$ and $T' = (N', D')$ are said to be *equivalent* if

$$P(N=N', \; D=D') = Q(N=N', \; D=D') = 1.$$

Definition 3.4. *A generalized sequential probability ratio test* (GSPRT) is a test $T = (N, D)$ of the form:

$$\begin{aligned} N &= \text{first } n \geqq 1 \text{ such that } \lambda_n \notin (A_n, B_n) \\ &= \infty \quad \text{if no such } n \text{ exists,} \\ D &= P \text{ on } [N < \infty, \; \lambda_N \leqq A_N], \text{ and} \\ &= Q \text{ on } [N < \infty, \; \lambda_N \geqq B_N] \end{aligned}$$

where A_n and B_n, $0 \leqq A_n \leqq B_n \leqq \infty$ are fixed constants (D is defined to be P if $\lambda_N = A_N = B_N = 0$ and as Q if $\lambda_N = A_N = B_N = \infty$. If $0 < \lambda_N = A_N = B_N < \infty$, D can be defined to be P or Q in an arbitrary manner).

We shall denote the above test by $S(A_N, B_N)$. If $A_n = A$ and $B_n = B$ with $A < B$, then the test $T = S(A, B)$ is called a *sequential probability ratio test* (SPRT).

Lemma 3.3. *Let* $T = (N, D)$ *be the GSPRT* $S(A_n, B_n)$. *Then, for each* $E \in \mathscr{F}_N$,

$$Q(E \cap [D=Q] \cap [\lambda_N \neq \infty]) \geqq \left(\inf_n B_n \right) P(E \cap [D=Q]) \tag{65}$$

and

$$P(E \cap [D=P] \cap [\lambda_N \neq 0]) \geqq \left(\sup_n A_n\right)^{-1} Q(E \cap [D=P]). \tag{66}$$

Proof. Let $\mathscr{E} = \mathscr{F}_N$ in relation (64). Then $E \cap [D=Q] \in \mathscr{E}$ and

$$Q(E \cap [D=Q] \cap [\lambda_N \neq \infty]) = \int_{E \cap [D=Q]} \lambda_N \, dP$$

$$\geqq \left(\inf_n B_n\right) P(E \cap [D=Q]).$$

The proof of (66) is similar.

Theorem 3.5. *Let $T=(N, D)$ be the GSPRT $S(A_n, B_n)$ and $T'=(N', D')$ be any other test for which $N' \leqq N$ a.s. under P and Q. Further, suppose* $\sup A_n \leqq \inf B_n$. *Then, for every $c \in [\sup A_n, \inf B_n]$*

$$c(\alpha' - \alpha - p) + (\beta' - \beta - q) \geqq 0 \tag{67}$$

and

$$(\alpha' - \alpha - p) + c^{-1}(\beta' - \beta - q) \geqq 0 \tag{68}$$

where

$$p = P(D=\Delta,\ D'=Q) + P(D=P,\ \lambda_N=0,\ D'=Q), \tag{69}$$

$$q = Q(D=\Delta,\ D'=P) + Q(D=Q,\ \lambda_N=\infty,\ D'=P), \tag{70}$$

and α', β' are error probabilities for the test T'.

Proof. Clearly

$$\alpha = P(D=Q) = P(D=D'=Q) + P(D=Q \neq D')$$

and

$$\alpha' = P(D'=Q) = P(D=D'=Q) + P(D \neq Q = D').$$

Since $N' \leqq N$, $D' \neq \Delta$ when $D=Q$. Hence

$$\alpha' - \alpha = P(D=P,\ D'=Q) + P(D=\Delta,\ D'=Q) - P(D=Q,\ D'=P)$$

and

$$\alpha' - \alpha - p = P(D=P,\ \lambda_N \neq 0,\ D'=Q) - P(D=Q,\ D'=P). \tag{71}$$

Similarly

$$\beta' - \beta - q = Q(D=Q,\ \lambda_N \neq \infty,\ D'=P) - Q(D=P,\ D'=Q). \tag{72}$$

Set $E=[D'=P]$ in (65) and $E=[D'=Q]$ in (66). Then it follows that

$$(\beta'-\beta-q)+Q(D=P,\ D'=Q)\geqq(\inf B_n)P(D=Q,\ D'=P) \tag{73}$$

and

$$(\alpha'-\alpha-p)+P(D=Q,\ D'=P)\geqq(\sup A_n)^{-1}Q(D=P,\ D'=Q). \tag{74}$$

These inequalities prove (67) for $0\leqq c<\infty$ and (68) for $0<c\leqq\infty$. Suppose $c=\infty$. Then $\inf R_n=\infty$, $\alpha=0$ and from (63),

$$\alpha'-\alpha-p=P(D=P,\ \lambda_N\neq 0,\ D'=Q)\geqq 0.$$

If $\alpha'-\alpha-p>0$, then (67) is obvious for $c=\infty$. If $\alpha'-\alpha-p=0$, then

$$P(D=P,\ \lambda_N\neq 0,\ D'=Q)=0$$

and by (64) for $\mathscr{E}=\mathscr{F}_N$,

$$Q(D=P,\ D'=Q)=Q(D=P,\ D'=Q,\ \lambda_N\neq\infty)$$

$$=\int_{[D=P,D'=Q]}\lambda_N\ \mathrm{d}P=0.$$

Hence (72) implies

$$\beta'-\beta-q=\geqq 0$$

and we get relation (67). Similar proof establishes (68) if $c=0$.

The above theorem has the following two simple but important corollaries.

Corollary 3.1. *Under the assumptions of Theorem 3.5, if $\alpha'\leqq\alpha$ and $\beta'\leqq\beta$, then*

$$\alpha'=\alpha \text{ and } \beta'=\beta.$$

Corollary 3.2. *If, in addition to the assumption of Theorem* 3.5, $\sup A_n\leqq 1\leqq\inf B_n$. then

$$\alpha'+\beta'\geqq\alpha+\beta.$$

Note that Corollary 3.1 implies in particular that

$$\alpha'<\alpha \text{ implies } \beta'>\beta \text{ and } \beta'<\beta \text{ implies } \alpha'>\alpha.$$

The assertion of Corollary 3.1 does not imply that the test T' cannot be superior to the GSPRT T. It is possible that $\alpha'=\alpha$, $\beta'=\beta$ and $N'<N$ with probability one both with respect to P and Q. Further results of this nature

are discussed in Eisenberg *et al.* (1976). They have shown that if an SPRT T' has at least one error probability strictly less than that of another SPRT T with one other error probability no larger, then T' require strictly more observations than T some of the time, under both P and Q, and never fewer observations. This generalizes some results of Wijsman (1960, 1963).

3.4 Sequential Probability Ratio Tests for Continuous Time Markov Processes

We shall briefly discuss some results on sequential probability ratio tests for continuous time Markov processes.

Let us consider a homogeneous Markov process with continuous time defined on a probability space $(\Omega, \mathscr{A}, P)$ with state space $(\mathscr{X}, \mathscr{B})$ where $\mathscr{X}$ is a Borel set in $\mathbb{R}^k$ and $\mathscr{B}$ is σ-field of subsets of $\mathscr{X}$ and suppose the transition probability function P_t, $t>0$ depends on a parameter $\theta\in\Theta$. Further suppose that, for each $\theta\in\Theta$

$$\lim_{t\to 0} P_t(x, \{x\}; \theta)=1 \text{ uniformly in } x. \tag{75}$$

Theorems 2·2; 2·3 of Chapter VI of Doob (1953) imply that, for each $\theta\in\Theta$,

$$\lim_{t\to 0} \frac{1-P_t(x, \{x\}; \theta)}{t}=Q(x, \theta)\leqq V(\theta)<\infty \tag{76}$$

this limit being uniform in x; for $x\notin B$,

$$\lim_{t\to 0} \frac{1}{t} P_t(x, B; \theta)=Q(x, B; \theta)\leqq Q(x; \theta) \tag{77}$$

this limit being uniform in $B\subset \mathscr{X}-\{x\}$.

Note that $Q(x, \mathscr{X}-\{x\}; \theta)=Q(x, \theta)$. By putting $Q(x, \{x\}; \theta)=0$, we can write $Q(x, \mathscr{X}; \theta)=Q(x; \theta)$. We shall assume that for every $x\in\mathscr{X}$, the measure $Q(x,\cdot; \theta)$ is absolutely continuous with respect to a σ-finite measure μ and we shall denote the $\mathscr{B}\times\mathscr{B}$-measurable version of its density by $q(x, y; \theta)$. Further suppose that, for every $\theta\in\Theta$, there exists $K(\theta)$ and $\rho(\theta)$ with $0<\rho(\theta)<1$ such that for every $t>0$,

$$|P_t(x, B; \theta)-\pi(B; \theta)|\leqq K(\theta)\, \rho(\theta)^t \tag{78}$$

for all $x\in\mathscr{X}$ and $B\in\mathscr{B}$, for some measure $\pi(\cdot; \theta)$. Finally, assume that

$$\operatorname*{ess\,sup}_{(x, y)\in\mathscr{X}\times\mathscr{X}} \left|\log \frac{q(x, y; \theta_1)}{q(x, y; \theta_0)}\right| <\infty \tag{79}$$

the essential supremum being taken relative to the measure $\pi(dx; \theta)\, Q(x, dy; \theta)$.

We shall suppose that the initial distribution is $\pi(\cdot;\theta)$ and the sample paths of the process are right continuous.

Let $\tau_T(B;\omega)$ denote the total duration of the time, the process stays in B during the interval $[0, T]$ for any $B\in\mathscr{B}$ and let $N_T(B_1\times B_2;\omega)$ be the number of passages from B_1 into B_2 during time $[0, T]$ where $B_1\times B_2$ is $\mathscr{B}\times\mathscr{B}$ measurable. It can be shown that $\tau_T(\cdot;\omega)$ is a positive measure on $\mathscr{B}$ and $N_T(\cdot;\omega)$ is a positive measure on $\mathscr{B}\times\mathscr{B}$ and $\tau_T(B;\cdot)$ and $N_T(B_1\times B_2;.)$ are random variables for fixed $B\in\mathscr{B}$ and $B_1\times B_2\in\mathscr{B}\times\mathscr{B}$ Define

$$\xi_T(\omega)=-\int[Q(x;\theta_1)-Q(x;\theta_0)]\,\mathrm{d}\tau_T(x;\omega)$$
$$+\int_{\mathscr{X}\times\mathscr{X}}\log\frac{q(x,y;\theta_1)}{q(x,y;\theta_0)}\,\mathrm{d}\,N_T(x,y;\omega), \tag{80}$$

for any $\theta_1\neq\theta_0$ in Θ. We shall call ξ_T the log likelihood ratio. One cannot write down the usual likelihood ratio for Markov processes with continuous time as defined above. The above definition is motivated by the fact that for each $\theta\in\Theta$, the likelihood equation can be written in the form

$$-\int_{\mathscr{X}}\frac{\partial}{\partial\theta}Q(x;\theta)\,\mathrm{d}\,\tau_T(x;\omega)+\int_{\mathscr{X}\times\mathscr{X}}\frac{\partial}{\partial\theta}\log q(x,y;\theta)\mathrm{d}N_T(x,y;\omega)=0.$$

Let us now consider the null hypothesis $H_0{:}\theta=\theta_0$ and the alternative $H_1{:}\theta=\theta_1$, $\theta_0\neq\theta_1$. We propose the following sequential procedure for testing H_0 against H_1. Let $0<b<1<a<\infty$.

If $\xi_T(\omega)\leqq\log b$, accept H_0;

if $\xi_T(\omega)\geqq\log a$, accept H_1;

if $\log b<\xi_T(\omega)<\log a$, continue observation.

In the last case, we continue the observation until the time T' such that $N_{T'}(\mathscr{X}\times\mathscr{X};\omega)=N_T(\mathscr{X}\times\mathscr{X};\omega)+1$ i.e. until the time the process reaches a new state. Let $S(\omega)$ be the stopping time. It is clear that $\xi_{S(\omega)}\leqq\log b$ or $\xi_{S(\omega)}\geqq\log a$ and $S(\omega)=+\infty$ if and only if for every $T>0$,

$$\log b<\xi_T(\omega)<\log a.$$

Let

$$C(\theta)=\int_{\mathscr{X}\times\mathscr{X}}\left[\log\frac{q(x,y;\theta_1)}{q(x,y;\theta_0)}\right]^2\pi(\mathrm{d}x;\theta)Q(x,\mathrm{d}y;\theta)+\int_0^\infty G(t;\theta)\mathrm{d}t \tag{81}$$

where

$$\frac{1}{2}\,G(t;\,\theta)=\int_{\mathscr{X}\times\mathscr{X}}[Q(x;\,\theta_1)-Q(x;\,\theta_0)][Q(u;\,\theta_1)-Q(u;\,\theta_0)]\pi(\mathrm{d}x;\,\theta)\times$$
$$[P_t(x,\,du;\,\theta)-\pi(\mathrm{d}u;\,\theta)]$$
$$-\int_{\mathscr{X}\times\mathscr{X}\times\mathscr{X}}\log\frac{q(x,\,y;\,\theta_1)}{q(x,\,y;\,\theta_0)}\,[Q(u;\,\theta_1)-Q(u;\,\theta_0)]\pi(\mathrm{d}x;\,\theta)Q(x,\,\mathrm{d}y;\,\theta)\times$$
$$[P_t(y,\,\mathrm{d}u;\,\theta)-\pi(\mathrm{d}u;\,\theta)]$$
$$-\int_{\mathscr{X}\times\mathscr{X}\times\mathscr{X}}[Q(u;\,\theta_1)-Q(u;\,\theta_0)]\,\log\frac{q(x,\,y;\,\theta_1)}{q(x,\,y;\,\theta_0)}\,\pi(\mathrm{d}u;\,\theta)\times \tag{82}$$
$$[P_t(u,\,\mathrm{d}x;\,\theta)-\pi(\mathrm{d}x;\,\theta)]Q(x,\,\mathrm{d}y;\,\theta)$$
$$+\int_{\mathscr{X}\times\mathscr{X}\times\mathscr{X}\times\mathscr{X}}\log\frac{q(x,\,y;\,\theta_1)}{q(x,\,y;\,\theta_0)}\,\log\frac{q(u,\,v;\,\theta_1)}{q(u,\,v;\,\theta_0)}\,\pi(\mathrm{d}x;\,\theta)\,Q(x,\,\mathrm{d}y;\,\theta)\times$$
$$[P_t(y,\,\mathrm{d}u;\,\theta)-\pi(\mathrm{d}u;\,\theta)]\,Q(u,\,\mathrm{d}v;\,\theta).$$

Under assumptions (78) and (79), it can be shown that

$$\int_0^\infty G(t;\,\theta)\mathrm{d}\theta$$

is absolutely convergent. We shall call the problem degenerate if

$$C(\theta)=0 \tag{83}$$

and

$$\int_{\mathscr{X}}[Q(x;\,\theta_1)-Q(x;\,\theta_0)]\pi(\mathrm{d}x;\,\theta)$$
$$=\int_{\mathscr{X}\times\mathscr{X}}\log\frac{q(x,\,y;\,\theta_1)}{q(x,\,y;\,\theta_0)}\,\pi(\mathrm{d}x;\,\theta)Q(x,\,\mathrm{d}y;\,\theta). \tag{84}$$

Lemma 3.4. *In the non-degenerate case, there exists $c>0$ such that*

$$P_\theta[|\xi_c|<\gamma]<1 \tag{85}$$

where

$$\gamma=\log a+|\log b|. \tag{86}$$

The proof consists of showing that $E(\xi_T^2)\to\infty$ as $T\to\infty$. We shall leave it to the reader.

Let $M(T)$ and $d(T)$ be integer-valued functions of T tending to ∞ as $T\to\infty$ such that

$$M(T)(c+d(T))\leqq T \tag{87}$$

and

$$\lim_{T\to\infty} M(T)\ \rho(\theta)^{d(T)}=0 \tag{88}$$

where c is chosen as in Lemma 3.4. One can choose $d(T)$ as the largest integer such that $d^2(T)\leqq T$ and

$$M(T)=\left[\frac{T}{c+d(T)}\right]\sim d(T).$$

Let us partition the interval $[0, T]$ into sub-intervals of length alternatively equal to c and to $d(T)$ i.e. into intervals

$$[m(c+d(T)),\ m(c+d(T))+c] \text{ and}$$
$$((m(c+d(T))+c,\ (m+1)(c+d(T))),$$

$m=0, 1, \ldots, M(T)-1$ and a residual interval $[M(T)\ (c+d(T)),\ T]$. Let us write

$$\begin{aligned} m_{\xi_c}(\omega)= &-\int_{\mathscr{X}}[Q(x;\theta_1)-Q(x;\theta_0)]\mathrm{d}\tau_{t',t''}(x;\omega) \\ &+\int_{\mathscr{X}\times\mathscr{X}}\log\frac{q(x,\ y;\ \theta_1)}{q(x,\ y;\ \theta_0)}\,\mathrm{d}N_{t',t''}(x;\omega) \end{aligned} \tag{89}$$

where

$$t'=m(c+d(T)) \text{ and } t''=m(c+d(T))+c$$

and $\tau_{t',t''}$ and $N_{t',t''}$ correspond to $[t', t'']$ as τ_T and N_T correspond to $[0, T]$.

The random variables m_{ξ_c}, $m=0, 1, \ldots, M(T)-1$ have the same distribution as that of ξ_c and as a consequence

$$E_\theta(m_{\xi_c}^2)=E_\theta(\xi_c^2).$$

The following lemma can be proved.

Lemma 3.5. $$\left|P_\theta\left(\bigcap_{m=0}^{M(T)-1}[m_{\xi_c}^2<\gamma^2]\right)-\prod_{m=0}^{M(T)-1}P_\theta(m_{\xi_c}^2<\gamma^2)\right|$$
$$\leqq P_\theta(\xi_c^2<\gamma^2)K(\theta)(M(T)-1)\rho(\theta)^{d(T)}.$$

As a consequence of lemmas 3.4 and 3.5, we have the following theorem.

Theorem 3.6. $P_\theta[S(\omega)<\infty]=1.$

Proof. Suppose $E_\theta(\xi_T)\neq 0$ for every T. Then $S<\infty$ with P_θ probability one since

$$\frac{1}{T}\,\xi_T\to\int_{\mathscr{X}}[Q(x;\,\theta_1)-Q(x;\,\theta_0)]\pi(dx;\,\theta)$$
$$+\int_{\mathscr{X}\times\mathscr{X}}\log\frac{q(x,\,y;\,\theta_1)}{q(x,\,y;\,\theta_0)}\,\pi(dx;\,\theta)Q(x,\,dy;\,\theta)$$

as $T\to\infty$ a.s. P_θ. This is due to the fact that

$$\frac{1}{T}\,\tau_T(B)\text{ and }\frac{1}{T}\,N_T(B_1\times B_2)$$

converge almost surely as $T\to\infty$. In order to include the general case when $E_\theta(T)=0$, we can prove the result in the following manner.
Suppose that for every $T\in\mathbb{R}_+$,

$$\log b<\xi_T(\omega)<\log a.$$

Then, for all $m=0,\,1,\ldots,\,M(T)-1$,

$$\log b<\xi_{m(c+d(T))+c}(\omega)<\log a,$$

and

$$\log b<\xi_{m(c+d(T))}(\omega)<\log a$$

from which we obtain by subtraction

$$-\gamma<m_{\xi_c}(\omega)<\gamma$$

where $\gamma=\log a+|\log b|$. As a consequence of Lemma 3.5, we have

$$P_\theta(S(\omega)=+\infty)\leqq\lim_{T\to\infty}P_\theta\left[\bigcap_{m=0}^{M(T)-1}[m_{\xi_c}^2<\gamma^2]\right]$$
$$\leqq\lim_{T\to\infty}[P_\theta(\xi_c^2<\gamma^2)]^{M(T)}$$
$$+K(\theta)P_\theta(\xi_c^2<\gamma^2)\cdot\lim_{T\to\infty}[M(T)-1]\,\rho(\theta)^{d(T)}$$
$$=0$$

by Lemma 3.4. Hence $P_\theta[S(\omega)<\infty]=1$.

Andrieu (1975) discussed similar problems for Markov processes with discrete time and continuous state space.

4. SEQUENTIAL TESTS FOR THE DRIFT OF WIENER PROCESS

Sequential tests for diffusion processes in general and Wiener process in particular, have been discussed in Chapter 9. In this section, we shall discuss some further sequential tests for the Wiener process.

4.1 Sequential Bayes Tests for the Sign of Drift for the Wiener Process

We first discuss the problem of testing whether the drift μ of a Wiener process is positive or negative given an *a priori* normal distribution. The problem is a continuous version of the discrete version of obtaining an asymptotically optimal Bayes solution (as the sampling cost approaches zero) of the problem of sequentially testing whether the unknown mean μ of a normal distribution with known variance is positive or negative when the prior is normal with fixed mean μ_0 and variance σ_0^2. The nature of the Bayes solution will depend on the loss function. We suppose that the loss due to taking the wrong action in the discrete case is

$$r(\mu)=k|\mu|+o(1)$$

as $\mu\to 0$. This problem in the continuous case reduces to the following. The problem is to test whether the drift μ of a Wiener process with variance σ^2 per unit time is positive or negative when the loss function is

$$r(\mu)=k|\mu| \tag{90}$$

and μ has an *a priori* distribution $N(\mu_0, \sigma_0^2)$ and there is a cost c per unit time of sampling. The rest of the section will deal with the reduction of this problem to the solution of a problem involving the diffusion equation.

At first, we shall normalize the problem of testing for the drift of a Wiener process to the case where the constants k, σ^2 *and* c are replaced by one. Let

$$X_t=\mu t+Z_t \tag{91}$$

where X_t is the observed Wiener process. Then Z_t is a continuous process of independent Gaussian increments with mean 0 and variance $E[Z_t^2]=\sigma^2 t$. Let

$$X_{t^*}^*=\alpha X_t,\ \mu^*=\beta\mu,\ t^*=\gamma t. \tag{92}$$

Then $X_{t^*}^*$ is a Wiener process with drift μ^* per unit time (measured in t^*) if

$$\frac{\alpha}{\beta\gamma}=1 \tag{93}$$

and $X_{t^*}^*$ has unit variance per unit time if

$$\frac{\alpha^2\sigma^2}{\gamma}=1. \tag{94}$$

Finally the Bayes risk for an arbitrary procedure in the original problem is

$$B=E[cT+\varepsilon(\mu)k|\mu|] \tag{95}$$

where T is the time of sampling, $\varepsilon(\mu)$ is the probability of error when μ is the mean drift with the given prior. Then

$$B=\frac{c}{\gamma}\,B^* \tag{96}$$

where

$$B^*=E\left[T^*+\varepsilon^*(\mu^*)\,\frac{\gamma k}{c\beta}\,|\mu^*|\right],\quad T^*=\gamma T, \tag{97}$$

$$\varepsilon^*(\mu^*)=\varepsilon(\mu).$$

If

$$\frac{\gamma k}{c\beta}=1, \tag{98}$$

the problem of minimizing B has been reduced to that of solving a corresponding problem with the constants c, σ and k replaced by ones. Note that the mean μ_0 and variance σ_0^2 of the prior are replaced by

$$\mu_0^*=\beta\mu_0,\quad \sigma_0^{*2}=\beta^2\sigma_0^2. \tag{99}$$

The solution of equations (93), (94) and (98) yields

$$\begin{aligned}\alpha&=c^{1/3}k^{-1/3}\sigma^{-4/3},\\ \beta&=c^{-1/3}k^{1/3}\sigma^{-2/3},\\ \gamma&=c^{2/3}k^{-2/3}\sigma^{-2/3},\end{aligned} \tag{100}$$

and hence

$$B=c^{1/3}k^{2/3}\sigma^{2/3}B^*. \tag{101}$$

The original problem has been reduced to a normalized one where the prior has large mean and standard deviation. Hereafter we shall treat the normalized problem.

Since the processes $\{X_s,\ 0\leqq s\leqq t\}$ for different values of μ are absolutely continuous with respect to each other (cf. Cameron and Martin, 1945) and since X_t is a sufficient statistic, the posterior distribution of μ given X_s on

$[0, t]$ may be computed in terms of X_t using the data that the conditional distribution of X_t given μ is normal with mean μt and variance t and the distribution of μ is normal with mean μ_0 and variance σ_0^2. It is easily seen that the conditional distribution of μ given $X_t = x$ is

$$N\left(\frac{x+\dfrac{\mu_0}{\sigma_0^2}}{t+\dfrac{1}{\sigma_0^2}}, \frac{1}{t+\dfrac{1}{\sigma_0^2}}\right).$$

Let $H_1{:}\mu > 0$ and $H_2{:}\mu \leqq 0$. The optimal Bayes procedure for testing H_1 versus H_2 is as follows. Using the *a posteriori* distribution at time t, compute B_{it}, the conditional risk due to stopping at time t and accepting H_i. Compute B_t, the infimum of risks corresponding to all sequential tests based on data $X_{t'} - X_t$ with $t' \geqq t$. Continue sampling if $B_t < \min (B_{1t}, B_{2t})$. Otherwise stop and accept the hypothesis corresponding to the smaller stopping risk. If $B_{1t} = B_{2t}$, any measurable procedure can be used to decide between H_1 and H_2.

Since the conditional distribution depends on the past only through X_t, this procedure can be represented by continuation and stopping sets in the (x, t) plane. Let

$$X'_{t'} = X_t + \frac{\mu_0}{\sigma_0^2}, \quad t' = t + \frac{1}{\sigma_0^2} \quad \cdot \tag{102}$$

Then the posterior distribution of μ given $X'_{t'} = x'$ is $N(x'/t', 1/t')$.

By this transformation, optimal procedures for all versions of the problem corresponding to normal prior distribution can be represented by two stopping regions and one continuation region in the (x', t') plane. Each version is distinguished by the point $(\mu_0/\sigma_0^2, 1/\sigma_0^2)$ from which the process starts. Thus the Bayes risk for the optimal procedure can be represented by a function $B_0(x', t')$ evaluated at (x', t'), $x' = \mu_0/\sigma_0^2$, $t' = 1/\sigma_0^2$. Similarly the conditional Bayes risk given X_s, $0 \leqq s \leqq t$, is $B_0(X_t + (\mu_0/\sigma_0^2), t + 1/\sigma_0^2)$.

The stopping regions and continuation regions in the (x', t') plane can be regarded as determining the Bayes solution for the testing problem when the prior distribution is replaced by the Lebesgue measure on $(-\infty, \infty)$ which is an improper prior.

We shall now treat the normalized problem in the (x', t') plane and delete the primes. Corresponding to each point (x, t), the *a posteriori* distribution of μ given $X_t = x$ will be $N(x/t, 1/t)$ and the conditional risk of stopping at t and accepting H_1 given $X_t = x$ is

$$D_1(x, t) = \int_{-\infty}^{0} |\mu| Q(\mu{:}\ x/t, 1/t) d\mu \tag{103}$$

where $Q(y; a, b)$ is the normal density with mean a and variance b. Then

$$D_1(x, t) = \frac{1}{\sqrt{t}} \left\{ Q\left(\frac{x}{\sqrt{t}}\right) - \frac{x}{\sqrt{t}} \left[1 - \phi\left(\frac{x}{\sqrt{t}}\right) \right] \right\} \tag{104}$$

where $Q(y) = Q(y; 0, 1)$ and $\phi(y)$ is the standard normal distribution function. Thus, for the specified loss function $r(\mu) = |\mu|$, if one should stop at time t with $X_t = x$, he should select the hypothesis according to the sign of x and his stopping risk is

$$D(x, t) = t^{-1/2} \psi(x; t^{-1/2}) \tag{105}$$

where

$$\psi(y) = \begin{cases} Q(y) - y[1 - \phi(y)], & y > 0 \\ Q(y) - y\phi(y), & y < 0. \end{cases} \tag{106}$$

Observe that ψ is a symmetric function and ψ has right and left derivatives as $-1/2$ and $+1/2$ respectively at $y = 0$.

Let us consider now an arbitrary measurable set where the continuation rule can be represented by an open set in the (x, t) plane and the conditional Bayes risk $B(x, t)$ is continuous everywhere and sufficiently differentiable for every point in the continuation set. We shall assume that, upon stopping, the best terminal decision is made so that $B(x, t) = D(x, t)$ on the stopping set. $B(x, t)$ represents the conditional Bayes risk which takes into account the probability of error and the expected cost of sampling further. Observe that if $X_t = x$, given μ,

$$X_{t+h} = x + h\mu + \sqrt{h}\, \eta_1 \tag{107}$$

where η_1 is $N(0, 1)$ and

$$\mu = \frac{x}{t} + \frac{1}{\sqrt{t}} \eta_2 \tag{108}$$

where η_2 is $N(0, 1)$ and independent of η_1. Hence

$$X_{t+h} = x\left(1 + \frac{h}{t}\right) + \left\{ h\left(1 + \frac{h}{t}\right) \right\}^{1/2} \eta$$

where η is $N(0, 1)$. It is not difficult to see that

$$P(|X_s - x| > a \text{ for some } s \text{ in } (t, t+h)) = o(\exp - a^2/4h).$$

Hence if (x, t) is an interior point of the continuation set, then there is an a such that the probability of termination within δ units of time is

$o(\exp - a^2/4\delta)$ as $\delta \to 0$ and

$$B(x, t) = \delta + \int_{x-a}^{x+a} B(u, t+\delta)\, Q\left(u; x\left(1+\frac{\delta}{t}\right), \delta\left(1+\frac{\delta}{t}\right)\right) du \qquad (109)$$

$$+ o(\exp - a^2/4\delta).$$

This relation implies that

$$B(x, t) = \delta + E[B + \delta B_t + Y B_x + \tfrac{1}{2} Y^2 B_{xx}] + o(\delta) \qquad (110)$$

where

$$Y = \frac{\delta x}{t} + \left\{\delta\left(1+\frac{\delta}{t}\right)^{1/2}\right\}\eta \qquad (111)$$

and the expectation is with respect to the standard normal distribution of η. Therefore

$$B = \delta + B + \delta B_t + \frac{\delta x}{t} B_x + \frac{\delta B_{xx}}{2} + o(\delta)\,.$$

Taking limit as $\delta \to 0$, we obtain the equation

$$1 + B_t + \frac{x}{t} B_x + \frac{B_{xx}}{2} = 0 \qquad (112)$$

with the boundary condition $B = D$ from the continuity of B. It can be shown that $B_x = D_x$ on the boundary for the optimal procedure. Here B_x, B_t denote the first partial derivatives of B with respect to x and t respectively and B_{xx} denotes the second partial derivative of B with respect to x.

Equation (112) can be reduced to the heat equation by the transformation

$$y = \frac{x}{t}, \quad s = \frac{1}{t}, 0 < t < \infty. \qquad (113)$$

If

$$\tilde{B}(y, s) = B(x, t) + t^{-1}$$

where $\tilde{B}$ represent the conditional risk including the amount already spent on sampling, we have

$$\tilde{B}_s = \tfrac{1}{2} \tilde{B}_{yy} \qquad (114)$$

with the boundary condition

$$\tilde{B}=D\left(\frac{y}{s},\frac{1}{s}\right)+\frac{1}{s}=\sqrt{s}\,\psi\left(\frac{y}{\sqrt{s}}\right)+\frac{1}{s}=\tilde{D}(y, s).$$

Optimal boundary is determined by the condition $\tilde{B}_y=-1-\phi(ys^{-1/2})$ on the boundary. Equation (114) is the heat equation or the diffusion equation.

Further discussion on the solution of equation (114) and the behaviour of $B(x, t)$ for large t etc. is discussed in Breakwell and Chernoff (1964), and Chernoff (1965a, b).

4.2 Sequential Detection of a Drift Change in a Wiener Process

Let $\{W(t), 0\leqq t<\infty\}$ be a Wiener process with variance σ^2 per unit time. In this section, we consider a continuous time analoge of a statistic due to Page (1954) for the sequential detection of an increase in the drift parameter of a Wiener process.

Without loss of generality, we can suppose that the initial drift is zero. In order to detect a positive shift in drift, one may subtract a positive reference value k to bias the path downward. The statistic suggested by Bagshaw and Johnson (1975) is, that there is a positive shift in drift at time t if

$$W(t)-kt-\inf_{s\leqq t}[W(s)-sk]>h \tag{115}$$

where h is selected to give a specified expected first passage time when $E[W(t)]=0$. Let μ be defined by $E[W(t)-kt]=\mu t$.

The functional (115) is equivalent to restraining $W(t)-kt$ to move between an absorbing barrier h and a reflecting barrier 0. Bagshaw and Johnson (1975) have proved that for any Wiener process with initial position w_0, drift parameter μ and variance σ^2 per unit time, the density of the first passage time to h, when there is a reflecting barrier at 0, is given by

$$g(t, w_0)=\frac{\sigma^2}{h}\exp\left\{\mu\sigma^{-2}(h-\omega_0)-\frac{1}{2}\mu^2t\sigma^{-2}\right\}$$
$$\left\{\sum_{n=1}^{\infty}\frac{\lambda_n\sin\lambda_n h}{1+\dfrac{\mu}{h\sigma^2}\dfrac{\sin^2\lambda_n h}{\lambda_n^2}}\exp(-\tfrac{1}{2}\lambda_n^2\sigma^2)Y_n(w)\right\} \tag{116}$$
$$+r(t)$$

where

$$
r(t)=\begin{cases} 0 & \text{if } -\mu h/\sigma^2<1 \\ \dfrac{3\sigma^2}{2h^3}(h-w_0)\exp\left(\dfrac{\mu^2 t}{2\sigma^2}+\dfrac{\mu}{\sigma^2}(h-w_0)\right) & \text{if } -\mu h/\sigma^2=1 \\ \dfrac{\sigma^2}{h}\left(-\left(\dfrac{\mu^2}{2\sigma^2}-\dfrac{q^2\sigma^2}{2}\right)t+\dfrac{\mu}{\sigma^2}(h-w_0)\right)\dfrac{q\sinh qh}{-1-\dfrac{-\mu\sinh^2 qh}{\sigma^2 hq^2}}v(w_0) & \text{if } -\dfrac{\mu h}{\sigma^2}>1. \end{cases}
$$

$$
Y_n(w)=\cos\lambda_n w+\frac{\mu}{\sigma^2\lambda_n}\sin\lambda_n w,
$$

and

$$
v(w)=\cosh qw+\frac{\mu}{\sigma^2 q}\sinh qw.
$$

Here $0<\lambda_1<\lambda_2<\ldots$ are solutions of $\tan\lambda h=-\lambda\sigma^2/\mu$ and q is the solution of $\tanh qh=-q\sigma^2/\mu$.

5. FURTHER TOPICS

In Section 2, we have discussed sequential estimation for processes with independent increments. Sequential estimation plans for binomial, Poisson and Wiener processes are studied in Kagan, *et al.* (1973) (Chapter 12). A short survey of sequential estimation for process with independent increments is given in Winkler (1977); he studied exponential type processes with multi-dimensional parameter.

Sequential probability ratio tests for Markov processes with discrete time are studied in Schmitz (1968, 1970) and Andrieu (1975). Rene Flavigny (1975) discussed the case of ϕ-mixing processes. Bickel and Yahav (1972) are concerned with the problem of testing $H_0{:}\mu\leqq 0$ versus $H_1{:}\mu>0$ by sequential sampling from a one parameter exponential family of densities $f(x,\mu)$ with cost c per observation and zero-one loss structure. They showed that if there is a bounded continuous prior density ψ on the parameter space and if an observation has mean zero and variance one if $\mu=0$, then this problem is equivalent to testing that the drift $\mu=0$ for the Wiener process with the same loss and cost structure and with the improper prior density identically equal to $\psi(0)$. These results are extensions of those discussed in Section 4.1 due to Chernoff (1961).

Finally, we mention that sequential procedures for detecting parameter changes in time series models have been developed by Bagshaw and Johnson (1975a). Muddapur (1974, 1976) developed control charts for a certain class of stochastic processes using "*V*-Mask" procedures suggested by Barnard (1959).

BIBLIOGRAPHICAL NOTES

Sequential estimation and testing for continuous time stationary processes with independent increments was originally investigated by Dvoretzky *et al.* (1953a,b). Magiera (1974) and Franz and Winkler (1976) studied efficient sequential estimation for exponential type of stationary processes with independent increments extending the earlier work of Trybula (1968). Hall (1969), Eisenberg, *et al.* (1976) and Andrieu (1975) studied sequential probability ratio tests for discrete parameter stochastic processes. Sequential Bayes tests for the sign of drift of a Wiener process were discussed by Chernoff (1961) in a series of papers. Bagshaw and Johnson (1975) gave tests for sequential detection of increase in drift for a Wiener process. Large sample theory of sequential estimation is due to Anscombe (1952, 1953) for i.i.d. observations and Sarma (1976) for discrete time stationary Markov processes.

Results in Section 2.1 are due to Prakasa Rao (1974). Sections 2.2 and 3.2 contain the work of Dvoretzky *et al.* (1953a, b). Results of Magiera (1974) are given in Section 2.3. Sections 3.1 and 3.3 contain the results developed by Hall (1969, 1970). Theorem 3.1 is due to Shepp (1969). Part of the material in Section 3.3 is due to Eisenberg, *et al.* (1976); Theorem 3.5 and the associated lemmas are due to them. Section 3.4 discusses the work Andrieu (1975). Results studied in Section 4.1 are due to Chernoff (1961). Section 4.2 contains the work of Bagshaw and Johnson (1975).

COMPLEMENTS

1. For the exponential type of processes discussed in Section 2, show that an unbiased sequential plan (τ, g, f) is efficient if and only if there exist constants K_i, $1 \leqq i \leqq 4$ such that

$$g(\theta)=\frac{K_1 W_1'(\theta)+K_2 W_2'(\theta)}{K_3 W_1'(\theta)+K_4 W_2'(\theta)}.$$

(Magiera, 1974)

2. Obtain a sequential test for testing a simple hypothesis about the value $p(0<p<1)$ against a simple alternative in the following process with stationary independent increments (called the negative binomial): the probability that $X(t)=k$ for every non-negative integer k is

$$\Gamma(t+k)p^t(1-p)^k/\Gamma(k+1)\Gamma(t).$$

(Dvoretzky *et al.*, 1953a)

3. Obtain a sequential test for testing about the value of θ $(\theta>0)$ in the following process with stationary independent increments (called the gamma process): the probability density of $X(t)$ at $x(x\geqq 0)$ is given by

$$x^{t-1}e^{-x/\theta}/\Gamma(t)\theta^t.$$

(Dvoretzky *et al.*, 1953a)

4. Obtain the power function of the test in Example 3.2.

(Dvoretzky *et al.*, 1953a)

5. Let $Z(\cdot)=(Z_1(\cdot), \ldots, Z_k(\cdot))$ be a k-dimensional homogeneous process with independent increments. Suppose $\delta_i=EZ_i(1)$ exists for all i. Under appropriate moment conditions, show that, for any Markov time T,

$$E\left(\sum_{i=1}^{k} Z_i(T)\right)=ET\left(\sum_{i=1}^{k}\delta_i\right)$$

whenever the right-hand side is well-defined, and if $\delta_i=0$ for all i, then

$$E\left(\sum_{i=1}^{k} Z_i(T)\right)^2=ET\left\{\sum_{i=1}^{k}\text{var}\,(Z_i(1))\right\}$$

whenever the right-hand side is finite.

(Hall, 1970)

6. *Testing of a hypothesis in a two-state Markov chain.*
Let the transition probability matrices under H_1 and H_2 be

$$P_1=\begin{pmatrix}1 & 0\\ p_1 & p_2\end{pmatrix} \text{ and } P_2=\begin{pmatrix}1 & 0\\ p_1' & p_2'\end{pmatrix}$$

respectively. Let w_{ij} be the loss incurred by accepting H_j when H_i is true, which is positive for $i\neq j$ and is zero for $i=j(i,j=1,2)$. Let $c_1>0$ and $c_2\geqq 0$ be the costs under H_1 and $c_1'>0$ and $c_2'\geqq 0$ under H_2 of observation belonging to the first and second state respectively. Obtain a Bayes solution for the sequential decision problem stated.

(Bhat, 1964)

7. Consider the negative binomial process discussed in Exercise 3. Let $w=(1-p)/p$. Find the minimax estimate of w corresponding to the loss function

$$W(\delta, w)=\frac{(\delta-w)^2}{w(1+w)}.$$

(Dvoretzky *et al.*, 1953b)

8. Consider the gamma process discussed in Exercise 3. Find the minimax

estimates of θ corresponding to the weight functions

$$W_1(\theta, \delta)=\left\{\left(\frac{\delta}{\theta}\right)^{\gamma}-1\right\}^2$$

where $\gamma>0$ is known and

$$W_2(\theta, \delta)=\left(\log\frac{\delta}{\theta}\right)^2.$$

(Dvoretzky *et al.*, 1953b)

9. *Sequential test for* φ*-mixing processes.* Suppose $\{X_n, n\geqq 1\}$ is a φ-mixing process (cf. Iosifescu and Theoderescu, 1969) such that the joint density of $(X_0, \ldots, X_n)$, with respect to a σ-finite measure μ_n, depends on a parameter $\theta\in\Theta$ and is given by

$$f_n(x_0, \ldots, x_n; \theta)=C_n(\theta)\left(\exp\left[\sum_{i=0}^{n-1} v_i(x_i, x_{i+1}; \theta)\right]\right)g_n(x_0, \ldots, x_n)$$

with $0<K_1(\theta)<C_n(\theta)<K_2(\theta)<\infty$, $K_1(\theta)$, $K_2(\theta)$ being independent of n. Let

$$R_n=\log\frac{C_n(\theta_1)}{C_n(\theta_0)}+\sum_{i=0}^{n-1}\{v_i(X_i, X_{i+1}; \theta_1)-v_i(X_i, X_{i+1}; \theta_0)\}$$

for $\theta_1\neq\theta_0$ in Θ. Let $0<b<1<a$. Consider the sequential test δ namely if $R_n\leqq\log b$, accept the hypothesis $H_0:\theta=\theta_0$; if $R_n\geqq\log a$, accept the hypothesis $H_1:\theta=\theta_1$; finally if $\log b<R_n<\log a$, continue the observations until one of the other two situations occur. Let v be the time of termination of the test. Show that if $\Sigma_{n=1}^{\infty}\varphi^{1/2}(n)<\infty$, then

(i) $P_\theta(v=+\infty)=0$,

(ii) $a\leqq\dfrac{1-\beta(\delta)}{\alpha(\delta)}$ and $b\geqq\dfrac{\beta(\delta)}{1-\alpha(\delta)}$

where $\alpha(\delta)$ and $\beta(\delta)$ are type I and type II errors corresponding to the test δ and

(iii) if $\alpha+\beta<1$, then the sequential test δ defined by the reals a and b defined by $a=(1-\beta)/\alpha$ and $b=\beta/(1-\alpha)$ is such that

$$\alpha(\delta)+\beta(\delta)\leqq\alpha+\beta.$$

(Rene' Flavigny, 1975)

10. (V-Mask Procedure). For the negative binomial process discussed in Exercise 3, derive the V-Mask procedure for controlling the changes in the intensity parameter $w=(1-p)p^{-1}$.

(Muddapur, 1974)

11. Derive Wald's fundamental identity from (64).

(Eisenberg, Ghosh and Simons 1976)

Appendix 1
Martingales

1. MARTINGALES AND LIMIT THEOREMS

In this section, we shall give a brief survey of martingales and some limit theorems for martingales. Doob (1953), Gikhman and Skorokhod (1974), Loeve (1963) and other treat these extensively in their books.

Definition. Let $(\Omega, \mathscr{F}, P)$ be a probability space and $\{\mathscr{F}_t, t \subset T\}$ be an increasing sequence of σ-algebras contained in $\mathscr{F}$. Let $\{\xi_t, \mathscr{F}_t, t \in T\}$ be a family of real-valued random variables such that ξ_t is $\mathscr{F}_t$-measurable and $E|\xi_t| < \infty$ for every $t \in T$. The family $\{\xi_t, t \in T\}$ is called a *martingale* if

$$E[\xi_t | \mathscr{F}_s] = \xi_s, \; s \leqq t, \; s, \; t \in T.$$

The family is called a *submartingale* if

$$E[\xi_t | \mathscr{F}_s] \geqq \xi(s), \; s \leqq t, \; s, \; t \in T$$

and it is called a *supermartingale* if

$$E[\xi_t | \mathscr{F}_s] \leqq \xi(s), \; s \leqq t, \; s, \; t \in T.$$

We have the following elementary results. If $\{\xi_t, \mathscr{F}_t, t \in T\}$ is a submartingale, then $E\xi_t$ is a monotonically nondecreasing function of t and if $\{\xi_t, \mathscr{F}_t, t \in T\}$ is a martingale, then $E(\xi_t)$ is a constant and $E|\xi_t|$ is monotonically non-decreasing in t.

A random variable τ is called a *Markov time* (*stopping time*) with respect to the family of σ-algebras $\{\mathscr{F}_t, t \in T\}$ if for arbitrary $t \in T$, the event $[\tau > t]$ belongs to $\mathscr{F}_t$. For any Markov time τ, $\mathscr{F}_\tau$ is the smallest σ-algebra containing sets of the form $F_t \cap [\tau > t]$, $F_t \in \mathscr{F}_t$, $t \in T$.

Suppose that $T = \{0, 1, 2, \ldots, n\}$ and $\{\mathscr{F}_k, 0 \leqq k \leqq n\}$ is a non-decreasing sequence of σ-algebras contained in $\mathscr{F}$. Let $\{\zeta_k, \mathscr{F}_k, k \in T\}$ be a sub-martingale and $\tau_1, \tau_2, \ldots, \tau_s$ be Markov times on $\{\mathscr{F}_k, k \in T\}$ such that $\tau_1 \leqq \tau_2 \leqq \ldots \leqq \tau_s$. Then $\{\eta_k, \mathscr{F}_{\tau_k}, k = 1, 2, \ldots, s\}$ forms a submartingale where $\eta_k = \zeta_{\tau_k}$. If, however, $\{\zeta_k, \mathscr{F}_k, k \in T\}$ is a martingale, then $\{\eta_k, \mathscr{F}_{\tau_k}, k = 1, 2, \ldots, s\}$ is a martingale.

In other words martingales observable at random instants of time are martingales and the same holds for submartingale or supermartingales. The following theorems hold.

Theorem 1.1. *Let $\{\zeta_k, \mathscr{F}_k, 0 \leqq k \leqq n\}$ be a sub-martingale. Then for any $\varepsilon > 0$,*

$$P\left[\max_{0 \leqq k \leqq n} \zeta_k \geqq \varepsilon\right] \leqq \frac{1}{\varepsilon} E[\zeta_n^+].$$

An important example of a martingale is $\{\zeta_t, \mathscr{F}_t, t \geqq 0\}$ where $\zeta_t = \exp[\alpha W_t - \alpha^2 t/2]$, $\alpha > 0$ where $\{W_t, t \geqq 0\}$ is the standard Wiener process and $\mathscr{F}_t = \sigma\{W(s), 0 \leqq s \leqq t\}$.

Theorem 1.2. *Let $\{\zeta_n, \mathscr{F}_n, n \geqq 1\}$ be a martingale. Then the following conditions are equivalent:*

(i) *the family $\{\zeta_n, n = 1, 2, \ldots\}$ is uniformly integrable.*

(ii) $E|\zeta_n - \zeta_{n'}| \to 0$ *as* $n, n \to \infty$.

Further if one of these conditions is satisfied, then the limit $\zeta_n = \zeta$ exists with probability one. If for some $p > 1$, $E|\zeta_n|^p \leqq C < \infty$, then (i) and (ii) are valid and $\zeta = \lim \zeta_n$ in the sense of convergence of L_p.

See Doob (1953) or Gikhman and Skorokhod (1974) for proofs of Theorems 1.1 and 1.2. As an application of the above theorem, one can obtain the following theorem which deals with likelihood ratios.

Theorem 1.3. *Let $(\mathscr{X}, \mathscr{B}, \mu)$ be a probability space and Q be a totally finite measure on $(\mathscr{X}, \mathscr{B})$ absolutely continuous with respect to μ. Let $\{A_{nk}, k \geqq 1\}$ be a measurable partition of $\mathscr{X}$ for each $n \geqq 1$. Let*

$$g_n(x) = \frac{Q(A_{nk}(x))}{\mu(A_{nk}(x))}$$

for $\mu(A_{nk}(x)) > 0$, where $A_{nk}(x)$ is that set of the sequence $\{A_{nk}, k = 1, 2, \ldots\}$

which contains x. If $\mu(A_{nk}(x))=0$, *let* $g_n(x)=0$. *Then*

(i) *the sequence* $\{g_n, \mathscr{F}_n, n \geqq 1\}$ *is a martingale where* $\mathscr{F}_n=\sigma(A_{n1}, A_{n2}, \ldots,)$;

(ii) *there exists the limit* $g(x)=\lim_{n\to\infty} g_n(x)$ *a.s.* $[\mu]$ *independent of the choice of the sequence of partitions;*

(iii) *for arbitrary* $B \in \beta$,

$$Q(B)=\int_B g(x)\mu(\mathrm{d}x).$$

Note that the above theorem gives a method for calculation of the Radon–Nikodym derivative. For further results, see Chapter 7, Gikhman and Skorokod (1974).

The following central limit theorem was proved for martingales by Billingsley (1961) and by Ibragimov (1963).

Theorem 1.4. *Let* $\{Z_n, n \geqq 1\}$ *be a strictly stationary ergodic process such that* $E(Z_1^2)$ *is finite and* $E[Z_n|Z_1, \ldots, Z_{n-1}]=0$ *a.s. for* $n>1$ *and* $E[Z_1]=0$. *Then the distribution of* $n^{-1/2}\Sigma_{k=1}^n Z_k$ *converges weakly to the normal distribution with mean* 0 *and variance* $E(Z_1^2)$.

This theorem is used in proving asymptotic normality of maximum likelihood estimators for parameters of stationary Markov processes.

We shall now state and prove another limit theorem due to Basawa and Scott (1977).

Theorem 1.5. *Let* $\{Z_k, k \geqq 1\}$ *be a sequence of random variables such that* $E(Z_k|Z_1, \ldots, Z_{k-1})=0$ a.s. *for* $k \geqq 1$ *and* $E(Z_1)=0$ *and* $E(Z_k^2)<\infty$, $k \geqq 1$. *Let*

$$S_n=Z_1+\ldots+Z_n \text{ and } I_n=\sum_{k=1}^n E(Z_k^2).$$

If, for the sequence of constants $\{I_n\}$,

(i) $I_n^{-1}\Sigma_{k=1}^n Z_k^2 \xrightarrow{p} \eta$ *where* η *is a random variable with* $P(\eta>0)=1$;

(ii) $I_n^{-1}E\{\max_{1 \leqq k \leqq n} Z_k^2\} \to 0$ *as* $n \to \infty$,

then

$$I_n^{-1/2}S_n \xrightarrow{\mathscr{L}} Z^*$$

where $Z^*=\eta^{1/2}Z$, *and* Z *is standard normal and independent of* η.

Proof. Since $\{S_n, n \geqq 1\}$ forms a martingale corresponding to family of σ-

algebras $\mathcal{F}_n = \sigma(Z_1, \ldots, Z_n)$, the sequence $\{Z_n, n \geqq 1\}$ is called a sequence of martingale differences. We may truncate the martingale differences and retain the martingale difference property by setting for any positive c,

$$Z_k^c = Z_k \chi_{[I_n^{-1}\Sigma_{j=1}^{k-1} Z_j^2 < c]}, \; S_k^c = \sum_{j=1}^{k} Z_j^c \tag{1}$$

and $S_0^c = Z_0^c = 0$ where χ_B denotes the indicator function of set B. We observe that $\{S_k^c, \mathcal{F}_k, k = 1, 2, \ldots, n\}$ is again a martingale sequence and satisfies

$$I_n^{-1} E\left\{\max_{1 \leq k \leq n} (Z_k^c)^2\right\} \to 0 \tag{2}$$

and

$$I_n^{-1} \sum_{k=1}^{n} (Z_k^c)^2 \xrightarrow{p} \eta^c \equiv \eta \chi_{[\eta < c]} + c \chi_{[\eta \geqq c]}. \tag{3}$$

From Theorem 4.2 of Billingsley (1968), the theorem follows provided we can show that

$$I_n^{-1/2} S_n^c \xrightarrow{\mathcal{L}} Z^{*c}$$

where $Z^{*c} = (\eta^c)^{1/2} Z$. Let $\bar{S}_n^c = I_n^{-1/2} S_n^c$ and $\bar{Z}_k^c = I_n^{-1/2} Z_k^c$. Applying the relation

$$e^x = (1 + x) \exp(x^2/2 + R(x))$$

where $|R(x)| \leqq |x|^3$ for any complex x such that $|x| = 1$, we obtain that for any real t,

$$E[\exp(it S_n^c)] = E[\phi_n + \{\phi_n \exp(R_n) - \phi_n\}] \tag{4}$$

where

$$\phi_n = \pi_n \exp\left(-\frac{t^2}{2} \sum_{k=1}^{n} (\bar{Z}_k^c)^2\right),$$

$$\pi_n = \sum_{k=1}^{n} (1 + it \bar{Z}_k^c),$$

and

$$R_n = \sum_{k=1}^{n} R(it \bar{Z}_k^c).$$

We shall prove that

$$E|\phi_n e^{R_n} - \phi_n| \to 0 \tag{5}$$

and

$$E(\phi_n) \to E(\exp(-t^2\eta^c/2)) \tag{6}$$

which together imply that

$$E(\exp(it\bar{S}_n^c)) \to E(\exp(-t^2\eta^c/2)) \tag{7}$$

which in turn shows that

$$I_n^{-1/2} S_n^c \xrightarrow{\mathscr{L}} Z^{*c}$$

by the continuity theorem for characteristic functions. Note that

$$|\phi_n| \leqq 1$$

since

$$|\pi_n| = \prod_{k=1}^{n} [1 + t^2(\bar{Z}_k^c)^2]^{1/2} \leqq \exp\left(t^2/2 \sum_{k=1}^{n} (\bar{Z}_k^c)^2\right)$$

and

$$|\phi_n e^{R_n}| = |\exp\ (it\bar{S}_n^c)| = 1$$

so that

$$|R_n| \leqq |t^3| \sum_{k=1}^{n} |\bar{Z}_k^c|^3 \leqq |t^3| \max_{1 \leqq k \leqq n} |\bar{Z}_k^c| \sum_{k=1}^{n} |\bar{Z}_k^c|^2$$

and the last term tends to zero in probability. The Bounded Convergence Theorem now gives (5). We can prove (6) as in Theorem 2 of Chatterji (1974) by showing that

$$\int \pi_n h \mathrm{d}P \to \int h \mathrm{d}P \tag{8}$$

for any square integrable h and observing that (3) implies

$$E\left| \exp\left(-\frac{t^2}{2} \sum_{k=1}^{n} (\bar{Z}_k^c)^2\right) - \exp\ (-t^2\eta^c/2)\right|^2 \to 0$$

since

$$0 \leqq e^{-x} \leqq 1$$

for $x > 0$.

Theorem 1.6. *Suppose the sequence $\{Z_k\}$ is a sequence of martingale differences as stated in Theorem 1.5. Let*

$$\xi_n = \sum_{k=1}^{n} E(Z_k^2 | \mathscr{F}_{k-1}).$$

If

(i) $I_n^{-1}\xi_n \xrightarrow{P} \eta$ *where* $P(\eta>0)=1$;

and

(ii) $I_n^{-1} \sum_{k=1}^{n} E(Z_k^2 \chi_{[|Z_k| \geqq \varepsilon I_n^{1/2}]} | \mathscr{F}_{k-1}) \xrightarrow{P} 0$ *for all* $\varepsilon > 0$,

then

$$I_n^{1/2} S_n \xrightarrow{\mathscr{L}} Z^*$$

and

$$\xi_n^{-1/2} S_n \xrightarrow{\mathscr{L}} Z.$$

Proof. See Hall (1977).

2. A RANDOM CENTRAL LIMIT THEOREM FOR MARTINGALES

Let $\xi_1, \xi_2, \ldots$ be random variables on some probability space $(\Omega, \mathscr{B}, P)$ and $\{\mathscr{F}_n, n \geqq 0\}$ be an increasing sequence of σ-algebras in $\mathscr{B}$ such that $\mathscr{F}_0 = \{\Omega, \phi\}$. Further suppose that ξ_n is $\mathscr{F}_n$-measurable and

$$E(\xi_n | \mathscr{F}_{n-1}) = 0, \; n \geqq 1 \tag{9}$$

with probability one. Define

$$S_n = \xi_1 + \ldots + \xi_n.$$

Then $\{S_n, \mathscr{F}_n, n \geqq 1\}$ is a martingale. Further, suppose there exists a $\delta > 0$ such that $E|\xi_k|^{2+\delta} \leqq M < \infty$ for all $k \geqq 1$ and

$$\lim_{n\to\infty} n^{-1} \sum_{k=1}^{n} E(\xi_k^2 | \mathscr{F}_{k-1}) = \sigma^2 > 0 \text{ a.s.}$$

Under the above conditions, it is known that $n^{-1/2} S_n \xrightarrow{\mathscr{L}} N(0, \sigma^2)$. Let v_n be a sequence of positive, integer valued, random variables such that

$$\frac{v_n}{n} \xrightarrow{P} c > 0 \tag{10}$$

where c is a constant a.s. We shall now prove that

$$v_n^{-1/2} S_{v_n} \xrightarrow{\mathscr{L}} N(0, \sigma^2). \tag{11}$$

Since $v_n/[nc] \xrightarrow{p} 1$ as $n \to \infty$, where $[x]$ denotes the largest integer less than or equal to x, writing

$$\frac{S_{v_n}}{v_n^{1/2}} = \left\{\frac{[nc]}{n}\right\}^{1/2} \left\{\frac{S_{[nc]}}{[nc]^{1/2}} + \frac{S_{v_n} - S_{[nc]}}{[nc]^{1/2}}\right\}$$

it follows that

$$\frac{S_{[nc]}}{[nc]^{1/2}} \xrightarrow{\mathscr{L}} N(0, \sigma^2)$$

and the problem reduces to showing that

$$\frac{S_{v_n} - S_{[nc]}}{[nc]^{1/2}} \xrightarrow{p} 0. \tag{12}$$

Let $\varepsilon > 0$. There exists an integer $n_0 > 0$ such that for every $n \geqq n_0$,

$$P(|v_n - [nc]| > \varepsilon[nc]) \leqq \varepsilon/2.$$

Let $L_1 = [(1-\varepsilon)nc]$ and $L_2 = [(1+\varepsilon)nc]$. Clearly, for $n \geqq n_0$,

$$\begin{aligned}
&P(|S_{v_n} - S_{[nc]}| > \varepsilon[nc]^{1/2})\\
&\leqq P(|S_{v_n} - S_{[nc]}| > \varepsilon[nc]^{1/2};\ |v_n - [nc]| \leqq \varepsilon[nc])\\
&\quad + P(|v_n - [nc]| > \varepsilon[nc])\\
&\leqq P\left(\sup_{L_1 \leqq n \leqq L_2} |S_n - S_{[nc]}| > \varepsilon[nc]^{1/2}\right) + \varepsilon/2\\
&\leqq P\left(\sup_{L_1 \leqq n \leqq [nc]} |S_n - S_{[nc]}| > \varepsilon[nc]^{1/2}\right)\\
&\quad + P\left(\sup_{[nc] < n \leqq L_2} |S_n - S_{[nc]}| > \varepsilon[nc]^{1/2}\right) + \varepsilon/2 \qquad (13)\\
&\leqq P\left(\sup_{L_1 \leqq n \leqq [nc]} |S_n - S_{L_1}| > \frac{\varepsilon}{2}[nc]^{1/2}\right)\\
&\quad + P\left(\sup_{[nc] < n \leqq L_2} |S_n - S_{[nc]}| > \varepsilon[nc]^{1/2}\right) + \varepsilon/2\\
&\leqq \frac{4}{\varepsilon^2[nc]} E(S_{[nc]} - S_{L_1})^2 + \frac{1}{\varepsilon^2[nc]} E(S_{[nc]} - S_{L_2})^2 + \varepsilon/2\\
&\leqq \frac{4}{\varepsilon^2[nc]} [E(S_{[nc]}^2) - E(S_{L1}^2) + E(S_{L2}^2) - E(S_{[nc]}^2)] + \varepsilon/2.
\end{aligned}$$

Suppose that

$$\lim_{\varepsilon\to\pm 0}\ \limsup_{n} \frac{E(S^2_{[nc]})-E(S^2_{[nc(1\pm\varepsilon)]})}{n\sigma^2}=c. \tag{14}$$

If (14) holds, it is easy to see that

$$\frac{S_{v_n}-S_{[nc]}}{[nc]^{1/2}}\xrightarrow{p}0$$

from relation (13). Hence we have the following Theorem.

Theorem 2.1. *Let $\{\xi_n, n\geqq 1\}$ be a sequence of random variables defined on a probability space $(\Omega, \mathscr{F}, P)$ and $\{\mathscr{F}_n, n\geqq 0\}$ be an increasing sequence of σ-algebras contained in $\mathscr{F}$ such that $\mathscr{F}_0=\{\Omega, \phi\}$ and $E(\xi_n|\mathscr{F}_{n-1})=0$, $n\geqq 1$ a.s. Further suppose that $n^{-1/2}S_n\xrightarrow{\mathscr{L}}N(0, \sigma^2)$ where $S_n=\xi_1+\ldots+\xi_n$ and $\{v_n\}$ is a sequence of positive integer valued random variables such that $n^{-1}v_n\xrightarrow{p}c$ where c is a constant. Then $v_n^{-1/2}S_{v_n}\xrightarrow{\mathscr{L}}N(0, \sigma^2)$ provided (14) holds.*

Random central limit theorem for independent random variables was proved by Renyi (1960) and Blum *et al.* (1963). This was generalized to strictly stationary-mixing processes whose partial sums form a martingale by Prakasa Rao (1969). Ghosh and Babu (1975) weakened the conditions assumed in Prakasa Rao (1969). Sarma (1976) proved that the mixing condition assumed by Prakasa Rao (1969) can be removed in case $v_n/n\xrightarrow{p}c$, constant. Sarma (1976) also assumed stationarity. It is clear from the proof given above that one can replace the assumption $n^{-1}v_n\xrightarrow{p}c$ by the assumption $v_n/f(n)\xrightarrow{p}1$ where $f(n)\uparrow\infty$ as $n\to\infty$. Theorem 2.1 is from Prakasa Rao (1974).

By an application of the Cramer–Wold technique, one can prove the following multivariate version of the random central limit theorem given above.

Theorem 2.2. *Let* $\xi_k'=(\xi_{k1}, \ldots, \xi_{kp})$, $k\geqq 1$ be a sequence of p-dimensional random vectors and $\{\mathscr{F}_n, n\geqq 0\}$ be an increasing sequence of σ-algebras such that $\mathscr{F}_0=\{\Omega, \phi\}$ and $E(\xi_{ki}|\mathscr{F}_{k-1})=0$ a.s. $1\leqq i\leqq p$, $k\geqq 1$. Further suppose that ξ_k is $\mathscr{F}_k$-measurable and $n^{-1/2}\Sigma_{k=1}^n\xi_k\xrightarrow{\mathscr{L}}N(0, \Gamma)$ where Γ is a positive definite matrix. Let $\{v_n\}$ be a sequence of positive, integer valued, random variables such that $n^{-1}v_n\xrightarrow{p}c>0$ where c is a constant. Then

$$v_n^{-1/2}\sum_{k=1}^{v_n}\xi_k\xrightarrow{\mathscr{L}}N(0, \Gamma)$$

provided that for every $\boldsymbol{\lambda} \neq 0$ in $\mathbb{R}^p$,

$$\lim_{\varepsilon \to \pm 0} \overline{\lim_{n}} \frac{E\left(\sum_{k=1}^{[nc]} \boldsymbol{\lambda}'\boldsymbol{\xi}_k\right)^2 - E\left(\sum_{k=1}^{[nc(1\pm\varepsilon)]} \boldsymbol{\lambda}'\boldsymbol{\xi}_k\right)^2}{n\boldsymbol{\lambda}'\Gamma\boldsymbol{\lambda}} = c.$$

3. EMBEDDING SUBMARTINGALES IN WIENER PROCESS WITH DRIFT

The Skorokhod (martingale) embedding theorem (Skorokhod, 1961) states that if $\{Y_n\}$ is a zero-mean martingale sequence and if W is a Wiener process without drift, then there exists a sequence $\{T_n\}$ of Markov times on $W(\cdot)$ so that the two sequences $\{W(T_n)\}$ and $\{Y_n\}$ have the same law. Hall (1969) has generalized this theorem to submartingales and Wiener processes with drift. He proved the following basic theorem.

Theorem 3.1. *Suppose $\delta > 0$ and $W_\delta(\cdot) = \{W_\delta(t),\ t \geqq 0;\ W_\delta(0) = 0\}$ is a Wiener process with drift δ and variance one per unit time. There exists a Markov time T on $W_\delta(\cdot)$ for which $W_\delta(T)$ and X have the same distribution if and only if*

$$E(e^{-2\delta X}) \leqq 1.$$

Moreover $ET = (EX)/\delta$.

The proof of this theorem can be found in Hall (1969). As a consequence of this theorem, one can obtain the following results.

(i) If $X \leqq a$ a.s. then $W_\delta(t) \leqq a$ for all $t \leqq T$.

(ii) If $E(e^{-2\delta X}) = 1$ and $X \geqq b$ a.s. then $W_\delta(t) \geqq b$ for all $t \leqq T$.

(iii) If $\mu = EX$, $\sigma^2 = \text{var } X$, then

$$(\sigma - \sqrt{\mu/\delta})^2 \leqq \delta^2 \text{ var } (T) \leqq (\sigma + \sqrt{\mu/\delta})^2$$

as long as var $(T) < \infty$.

Let $\{Y_n, n \geqq 1\}$ be a sequence of random variables defined on a probability space $(\Omega, \mathscr{F}, P)$ and $\{\mathscr{F}_n\}$ be an increasing sequence of sub-σ-algebras of $\mathscr{F}$ such that Y_n is $\mathscr{F}_n$-measurable. Let $\mathscr{F}_0 = \mathscr{F}$. We shall say that $\{Y_n, \mathscr{F}_n\}$ is an *embeddable submartingale* (supermartingale) if there exists a $\theta > 0 (\theta < 0)$ for which, with $Z_n = e^{-\theta Y_n}$, $n > 0$, $Z_0 = 1$, $\{Z_n, \mathscr{F}_n, n \geqq 0\}$ is a supermartingale i.e. $E(Z_{n+1} | \mathscr{F}_n) \leqq Z_n$ a.s. for all $n \geqq 0$. Any such θ is an *exponent* of the sequence.

Theorem 3.2. *An embeddable submartingale is a submartingale.*

Proof. Since $\{Z_n\}$ is a positive supermartingale, $EZ_n < \infty$ by definition. Hence Y_n^- has a moment generating function and therefore a finite mean. Since $Y_n = \theta^{-1} \log Z_n$,

$$E(Y_{n+1} | \mathscr{F}_n) = -\theta^{-1} E(\log Z_{n+1} | \mathscr{F}_n) \geqq -\theta^{-1} \log E(Z_{n+1} | \mathscr{F}_n)$$

by Jensen's inequality and the last term is greater than or equal to $-\theta^{-1} \log Z_n = Y_n$ since $\{Z_n\}$ is a supermartingale.

An embeddable submartingale is not a martingale (except in degenerate cases) in view of the strict nature of Jensen's inequality.

Let $X_n = Y_n - Y_{n-1}$, $n > 1$ and $X_1 = 1$. It is easy to see that $\{Y_n, \mathscr{F}_n\}$ is an embeddable submartingale if and only if there exists $\theta > 0$ for which

$$E(e^{-\theta X_n} | \mathscr{F}_{n-1}) \leqq 1 \text{ a.s., } n \geqq 1. \tag{16}$$

Let θ_n be the non-zero root, if any, of $E(e^{-\theta X_{n-1}} \mathscr{F}_n) = 1$ *and* $\theta_n = \infty$ if $Pr[X_{n+1} \geqq 0 | \mathscr{F}_n] = 1$. Hence θ_n is $\mathscr{F}_n$-measurable. If θ_n is bounded away from zero a.s. for all n, then (and only then) $\{Y_n\}$ is an embeddable submartingale.

Theorem 3.3 *If $\{Y_n, \mathscr{F}_n\}$ is an embeddable submartingale with exponent θ, then $Y = \lim_{n\to\infty} Y_n$ exists almost surely, $-\infty < Y \leqq +\infty$ and $E(e^{-\theta Y}) < \infty$.*

Proof. This is an immediate consequence of Theorem 4.1 of Doob (1953), p. 324 applied to the non-positive submartingale $\{-e^{-\theta Y_n}, \mathscr{F}_n\}$.

Since a sequence of likelihood ratios forms a martingale (the correct hypothesis being represented in the denominator and the distributions being mutually absolutely continuous), the sequence of log likelihood ratios is an embeddable supermartingale with exponent $\theta = -1$. If $X_n = Y_n - Y_{n-1} \geqq 0$ a.s. for all n, then $\{Y_n, \mathscr{F}_n\}$ is an embeddable submartingale with arbitrary positive exponent.

We shall now state the main submartingale embedding theorem. Theorem 3.3, Chapter 12 is an application of this theorem.

Theorem 3.4. *Suppose $\delta > 0$ (< 0 respectively) and $W_\delta(\cdot) = \{W_\delta(t);\ t = 0, W_\delta(0) = 0\}$ be a Wiener process with drift δ per unit time. Then there exists a sequence $\{T_n\}$ of successive Markov times on $W_\delta(\cdot)$ for which*

$$\{W_\delta(T_n),\ n = 1, 2, \ldots\}$$

and $\{Y_n, n = 1, 2, \ldots\}$ have the same distributions if and only if $\{Y_n, \mathscr{F}_n\}$ is an embeddable submartingale (supermartingale, respectively) for some $\{\mathscr{F}_n\}$ with *exponent* 2δ. *Moreover*

$$E(T_n) = EY_n/\delta (\leqq \infty).$$

The necessity part of the proof is based on a similar proof of Theorem 3.1. We omit it. For the sufficiency part, the sequence $\{T_n\}$ is constructed recursively by applying Theorem 3.1 to the conditional distribution of X_n given $\mathscr{F}_{n-1}$, with the origin in (t, W_δ) plane shifted to $(T_{n-1}, W_\delta(T_{n-1}))$. We shall now state some properties of the sequence $\{T_n\}$ so constructed. If $\{Y_n\}$ has independent increments, then $\{T_n\}$ may be also chosen to have independent increments; if $\limsup E(X_n | \mathscr{F}_{n-1}) \geqq \varepsilon > 0$ a.s. then $T_n \to \infty$ a.s. As a consequence of the Submartingale Embedding Theorem, we have the following result.

Theorem 3.5. *Suppose* $\{Y_n, \mathscr{F}_n\}$ *is embeddable in* $W_\delta(\cdot)(\delta > 0)$. *Then, for any* $b < 0$,

$$Pr\left\{\inf_n Y_n \leqq b\right\} \leqq e^{2\delta b} E(e^{-2\delta Y_1}) \leqq e^{2\delta b}.$$

Proof. We can embed $Y_n - c$ in $W_\delta(t)$; $t \geqq 0$, $W_\delta(0) = 0$ where c satisfies $Ee^{-2\delta Y_1} = e^{-2\delta c}$. Then

$$\begin{aligned} Pr\left[\inf_n (Y_n - c) \leqq b - c\right] &= Pr\left[\inf_n W_\delta(T_n) \leqq b - c\right] \\ &\leqq Pr\left[\inf_t W_\delta(t) \leqq b - c\right] \\ &= \exp(2\delta(b - c)) \end{aligned}$$

by Lemma 3.2.

Further applications of the embedding theorem to random walks and sequential analysis are given in Hall (1969). Results in this section are due to Hall (1969, 1970).

4. STRUCTURE OF CONTINUOUS PARAMETER MARTINGALES

We shall now give briefly some results on the theory and structure of continuous parameter martingales.

Let $\{\mathscr{F}_t, t \geqq 0\}$ be a right-continuous increasing sequence of sub-σ-fields of $\mathscr{F}$, $\mathscr{F}_0$ complete with respect to a probability measure P of a probability space $(\Omega, \mathscr{F}, P)$. Assume further that $\{V_t, \mathscr{F}_t; t \geqq 0\}$ is a square integrable martingale with zero mean and that almost all the sample paths of $V = \{V_t; t \geqq 0\}$ are right continuous. Thus, for all $t \geqq 0$,

$$E[V_t] = 0, \; E[V_t^2] < \infty, \tag{17}$$

$$E[V_t | \mathscr{F}_s] = V_s \text{ a.s., } s \leqq t, \tag{18}$$

and V_t is $\mathscr{F}_t$-measurable. Results of Meyer (1962) indicate that there exists a right continuous increasing process $\{I_t; t \geqq 0\}$ adapted to $\{\mathscr{F}_t, t \geqq 0\}$ (i.e. I_t is $\mathscr{F}_t$-measurable) satisfying

$$E\{(V_t - V_s)^2 | \mathscr{F}_s\} = E\{I_t - I_s | \mathscr{F}_s\} \text{ a.s. } 0 \leqq s \leqq t. \tag{19}$$

The process $\{I_t, t \geqq 0\}$ is the continuous analogue of the conditional variance

$$I_n = \sum_{j=1}^{n} E(X_j^2 | \zeta_{j-1}), \; n = 1, 2 \ldots \tag{20}$$

for the discrete parameter square integrable martingale $\{S_n = \Sigma_{j=1}^n X_j, \zeta_n, n \geqq 1\}$. In analogy with the definition of I_n, we can give the formal definition of I_t:

$$I_t = \int_0^t E[(dV_s)^2 | \mathscr{F}_s]. \tag{21}$$

For some applications, one can use this formal definition as a check in computing I_t.

We shall assume that the following condition is satisfied.

There exists a separable and measurable random function $f = \{f_t, t \geqq 0\}$ adapted to $\{\mathscr{F}_t, t \geqq 0\}$ for which

$$I_t = \int_0^t f_u^2 du \text{ a.s.} \tag{22}$$

For a definition of separability and measurability of a random function, the reader is referred to Doob (1953). The analogue for continuous time martingales of the strong law can now be stated as follows.

Theorem 4.1. *If* $\{V_t, \mathscr{F}_t, t \geqq 0\}$ *satisfies relations* (17) *and* (19) *and condition* (22) *holds then*

$$I_t^{-1} V_t \to 0 \text{ a.s. } on \; [I_t \to \infty]. \tag{23}$$

Before we state a central limit theorem for our purpose, we shall discuss some further results.

Theorem 4.2. *Let* $V_t = Y_{\tau_t}$ a.s. *where* Y *is a stationary process with independent increments satisfying* $E(Y_t^2) = t$, $t \geqq 0$ *and* $\tau = \{\tau_t, t \geqq 0\}$ *satisfies*

(i) τ_t *is a non-decreasing function of* t a.s.

(ii) *for some function* $m(t) \uparrow \infty$ *as* $t \to \infty$,

$$\frac{\tau_t}{m(t)} \xrightarrow{p} \eta^2 \tag{24}$$

where $P(\eta^2 > 0) > 0$. *Let* $V_t(a) = \tau_t^{-1/2} Y_{a\tau_t}$, $0 \leqq a \leqq 1$. *Then the stochastic process* V_t *converges in distribution to Wiener process on* $[0, 1]$ *and*

$$V_t(1) = \tau_t^{-1/2} V_t \xrightarrow{\mathcal{L}} N(0, 1). \tag{25}$$

Convergence here holds with respect to any probability measure μ *absolutely continuous with respect to* $P_B(\cdot) = P(\cdot | B)$ *where* $B = [\eta^2 > 0]$.

Billingsley (1968) gives a discussion of the concept convergence in distribution for stochastic processes. The above theorem is an invariance principle and is proved by method similar to that of Theorem 17.2 of Billingsley (1968). Dion (1973) discussed a discrete time version of the above theorem.

The next theorem identifies τ_t of Theorem 4.2 with I_t.

Theorem 4.3. *Let* $\{V_t, \mathscr{F}_t, t \geqq 0\}$ *satisfy relations* (17) *and* (18) *and have continuous sample paths almost surely. Then there exists a standard Wiener process* $\{W_t, t \geqq 0\}$ *such that*

$$V_t = W_{I_t} \text{ a.s.}, \ t \geqq 0.$$

The above theorem is due to Kunita and Watanabe (1967). To prove a central limit theorem for U_t using the norming $I_t^{-1/2}$, one has to find a process Y as in Theorem 4.2 and check whether relation (24) holds or not.

Appendix 2

Stochastic Differential Equations

1. STOCHASTIC INTEGRALS

In this section, we shall give a brief survey of the theory of stochastic integrals. An excellent discussion of this topic is given in Gikhman and Skorokhod (1972).

Let $\{W(t),\ t \geqq 0\}$ be a Wiener process i.e. a homogeneous Gaussian process with independent increments for which $W(0)=0$, $EW(t)=0$ and var $W(t)=t$. It is known that

$$P\left[\sup_{0 \leqq t \leqq T} W(t) > x\right] = \sqrt{2/\pi} \int_x^\infty \exp(-u^2/2T)\mathrm{d}u \tag{1}$$

for any $T>0$ and $x>0$ and for each $\varepsilon>0$ and $m>0$, then exists a constant $L_m(\varepsilon)$ such that

$$P\left[\sup_{0 \leqq s \leqq t} |W(s)| > \varepsilon\right] \leqq \mathrm{L}_m(\varepsilon)t^m. \tag{2}$$

Doob (1953) has given the following characterization of Wiener process: let $\{\xi(t),\ t \geqq 0\}$ be a stochastic process with continuous sample paths with probability one and suppose that $\{\mathscr{F}_t,\ t \geqq 0\}$ is a sequence of σ-algebras such that $\mathscr{F}_{t_1} \subset \mathscr{F}_{t_2}$ if $t_1 < t_2$. Further suppose that

(i) for all $t \geqq 0$, $\xi(t)$ is $\mathscr{F}_t$-measurable,
(ii) $E[\xi(t+h)-\xi(t)|\mathscr{F}_t]=0$ a.s. for all $t \geqq 0$, $h>0$, and
(iii) $E[(\xi(t+h)-\xi(t))^2|\mathscr{F}_t]=h$ a.s. for all $t \geqq 0$, $h>0$.

Then $\{\xi(t), t \geqq 0\}$ is a Wiener process.

We shall first define the stochastic integral

$$\int f(t)\mathrm{d}W(t) \tag{3}$$

with respect to suitable class of random functions $f(t)$.

For all $t \in [0, T]$, let $\{\mathscr{F}_t\}$ be a family of σ-algebras satisfying the properties (i) for $t_1 < t_2$, $\mathscr{F}_{t_1} \subset \mathscr{F}_{t_2}$, (ii) $W(t)$ is $\mathscr{F}_t$-measurable and (iii) the process $W_t(s) = W(t+s) - W(t)$, $s \geqq 0$ is independent of $\mathscr{F}_t$ for any fixed t.

Let $H[0, T]$ be the class of the random functions defined on $[0, T]$, $\mathscr{F}_t$-measurable for each t and for which

$$\int_0^T Ef^2(t)\mathrm{d}t < \infty$$

and let $H_2[0, T]$ be the class of all random functions $f(t)$ defined on $[0, T]$, $\mathscr{F}_t$-measurable for each t and for which

$$\int_0^T f^2(t)\mathrm{d}t$$

is finite with probability one. If f is a step function in $H_2[0, T]$ i.e. a function f for which there exists a partition of $[0, T]$ $(0 = t_0 < t_1 < \ldots < t_m = T)$ such that $f(t) = f(t_k)$, $t_k \leqq t < t_{k+1}$, define the integral

$$\int_0^T f(t)\mathrm{d}W(t)$$

as

$$\sum_{k=0}^{m-1} f(t_k)[W(t_{k+1}) - W(t_k)].$$

It is not difficult to show that the following properties hold for step functions.

(i) If f_1 and $f_2 \in H_2[0, T]$ and α_1, α_2 are random variables for which

$\alpha_1 f_1(t)+\alpha_2 f_2(t) \in H_2[0, T]$, then

$$\int_0^T [\alpha_1 f_1(s)+\alpha_2 f_2(s)]\mathrm{d}W(s)=$$

$$\alpha_1 \int_0^T f_1(s)\mathrm{d}W(s)+\alpha_2 \int_0^T f_2(s)\mathrm{d}W(s).$$

(ii) If $\chi_{[\alpha,\beta]}$ is the indicator function of the interval $[\alpha, \beta]$ contained in $[0, T]$, then

$$\int_0^T \chi_{[\alpha,\beta]}(t)\mathrm{d}W(t)=W(\beta)-W(\alpha).$$

(iii) If $f \in H[0, T]$, then

$$E\left[\int_0^T f(t)\mathrm{d}W(t)\right]=0,\ E\left[\int_0^T f(t)\mathrm{d}W(t)\right]^2=\int_0^T E[f^2(t)]\mathrm{d}t.$$

(iv) For $f \in H_2[0, T]$, $C>0$ and $N>0$,

$$P\left\{\left|\int_0^T f(t)\mathrm{d}W(t)\right|>C\right\} \leq P\left\{\int_0^T f^2(t)\mathrm{d}t>N\right\}+\frac{N}{C^2}.$$

Now let $f \in H[0, T]$ be such that there exists a sequence of step functions $g_n \in H_2[0, T]$ for which

$$\lim_{n\to\infty} E\left[\int_0^T \{f(t)-g_n(t)\}^2\mathrm{d}t\right]=0$$

(see Lemma 4.4 of Lipcer and Sirjaev, 1977). Then the sequence of random variables

$$\int_0^T g_n(t)\mathrm{d}W(t)$$

is fundamental in the sense of convergence in the mean square. Hence it converges to some limit which we will denote by

$$\int_0^T f(t)\mathrm{d}W(t).$$

It can be shown that this integral is independent of the choice of the sequence $\{g_n\}$ and is well defined almost surely.

Now assume that $f \in H_2[0, T]$. Then we can construct a sequence of functions $g_n(t)$ for which

$$\int_0^T \{g_n(t)-f(t)\}^2\mathrm{d}t \xrightarrow{P} 0$$

which in turn implies that (see Lemma 4.5 of Lipcer and Sirjaev, 1977)

$$\int_0^T g_n(t)\mathrm{d}W(t)$$

is fundamental in the sense of convergence in probability by (iv). Hence it converges (in probability) to some limit which is independent of the choice of $g_n(t)$. This limit is denoted by

$$\int_0^T f(t)\mathrm{d}W(t).$$

Properties (i)–(iv) for the general case follow from their validity for step functions and the possibility of proceeding to the limiting case.

For any $f \in H_2[0, T]$ and $\chi_{[\alpha,\beta]}(t)$ the indicator function of the interval $[\alpha, \beta] \subset [0, T]$, the symbol

$$\int_\alpha^\beta f(t)\mathrm{d}W(t)$$

will denote the stochastic integral

$$\int_0^T f(t)\chi_{[\alpha,\beta]}(t)\mathrm{d}W(t).$$

The following sharper form of property (iii) is also true:

(iii)*. If $f \in H[0, T]$ then, for $[\alpha, \beta] \subset [0, T]$,

$$E\left[\int_\alpha^\beta f(t)\mathrm{d}W(t) \mid \mathscr{F}_\alpha\right] = 0 \text{ a.s.}$$

and

$$E\left\{\left[\int_\alpha^\beta f(t)\mathrm{d}W(t)\right]^2 \middle| \mathscr{F}_\alpha\right\} = \int_\alpha^\beta E(f^2(t) \mid \mathscr{F}_\alpha)\mathrm{d}t \text{ a.s.}$$

For any $f \in H_2[0, T]$, define

$$I(t) = \int_0^t f(s)\mathrm{d}W(s)$$

The process $I(t)$ is defined up to stochastic equivalence. Since every process is stochastically equivalent to a separable process, we will assume that $I(t)$ is separable. The following theorem holds for the stochastic process $I(t)$.

Theorem 1.1. *Let $f \in H_2[0, T]$. Then the separable process*

$$I(t) = \int_0^t f(s)\mathrm{d}W(s)$$

is continuous with probability one. Moreover for any $C > 0$, $N > 0$

$$P\left\{\sup_{0 \leqq t \leqq T} \left| \int_0^t f(s)\mathrm{d}W(s) \right| > C\right\} \leqq P\left\{\int_0^T f^2(t)\mathrm{d}t > N\right\} + \frac{N}{C^2}.$$

In particular, if

$$\int_0^T E[f^2(t)]\mathrm{d}t < \infty$$

then for any $a > 0$

$$P\left\{\sup_{0 \leqq t \leqq l} \left| \int_0^t f(s)\mathrm{d}W(s) \right| > a\right\} \leqq \frac{1}{a^2} \int_0^T E[f^2(t)]\mathrm{d}t,$$

and

$$E\left\{\sup_{0 \leqq t \leqq T} \left| \int_0^t f(s)\mathrm{d}W(s) \right|\right\} \leqq 4 \int_0^T E[f^2(t)]\mathrm{d}t.$$

Proof. See Gikhman and Skorokhod (1972), p. 20.

Let the process $\{\zeta(t), t \in [0, T]\}$ satisfy for all $0 \leqq t_1 < t_2 \leqq T$, the relation

$$\zeta(t_2) - \zeta(t_1) = \int_{t_1}^{t_2} a(t)\mathrm{d}t + \int_{t_1}^{t_2} b(t)\mathrm{d}W(t)$$

where $|a(t)|^{1/2} \in H_2[0, T]$ and $b(t) \in H_2[0, T]$. We will then say that the process $\zeta(t)$ has a stochastic differential on $[0, T]$:

$$\mathrm{d}\zeta(t) = a(t)\mathrm{d}t + b(t)\mathrm{d}W(t).$$

The following theorems can be proved by using the above notion of stochastic differential. For proofs, see Gikhman and Skorokhod (1972).

Theorem 1.2. *If f_1 and f_2 are in $H[0, T]$ then*

$$E\left\{\int_0^T f_1(t)\mathrm{d}W(t) \int_0^T f_2(t)\mathrm{d}W(t)\right\} = \int_0^T E[f_1(t) f_2(t)]\mathrm{d}t$$

and if $f \in H[0, T]$ and for some integer $m \geqq 1$

$$\int_0^T E f^{2m}(t)\mathrm{d}t < \infty$$

then

$$E\left[\int_0^T f(t)\mathrm{d}W(t)\right]^{2m} \leqq [m(2m-1)]^{m-1}T^{m-1}\int_0^T Ef^{2m}(t)\mathrm{d}t.$$

Theorem 1.3. *If f_1 and f_2 are in $H[0, T]$ and $W_1(s)$ and $W_2(s)$ are independent Wiener processes, then*

$$E\left\{\int_0^T f_1(s)\mathrm{d}W_1(s)\int_0^T f_2(s)\mathrm{d}W_2(s)\right\}=0.$$

If $f \in H_2[0, T]$ and ζ is a random variable for which $P[0 \leqq \zeta \leqq T]=1$, then by the integral

$$\int_0^\zeta f(t)\mathrm{d}W(t)$$

we will understand the variables $I(\zeta)$ where

$$I(t)=\int_0^t f(s)\mathrm{d}W(s)$$

is a continuous process with probability one. $I(\zeta)$ satisfies properties (i) and (ii) but need not satisfy (iii) in general. We can, however, define a class of random variables for which property (iii) holds. This class of random variables is the class of Markov times with respect to the collection of σ-algebras $\{\mathscr{F}_t, t \geqq 0\}$.

Typical examples of Markov times are the first passage times. For example,

$$\zeta_1=\inf\{t: W(t)=a\},$$

$$\zeta_2=\inf\left\{t: \int_0^t f(u)\mathrm{d}W(u)=a\right\}$$

are Markov times. If ζ_1 and ζ_2 are random variables such that $P(0 \leqq \zeta_1 \leqq \zeta_2 \leqq T)=1$, then we define

$$\int_{\zeta_1}^{\zeta_2} f(t)\mathrm{d}W(t)=\int_0^{\zeta_2} f(t)\mathrm{d}W(t)-\int_0^{\zeta_1} f(t)\mathrm{d}W(t).$$

The following theorem holds for stochastic integrals with limits as Markov times.

Theorem 1.4. *Assume that $f \in H[0, T]$ and that ζ_1 and ζ_2 are Markov times* *for which* $P[0 \leqq \zeta_1 \leqq \zeta_2 \leqq T] = 1$. *Then*

$$E\left[\int_{\zeta_1}^{\zeta_2} f(t)\mathrm{d}W(t)\right] = 0,$$

and

$$E\left[\int_{\zeta_1}^{\zeta_2} f(t)\mathrm{d}W(t)\right]^2 = E\left[\int_{\zeta_1}^{\zeta_2} f^2(t)\mathrm{d}t\right].$$

Furthermore

$$E\left[\int_{\zeta_1}^{\zeta_2} f(t)\mathrm{d}W(t) | \mathscr{F}_{\zeta_1}\right] = 0 \text{ a.s.}$$

and

$$E\left[\left\{\int_{\zeta_1}^{\zeta_2} f(t)\mathrm{d}W(t)\right\}^2 \,\middle|\, \mathscr{F}_{\zeta_1}\right] = E\left[\int_{\zeta_1}^{\zeta_2} f^2(t)\mathrm{d}t \,\middle|\, \mathscr{F}_{\zeta_1}\right] \text{ a.s.}$$

Theorem 1.5. *If* f_1 *and* f_2 *belong to* $H[0, T]$ *for every* $T > 0$, $W_1(s)$ *and* $W_2(s)$ *are independent Wiener processes and* τ_1 *and* τ_2 *are Markov times for which* $P[0 \leqq \tau_1 \leqq \tau_2] = 1$, *then*

$$E\left\{\int_{\tau_1}^{\tau_2} f_1(s)\mathrm{d}W_1(s) \int_{\tau_1}^{\tau_2} f_2(s)\mathrm{d}W_2(s) \,\middle|\, \mathscr{F}_{\tau_1}\right\} = 0 \quad \text{a.s.}$$

Theorem 1.6. *If* τ *is a Markov time, then the process* $\tilde{W}(t) = W(t+\tau) - W(\zeta)$ *is also a Wiener process and is independent of* $\mathscr{F}_\tau$.

This property of the Wiener process is called the "strong Markov Property."

Theorem 1.7. *Let* $\mathbf{G}(s) = (G_1(s), \ldots, G_m(s))$ *be a vector valued random function with components in* $H[0, T]$ *for every* $T > 0$ *and* $W(s) = (W_1(s), \ldots, W_m(s))$ *be the Wiener process with independent components. Further suppose that*

$$\int_0^\infty (\mathbf{G}(s), \mathbf{G}(s))\mathrm{d}s = \infty \tag{4}$$

almost surely. Let

$$\tau_t = \inf\left\{u: \int_0^u (\mathbf{G}(s), \mathbf{G}(s))\mathrm{d}s \geqq t\right\}, \tag{5}$$

then the process

$$\zeta_t = \int_0^{\tau_t} (\mathbf{G}(s), \mathrm{d}\mathbf{W}(s)) \tag{6}$$

is Wiener.

Proof. For each $t \geqq 0$, τ_t is a Markov time with respect to the σ-algebra $\{\mathscr{F}_t, t \geqq 0\}$ and $\tau_{t_1} < \tau_{t_2}$ for $t_1 < t_2$. Since

$$\int_0^{\tau_t} (\mathbf{G}(s), \mathbf{G}(s))\mathrm{d}s = t$$

it follows that

$$E\left\{\int_0^{\tau_t} (\mathbf{G}(s), \mathbf{G}(s))\mathrm{d}s\right\} < \infty.$$

By properties of stochastic integrals with random limits, we find that

$$E\left\{\int_{\tau_{t_1}}^{\tau_{t_2}} (G(s), \mathrm{d}\mathbf{W}(s)) | \mathscr{F}_{\tau_{t_1}}\right\} = 0 \text{ a.s.} \tag{7}$$

and

$$E\left\{\left[\int_{\tau_{t_1}}^{\tau_{t_2}} (\mathbf{G}(s), \mathrm{d}\mathbf{W}(s))\right]^2 | \mathscr{F}_{\tau_{t_1}}\right\}$$

$$= \sum_{k=1}^{m} \sum_{j=1}^{m} E\left\{\int_{\tau_{t_1}}^{\tau_{t_2}} G_k(s)\mathrm{d}W_k(s) \int_{\tau_{t_1}}^{\tau_{t_2}} G_j(s)\mathrm{d}W_j(s) | \mathscr{F}_{\tau_{t_1}}\right\}$$

$$= E\left\{\int_{\tau_{t_1}}^{\tau_{t_2}} (\mathbf{G}(s), \mathbf{G}(s))\mathrm{d}s | \mathscr{F}_{\tau_{t_1}}\right\} = t_2 - t_1. \tag{8}$$

The continuity of the process $(\zeta_t, t \geqq 0)$ follows from general properties of stochastic integrals (cf. Gikhman and Skorokhod, 1972, p. 20). Relations (7) and (8) together imply that $\{\zeta(t), t \geqq 0\}$ is a Wiener process by a theorem of Doob (cf. Gikhman and Skorokhod, 1972, p. 8). This completes the proof of the theorem.

An alternative proof of an improved version of Theorem 1.7 is given below.

Theorem 1.8. *Let* $\mathbf{G}(s) = (G_1(s), \ldots, G_m(s))$ *be a vector valued random function with components in* $H_2[0, T]$ *for every* $T > 0$ *and*

$$\int_0^{\infty} (\mathbf{G}(s), \mathbf{G}(s))\,\mathrm{d}s = \infty \text{ a.s.}$$

Let

$$\tau_T = \inf\left\{t: \int_0^t (\mathbf{G}(s), \mathbf{G}(s))\mathrm{d}s = Tb^2\right\}, \tag{9}$$

and.

$$\xi = \frac{1}{\sqrt{T}} \int_0^{\tau_T} (\mathbf{G}(s), \mathrm{d}\mathbf{W}(s)). \tag{10}$$

Then ξ is normal with mean zero and variance b^2.

Proof. The proof follows from the identity

$$E\left\{\exp\left[i \int_0^{\infty} (\mathbf{f}(s), \mathrm{d}\mathbf{W}(s)) + \frac{1}{2} \int_0^{\infty} (\mathbf{f}(s), \mathbf{f}(s))\mathrm{d}s\right]\right\} = 1, \tag{11}$$

(cf. McKean, 1969) whenever

$$E\left[\exp\left\{\frac{1}{2} \int_0^{\infty} (\mathbf{f}(s), \mathbf{f}(s))\mathrm{d}s\right\}\right] < \infty. \tag{12}$$

Let $\mathbf{f}(s) = \lambda T^{-1/2} \chi_{[\tau \geqq t]} \mathbf{G}(s)$. It is clear that (12) holds for $\mathbf{f}(s)$ from the definition of τ given in (9). Hence

$$E\left\{\exp\left[i\lambda\xi + \frac{1}{2}\frac{\lambda^2}{T} \int_0^{\tau} (\mathbf{G}(s), \mathbf{G}(s))\mathrm{d}s\right]\right\} = 1$$

for all $\lambda \in \mathbb{R}$. Therefore

$$E[\mathrm{e}^{i\lambda\xi}] = \mathrm{e}^{-1/2\lambda^2 b^2} \tag{13}$$

which proves that ξ is normal with mean zero and variance b^2.

Theorem 1.7 is due to Taraskin (1974) and Theorem 1.8 is from Kutoyants (1975). The following inequality is useful in obtaining some bounds.

Theorem 1.9. *If $f \in H_2[0, 1]$, then, for any $\alpha > 0$ and $\beta > 0$,*

$$P\left[\sup_{0 \leqq t \leqq 1} \left\{\int_0^t f(s)\mathrm{d}W(s) - \frac{\alpha}{2} \int_0^t f^2(s)\mathrm{d}s\right\} > \beta\right] \leqq \mathrm{e}^{-\alpha\beta}.$$

Proof of this theorem can be obtained from McKean (1969). It follows from the fact that

$$Z(t) = \exp\left\{\int_0^t f(s)\mathrm{d}W(s) - \frac{1}{2}\int_0^t f^2(s)\mathrm{d}s\right\}, \quad 0 \leqq t \leqq 1$$

is a supermartingale with respect to the family: $\{\mathscr{F}_t, 0 \leqq t \leqq 1\}$, $\mathscr{F}_t = \sigma\{W(s)$: $0 \leqq s \leqq t\}$.

2. CENTRAL LIMIT THEOREM FOR VECTOR-VALUED STOCHASTIC INTEGRALS

We shall now state and prove a central limit theorem for stochastic integrals due to Kutoyants (1975).

Suppose that $F(s)=((f_{kj}(s)))_{n\times m}$ is a random matrix valued function such that its elements $f_{kj}\in H[0,T]$ for all $T>0$. Then, it can be shown (Gikhman and Skorokhod, 1972)) that the stochastic integral

$$\int_0^T F(s)\mathrm{d}\mathbf{W}(s) \tag{14}$$

is defined a.s. for every $T>0$ and is a vector valued random variable. We are interested in conditions under which

$$T^{-1/2}\int_0^T F(s)\mathrm{d}\mathbf{W}(s) \tag{15}$$

is asymptotically normal as $T\to\infty$. Let us set $\mathbf{f}_k(s)=f_{kl}(s),\ \ldots,\ f_{km}(s)$, $1\leqq k\leqq n$. Then

$$\int_0^T F(s)\mathrm{d}\mathbf{W}(s)=\left\{\int_0^T(\mathbf{f}_1(s),\ \mathrm{d}\mathbf{W}(s)),\ \ldots,\ \int_0^T(\mathbf{f}_n(s),\ \mathrm{d}\mathbf{W}(s))\right\}.$$

Theorem 2.1. *Suppose the random matrix valued function $F(s)$ satisfies the condition*

$$T^{-1}\int_0^T(\mathbf{f}_k(s),\ \mathbf{f}_j(s))\mathrm{d}s\xrightarrow{p}c_{kj}\ \ as\ \ T\to\infty \tag{16}$$

where c_{kj}, $1\leqq k\leqq n$, $1\leqq j\leqq n$ are finite. Then the distribution of

$$T^{-1/2}\int_0^T F(s)\mathrm{d}\mathbf{W}(s) \tag{17}$$

converges to the normal distribution with mean zero and covariance matrix $C=((c_{kj}))$ as $T\to\infty$.

Proof. Clearly C is symmetric and non-negative definite. We shall prove that for any $\lambda=(\lambda_1,\ \ldots,\ \lambda_n)\in\mathbb{R}^n$ such that $(C\lambda,\ \lambda)\neq 0$,

$$T^{-1/2}\sum_{k=1}^{n}\lambda_k\int_0^T(\mathbf{f}_k(s),\ \mathrm{d}\mathbf{W}(s))$$

is asymptotically normal as $T\to\infty$. Let $\mathbf{f}_\lambda(s)=\Sigma_{k=1}^n\ \lambda_k\mathbf{f}_k(s)$. Condition (16) of the theorem implies that

$$T^{-1}\int_0^T (\mathbf{f}_\lambda(s), \mathbf{f}_\lambda(s))ds \xrightarrow{p} (C\lambda, \lambda). \tag{18}$$

Define

$$\begin{aligned} \mathbf{f}_\lambda^*(s) &= \mathbf{f}_\lambda(s) \quad \text{for } 0 \leqq s \leqq T \\ &= \mathbf{1} \qquad\quad \text{for } T < s \leqq T^* \end{aligned} \tag{19}$$

where $T^* = T\{1 + m^{-1}(C\lambda, \lambda)\}$. Further let

$$\tau_T = \inf\left\{u: \int_0^u (\mathbf{f}_\lambda^*(s), \mathbf{f}_\lambda^*(s))ds = T(C\lambda, \lambda)\right\}.$$

It follows from Theorem 1.8 that

$$\xi_T = T^{-1/2}\int_0^{\tau_T} (\mathbf{f}_\lambda^*(s), d\mathbf{W}(s))$$

is normal with mean 0 and variance $(C\lambda, \lambda)$. Let

$$\eta_T = T^{-1/2}\int_0^T (\mathbf{f}_\lambda(s), d\mathbf{W}(s)).$$

Now, for any $\delta > 0$ and $\alpha > 0$,

$$\begin{aligned} &P\{|\eta_T - \xi_T| > \delta\} \\ &= P\left\{T^{-1/2}\left|\int_0^T (\mathbf{f}_\lambda(s), d\mathbf{W}(s)) - \int_0^{\tau_T} (\mathbf{f}_\lambda^*(s), d\mathbf{W}(s))\right| > \delta\right\} \\ &= P\left\{T^{-1/2}\left|\int_0^{T^*} (\chi_{[t \leqq T]} - \chi_{[t \leqq \tau_T]})(\mathbf{f}_\lambda^*(s), d\mathbf{W}(s))\right| > \delta\right\} \\ &\leqq \frac{\alpha}{\delta^2} + P\left\{\frac{1}{T}\int_0^{T^*} \left|\chi_{[t \leqq T]} - \chi_{[t \leqq \tau_T]}\right| (\mathbf{f}_\lambda^*(s), \mathbf{f}_\lambda^*(s))ds \geqq \alpha\right\}. \end{aligned} \tag{20}$$

Here χ_B denotes the indicator function of set B. Let

$$A = \left\{w: \frac{1}{T}\int_0^{T^*} \left|\chi_{[t \leqq T]} - \chi_{[t \leqq \tau_T]}\right| (\mathbf{f}_\lambda^*(s), \mathbf{f}_\lambda^*(s))ds \geqq \alpha\right\}. \tag{21}$$

By the definition of Markov time τ_T, $P(A \cap [\tau_T = T]) = 0$. Furthermore

$$\begin{aligned} &P(A \cap [\tau_T < T]) \\ &= P\left\{\frac{1}{T}\int_0^{T^*} (\chi_{[t \leqq T]} - \chi_{[t \leqq \tau_T]})(\mathbf{f}_\lambda^*(s), \mathbf{f}_\lambda^*(s))ds \geqq \alpha,\ \tau_T < T\right\} \end{aligned}$$

$$=P\left\{\frac{1}{T}\int_0^T(\mathbf{f}_\lambda(s),\ \mathbf{f}_\lambda(s))ds-\frac{1}{T}\int_0^{\zeta_T}(\mathbf{f}_\lambda^*(s),\ \mathbf{f}_\lambda^*(s))ds\geqq\alpha,\ \tau_T<T\right\}$$

$$=P\left\{\frac{1}{T}\int_0^T(\mathbf{f}_\lambda(s),\ \mathbf{f}_\lambda(s))ds-(C\lambda,\ \lambda)\geqq\alpha,\ \tau_T<T\right\}.$$

Similarly we can show that

$$P(A\cap[\tau_T>T])=P\left\{(C\lambda,\ \lambda)-\frac{1}{T}\int_0^T(\mathbf{f}_\lambda(s),\ \mathbf{f}_\lambda(s))ds\geqq\alpha,\tau_T>T\right\}.$$

Hence

$$P(A)\leqq P\left\{\ \frac{1}{T}\int_0^T(\mathbf{f}_\lambda(s),\ \mathbf{f}_\lambda(s))ds-(C\lambda,\ \lambda)\ \ \geqq\alpha\right\} \tag{22}$$

and the term on the right-hand side of (22) tends to zero by (18). Hence it follows from (20) that $\eta_T-\xi_T\xrightarrow{p}0$ as $T\to\infty$. But ξ_T is normal with mean 0 and variance $(C\lambda,\ \lambda)$ for every $T>0$. Therefore

$$\eta_T\xrightarrow{\mathscr{L}}N(0,\ (C\lambda,\ \lambda)). \tag{23}$$

This completes the proof of the theorem.

We have the following corollary for diffusion processes.

Corollary 2.1. *Let $\{\xi(t),\ t\geqq0\}$ be a recurrent diffusion process with values in $\mathbb{R}^m$ and suppose that it is ergodic with stationary measure $G(\mathbf{x})$. Let $\phi(\mathbf{x})$ be a matrix valued function defined on $\mathbb{R}^m$ with bounded entries $\varphi_{kj}(\mathbf{x})$, $1\leqq k\leqq n$, $1\leqq j\leqq m$.*
Then

$$T^{-1/2}\int_0^T\phi(\xi(s))d\mathbf{W}(s)$$

is asymptotically normal with mean zero and covariance matrix

$$\left(\left(\int_{\mathbb{R}^m}(\boldsymbol{\phi}_k(\mathbf{x}),\ \boldsymbol{\phi}_j(\mathbf{x}))G(d\mathbf{x})\right)\right)$$

where

$$\boldsymbol{\phi}_k(\mathbf{x})=\{\boldsymbol{\phi}_{k1}(\mathbf{x}),\ldots,\boldsymbol{\phi}_{km}(\mathbf{x})\}.$$

The corollary follows from theorem 2.1 in view of the fact that

$$\frac{1}{T}\int_0^T(\boldsymbol{\phi}_k(\xi(s)),\ \boldsymbol{\phi}_j(\xi(s)))ds\to\int_{\mathbb{R}^m}(\boldsymbol{\phi}_k(\mathbf{x}),\ \boldsymbol{\phi}_j(\mathbf{x}))G(d\mathbf{x})\ \text{a.s.}$$

by the result in Khasminskii (1960) for recurrent diffusion processes.

3 STOCHASTIC DIFFERENTIAL EQUATIONS

Consider the stochastic differential equation

$$\mathrm{d}\eta(t)=a(t,\ \eta(t))\mathrm{d}t+\sigma(t,\ \eta(t))\mathrm{d}W(t) \tag{24}$$

where $a(t;\ x)$ and $\sigma(t,\ x)$, defined and measurable for $t\in[0,\ T]$, $x\in\mathbb{R}$, are called coefficients. The random process $\{\eta(t),\ 0\leqq t\leqq T\}$ is called a solution of (24) on $[0,\ T]$ if it satisfies the following conditions:

(i) if $\mathscr{F}_t$ denotes the smallest σ-algebra with respect to which the variables $\eta(s)$ for $s\leqq t$ and $W(s)$ for $s\leqq t$ are measurable the process $W_t(s)=W(t+s)-W(t)$ is independent of $\mathscr{F}_t$;

(ii) if $H_2[0,\ T]$ denotes the space of measurable random functions $\varphi(t)$ which, for each $t\in[0,\ T]$, are $\mathscr{F}_t$-measurable and for which the integral

$$\int_0^T \varphi^2(t)\mathrm{d}t<\infty \text{ a.s.}$$

then $|a(t,\ \eta(t))|^{1/2}$ and $\sigma(t,\ \eta(t))$ belong to $H_2[0,\ T]$.

(iii) the process $\eta(t)$ has on $[0,\ T]$, the stochastic differential

$$\mathrm{d}\eta(t)=\overline{a}(t)\mathrm{d}t+\overline{\sigma}(t)\mathrm{d}W(t)$$

where $\overline{a}(t)=a(t,\ \eta(t))$ a.s. and $\overline{\sigma}(t)=\sigma(t,\ \eta(t))$ a.s. If $\eta(t)$ is a solution of (24), then $\eta(0)$ will be independent of $W(t)-W(0)$, $t>0$ and

$$\eta(t)=\eta(0)+\int_0^t a(s,\ \eta(s))\mathrm{d}s+\int_0^t \sigma(s,\ \eta(s))\mathrm{d}W(s). \tag{25}$$

Equations (24) and (25) are equivalent and are to be solved for given $\eta(0)$ with $\eta(0)$ chosen independent of $W(t)$. If $\sigma(t,\ x)=0$, equation (24) can be considered as an ordinary differential equation.

Theorem 3.1. gives sufficient conditions for the existence and uniqueness of solutions of (24).

Theorem 3.1. *Suppose the following assumptions are satisfied:*

(i) *the functions $a(t,\ x)$ and $\sigma(t,\ x)$ are defined for $t\in[0,\ T]$ and $x\in\mathbb{R}$, and measurable with respect to all the arguments;*

(ii) *there exists a constant K such that for $t\in[0,\ T]$ and $x,\ y\in\mathbb{R}$,*

$$|a(t,\ x)-a(t,\ y)|+|\sigma(t,\ x)-\sigma(t,\ y)|\leqq K|x-y|$$

$$|a(t,\ x)|^2+|\sigma(t,\ x)|^2\leqq K^2(1+|x|^2)$$

(iii) *$\eta(0)$ is independent of $W(t)$ and $E\ \eta^2(0)<\infty$.*

Then there exists a solution of (24) satisfying the conditions:

(a) $\eta(t)$ *is continuous a.s. and* $\eta(t)=\eta(0)$ *for* $t=0$
(b) $\sup\{E\,\eta^2(t);\ 0\leqq t\leqq T\}<\infty$.

If $\eta_1(t)$ *and* $\eta_2(t)$ *are two solutions of* (24) *satisfying* (a) *and* (b), *then*

$$P\left\{\sup_{0\leqq t\leqq T}|\eta_1(t)-\eta_2(t)|=0\right\}=1.$$

Theorem 3.2. *In addition to the assumptions* (a) *and* (b) *suppose that* $E[\eta(0)]^{2m}<\infty$ *for some integer* $m>0$. *Then there exists a constant* C *depending only on* m, K *and* T *for which*

$$E[\eta(t)]^{2m}\leqq E[1+(\eta(0))^{2m}]\ \mathrm{e}^{Ct}.$$

The next theorem shows the connection between diffusion processes and stochastic differential equations.

Theorem 3.3. *Assume that* $\eta(t)$ *is a solution of*

$$\mathrm{d}\eta(t)=a(t,\,\eta(t))\mathrm{d}t+\sigma(t,\,\eta(t))\mathrm{d}W(t) \tag{26}$$

where coefficients satisfy the conditions for existence and uniqueness of solution. Then $\eta(t)$ *will be a Markov process whose transition probability is defined by*

$$P(t,\ x;\ s,\ A)=P(\eta_{x,t}(s)\in A)$$

where $\eta_{x,t}(s)$ *is a solution of*

$$\eta_{x,t}(s)=x+\int_t^s a(u,\ \eta_{x,t}(u))\mathrm{d}u+\int_t^s \sigma(u,\ \eta_{x,t}(u))\mathrm{d}W(u) \tag{27}$$

on the interval $[t,\ T]$.

Theorem 3.4. *Let* $a(t,\ x)$ *and* $\sigma(t,\ x)$ *be continuous in both arguments and assume that for some* K

$$|a(t,\ x)|^2+|\sigma(t,\ x)|^2\leqq K(1+|x|^2)$$

and that for each N, *there exists* L_N *with* $|x|\leqq N$, $|y|\leqq N$ *for which*

$$|a(t,\ x)-a(t,\ y)|+|\sigma(t,\ x)-\sigma(t,\ y)|\leqq L_N|x-y|.$$

Then the process $\eta(t)$ *as a solution of* (26) *will be a diffusion with diffusion coefficient* $b(t,\ x)=\sigma^2(t,\ x)$ *and drift coefficient* $a(t,\ x)$.

Theorem 3.5. *Let* $\xi(t)$ *be a diffusion process on* $[0,\ T]$ *with coefficients* $a(t,\ x)$ *and* $b(t,\ x)$ *which satisfy the following properties.*

(i) *$a(t, x)$ is continuous in both arguments and for some K satisfies*

$$|a(t, x)| \leqq K(1+|x|);$$

(ii) *$b(t, x)$ is continuous in both arguments and has continuous bounded derivatives $\partial/\partial t\, b(t, x)$* and *$\partial/\partial x\, b(t, x)$, also $1/b(t, x)$ is bounded;*

(iii) *there exists a function $\psi(x)$, independent of t and Δ, for which*

$$\psi(x) > 1+|x|, \quad \sup_{0 \leqq t \leqq T} E[\psi(\xi(t))] < \infty$$

and

$$\left|\int (y-x)P(t, x; t+\Delta, y)\right| + \int (y-x)^2 P(t, x; t+\Delta, \mathrm{d}y) \leqq \psi(x)\Delta,$$

$$\int (|y|+y^2)P(t, x; t+\Delta, \mathrm{d}y) \leqq \psi(x).$$

Then there exists a Wiener process $W(t)$ for which $\xi(t)$ satisfies the stochastic differential equation

$$\mathrm{d}\xi(t) = a(t, \xi(t))\mathrm{d}t + (b(t, \xi(t)))^{1/2}\mathrm{d}W(t).$$

In view of the above theorems, the solutions of stochastic differential equations and diffusion processes comprise one and the same class of processes under some conditions. Proofs of all the theorems stated in this section and further material on stochastic differential equations can be found in Gikhman and Skorokhod (1972), Arnold (1974) and Friedman (1975).

Appendix 3

Proof of Sudakov's Lemma (Theorem 2.3 of Chapter 12)

The existence of the densities $p_{\theta_0}(t, x; \theta)$ is guaranteed by the absolute continuity of the processes on any finite interval. We consider a sequence of Markov times $\tau_n(\omega)$ defined as follows. Let $t_k^{(n)}=k2^{-n}$, $k, n=1, 2, \ldots$. We set

$$\tau_n(w)=-2^{-n}[-2^n\tau(w)]$$

where $\lfloor x \rfloor$ denotes the largest integer less than or equal to x. It is clear that $x(\tau_n(w), w)$ is measurable and $\tau_n(w) \downarrow \tau(w))$ and from the assumption that the sample paths are right continuous, it follows that

$$\lim_{n\to\infty} x(\tau_n(w), w)=x(\tau(w), w) \text{ a.s. } \mu_{\theta_0}$$

proving the measurability of $x(\tau(w), w)$.

We first show that the theorem holds if $\tau(w)$ is replaced by $\tau_n(w)$; it is easy to check that $\{\tau_n\}$ are Markov times. The measures $m_\theta^{(n)}$, generated by τ_n, are concentrated on the subsets of the set $T\times \mathscr{X}$ given by the condition $t=t_k^{(n)}$, $k=1, 2, \ldots$ We fix any one of the points $t_k^{(n)}$ and consider the restriction $m_{k,\theta}^{(n)}$ of $m_\theta^{(n)}$ on the subset $t=t_k^{(n)}$.

Let $Y_k^{(n)}=\{w:\tau_n(w)=t_k^{(n)}\}$. Consider the σ-algebra on Ω generated by the functions $\xi(w, t)$, $0\leq t\leq t_k^{(n)}$. Since τ_n is a Markov time, $Y_k^{(n)}$ is measurable with respect to that σ-algebra. On the same σ-algebra, the measure μ_θ is absolutely continuous with respect to the measure μ_{θ_0}; moreover, from the fact that the end time t determines a sufficient statistic, it follows that the density of μ_θ relative to μ_{θ_0} depends only on $t_k^{(n)}$ and hence is equal to $p_{\theta_0}(t_k^{(n)}, x(t_k^{(n)}, w); \theta)$. The measure $m_{k,\theta_0}^{(n)}$ is the image of the restriction $(\mu_{\theta_0}|Y_k^{(n)})$, of the measure μ_{θ_0} to the set $Y_k^{(n)}$, under the mapping

$\pi: w \to (t_k^{(n)}, x(t_k^{(n)}, w)) \in T \times \mathscr{X}$ and similarly for $m_{k,\theta}^{(n)}$. Hence

$$m_{k,\theta_0}^{(n)}(A) = \mu_{\theta_0}(\pi^{-1} A \cap Y_k^{(n)}) = (\mu_{\theta_0}|Y_k^{(n)})(\pi^{-1} A)$$

and

$$m_{k,\theta}^{(n)}(A) = (\mu_\theta|Y_k^{(n)})(\pi^{-1} A) = \int_{\pi^{-1} A} p_{\theta_0}(t_k^{(n)}, x(t_k^{(n)}, w); \theta) \mathrm{d}\mu_{\theta_0}$$

$$= \int_A p_{\theta_0}(t_k^{(n)}, x; \theta) \mathrm{d}m_{k,\theta_0}^{(n)}$$

for any measurable set A. Therefore $(\mathrm{d}m_{k,\theta}^{(n)}/\mathrm{d}m_{k,\theta_0}^{(m)}) = p_{\theta_0}(t_k^{(n)}, x; \theta)$ and hence $(\mathrm{d}m_\theta^{(n)}/\mathrm{d}m_{\theta_0}^{(n)}) = p_{\theta_0}(t, x; \theta)$ if $\tau = \tau_n$.

To conclude the proof, observe that

$$x(\tau_n(w), w) \to x(\tau(w), w) \text{ a.s.}$$

implies that the measures $m_\theta^{(n)}$ generated by τ_n converge to measure m_θ in the sense of weak convergence of measures on $[0, \infty) \times \mathscr{X} \subset \mathbb{R}^{m+1}$ for any fixed $\theta \in \Theta$. We then verify the following assertion.

Let $\{m_n\}$ be a sequence of measures on $\mathbb{R}^m$ converging weakly to a measure m and $p(t, x)$ be a non-negative continuous function. Then the sequence $\{m'_n\}$ of measure having the same density $p(t, x)$ with respect to the corresponding members of the sequence $\{m_n\}$ (p is independent of n), converge weakly to a measure m' having the same density with respect to m. This assertion follows immediately from the definition of weak convergence: for any continuous function $f(t, x)$ with compact support, we have

$$\int f(t, x) \mathrm{d}m'_n = \int p(t, x) f(t, x) \mathrm{d}m_n \to \int p(t, x) f(t, x) \mathrm{d}m = \int f(t, x) \mathrm{d}m'.$$

This concludes the proof.

This proof is from Kagan *et al.* (1973) and is due to Sudakov (1969).

Appendix 4

Generalized Functions and Generalized Stochastic Processes

Most of the theory relevant to us in connection with generalized functions and generalized stochastic processes can be found in Gelfand and Vilenkin (1964), Gelfand and Shilov (1964) and Yosida (1966).

Let us define $K = K_1$ as in Section 2, Chapter 11 and endow the space K with the topology given for $\mathscr{D}(\mathbb{R})$. Let K' denote the topological dual of K with topology as the weak*-topology. Denote by $\mathscr{B}$ the σ-field generated by open sets of K'. Elements of K' are called generalized functions.

A generalized function φ^* is said to be *regular* if there exists a measurable integrable function t such that

$$\langle \varphi^*, \varphi \rangle = \int t(x)\varphi(x)\mathrm{d}x$$

for all $\varphi \in K$. All other generalized functions are said to be *singular*. The generalized function δ defined by $\langle \delta, \varphi \rangle = \varphi(0)$ for $\varphi \in K$ is called a *δ-function* at zero.

A sequence of functions g_n is called a *δ-sequence* if for every integrable continuous function t

$$\lim_{n\to\infty} \int t(x)\, g_n(x-y)\mathrm{d}x = t(y).$$

A probability measure μ on $(K, \mathscr{B})$ is called a *generalized stochastic process* (GSP). The characteristic functional of a GSP μ is defined as

$$C(\varphi) = \int_{K'} e^{i<\theta^*,\varphi>} d\mu$$

for $\varphi \in K$. It can be shown that $C(0)=1$, C is continuous in topology on K and C is positive definite i.e. for every positive integer m, for arbitrary $\varphi_1, \ldots, \varphi_m$ in K and arbitrary complex numbers $b_1, \ldots, b_m$, we have

$$\sum_{i,j} b_i \overline{b_j}\, C(\varphi_i - \varphi_j) \geqq 0.$$

By a well known result (cf. Gelfand and Vilenkin, (1964), p. 262), every functional on $K(a)$, the space of infinitely differentiable functions vanishing outside $[-a, a]$, satisfying the above properties is a characteristic functional i.e. every such functional is the Fourier transform of a countably additive probability measure on the topological dual $K'(a)$.

A sequence $\{\mu_n\}$ of GSPs' is said to converge weakly to a GSP μ if the characteristic functionals $C_{\mu_n}(\varphi)$ converge to $C_\mu(\varphi)$ for each $\varphi \in K$. We denote it by $\mu_n \overset{\mathscr{L}}{\to} \mu$.

A GSP μ is said to be *Gaussian* if $C_\mu(\varphi) = e^{-1/2\,B(\varphi,\varphi)}$ where $B(\varphi, \psi)$ is a positive definite continuous bilinear functional on $K \times K$.

Let $\varphi^* \in K'$. The derivative of φ^*(denoted by $D\varphi^*$) is defined by

$$\langle D\varphi^*, \varphi \rangle = -\langle \varphi^*, D\varphi \rangle$$

for all $\varphi \in K$, where $D\varphi$ is the derivative of φ.

The derivative of a GSP μ(denoted by $D\mu$), if it exists, is a process such that

$$C_{D\mu}(\varphi = \overline{C_\mu(D\varphi)}$$

where bar denotes complex conjugate.

For any φ, ψ in K, define

$$\varphi * \Psi = \int \varphi(t)\psi(x-t)dt.$$

It is clear that $\varphi * \psi \in K$. Let φ^* be a random generalized function. $E\varphi^* = \psi^*$, if it exists, is a function ψ^* such that

$$E\langle \varphi^*, \varphi \rangle = \langle \psi^*, \varphi \rangle$$

for all $\varphi \in K$.

References

Aalen, O. (1975). Statistical Inference for a family of counting processes. Ph.D. thesis, Univ. California, Berkeley.

Aalen, O. (1978). Nonparametric inference for a family of counting processes. *Ann. Statist.* **6**, 701–726.

Ahmad, I. (1977a). Nonparametric density estimation for dependent variables using delta sequences. Technical Report, McMaster University.

Ahmad, I. (1977b). Strong consistency of density estimation by orthogonal series methods for dependent variables with applications. Technical Report, McMaster University.

Ahmad, R. (1974). Distribution-free statistical hypotheses testing for stochastic processes. *Math. Operationsforch. Statist.* **5**, 643–656.

Ahmad, R. (1975). Some characterizations of the exchangeable processes and distribution-free tests. *In* "Statistical Distributions in Scientific Work" Vol. 3. D. Reidel Publishing Company. Dordrecht, Holland.

Akritas, M. G. (1978). Contiguity of probability measures associated with continuous time stochastic processes. Ph.D. Thesis, University of Wisconsin, Madison.

Akritas, M. G. and Johnson, R. A. (1978) Asymptotic inference in Lévy processes of the discontinuous type. Technical report, Univ. Wisconsin, Madison.

Akritas, M. G. and Roussas, G. G. (1978). Asymptotic expansion of the loglikelihood function based on stopping times defined on a Markov process. (Preprint) Univ. Patras.

Albers, W. (1978). Testing the mean of a normal population under dependence *Ann. Statist.* **6**, 1337–1344.

Albers, W. (1978). One sample rank tests under Autoregressive dependence. *Ann. Statist.* **6**, 836–845.

Albert, A. (1962). Estimating the infinitesimal generator of a continuous time finite state Markov process. *Ann. Math. Statist.* **38**, 727–753.

Anderson, T. W. (1959). On asymptotic distributions of estimates of parameters of stochastic difference equations. *Ann. Math. Statist.* **30**, 676–687.
Anderson, T. W. (1971). "The Statistical Analysis of Time Series" Wiley, New York.
Anderson, T. W. and Goodman, L. A. (1957). Statistical inference about Markov chains. *Ann. Math. Statist* **28**, 89–110.
Andreieu, C. (1975). Une formulation, et la finitude, du test sequential du rapport des vraisemblaces dans le cas Markovien a temps continu. *Z. Wahrschienlichkeits-theorie und Verw. Gebiete*, **32**, 333–344.
Anscombe, F. J. (1952). Large sample theory of sequential estimation. *Proc. Cambridge Philos. Soc.* **40**, 600–607.
Anscombe, F. J. (1953). Sequential estimation, *J. Roy. Statist. Soc. Ser. B.* **15**, 1–29.
Arato, M., Kolmogorov, A. N. and Sinai, Ya. G. (1962). Evaluation of the parameters of a complex stationary Gauss–Markov process. *Soviet. Math. Dokl.* **3**, 1368–1371.
Arnold, L. (1974). "Stochastic Differential Equations. Theory and Applications" Wiley-Interscience, New York.
Athreya, K. B. and Keiding, N. (1977). Estimation theory for continuous time branching processes. *Sankhyá, Ser. A*, **39**, 101–123.
Athreya, K. D. and Ney, P. E. (1972). "Branching Process" Springer-Verlag, Berlin.
Bagshaw, B. and Johnson, R. A. (1975). Sequential detection of a drift change in a Wiener process. *Comm. Statistics*, **4**, 787–796.
Bagshaw, M. and Johnson, R. A. (1975a). Sequential procedures for detecting parameter changes in a time series model. Technical Report. University of Wisconsin, Madison.
Bahadur, R. (1958). A note on the fundamental identity of sequential analysis. *Ann. Math. Statist.* **29**, 534–543.
Bailey, N. T. J. (1975). "The Mathematical Theory of Infectious Diseases" (second edition) Griffin, London.
Banon, G. (1978). Nonparametric identification for diffusion processes. *SIAM J. Control and Optimization*, **16**, 380–395.
Barlow, R. E. and Proschan, F. (1965). "Mathematical Theory of Reliability" Wiley, New York.
Barnard, G. A. (1959). Control charts and stochastic processes. *J. Roy. Statist. Soc. Ser. B*, **21**, 239–257.
Bar-Shalom, Y. (1971). On the asymptotic properties of maximum likelihood estimate obtained from dependent observations. *J. Roy Statist. Soc. Ser. B*, **33**, 72–77.
Bartlett, M. S. and Diananda, P. H. (1950). Extensions of Quenouille's test for auto-regressive schemes. *J. Roy. Statist. Soc. Ser. B*, **12**, 108–115.
Bartlett, M. S. (1963). Statistical estimation of density functions. *Sankhya Ser. A*, **25**, 245–254.
Bartoo, J. B. and Puri, P. S. (1967). On optimal tests of composite statistical hypotheses. *Ann. Math. Statist.* **38**, 1845–1852.
Basawa, I. V. (1972). Estimation of the autocorrelation coefficient in simple Markov chains. *Biometrika*, **59**, 85–89.
Basawa, I. V. (1974). Maximum likelihood estimation of parameters in renewal and Markov-renewal processes. *J. Aust. Statist. Soc.* **16**, 33–43.
Basawa, I. V. (1980). Remarks on Bahadur efficiency of conditional tests. *Ann. Statist.* (to appear).

Basawa, I. V. and Brockwell, P. J. (1978). Inference for gamma and stable processes. *Biometrika*, **65**, 129–133.
Basawa, I. V. and Scott, D. J. (1976). Efficient tests for branching processes. *Biometrika* **63**, 531–536.
Basawa, I. V. and Scott, D. J. (1977). Efficient tests for stochastic processes. *Sankhya Ser. A*, **39**, 21–31.
Basawa, I. V., Feigin, P. D. and Heyde, C. C. (1976). Asymptotic properties of maximum likelihood estimators for stochastic processes. *Sankhya Ser. A*, **38**, 259–270.
Becker, N. G. (1974). On parametric estimation for mortal branching processes. *Biometrika* **61**, 393–399.
Bell, C. (1960). On the structure of distribution-free statistics. *Ann. Math. Statist.*, **31**, 703–709.
Bell, C. (1964). Some basic theorems of distribution-free statistics. *Ann. Math. Statist.*, **35**, 150–156.
Bell, C. and Doksum, K. (1965). Some new distribution-free statistics. *Ann. Math. Statist.*, **36**, 203–214.
Bell, C. and Donoghue, J. F. (1969). Distribution-free tests of randomness. *Sankhya Ser. A*, **31**, 157–176.
Bell, C. and Smith, P. J. (1969). Some nonparametric tests for the multivariate goodness-of-fit, multisample, independence and symmetry problems. *In* "Multivariate Analysis", Academic Press, New York.
Bell, C., Woodroofe, M. and Avadhani, T. V. (1970). Some nonparametric tests for stochastic processes. *In* "Nonparametric Techniques in Statistical Inference" Cambridge University Press.
Berger, A. and Wald, A. (1949). On distinct hypotheses. *Ann. Math. Statist.* **20**, 104–109.
Besag, J. (1974). Spatial interaction and the statistical analysis of lattice systems. *J. Roy. Statist. Soc. B* **36**, 192–236.
Bhat, B. R. (1964). Bayes solution of sequential decision problem for Markov dependent observations. *Ann. Math. Statist.* **35**, 1656–1662.
Bhat, B. R. (1974). On the method of maximum likelihood for dependent observations. *J. Roy. Statist. Soc. B* **36**, 48–53.
Bhat, B. R. and Kulkarni, S. R. (1972). Optimal asymptotic tests of composite hypotheses for stochastic processes. *J. Karnatak Univ.* **27**, 73–89.
Bhat, B. R. and Nagnur, B. N. (1965). Locally asymptotically most stringent tests and Lagrangian multiplier tests of linear hypotheses. *Biometrika*, **52**. 459–468.
Bickel, P. J. and Lehmann, E. (1969). Unbiased estimation in convex families. *Ann. Math. Statist.* **40**, 1523–1535.
Bickel, P. J. and Yahav, J. (1967). Asymptotically pointwise optimal procedures in sequential analysis. In "Proceedings Fifth Berkeley Symposium Mathematical Statistics and Probability" **1**, 401–413, University of California Press.
Bickel, P., and Yahav, J. (1969). Some contributions to the asymptotic theory of Bayes solutions. *Z. Wahrscheinlichkeitstheorie und verw. Gebiete* **11**, 257–275.
Bickel, P. and Yahav, J. (1972). On the Wiener process approximation to Bayesian sequential testing problems. *In* "Proceedings Sixth Berkeley Symposium Mathematical Statistics and Probability" Vol. 1, 57–83, University of California Press.
Billingsley, P. (1961). The Lindeberg–Levy theorem for martingales. *Proc. Amer. Math. Soc.* **12**, 788–792.
Billingsley, P. (1961a) "Statistical Inference for Markov Processes." University of Chicago Press.

Billingsley, P. (1961b). Statistical Methods for Markov Chains, *Ann. Math. Statist.* **32**, 12–40.

Billingsley, P. (1968). "Convergence of Probability Measures" Wiley, New York.

Birnbaum, Z. W. and Rubin, H. (1954). On distribution-free statistics. *Ann. Math. Statist.* **25**, 593–598.

Blum, J. R., Hanson, D. L. and Rosenblatt, J. I. (1963). On the central limit theorem for the sum of a random number of independent random variables. *Z. Wahrscheinlichkeitstheorie Verw. Gebiete*, **1**, 389–393.

Bochner, S. (1955). "Harmonic Analysis and the Theory of Probability" University of Chicago Press.

Borwanker, J. D. (1967). Some Asymptotic Results for Stationary Processes. Ph.D. Thesis. Univ. of Minnesota.

Borwanker, J. D. (1971). Asymptotic theory of density estimation. *Z. Wahrscheinlichkeitstheorie und verw Gebiete* **20**, 182–188.

Borwanker, J. D., Kallianpur, G. and Prakasa Rao, B. L. S. (1971). The Bernstein–von Mises theorem for Markov processes. *Ann. Math. Statist.* **42**, 1241–1253.

Bosq, D. (1970). Contribution a la Theorie de L'estimation Fonctionelle. Publications de l'Institut de Statistique de l'Universite de Paris, **19**.

Bosq, D. (1973). Sur l'estimation de la densite d'un processus stationnaire et milangeant. *C.R. Acad. Sci. Paris Ser. A* **277**, 535–538.

Bosq, D. (1975). Inegalite de Bernstein pour les processus stationnaires et melangenats. Applications. *C.R. Acad. Sci. Paris Ser. A* **281**, 1095–1098.

Boswell, M. T. (1966). Estimating and testing in a stochastic process of Poisson type. *Ann. Math. Statist.* **37**, 1564–1573.

Box, G. E. P. and Jenkins, G. M. (1976). "Time series analysis" (second ed.), Holden-Day, San-Francisco.

Bradley, R. A. (1973). Asymptotic properties of likelihood estimators for associated populations and their applications. "Proceedings of the Prague Symposium on Asymptotic Statistics" Vol. 1, 37–54.

Bradley, R. A. and Gart, J. J. (1962). The asymptotic properties of ML estimators when sampling from associated populations. *Biometrika*, **49**, 205–214.

Breakwell, J., and Chernoff, H. (1964). Sequential tests for the mean of a normal distribution II (Large t). *Ann. Math. Statist.* **35**, 162–173.

Breiman, L. (1968). "Probability". Addison-Wesley: Reading.

Brillinger, D. (1975). Statistical inference for stationary point processes. *In* "Stochastic Processes and Related Topics" Vol. 1, Academic Press, New York.

Brown, B. M. (1971). Martingale central limit theorems. *Ann. Math. Statist.* **42**, 59–66.

Brown, B. M. (1974a). A sequential procedure for diffusion processes. *In* "Studies in Probability and Statistics" (Papers in honour of E. J. G. Pitman) Academic Press.

Brown, B. M. (1974b). A restricted sequential test. *J. Roy. Statist. Soc. Ser. B* **36**, 455–465.

Brown, B. M., and Hewitt, J. I. (1975). Asymptotic likelihood theory for diffusion processes. *J. Appl. Prob.* **12**, 228–238.

Brown, M. (1971). Discrimination of Poisson process. *Ann. Math. Statist.* **42**, 773–776.

Brown, M. (1972). Statistical analysis of nonhomogeneous Poisson processes. *In* "Stochastic Point Processes" Wiley, New York.

Buhler, W. G. and Puri, P. S. (1966). On optimal asymptotic tests of composite hypotheses with several constraints. *Z. Wahrscheinlichkeitstheorie Verw. Gebiete* **5**, 71–88.

Caines, P. E. (1975). A note on the consistency of maximum likelihood estimates for finite families of stochastic processes. *Ann. Math Statist.* **3**, 539–546.

Cameron, R. H., and Martin, W. T. (1945). Transformation of Wiener integrals under a general class of linear transformations. *Trans. Amer. Math. Soc.* **58**, 184–219.

Chanda, K. C. (1974). Strong mixing properties of linear stochastic processes. *J. Appl. Prob.* **11**, 401–408.

Chao, M. T. (1970). The asymptotic behaviour of Bayes estimates. *Ann. Math. Statist.* **41**, 601–609.

Chatfield, C. (1973). Statistical inference regarding Markov chain models. *Applied Statistics*, **22**, 7–20.

Chatterjee, S. (1974). A principle of subsequences in probability theory, the central limit theorem. *Advances in Math.* **13**, 31–53.

Chernoff, H. (1952). A measure of asymptotic efficiency for tests of a hypothesis based on the sum of observations. *Ann. Math. Statist.* **23**, 493–507.

Chernoff, H. (1956). Large sample theory: parametric case. *Ann. Math. Statist.* **27**, 1–22.

Chernoff, H. (1961). Sequential tests for the mean of a normal distribution. *In* "Proceedings Fourth Berkeley Symposium Mathematical Statistics and Probability" 1, 79–91, University of California Press.

Chernoff, H. (1965a). Sequential tests for the mean of a normal distribution IV (discrete case). *Ann. Math. Statist.* **36**, 55–68.

Chernoff, H. (1965b). Sequential tests for the mean of a normal distribution III (small t). *Ann. Math. Statist.* **36**, 28–54.

Chernoff, H., and Rubin, H. (1956). The estimation of the location of a discontinuity of a density. *In* "Proceedings Third Berkeley Symposium Mathematical Statistics and Probability, 1, 19–37. University of California Press.

Chung, K. L. (1967). "Markov Chains with Stationary Transition Probabilities" (second edition). Springer-Verlag. Berlin.

Cliff, A. D. and Ord, J. K. (1975). Model building and the analysis of spatial pattern in human geography. *J. Roy. Statist. Soc. B*, **37**, 297–348.

Cox, D. R. and Smith, W. L. (1961). "Queues", Methuen, London.

Cox, D. R. and Miller, H. D. (1965). "The Theory of Stochastic Processes" Methuen, London.

Cox, D. R. and Lewis, P. A. W. (1966). "Statistical Analysis of Series of Events" Methuen, London.

Cox, D. R. and Hinkley, D. V. (1974). "Theoretical Statistics" Chapman and Hall, London.

Cramer, H. (1946). "Mathematical Methods of Statistics" Princeton University Press.

Crow, J. F. and Kimura, M. (1970). "An Introduction to Population Genetics Theory" Harper and Row, New York.

Crowder, M. J. (1976). Maximum likelihood estimation for dependent observations. *J. Roy. Statist. Soc. Ser. B* **38**, 45–53.

Darling, D. A. (1955). The Cramer–von Mises statistic in the parametric case. *Ann. Math. Statist.* **26**, 1–20.

Darroch, J. M. and Seneta, E. (1965). On quasi-stationary distributions in absorbing discrete time finite Markov chains. *J. Appl. Prob.* **2**, 88–100.

Dawid, A. P. (1970). On the limiting normality of posterior distribution. *Proc. Camb. Philosp. Soc.* **67**, 625–633.

Delecroix, M. (1975). Sur l'estimation des densites marginales et de transition d'un processus stationnaire et melangeant. Doctoral thesis (3rd cycle), Université des Sciences et Techniques de Lille.

Devore, J. L. (1973). The naive rule for reconstructing a noisy Markov chain. *Biometrika* **60**, 227–233.

Dieudonne, J. (1972). "Treatise on Analysis." Vol. 3, Academic Press, New York and London.

Dion, J. P. (1972). Estimation des probabilites initiales et de la moyenne d'un processus de Galton-Watson. These de doctorat, Université de Montreal.

Dion, J. P. (1973). Theoremes limites fonctionnels pour une somme d'un nombre aleatoire de variables aleatoires dependantes. (preprint). University of Quebec at Montreal.

Dion, J. (1974). Estimation of the mean and initial probabilities of a branching process. *J. Appl. Probability* **11**, 687–694.

Dion, J. P. (1975). Estimation of the variance of a branching process. *Ann. Statist.* **3**, 1183–1187.

Doob, J. L. (1949). Applications of the theory of martingales. *Col. Int. du C.N.R.S. Paris*, pp. 22–28.

Doob, J. L. (1953). "Stochastic Processes" Wiley, New York.

Duncan, T. E. (1967). Probability Densities for Diffusion Processes with Applications to Non-linear Filtering Theory and Detection Theory, Ph.D. dissertation. Dept. of Elect. Eng. Stanford University.

Dunford, N., and Schwartz, J. T. (1958). "Linear Operators", Part I. Interscience, New York.

Dvoretzky, A., Kiefer, J. and Wolfowitz, J. (1953a). Sequential decision problems with continuous time parameter. Testing hypotheses. *Ann. Math. Statist.* **24**, 254–264.

Dvoretzky, A., Kiefer, J. and Wolfowitz, J. (1953b). Sequential decison problems for processes with continuous time parameter. Problems of estimation. *Ann. Math. Statist.* **24**, 403–415.

Dynkin, E. (1965). "Markov Processes" Vol. I, Springer-Verlag, Berlin.

Eisenberg, B., Ghosh, B. K., and Simons, G. (1976). Properties of generalized sequential probability ratio tests. *Ann. Statist.* **4**, 237–251.

Epstein, B. and Sobel, M. (1955). Sequential life tests in the exponential case. *Ann. Math. Statist.* **26**, 82–93.

Ershov, M. P. (1971). Sequential estimation of diffusion processes. *Theory Probability Appl.* **15**, 688–700.

Ewens, W. J. (1969). "Population Genetics" Methuen, London.

Feigin, P. D. (1975). Maximum Likelihood Estimation for Stochastic Processes—A Martingale Approach. Ph.D. Thesis, Australian National University.

Feigin, P. D. (1978). Maximum likelihood estimation for continuous time stochastic processes. *Adv. Appl. Prob.* **8**, 712–736.

Feigin, P. D. (1978). The efficiency criteria problem for stochastic processes. *Stoch. Proc. Appl.* **6**, 115–127.

Feller, W. (1957). "Introduction to Probability Theory and its Applications," Vol. I, Wiley, New York.

Feller, W. (1966). "Introduction to Probability Theory and its Applications", Vol. II, Wiley, New York.

Ferguson, T. (1967). "Mathematical Statistics: A Decision Theoretic Approach" Academic Press, New York and London.

Fienberg, S. E. (1974). Stochastic models for single neuron firing trains: a survey. *Biometrics*, **30**, 399–427.

Flavigny, R. (1975). A propos d'un test sequential concernant une suite φ-milangeante. *C.R. Acad. Sc. Paris. Ser. A* **280**, 1633–1636.

Földes, A and Revesz, P. (1974). A general method of density estimation. *Studia Scient. Math. Hungar.* **9**, 81–92.

Foldes, A. (1974). Density estimation for dependent sample. *Studia Sci. Math. Hungar.* **9**, 443–452.

Foutz, R. (1974). Studies in large sample theory. Ph.D. Thesis. The Ohio State University.

Franz, J. and Winkler, W. (1975). Optimale schatzverfahren fur parameter bei einer speziellen klasse stochastischer prozesse. (Preprint) Technische Universitat, Dresden.

Franz, J. and Winkler, W. (1976). Uber stoppzeiten bei statistischen problemen , fur homogene prozesse mit unabhangigen zuwochsen. *Math. Nachr.* **70**, 37–53.

Freedman, D. A. (1963). On the asymptotic behaviour of Bayes estimates in the discrete case. *Ann. Math. Statist.* **34**, 1386–1403.

Friedman, A. (1975). "Stochastic Differential Equations and Applications". Vol. 1. Academic Press, New York and London.

Gani, J. (1955). Some theorems and sufficiency conditions for the maximum likelihood estimator of an unknown parameter in a simple Markov chain. *Biometrika*, **42**, 342–359.

Gastwirth, J. L. and Rubin, H. (1971). Effect of dependence on the level of some one-sample tests. *J. Amer. Statist. Assoc.* **66**, 816–820.

Gastwirth, J. L. and Rubin, H. (1975a). Asymptotic distribution theory of the empiric c.d.f. for mixing stochastic processes. *Ann. Statist.* **3**, 809–824.

Gastwirth, J. L. and Rubin, H. (1975b). The behavior of robust estimators on dependent data. *Ann. Statist.* **3**, 1070–1100.

Gastwirth, J. L., Rubin, H. and Wolff, S. S. (1967). The effective of autoregressive dependence on a nonparametric test. *IEEE Trans. Information Theory IT* **13**, 311–313.

Gelfand, I. M. and Shilov, G. E. (1964). "Generalized Functions" Vol. 1, Academic Press, New York and London.

Gelfand, I. M., and Vilenkin, N. Y. (1964). "Generalized Functions" Vol. 4, Academic Press, New York and London.

Ghosh, M., and Babu, G. J. (1976). A random functional central limit theorem for martingales. *Acta. Math. Acad. Sci. Hungar.* **27**, 301–306.

Gikhman, I. I. and Skorokhod, A. V. (1966). On densities of probability measures in functions spaces. *Russian Math. Surveys*, **21**, 83–156.

Gikhman, I. I. and Skorokhod, A. V. (1969). "Introduction to the Theory of Random Processes" W. B. Saunders, Philadelphia.

Gikhman, I. I. and Skorokhod, A. V. (1972). "Stochastic Differential Equations" Springer Verlag, Berlin.

Gikhman, I. I. and Skorokhod, A. V. (1974). "The Theory of Stochastic Processes" Vol. 1, Springer Verlag, Berlin.

Girsanov, I. V. (1960). On transforming a certain class of stochastic processes by absolutely continuous substitution of measures. *Theor. Probability Appl.* **5**, 285–301.

Gleser, L. (1969). On limiting distributions for sums of random number of independent random vectors. *Ann. Math. Statist.* **40**, 935–941.

Goyal, T. L. and Harris, C. M. (1972). Maximum likelihood estimates for queues with state-dependent service. *Sankhya Ser. A* **34**, 65–80.

Gnendenko, B. V., Belyayev, Yu. K. and Solovyev, A. D. (1965). "Mathematical Methods in Reliability Theory" (English translation) Academic Press, New York and London.

Grandell, J. (1976). "Doubly Stochastic Poisson Processes" Springer-Verlag, Berlin, Lecture Notes No. 529.

Grenander, U. V. (1951). Stochastic processes and statistical inference. *Ark. Mat.* **1**, 195–277.

Grenander, U. V. (1967). Eight lectures on statistical inference in stochastic processes. Technical Report No. 2, Division of Applied Mathematics, Brown University, Providence, Rhode Island.

Hajek, J. and Sidak, Z. (1967). "Theory of Rank Tests" Academia, Prague.

Hall, P. G. (1975). Martingale Limit Theory. M.Sc. Thesis, Australian National University.

Hall, P. G. (1977). Martingale invariance principles. *Ann. Probability* **5**, 875–887.

Hall, W. J. (1969). Embedding submartingales in Wiener processes with drift, with applications to sequential analysis. *J. Appl. Prob.* **6**, 612–362.

Hall, W. J. (1970). On Wald's equations in continuous time. *J. Appl. Prob.* **7**, 59–68.

Harris, T. E. (1948). Branching processes. *Ann. Math. Statist.* **19**, 474–494.

Harris, T. E. (1963). "Branching Processes" Springer-Verlag, Berlin.

Hewitt, E. and Savage, L. J. (1955). Symmetric measures on Cartesian products. *Trans. Amer. Math. Soc.* **80**, 470–501.

Heyde, C. C. (1975). Remarks on efficiency in estimation for Branching Processes. *Biometrika* **62**, 49–55.

Heyde, C. C. and Seneta, E. (1972). Estimation theory for growth and immigration rates in a multiplicative process. *J. Appl. Prob.* **9**, 235–256.

Heyde, C. C. and Scott, D. J. (1973). Invariance principles for the law of the iterated logarithm for martingales and processes with stationary increments. *Ann. Prob.* **1**, 428–436.

Heyde, C. C. (1978). An optimum property of the maximum likelihood estimator for stochastic processes. *Stoch. Proc. Appl.* **8**, 1–9.

Hipp, C. and Michel, R. (1976). On the Bernstein–von Mises approximation of posterior distributions. *Ann. Statist.* **4**, 972–980.

Hodges, J. L., Jr. and Lehmann, E. L. (1963). Estimates of location based on rank tests. *Ann. Math. Statist.* **34**, 598–611.

Holevo, A. S. (1967). Estimates of the drift parameters for a diffusion process by the method of stochastic approximation. *In* "Studies in the Theory of Self-adjusting Systems" Vycise. Centr. Akad. Nauk SSR, Moscow.

Hopper, J. and Brockwell, P. J. (1978). A stochastic model for cell populations with circadian rhythms. *Cell and Tissue Kinetics* (to appear).

Huber, P. J. (1964). Robust estimation of a location parameter. *Ann. Math. Statist.* **35**, 73–101.

Huber, P. J. (1965). The behaviour of maximum likelihood estimates under nonstandard conditions. "Proceedings Fifth Berkely Symposium Mathematical Statistics and Probability". **1**, 221–233. Universal of California Press.

Huber, P. J. (1972). Robust statistics: A review. *Ann. Math. Statist.* **43**, 1041–1067.

Ibragimov, I. A. (1962). Some limit theorems for stationary processes. *Theor. Probability Appl.* **7**, 349–382.

Ibragimov, I. A. (1963). A central limit theorem for a class of dependent random variables. *Theor. Probability Appl.* **8**, 83–89.

Ibragimov, I. A. and Linnik, Yu, V. (1971). "Independent and Stationary Sequences of Random Variables" Wolters-Noordhoff, Groningen.

Ibragimov, I. A. and Khas'minskii, R. Z. (1972). Asymptotic behaviour of statistical estimators in the smooth case. I. Study of the likelihood ratio. *Theor. Probability Appl.* **17**, 445–462.

Ibragimov, I. A. and Khas'minskii, R. Z. (1973a). Asymptotic behaviour of statistical estimators. II. Limit theorems for the a posterior density and Bayes estimators. *Theor. Probability Appl.* **18**, 76–91.

Ibragimov, I. A. and Khas'minskii, R. Z. (1973b). On moments of generalized Bayes estimators and maximum likelihood estimators. *Theor. Probability Appl.* **18**, 508–520.

Ibragimov, I. A. and Khas'minskii, R. Z. (1974a). On sequential estimation. *Theor. Probability Appl.* **19**, 233–244.

Ibragimov, I. A. and Khas'minskii, R. Z. (1974b). On sequential estimation of the location parameter for families of distributions with discontinuous densities. *Theor. Probability Appl.* **19**, 699–682.

Ibragimov, I. A. and Khas'minskii, R. Z. (1975a). Local asymptotic normality for non-identically distributed observations. *Theor. Probability Appl.* **20**, 246–260.

Ibragimov, I. A. and Khas'minskii, R. Z. (1975b). Properties of maximum likelihood and Bayes estimators for non-identically distributed observations. *Theor. Probability Appl.* **20**, 689–697.

Inagaki, N. (1973). Asymptotic relations between the likelihood estimating function and the maximum likelihood estimator. *Ann. Inst. Statist. Math.* **25**, 1–26.

Inagaki, N., and Ogata, Y. (1975). The weak convergence of likelihood ratio random fields and its applications. *Ann. Inst. Statist. Math.* **27**, 391–419.

Inagaki, N. and Ogata, Y. (1977). The weak convergence of likelihood ratio random field for Markov observations. *Ann. Inst. Statist. Math.* **29**, 165–187.

Inagaki, N. and Ogata, Y. (1976). The weak convergence of likelihood ratio random field for Markov observations II. Asymptotic analysis for statistical Markovian models. Research Memorandom No. 96. The Institute of Statistical Mathematics, Japan.

Iosifescu, M. and Theoderescu, R. (1969). "Random Processes" Springer-Verlag, Berlin.

Jagers, P. (1975) Branching Processes with Biological Applications" Wiley, London.

Johnson, R. A. (1970). Asymptotic expansions associated with posterior distributions. *Ann. Math. Statist.* **41**, 851–864.

Kagan, A. M., Linnik, Yu, V. and Rao, C. R. (1973). "Characterization Problems in Mathematical Statistics". Wiley, New York.

Kailath, T. C. (1971). The structure of Radon–Nikodym derivatives with respect to Wiener and related measures. *Ann. Math. Statist.* **42**, 1054–1067.

Kailath, T. and Zakai, M. (1971). Absolute continuity and Radon–Nikodym derivatives for certain measures relative to Wiener measure. *Ann. Math. Statist.* **42**, 130–140.

Karhunen, K. (1947). Ueber lineare methoden der wahrscheinlichkeitsrechnung. *Ann. Acad. Sci. Fenn.*, **37**.

Karlin, S. and Taylor, (1975). "A First Course in Stochastic Processes" (second edition). Academic Press, New York and London.

Keiding, N. (1974). Estimation in the birth process. *Biometrika*, **61**, 71–80.

Keiding, N. (1975). Maximum likelihood estimation in the birth and death process. *Ann. Statist.* **3**, 363–372.

Kemeny, J. G. and Snell, J. L. (1960). "Finite Markov Chains" Van Nostrand, Princeton, N. J.

Kendall, D. G. (1949). Stochastic processes and population growth. *J. Roy. Statist. Soc. Ser. B*, **11**, 230–264.

Kelker, D. (1968). Distribution theory of spherical distributions and some characterization theorems. Technical report RM-210. Michigan State University.

Khasminskii, R. Z. (1960). Ergodic properties of recurrent diffusion processes and stabilization of the solution of the Cauchy problem for parabolic equations. *Theor. Probability Appl.* **5**, 179–196.

Kiefer, J. (1957). Invariance minimax sequential estimation and continuous time processes. *Ann. Math. Statist.* **28**, 573–601.

Kiefer, J. and Weiss, L. (1957). Some properties of generalised sequential probability ratio tests. *Ann. Math. Statist.* **28**, 57–74.

Kingman, J. F. G. (1972). "Regenerative Phenomena" Wiley. New York.

Klotz, J. (1973). Statistical inference in Bernoulli trials with dependence. *Ann. Statist.* **1**, 373–379.

Kolmogorov, A. N. and Rozanov, Y. A. (1960). On strong mixing conditions for stationary Gaussian random processes. *Theor. Probability Appl.* **5**, 204–208.

Koul, H. A. (1975). Asymptotic normality of H.-L. estimators based on dependent data. *Ann. Inst. Statist. Math.* **27**, 429–441.

Koul, H. A. (1977). Behavior of robust estimators in the regression model with dependent errors. *Ann. Statist.* **5**, 681–699.

Kulinic, G. L. (1975). On an estimation of the drift parameter of a stochastic diffusion equation. *Theory. Probability Appl.* **20**, 384–387.

Kunita, H. and Watanabe, S. (1967). On square integrable martingales. *Nagoya Math. J.* **30**, 209–245.

Kutoyants Yu. A. (1975a). On a hypotheses testing problem and asymptotic normality of stochastic integrals. *Theory-Probability Appl.* **20**, 376–384.

Kutoyants, Yu. A. (1975b). Local asymptotical normality for the diffusion type processes. *Izvest Akad Nauk Arminski. SSR* **10**, 103–112.

Kutoyants, Yu. A. (1976). On asymptotic theory of signal detection. *Radio Technika and Electronica* **7**, 1458–1466.

Kutoyants Yu. A. (1977a). Estimation of the trend parameter of a diffusion process in the smooth case. *Teor. Veroyat iee Primenen* **22**, 409–415 (Russian).

Kutoyants, Yu A. (1977b). On the estimation of trend parameter *Izv. Akad. Nauk. Armeniski. SSR.* **12**, 245–251 (Russian).

Kutoyants, Yu. A. (1977c) Parameter estimation for processes of diffusion type. *Theor. Probability Appl.* (Russian) (To appear).

Kryukov, V. I. (1976). Wald's identity and random walk models for neuron firing. *Adv. Appl. Prob.* **8**, 257–277.

Lai, T. L. (1973a). Gaussian processes, moving averages and quick detection problems. *Ann. Probability*, **1**, 825–837.

Lai, T. L. (1973b). Optimal stopping and sequential tests which minimize the expected sample size. *Ann. Statistics.* **1**, 659–673.

Le Breton, A. (1974). Parameter estimation in a vector linear stochastic differential equation. *In* "Transactions of the Seventh Prague Conference on Information Theory. Statistical Decision functions and Random Processes" Vol A, 353–366.

Le Breton, A. (1976). On continuous and discrete sampling for parameter estimation in diffusion type processes. *In* "Mathematical Programming Studies" **5**, 124–144.

Le Cam, L. (1953). "On Some Asymptotic Properties of Maximum Likelihood and Related Bayes Estimates" University of California Publications in Statistics 1, 277–330.

Le Cam, L. (1955). An extension of Wald's theory of statistical decision functions. *Ann. Math. Statist.* **26**, 69–81.

Le Cam, L. (1958). Les proprietes asymptotiques des solutions de Bayes. *Publ. Inst. Statist. Univ. Paris* **8**, 17–35.

Le Cam, L. (1970). Remarks on the Bernstein–von Mises theorem. (Preprint). University of California, Berkeley.

Le Cam, L. (1970). On the assumptions used to prove asymptotic normality of maximum likelihood estimates. *Ann. Math. Statist.* **41**, 802–828.

Lee, E. B. and Markus, L. (1967). "Foundations of Optimal Control Theory," Wiley, New York.

Lee, T. C., Judge, G. G. and Zellner, A. (1970). "Estimating the Parameters of the Markov Probability Model from Aggregate Time Series Data". North-Holland, Amsterdam.

Lehmann, E. L. (1959). "Testing Statistical Hypotheses," Wiley, New York.

Lehmann, E. L. (1966). Some concepts of dependence. *Ann, Math. Statist.* **37**, 1137–1153.

Lehmann, E. L. and Stein, C. (1949). On the theory of some nonparametric hypotheses. *Ann. Math. Statist.* **20**, 28–45.

Levit, B. Ya. (1972). Generalised Bayes estimates for transmission in channels with feedback. *Prob. Peredachi. Inform.* **8**, 184–193 (Russian).

Levit, B. Ya. (1974). On the behaviour of the generalized Bayes estimators for Markov observations. *Theor. Probability Appl.* **19**, 332–346.

Linkov, Yu. V. (1975a). Asymptotic theory of Bayes estimates of parameters of shift diffusion processes. *In* "Theory of Random Processes." 3, Naukov Dumka, Kiev. (Russian) 50–54..

Linkov, Yu. V. (1975b). Generalized Bayes estimates of parameters of shift diffusion processes. *Theor. Probability Math. Statist.* **13**. Kiev. 92–99 (Russian).

Linkov, Yu. V. (1977). On statistical estimators for parameters of diffusion processes. *In* "Abstracts of communications. Second Vilnius Conference on Prob. Theory and Math. Statist." 1, 240–241 (Russian).

Lipcer, R. S. and Sirjaev, A. N. (1972). On the absolute continuity of measures corresponding to processes of diffusion type. *Math. USSR Izvestija* **6**, 839–882.

Lipcer, R. S. and Sirjaev, A. N. (1974). "Statistics of Random Processes" (Russian) Nauka, Moscow.

Lipcer, R. S. and Sirjaev, A. N. (1977). "Statistics of Random Processes I General Theory." Springer Verlag, Berlin.

Loeve, M. (1963). "Probability Theory" (3rd edition) Van Nostrand, Princeton.

Lorden, G. (1973). Open-ended tests for Koopman-Dormois families. *Ann. Statistics* **1**, 633–643.

Maclean, C. J. (1974). Estimation and testing of an exponential rate function within the nonstationary Poisson process. *Biometrika* **61**, 81–85.

Magiera, R. (1974). On the inequality of Cramer–Rao type in sequential estimation theory. *Zastosowania Matematyki Applicationes Mathematicae* **14**, 227–235.

Majone, G. D. (1966). Asymptotic behaviour of Bayes estimates in Borel sets. Ph.D. Thesis. University of California, Berkeley.

Mann, N. R., Schafer, R. E. and Singpurwalla, N. D. (1974). "Methods for Statistical Analysis of Reliability and Life Data" Wiley, New York.

Marc Moore (1976). Remarks on the asymptotic behaviour of posterior laws with applications. *Utilitas Math.* **11**, 237–259.

Mardia, K. V. (1972). "Statistics of directional data." Academic Press, London and New York.

Marshall, A. W. and Olkin, I. (1967). A multivariate exponential distribution. *J. Amer. Statist. Assoc.* **62**, 30–44.

Martin, J. J. (1967). "Bayesian Decision Problems and Markov Chains" Wiley, New York.

Matthews, J. P. (1970). A central limit theorem for absorbing Markov chains. *Biometrika* **57**, 129–139.

Matusita, K. (1955). Decision rules, based on the distance for problems of fit, two-samples and estimation. *Ann. Math. Statist.* **26**, 631–640.

McKean, H. P. (1969). "Stochastic Integrals" Academic Press, New York.

Meyer, P. A. (1962). A decomposition theorem for super martingales. *Illinois J. Math.* **6**, 193–205.

Mihoc, Gh. and Craiu, M. (1972). "Inferenta Statistica pentru Variable Dependente" Editura Acadermiei Republicii Socialiste Romania.

Modestino, J. W. (1969). Nonparametric and adaptive detecting on dependent data. Technical report 27. Dept. of Electrical Engineering. Princeton University.

Moore, E. H. and Pyke, R. (1968). Estimation of the transition distributions of a Markov Renewal process. *Ann. Inst. Statist. Math.* **20**, 411–424.

Moran, P. A. P. (1951). Estimation methods for evolutive processes. *J. Roy. Statist. Soc. Ser. B* **13**, 141–146.

Moran, P. A. P. (1953). The estimation of the parameters of a birth and death process. *J. Roy. Statist. Soc. Ser. B* **15**, 241–245.

Moran, P. A. P. (1959). "The Theory of Storage" Methuen, London.

Morgan, B. J. T. (1976). Stochastic models for grouping changes. *Adv. Appl. Prob.* **8**, 30–57.

Muddapur, M. V. (1974). V-Mask for the negative binomial process. *J. Indian Statist. Assoc.* **12**, 31–38.

Muddapur, M. V. (1976). Control charts for certain class of additive processes. (Preprint). Karnatak University, Dharwar (India).

Musiela, M. (1976). Risk of maximum likelihood estimators of parameters of an Ito–type process. *Bull. de,l'Acad. Polon. Sci. Ser. Math. Astron. et Phys.* **34**, 1035–1039.

Musiela, M. (1977a). Sequential estimation of parameters of a stochastic differential equation (unpublished).

Musiela, M. (1977b). Estimation of parameters of a diffusion type process (Preprint).

Nadas, A. (1973). Best test for zero drift based on first passage time in Brownian motion. *Technometrics* **15**, 125–132.

Neyman, J. (1959). Optimal asymptotic tests of composite statistical hypotheses. *In* "Probability and Statistics". (H. Cramer, ed.) Wiley, New York.

Neyman, J. and Scott, E. L. (1948). Consistent estimates based on partially consistent observations. *Econometrica*, **16**, 1–32.

Neyman, J. and Scott, E. L. (1965). Asymptotically optimal tests of composite hypotheses for randomized experiments with non-controlled predictor variables. *J. Amer. Statist. Assoc.* **60**, 699–721.

Norman, M. F. (1971). Statistical inference with dependent observations. Extensions of classical procedures. *J. Mathematical Psychology.* **8**, 444–451.

Novikov, A. A. (1972). Sequential estimation of the parameters of diffusion processes. *Mathematical Notes* **12**, 812–818.
Ord, J. K. (1975). Estimation methods for models of spatial interaction. *J. Amer. Statist. Assoc.* **70**.
Page, E. S. (1954). Continuous inspection schemes. *Biometrika* **41**, 100–115.
Parthasarathy, K. R. (1967). "Probability Measures on Metric Spaces" Academic Press, New York and London.
Parzen, E. (1962). On estimation of a probability density function and mode. *Ann. Math. Statist.* **33**, 1065–1076.
Phillip, W. (1969a). The remainder in the central limit theorem for mixing stochastic processes. *Ann. Math. Statist.* **40**, 601–609.
Phillip, W. (1969b). The central limit problem for mixing sequences of random variables. *Z. Wahrscheinlichkeitstheorie Verw. Gebiete.* **12**, 155–171.
Phillippou, A. (1977a). On the existence and consistency of the MLE in the dependent case. *In* "Proceedings of the International Conference on Information Sciences and Systems".
Phillippou, A. (1977b). Asymptotic normality of the maximum likelihood estimate in arbitrary stochastic processes. (Preprint) University of Patras.
Pitman, E. J. G. (1937). Significance tests which may be applied to samples from any population. *Suppl. J. Roy. Statist. Soc.* **4**, 119–130.
Portnoy, S. L. (1977). Robust estimation in dependent situations. *Ann. Statist.* **5**, 22–43.
Prabhu, N. U. (1965). "Queues and Inventories" Wiley, New York.
Prakasa Rao, B. L. S. (1968). Estimation of the location of the cusp of a continuous density. *Ann. Math. Statist.* **39**, 76–87.
Prakasa Rao, B. L. S. (1969). Random central limit theorems for martingales. *Acta. Math. Acad. Sci. Hungar.* **20**, 217–222.
Prakasa Rao, B. L. S. (1972). Maximum likelihood estimation for Markov processes. *Ann. Inst. Statist. Math.* **24**, 333–345.
Prakasa Rao, B. L. S. (1973). On the rate of convergence of estimators for Markov processes. *Z. Wahrscheinlichkeitstheorie Verw. Gebiete.* **26**. 141–152.
Prakasa Rao, B. L. S. (1974). Statistical inference for stochastic processes. Technical Report CRM-465, Centre de Recherches Mathematiques, Universite de Montreal To appear in Proc. Adv. Sem Prob. Theory and its Appl. (1977). Annamalai Univ. (India).
Prakasa Rao, B. L. S. (1975a). The Berry-Esseen bound for minimum contrast estimators in the independent not identically distributed case. *Metrika*, **22**, 225–239.
Prakasa Rao, B. L. S. (1975b). Tightness of probability measures generated by stochastic processes on metric spaces. *Bull. Inst. Math. Academia Sinica* **3**, 353–367.
Prakasa Rao, B. L. S. (1977). Bounds for the equivalence of Bayes and Maximum likelihood estimators for Markov processes. (Preprint). Indian Statistical Institute, Delhi.
Prakasa Rao, B. L. S. (1977). Berry-Esseen type bound for density estimators of stationary Markov processes. *Bull. Math. Statist.* **17**, 15–21.
Prakasa Rao, B. L. S. (1978a). On the rate of convergence of Bernstein-Von Mises approximation for Markov processes. *Serdica Bulg. Math. Pub.* **4**, 36–42.
Prakasa Rao, B. L. S. (1978b). Density estimation for Markov processes using Delta-sequences. *Ann. Inst. Statist. Math.* **30**, 321–328.

Prakasa Rao, B. L. S. (1978c). On some problems of inference for Markov processes. *In* "Proceedings of General Mathematics Seminar," University of Patras (Greece), (to appear).

Prakasa Rao, B. L. S. (1979). Nonparametric estimation for continuous time Markov processes via delta families *Publ. Inst. Statist. Univ. Paris.* **24** (To appear).

Prakasa Rao, B. L. S. (1981). The Bernstein – von Mises Theorem for a class of diffusion processes. *The Theory of Random Processes*, **9** (Russian) to appear.

Prasad, M. S. (1973). Some Contribution to the Theory of Maximum Likelihood Estimation for Dependent Random Variables. Ph.D. Thesis, Indian Institute of Technology, Kanpur.

Prasad, M. S. and Prakasa Rao, B. L. S. (1976). Maximum likelihood estimation for dependent random variables. *J. Ind. Stat. Assoc.* **14**, 75–97.

Pratt, J. W. (1960). On interchanging limits and integrals. *Ann. Math. Statist.* **31**, 74–77.

Preston, P. F. (1971). An emprirical Bayes problem with a Markovian parameter. *Biometrika* **58**, 535–543.

Prohorov, Yu. V. (1956). Convergence of random processes and limit theorems in probability theory. *Theor. Probability Appl.* **1**, 157–214.

Prohorov, Yu. V. and Rozanov, Yu. A. (1969). Probability Theory. Springer-Verlag, Berlin.

Quine, M. P. (1976). On the estimation of means in a subcritical branching process with immigration. *Ann. Probability* **4**, 319–325.

Ranneby, B. (1975a). Maximum likelihood estimation for continuous time Markov chains. Technical Report No. 1975–4, Institute of Mathematics and Statistics. University of Umea, Sweden.

Ranneby, B. (1975b). Robustness against dependance of maximum likelihood estimates in i.i.d. models. Technical report No. 1975–9. Institute of Mathematics and Statistics, University of Umea, Sweden.

Rao, C. R. (1948). Large sample tests of statistical hypothesis concerning several parameters with applications to problems of estimation. *Proc. Cambridge Phil. Soc.* **44**, 50–57.

Rao, C. R. (1958). Maximum likelihood estimation for the multinomial distribution with an infinite number of cells. *Sankhya* **20**, 211–218.

Rao, C. R. (1962). Efficient estimates and optimum inference procedure in large samples. *J. Roy. Statist. Soc. Ser. B* **24**, 46 72.

Rao, C. R. (1973). "Linear Statistical Inference and its Applications" (2nd edition) Wiley, New York.

Rao, C. R. and Mitra, S. K. (1972). "Generalized Inverse of Matrices and its Applications" Wiley, New York.

Rao, M. M. (1963). Inference in stochastic processes I. *Theor. Probability Appl.* **8**, 266–281.

Rao, M. M. (1966). Inference in stochastic processes II. *Z. Wahracheinlichkeitstheorie Verw. Gebiete*, **5**, 317–335.

Ray, R. M. (1973). Note on the consistency of $C(\alpha)$ tests. *IMS Bull.* (Abstract 140–20), **2**, 178.

Renyi, A. (1960). On the central limit theorem for the sum of a random number of independent random variables. *Acta. Math. Acad. Sci. Hungar.* **11**, 97–102.

Robbins, H. and Samuel, E. (1966). An extension of a lemma of Wald. *J. Appl. Prob.* **3**, 272–273.

Rosenblatt, M. (1952). Remarks on the multivariate transformation. *Ann. Math. Statist.* **23**, 470–472.

Rosenblatt, M. (1956). Remarks on some nonparametric estimates of a density function *Ann. Math. Statist.* **27**, 832–837.

Rosenblatt, M. (1970). Density estimates and Markov sequences. *In* "Nonparametric Techniques in Statistical Inference" Cambridge University Press.

Roussas, G. G. (1965). Extension to Markov processes of a result by a Wald about the consistency of the maximum likelihood estimate. *Z. Wahrscheinlichkeitstheorie Verw. Gebriete.* **4**, 69–73.

Roussas, G. G. (1968a). Asymptotic normality of the maximum likelihood estimate in Markov processes. *Metrika*, **14**, 62–70.

Roussas, G. G. (1968b). On some properties of nonparametric estimates of probability density functions. *Bull. Soc. Mathematique de Grece.* **2**, 29–43.

Roussas, G. G. (1969a). Nonparametric estimation in Markov processes. *Ann. Inst. Statist. Math.* **21**, 73–87.

Roussas, G. G. (1969b). Nonparametric estimation of the transition distribution function of a Markov process. *Ann. Math. Statist.* **40**, 1386–1400.

Roussas, G. G. (1972). "Contiguity of Probability Measures" Cambridge University Press.

Roussas, G. G. (1976). Asymptotic distribution of the log-likelihood function for stochastic processes. (Preprint). University of Patras, Greece.

Rubin, H. (1961). The estimation of discontinuities in multivariate densities and related problems in stochastic processes. *In* "Proceedings of the Fourth Berkely Symposium Mathematical Statistics and Probability". 563–574. University of California Press.

Rudin, W. (1964). "Principles of Mathematical Analysis" (second ed.) McGraw–Hill, New York.

Sagdar, D. (1974). On an approximate test of hypotheses about the correlation function of a Gaussian random process. *Theor. Probability Math. Stat.* **2**, 231–238.

Sarma, Y. R. (1968). Sur les tests et sur l'estimation de parametres pour certains processue stochastiques stationnaires, *Pub. Inst. Statist. Univ. Paris*, **17**, 1–124.

Sarma, Y. R. (1973). Large sample tests concerning several parameters in stationary Markov processes (Preprint). Tulane University.

Sarma, Y. R. (1976). Large sample theory of sequential estimation in stationary Markov processes. *Pub. Inst. Statist. Univ. Paris*, **21**, 57–70.

Scheffe, H. (1943). On a measure problem arising in the theory of nonparametric tests. *Ann. Math. Statist.* **14**, 227–233.

Schmitz, N. (1968). Likelihoodquotienten–Sequenztests bei homogenen Markovschen Ketten. *Biometriche Zeitschrift.* **10**, 231–347.

Schmitz, N. (1970). Existenz von sequentiellen tests zu vorgegbenen niveans. *Arch. Math.* **21**, 617–628.

Schonberg, I. J. (1938). Metric spaces and completely monotone functions. *Ann. of Math. II Ser.* **39**, 811–841.

Schwartz, L. (1965). On Bayes procedures. *Z. Wahrscheinlichkeitstheorie Verw. Gebiete.* **4**, 10–26.

Scott, D. J. (1973). Central limit theorems for martingales and for processes with stationary increments using a Skorokhood representation approach. *Adv. Appl. Prob.* **5**, 119–137.

Seal, H. L. (1969). "Stochastic Theory of Risk Business". Wiley, New York.

Sen, P. K. (1971). A note on weak convergence of empirical processes for sequences of φ-mixing random variables. *Ann. Math. Statist.* **42**, 2131–2133.
Sen, P. K. (1977). On Wiener process embedding for linear combination of order statistics. *Sankhya. Ser A* **39**, 138–143.
Sen, P. K. and Ghosh, M. (1973). A Chernoff Savage representation of rank order statistics for stationary φ-mixing processes. *Sankhya Ser. A* **35**, 153–172.
Serfling, R. J. (1968). The Wilcoxon two-sample statistic on strongly mixing processes. *Ann. Math. Statist.* **39**, 1202–1209.
Shanbhag, D. N. and Basawa, I. V. (1972). A note on serial correlation coefficients. *Cal. Stat. Assn. Bull.* **21**, 57–62.
Shepp, L. (1969). Explicit solutions to some problems of optimal stopping. *Ann. Math. Statist.* **40**, 993–1010.
Sherman, B. (1950). A random variable related to the spacing of sample values. *Ann. Math. Statist.* **21**, 339–361.
Silvey, S. D. (1961). A note on the maximum likelihood in the case of dependent random variables. *J. Roy. Statist. Soc. Ser. B.* **23**, 444–452.
Sirjaev, A. N. (1972). Statistics of diffusion type processes. *In* "Proc. Second Japan–USSR Symp. on Prob. Theory" 397–412, Springer-Verlag, Berlin Lecture Notes in Mathematics, 330.
Skorokhod, A. V. (1965). "Studies in the Theory of Random Processes" Addison-Wesley, Reading, Mass.
Smith, P. J. (1969). Structure of nonparametric tests of some multivariate hypotheses. Ph.D. Thesis, Case Western Reserve University.
Snyder, D. L. (1975). "Random Point Processes" Wiley-Interscience, New York.
Stepan, J. (1973). The weak convergence of Bayes estimators. *In* "Trans. Sixth Prague Conf. on Inf. Theory. Stat. Dec. Functions and Rand. Proc. 807–814.
Stigler, S. M. (1971). The estimation of the probability of extinction and other parameters associated with branching processes. *Biometrika* **58**, 499–508.
Straf, M. L. (1970). Weak convergence of random processes with several parameters. *In* "Proceedings Sixth Berkely Symposium" 187–221. University of California Press.
Strasser, H. (1973). On Bayes estimates. *J. Multivariate Analysis* **3**, 293–310.
Strasser, H. (1975). Asymptotic properties of posterior distributions. *Z. Wahrscheinlichkeitstheorie und verw Gebiete*, **35**, 269–282.
Strasser, H. (1977). Improved bounds for the equivalence of Bayes and maximum likelihood estimation *Teoria Varoyat. iee primenen* **22**, 358–370.
Striebel, C. T. (1959). Densities for stochastic processes. *Ann. Math. Statist.* **30**, 559–567.
Sudakov, V. N. (1969). On measures defined by Markovian moments. In "Investigations on the Theory of Random Processes" Vol. 12, Memoirs of the Scientific Seminars of the Leningrad section of the Steklov Math. Inst., 157–164.
Sweeting, T. J. (1978). On efficient tests for branching processes. *Biometrika* **65**, 123–127.
Taraskin, A. F. (1970). The asymptotic normality of stochastic integrals and estimates of the coefficient of diffusion process transfer. *In* "Mathematical Physics" no. 8 Naukova Dumka, Kiev, 149–163 (Russian).
Taraskin, A. F. (1971a). Statistical problems for a class of stochastic differential equations. *In* "Mathematical Physics" no. 10, Naukova Dumka, Kiev, 91–99 (Russian).
Taraskin, A. F. (1971b). Parameter estimation by method of maximum likelihood

for a stationary process. *In* "Mathematical Physics". no. 9, Naukova Dumka, Kiev, 123–131 (Russian).

Taraskin, A. F. (1974). On the asymptotic normality of vector-valued stochastic integrals and estimates of drift parameters of a multidimensional diffusion process. *Theor. Probability Math. Statist.* **2**, 209–224.

Taraskin, A. F. (1975). On confidence regions for parameters of diffusion type Markov processes. "Matematika" no. 1. Kyubushev. 3–9 (Russian).

Taylor, H. (1972). "Statistical Decomposition Analysis". North-Holland, Amsterdam.

Toe, D. T. (1976). Statistical Analysis of Markovian and Related Time Series. Ph.D. Thesis, La Trobe University.

Trybula, S. (1968). Sequential estimation in processes with independent increments. *Dissertationes Mathematicae*, **60**.

Wald, A. (1943). Tests of statistical hypotheses concerning several parameters when the number of observations is large *Trans. Amer. Math. Soc.* **54**, 426–482.

Wald, A. (1947). "Sequential Analysis" Wiley, New York.

Wald, A. (1948). Asymptotic properties of the maximum likelihood estimate of an unknown parameter of a discrete stochastic process. *Ann. Math. Statist.* **20**, 595–601.

Wald, A. (1949). Note on the consistency of the maximum likelihood estimate. *Ann. Math. Statist.* **20**, 595–601.

Wald, A. (1950). "Statistical Decision Functions" Wiley, New York.

Walker, A. M. (1969). On the asymptotic behaviour of posterior distributions. *J. Roy. Statist. Soc. Ser. B*, **31**, 80–88.

Walter, G. and Blum, J. R. (1976). Probability density estimation using delta-sequences. (Preprint) The University of Wisconsin, Milwaukee.

Weiss, L. (1971). Asymptotic properties of maximum likelihood estimators in some non-standard cases. *J. Amer. Statist. Assoc.* **66**, 345–350.

Weiss, L. (1973). Asymptotic properties of maximum likelihood estimates in some nonstandard cases II. *J. Amer. Statist. Assoc.* **68**, 428–430.

Wertz, W (1971). Empirische betrachtungen und normal approximation bei dichteschatzungen. *Operations Research-Verfahren* **8**, 430–440.

Wijsman, R. (1960). A monotonicity property of the sequential probability ratio test. *Ann. Math. Statist.* **31**, 677–684. (Correction. *Ann. Statistics* (1975) **3** 796).

Wijsman, R. (1963). Existence, uniqueness and monotonicity of sequential probability ratio tests. *Ann. Math. Statist.* **34** 1541–1548.

Winkler, W. (1977). Sequential estimation in processes with independent increments. *In* "Banach Center Publications. Vol. 6. (To appear).

Wolff, R. W. (1965). Problems of statistical inference for birth and death queueing models. *Operation Res.* **13**, 343–357.

Wolfowitz, J. (1950). Minimax estimates of the mean of a normal distribution with known variance. *Ann. Math. Statist.* **21**, 218–230.

Yamada, K. (1976). Asymptotic behavior of posterior distributions for random processes under incorrect models. *J. Math. Anal. and Appl.* **56**, 294–308.

Yang, G. L. (1968). Contagion in stochastic models for epidemics. *Ann. Math. Statist.* **39**, 1865–1889.

Yosida, K. (1968). Functional Analysis. Springer-Verlag, Berlin.

Index

Probability and Mathematical Statistics

A Series of Monographs and Textbooks

Thomas Ferguson. Mathematical Statistics: A Decision Theoretic Approach. 1967

Howard Tucker. A Graduate Course in Probability. 1967

K. R. Parthasarathy. Probability Measures on Metric Spaces. 1967

P. Révész. The Laws of Large Numbers. 1968

H. P. McKean, Jr. Stochastic Integrals. 1969

B. V. Gnedenko, Yu. K. Belyayev, and A. D. Solovyev. Mathematical Methods of Reliability Theory. 1969

Demetrios A. Kappos. Probability Algebras and Stochastic Spaces. 1969

Ivan N. Pesin. Classical and Modern Integration Theories. 1970

S. Vajda. Probabilistic Programming. 1972

Sheldon M. Ross. Introduction to Probability Models. 1972

Robert B. Ash. Real Analysis and Probability. 1972

V. V. Fedorov. Theory of Optimal Experiments. 1972

K. V. Mardia. Statistics of Directional Data. 1972

H. Dym and H. P. McKean. Fourier Series and Integrals. 1972

Tatsuo Kawata. Fourier Analysis in Probability Theory. 1972

Fritz Oberhettinger. Fourier Transforms of Distributions and Their Inverses: A Collection of Tables. 1973

Paul Erdös and Joel Spencer. Probabilistic Methods in Combinatorics. 1973

K. Sarkadi and I. Vincze. Mathematical Methods of Statistical Quality Control. 1973

Michael R. Anderberg. Cluster Analysis for Applications. 1973

W. Hengartner and R. Theodorescu. Concentration Functions. 1973

Kai Lai Chung. A Course in Probability Theory, Second Edition. 1974

L. H. Koopmans. The Spectral Analysis of Time Series. 1974

L. E. Maistrov. Probability Theory: A Historical Sketch. 1974

William F. Stout. Almost Sure Convergence. 1974

E. J. McShane. Stochastic Calculus and Stochastic Models. 1974

Robert B. Ash and Melvin F. Gardner. Topics in Stochastic Processes. 1975

Avner Friedman, Stochastic Differential Equations and Applications, Volume 1, 1975; Volume 2. 1975

Roger Cuppens. Decomposition of Multivariate Probabilities. 1975

Eugene Lukacs. Stochastic Convergence, Second Edition. 1975

H. Dym and H. P. McKean. Gaussian Processes, Function Theory, and the Inverse Spectral Problem. 1976

N. C. Giri. Multivariate Statistical Inference. 1977

Lloyd Fisher and John McDonald. Fixed Effects Analysis of Variance. 1978

Sidney C. Port and Charles J. Stone. Brownian Motion and Classical Potential Theory. 1978

Konrad Jacobs. Measure and Integral. 1978

K. V. Mardia, J. T. Kent, and J. M. Biddy. Multivariate Analysis. 1979

Sri Gopal Mohanty. Lattice Path Counting and Applications. 1979

Ishwar V. Basawa and B. L. S. Prakasa Rao, Statistical Inference for Stochastic Processes. 1980

in preparation

Michel Metivier and J. Pellaumail. Stochastic Integration. 1980

Y. L. Tong. Probability Inequalities in Multivariate Distributions. 1980

M. B. Priestly, Spectral Analysis and Time Series. 1980

Probability and Mathematical Statistics

A Series of Monographs and Textbooks

Editors ***Z. W. Birnbaum*** ***E. Lukacs***

University of Washington *Bowling Green State University*
Seattle, Washington *Bowling Green, Ohio*

Thomas Ferguson. Mathematical Statistics: A Decision Theoretic Approach. 1967

Howard Tucker. A Graduate Course in Probability. 1967

K. R. Parthasarathy. Probability Measures on Metric Spaces. 1967

P. Révész. The Laws of Large Numbers. 1968

H. P. McKean, Jr. Stochastic Integrals. 1969

B. V. Gnedenko, Yu. K. Belyayev, and A. D. Solovyev. Mathematical Methods of Reliability Theory. 1969

Demetrios A. Kappos. Probability Algebras and Stochastic Spaces. 1969

Ivan N. Pesin. Classical and Modern Integration Theories. 1970

S. Vajda. Probabilistic Programming. 1972

Sheldon M. Ross. Introduction to Probability Models. 1972

Robert B. Ash. Real Analysis and Probability. 1972

V. V. Fedorov. Theory of Optimal Experiments. 1972

K. V. Mardia. Statistics of Directional Data. 1972

H. Dym and H. P. McKean. Fourier Series and Integrals. 1972

Tatsuo Kawata. Fourier Analysis in Probability Theory. 1972

Fritz Oberhettinger. Fourier Transforms of Distributions and Their Inverses: A Collection of Tables. 1973

Paul Erdös and Joel Spencer. Probabilistic Methods in Combinatorics. 1973

K. Sarkadi and I. Vincze. Mathematical Methods of Statistical Quality Control. 1973

Michael R. Anderberg. Cluster Analysis for Applications. 1973

W. Hengartner and R. Theodorescu. Concentration Functions. 1973

Kai Lai Chung. A Course in Probability Theory, Second Edition. 1974

L. H. Koopmans. The Spectral Analysis of Time Series. 1974

L. E. Maistrov. Probability Theory: A Historical Sketch. 1974

William F. Stout. Almost Sure Convergence. 1974

E. J. McShane. Stochastic Calculus and Stochastic Models. 1974

Robert B. Ash and Melvin F. Gardner. Topics in Stochastic Processes. 1975

Avner Friedman, Stochastic Differential Equations and Applications, Volume 1, 1975; Volume 2. 1975

Roger Cuppens. Decomposition of Multivariate Probabilities. 1975

Eugene Lukacs. Stochastic Convergence, Second Edition. 1975

H. Dym and H. P. McKean. Gaussian Processes, Function Theory, and the Inverse Spectral Problem. 1976

N. C. Giri. Multivariate Statistical Inference. 1977

Lloyd Fisher and John McDonald. Fixed Effects Analysis of Variance. 1978

Sidney C. Port and Charles J. Stone. Brownian Motion and Classical Potential Theory. 1978

Konrad Jacobs. Measure and Integral. 1978

K. V. Mardia, J. T. Kent, and J. M. Biddy. Multivariate Analysis. 1979

Sri Gopal Mohanty. Lattice Path Counting and Applications. 1979

Ishwar V. Basawa and B. L. S. Prakasa Rao, Statistical Inference for Stochastic Processes. 1980

in preparation

Michel Metivier and J. Pellaumail. Stochastic Integration. 1980

Y. L. Tong. Probability Inequalities in Multivariate Distributions. 1980

M. B. Priestly, Spectral Analysis and Time Series. 1980